Algebra I

ALL-IN-ONE

by Mary Jane Sterling

Algebra I All-in-One For Dummies®

Published by: **John Wiley & Sons, Inc.**, 111 River Street, Hoboken, NJ 07030-5774, www.wiley.com

For general information on our other products and services, please contact our Customer Care Department within the U.S. at 877-762-2974, outside the U.S. at 317-572-3993, or fax 317-572-4002. For technical support, please visit https://hub.wiley.com/community/support/dummies.

Wiley publishes in a variety of print and electronic formats and by print-on-demand. Some material included with standard print versions of this book may not be included in e-books or in print-on-demand. If this book refers to media such as a CD or DVD that is not included in the version you purchased, you may download this material at http://booksupport.wiley.com. For more information about Wiley products, visit www.wiley.com.

Library of Congress Control Number: 2021947882

ISBN 978-1-119-84304-7 (pbk); ISBN 978-1-119-84305-4 (ebk); ISBN 978-1-119-84306-1 (ebk)

SKY10031265_110821

Contents at a Glance

Table of Contents

Introduction

"What is algebra?" "Is it really that important in the study of other math courses?" "Where did it come from?" And my favorite question from students: "What do I need this for?"

Algebra is really the basis of most courses that you take in high school and college. You can't do anything in calculus without a good algebra background. And there's lots of algebra in geometry. You even need algebra in computer science! Algebra was created, modified, and continues to be tweaked so that ideas and procedures can be shared by everyone. With all people speaking the same "language," there are fewer misinterpretations.

Algebra, or *al-jabr* in Arabic, means "a reunion of broken pieces." How appropriate!

About This Book

This book covers just about everything you'd ever want to know about basic algebra. And it provides opportunities for further discoveries. You'll find explanations, examples, practice opportunities, and problems to test your comprehension. This book starts with basic operations and terminology, gives you information on simplifying and organizing expressions, runs through equation-solving, introduces applications, and goes visual with the graphing. When finished, you should:

- Be familiar with notation and terminology.
- Have confidence in finding the correct answer.
- Look forward to more challenges with Algebra II and other courses.

Each new topic provides:

- Example problems with answers and solutions.
- Practice problems with answers and solutions.

Each chapter provides:

- A test with problems representing the topics covered.
- Solutions to the test problems.

Online quizzes are also available for even more practice and confidence-building.

This book also has a few conventions to keep in mind:

- New terms introduced in a chapter, as well as variables, are in *italics.*
- Keywords in lists and numbered steps are in **boldface.**
- Any websites appear in monofont.
- The final answers to problems appear in **bold.** Then the explanation follows.

Foolish Assumptions

You are reading this book to learn more about algebra, so I'm assuming that you have some of the other basic math skills coming in: familiarity with fractions and their operations, comfort with handling decimals and the operations involved, some experience with integers (signed whole numbers) and how they operate, and some graphing knowledge — how to place points on a graphing plane. If you don't have as much knowledge as you'd like related to some items mentioned, you might want to refer to some resources such as *Basic Math & Pre-Algebra For Dummies* or *Pre-Algebra Essentials For Dummies.*

I am also assuming that you're as excited about mathematics as I am. Oh, okay, you don't have to be that excited. But you're interested and eager and anxious to increase your mathematical abilities. That's the main thing you need.

Icons Used in This Book

You'll see the following five icons throughout the book:

Each example is an algebra question based on the discussion and explanation, followed by a solution. Work through these examples, and then refer to them to help you solve the practice problems that follow them as well as the quiz questions at the end of the chapter.

This icon points out important information that you need to focus on. Make sure you understand this information fully before moving on. You can skim through these icons when reading a chapter to make sure you remember the highlights.

Tips are hints that can help speed you along when answering a question. See whether you find them useful when working on practice problems.

This icon flags common mistakes that students make if they're not careful. Take note and proceed with caution!

When you see this icon, it's time to put on your thinking cap and work out a few practice problems on your own. The answers and detailed solutions are available so you can feel confident about your progress.

Beyond the Book

In addition to what you're reading right now, this book comes with a Cheat Sheet that provides quick access to some formulas and rules and processes that are frequently used. To get this Cheat Sheet, simply go to www.dummies.com and type **Algebra I All In One For Dummies Cheat Sheet** in the Search box.

You'll also have access to online quizzes related to each chapter. These quizzes provide a whole new set of problems for practice and confidence-building. To access the quizzes, follow these simple steps:

1. **Register your book or ebook at Dummies.com to get your PIN.** Go to www.dummies.com/go/getaccess.
2. **Select your product from the drop-down list on that page.**
3. **Follow the prompts to validate your product, and then check your email for a confirmation message that includes your PIN and instructions for logging in.**

If you do not receive this email within two hours, please check your spam folder before contacting us through our Technical Support website at http://support.wiley.com or by phone at 877-762-2974.

Now you're ready to go! You can come back to the practice material as often as you want — simply log on with the username and password you created during your initial login. No need to enter the access code a second time.

Your registration is good for one year from the day you activate your PIN.

Where to Go from Here

This book is organized so that you can safely move from whichever chapter you choose to start with and in whatever order you like. You can strengthen skills you feel less confident in or work on those that need some attention.

If you haven't worked on any algebra recently, I'd recommend that you start out with Chapter 1 and some other chapters in the first unit. It's important to know the vocabulary and basic notation so you understand what is being presented in later chapters.

If you're all set with the basic operations, then a good place to go is Chapter 11, where you find out about factoring different types of expressions. Factoring makes it possible to simplify expressions and solve equations.

When you're ready for the "What do I need this for?" question, go to Chapter 20 or Chapter 21, where you see some of the many types of applications that use algebraic expressions and solutions.

There are other resources, such as *Basic Math & Pre-Algebra For Dummies* and *Pre-Algebra Essentials For Dummies* (John Wiley & Sons, Inc.), if you think you need more background. And, of course, when you've finished here and are ready for the next challenge, be sure to check out *Algebra II For Dummies* and *Pre-Calculus For Dummies.* And that's just the beginning!

1

Starting Out with Numbers and Properties

Contents at a Glance

IN THIS CHAPTER

» Identifying the different types of numbers

» Placing numbers on a number line

» Becoming familiar with the vocabulary

» Recognizing the operations of algebra

Chapter 1
Assembling Your Tools: Number Systems

You've undoubtedly heard the word *algebra* on many occasions, and you know that it has something to do with mathematics. Perhaps you remember that algebra has enough stuff in it to require taking two separate high school algebra classes — Algebra I and Algebra II. But what exactly is algebra? What is it really used for?

This book answers these questions and more, providing the straight scoop on some of the contributions to algebra's development, what it's good for, how algebra is used, and what tools you need to make it happen. In this chapter, you find some of the basics necessary to make it easier to find your way through the different topics in this book.

In a nutshell, *algebra* is a way of generalizing arithmetic. Through the use of *variables* (letters representing numbers) and formulas or equations involving those variables, you solve problems. The problems may be in terms of practical applications, or they may be puzzles for the pure pleasure of the solving. Algebra uses positive and negative numbers, integers, fractions, operations, and symbols to analyze the relationships between values. It's a systematic study of numbers and their relationships, and it uses specific rules.

Identifying Numbers by Name

Where would mathematics and algebra be without numbers? Numbers aren't just a part of everyday life, they are the basic building blocks of algebra. Numbers give you a value to work with. Where would civilization be today if not for numbers? Without numbers to figure the distances, slants, heights, and directions, the pyramids would never have been built. Without numbers to figure out navigational points, the Vikings would never have left Scandinavia. Without numbers to examine distance in space, humankind could not have landed on the moon.

Even the simplest tasks and the most common of circumstances require a knowledge of numbers. Suppose that you wanted to figure the amount of gasoline it takes to get from home to work and back each day. You need a number for the total miles between your home and business, and another number for the total miles your car can run on a gallon of gasoline.

The different sets of numbers are important because what they look like and how they behave can set the scene for particular situations or help to solve particular problems. It's sometimes really convenient to declare, "I'm only going to look at whole-number answers," because whole numbers don't include fractions or negatives. You could easily end up with a fraction if you're working through a problem that involves a number of cars or people. Who wants half a car or, heaven forbid, a third of a person?

Algebra uses different sets of numbers, in different circumstances. I describe the different types of numbers here.

Realizing real numbers

Real numbers are just what the name implies. In contrast to imaginary numbers, they represent *real* values — no pretend or make-believe. Real numbers cover the gamut and can take on any form — fractions or whole numbers, decimal numbers that can go on forever and ever without end, positives and negatives. The variations on the theme are endless.

Counting on natural numbers

A *natural number* (also called a *counting number*) is a number that comes naturally. What numbers did you first use? Remember someone asking, "How old are you?" You proudly held up four fingers and said, "Four!" The natural numbers are the numbers starting with 1 and going up by ones: 1, 2, 3, 4, 5, 6, 7, and so on into infinity. You'll find lots of counting numbers in Chapter 8, where I discuss prime numbers and factorizations.

Whittling out whole numbers

Whole numbers aren't a whole lot different from natural numbers. Whole numbers are just all the natural numbers plus a 0: 0, 1, 2, 3, 4, 5, and so on into infinity.

Whole numbers act like natural numbers and are used when whole amounts (no fractions) are required. Zero can also indicate none. Algebraic problems often require you to round the answer to the nearest whole number. This makes perfect sense when the problem involves people, cars, animals, houses, or anything that shouldn't be cut into pieces.

Integrating integers

Integers allow you to broaden your horizons a bit. Integers incorporate all the qualities of whole numbers and their opposites (called their *additive inverses*). *Integers* can be described as being positive and negative whole numbers and zero: −3, −2, −1, 0, 1, 2, 3.

Integers are popular in algebra. When you solve a long, complicated problem and come up with an integer, you can be joyous because your answer is probably right. After all, it's not a fraction! This doesn't mean that answers in algebra can't be fractions or decimals. It's just that most textbooks and reference books try to stick with nice answers to increase the comfort level and avoid confusion. This is my plan in this book, too. After all, who wants a messy answer — even though, in real life, that's more often the case. I use integers in Chapter 14 and those later on, where you find out how to solve equations.

Being reasonable: Rational numbers

Rational numbers act rationally! What does that mean? In this case, acting rationally means that the decimal equivalent of the rational number behaves. The decimal eventually ends somewhere, or it has a repeating pattern to it. That's what constitutes "behaving."

Some rational numbers have decimals that end such as: 3.4, 5.77623, −4.5. Other rational numbers have decimals that repeat the same pattern, such as $3.164164\overline{164}$, or $0.66666666\overline{6}$. The horizontal bar over the 164 and the 6 lets you know that these numbers repeat forever.

In *all* cases, rational numbers can be written as fractions. Each rational number has a fraction that it's equal to. So one definition of a *rational number* is any number that can be written as a fraction, $\frac{p}{q}$, where p and q are integers (except q can't be 0). If a number can't be written as a fraction, then it isn't a rational number. Rational numbers appear in Chapter 16, where you see quadratic equations, and later, when the applications are presented.

Restraining irrational numbers

Irrational numbers are just what you may expect from their name: the opposite of rational numbers. An *irrational number* cannot be written as a fraction, and decimal values for irrationals never end and never have a nice pattern to them. Whew! Talk about irrational! For example, π, with its never-ending decimal places, is irrational. Irrational numbers are often created when using the quadratic formula, as you see in Chapter 16, because you find the square roots of numbers that are not perfect squares, such as: $\sqrt{6}$ and $\sqrt{85}$.

Picking out primes and composites

A number is considered to be *prime* if it can be divided evenly only by 1 and by itself. The prime numbers are: 2, 3, 5, 7, 11, 13, 17, 19, 23, 29, 31, and so on. The only prime number that's even is 2, the first prime number. Mathematicians have been studying prime numbers for centuries, and prime numbers have them stumped. No one has ever found a formula for producing all the primes. Mathematicians just assume that prime numbers go on forever.

A number is *composite* if it isn't prime — if it can be divided by at least one number other than 1 and itself. So the number 12 is composite because it's divisible by 1, 2, 3, 4, 6, and 12. Chapter 8 deals with primes, but you also see them throughout the chapters, where I show you how to factor primes out of expressions.

Numbers can be classified in more than one way, the same way that a person can be classified as male or female, tall or short, blonde or brunette, and so on. The number −3 is negative, it's an integer, it's an odd number, it's rational, and it's real. The number −3 is also a negative prime number. You should be familiar with all these classifications so that you can read mathematics correctly.

Zero: It's Complicated

Zero is a very special number. It wasn't really used in any of the earliest counting systems. In fact, there is no symbol for zero in the Roman numerals!

Zero is a very useful number, but it also comes with its challenges. You can't divide by zero, but you can add zero to a number and multiply a number by 0. You'll find zero popping up in the most interesting places!

Imagining imaginary numbers

Yes, there are imaginary numbers in mathematics. These numbers were actually created by mathematicians who didn't like not finishing a problem! They would be trying to solve a quadratic equation and be stumped by the situation where they needed the square root of a negative number. There was no way to deal with this.

So some clever mathematicians came up with a solution. They declared that $\sqrt{-1}$ must be equal to *i*. Yes, the *i* stands for "imaginary." You'll see how this works in Chapter 16.

Coping with complex numbers

A complex number isn't really all that mysterious. This is just a designation that allows for you to deal with both real and imaginary parts of a number. A complex number has some of each! Complex numbers have the general format of $a + bi$, where a and b are real numbers, and the i is that imaginary number, $\sqrt{-1}$.

EXAMPLE

Q. Using the choices: natural, whole, integer, rational, irrational, prime, and imaginary, which of these can be used to describe the number 8?

A. **Natural, whole, integer, rational.** The number 8 fits all of these descriptions. It is rational, because you can write it as a fraction such as $\frac{8}{1}$ or $\frac{24}{3}$.

Q. Using the choices: natural, whole, integer, rational, irrational, prime, and imaginary, which of these can be used to describe the number $-\frac{2}{3}$?

A. **Rational.** This is written as a fraction but cannot be reduced to create an integer.

Q. Using the choices: natural, whole, integer, rational, irrational, prime, and imaginary, which of these can be used to describe the number $\sqrt{17}$?

A. **Irrational.** The number 17 isn't a perfect square, so the decimal equivalence of $\sqrt{17}$ is a decimal that goes on forever without repeating or terminating.

Q. Using the choices: natural, whole, integer, rational, irrational, prime, and imaginary, which of these can be used to describe the number $\sqrt{-9}$?

A. **Imaginary.** Even though 9 is a perfect square, so you can write the number as $\sqrt{-1} \cdot \sqrt{9}$ and then simplify it to read $i \cdot 3$ or $3i$, this number stays imaginary.

YOUR TURN

1. Identify which of the following numbers are natural numbers:

$$-41,\ 15,\ -5.2,\ 11,\ 3.2121...,\ -\frac{12}{3},\ \frac{14}{11},\ \sqrt{-5},\ \sqrt{10},\ \sqrt{9}$$

2. Identify which of the following numbers are integers:

$$-41,\ 15,\ -5.2,\ 11,\ 3.2121...,\ -\frac{12}{3},\ \frac{14}{11},\ \sqrt{-5},\ \sqrt{10},\ \sqrt{9}$$

3. Identify which of the following numbers are rational numbers:

$$-41,\ 15,\ -5.2,\ 11,\ 3.2121...,\ -\frac{12}{3},\ \frac{14}{11},\ \sqrt{-5},\ \sqrt{10},\ \sqrt{9}$$

4. Identify which of the following numbers are irrational numbers:

$$-41,\ 15,\ -5.2,\ 11,\ 3.2121...,\ -\frac{12}{3},\ \frac{14}{11},\ \sqrt{-5},\ \sqrt{10},\ \sqrt{9}$$

5. Identify which of the following numbers are prime numbers:

$$-41,\ 15,\ -5.2,\ 11,\ 3.2121...,\ -\frac{12}{3},\ \frac{14}{11},\ \sqrt{-5},\ \sqrt{10},\ \sqrt{9}$$

6. Identify which of the following numbers are imaginary numbers:

$$-41,\ 15,\ -5.2,\ 11,\ 3.2121...,\ -\frac{12}{3},\ \frac{14}{11},\ \sqrt{-5},\ \sqrt{10},\ \sqrt{9}$$

Placing Numbers on the Number Line

A number line is labeled with numbers that increase as you move from left to right. And numbers are listed with an equal amount or value between any two consecutive numbers.

Numbers are placed on a number line to give you a visual picture of how they compare, how far apart they are, and what is missing between them. The two number lines shown here are examples of some versions that are possible. In Figure 1-1, you see the half-way mark indicated between units. And in Figure 1-2, the negative and positive integers are shown, with 0 in the middle.

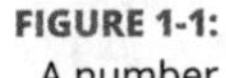

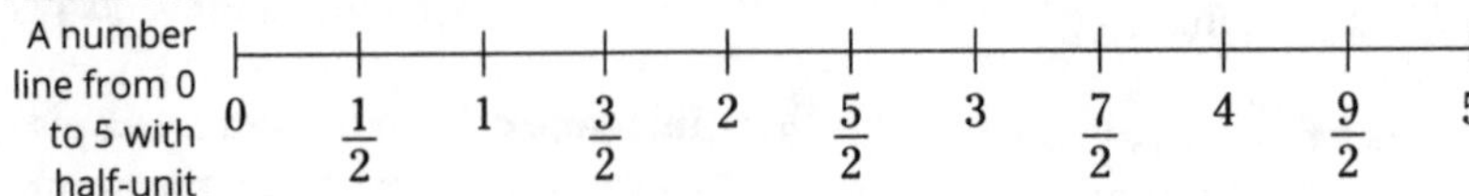

FIGURE 1-1: A number line from 0 to 5 with half-unit increments.

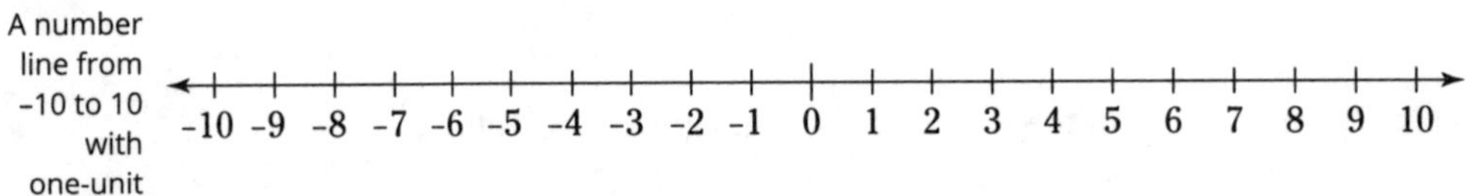

FIGURE 1-2: A number line from -10 to 10 with one-unit increments.

EXAMPLE

Q. Place the numbers 3, $-6, \frac{1}{2}, -2.6$ on a number line.

A.

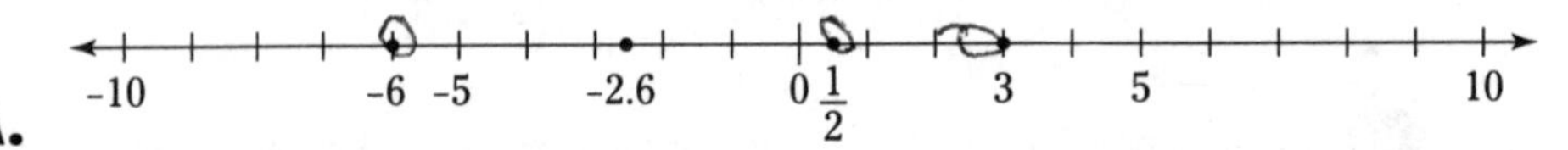

Note that the marks representing these numbers on the number line are marked with dots or points. The points for the fraction and decimal numbers are approximated, because the tickmarks for these numbers aren't on the number line to make them exact.

YOUR TURN

7 Place the following numbers on the number line: $-6, -1, 0.5, 2, 3.2$

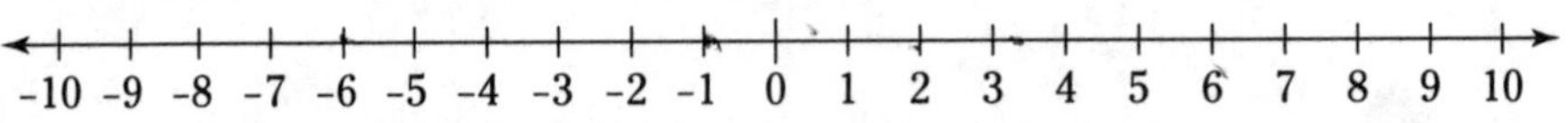

8 Place the following numbers on the number line: $-2\frac{2}{3}, \frac{1}{3}, 2\frac{1}{3}, 3\frac{2}{3}$

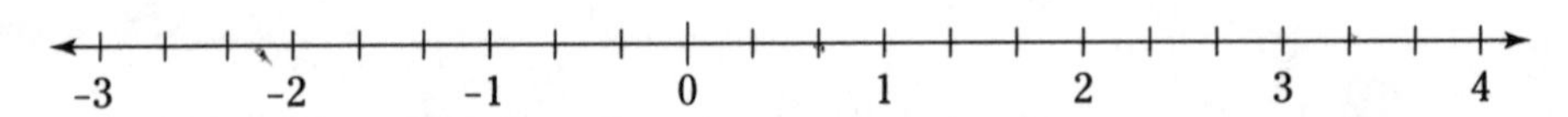

Speaking in Algebra

Algebra and symbols in algebra are like a foreign language. They all mean something and can be translated back and forth as needed. It's important to know the vocabulary in a foreign language; it's just as important in algebra.

Being precise with words

The words used in algebra are very informative. You need to know their exact meaning, because they convey what is happening.

- An *expression* is any combination of values and operations that can be used to show how things belong together and compare to one another. $2x^2 + 4x$ is an example of an expression. Think of an expression as being the equivalent of a phrase or part of a sentence; you have some subjects and conjugates, but no verbs. You see how items are distributed over expressions in Chapter 9.
- A *term,* such as 4*xy*, is a grouping together of one or more *factors* (variables and/or numbers) all connected by multiplication or division. In this case, multiplication is the only thing connecting the number with the variables. Addition and subtraction, on the other hand, separate terms from one another. For example, the expression $3xy + 5x - 6$ has three *terms.*
- An *equation* uses a sign to show a relationship — that two things are equal. By using an equation, tough problems can be reduced to easier problems and simpler answers. An example of an equation is $2x^2 + 4x = 7$. See Chapters 14 through 18 for more information on equations.
- An *operation* is an action performed upon one or two numbers to produce a resulting number. Operations include addition, subtraction, multiplication, division, square roots, and so on. See Chapter 7 for more on operations.
- A *variable* is a letter representing some unknown; a variable always represents a number, but it *varies* until it's written in an equation or inequality. (An *inequality* is a comparison of two values. For more on inequalities, turn to Chapter 19.) Then the fate of the variable is set — it can be solved for, and its value becomes the solution of the equation. By convention, mathematicians usually assign letters at the end of the alphabet to be variables to be solved for in a problem (such as *x*, *y*, and *z*).
- A *constant* is a value or number that never changes in an equation — it's constantly the same. The number 5 is a constant because it is what it is. A letter can represent a constant if it is assigned a definite value. Usually, a letter representing a constant is one of the first letters in the alphabet. In the equation $ax^2 + bx + c = 0$, *c* is a constant and *x* is the variable.
- A *coefficient* is another type of constant. It is a multiplier of a variable. In the equation $ax^2 + bx + c = 0$, *a* and *b* are coefficients. They have constant, assigned values and are factors, but they have the special role of multiplying variables.
- An *exponent* is a small number written slightly above and to the right of a variable or number, such as the 2 in the expression 3^2. It's used to show repeated multiplication. An exponent is also called the *power* of the value. For more on exponents, see Chapter 5.

EXAMPLE

Q. Identify the terms, coefficients, factors, exponents, and constants in the expression $4x^2 - 3x + 2$.

A. There are three terms, separated by the subtraction and addition symbols. In the first term, the 4 is the coefficient, and the 4 and x^2 are factors. The 2 is the exponent. In the second term, the 3 and the x are factors. The exponent 1 isn't shown on the x; it's just assumed. And the final term, the 2, is a constant.

Q. Identify the terms, coefficients, factors, exponents, and constants in the expression $P\left(1+\frac{r}{n}\right)^{nt}$.

A. This expression has just one term. The P is a factor, and the parentheses form the other factor. There are two terms in the parentheses, and the exponent on the parentheses is nt.

YOUR TURN

9. How many terms are there in the expression: $4x - 3x^3 + 11$?

10. How many factors are found in the expression: $3xy + 2z$?

11. Which are the variables and which are the constants in the expression: $\frac{(x-h)^2}{a} + \frac{(y-k)^2}{b} = 1$?

12. Which are the exponents in the expression: $z^2 + z^{1/2} - z$?

Describing the size of an expression

An expression is a combination of terms and operations and can take on many different formats. In Chapters 11 through 16, you see many types of expressions and several types of equations that are created from the expressions. Many of these expressions and equations have very precise and descriptive names.

- A *polynomial* is an expression containing variables and constants. It consists of one or more terms. The terms are separated by addition and subtraction. And the exponents on the variable terms are always whole numbers, never fractions or negative numbers.

  Ex: $5x^4 - 2x^3 + x - 13$

- A *monomial* is a polynomial consisting of exactly 1 term.

  Ex: $15y$

- A *binomial* is a polynomial consisting of exactly 2 terms.

  Ex: $x^2 - 1$

- A *trinomial* is a polynomial consisting of exactly 3 terms.

  Ex: $3x^2 - 4x + 15$

- A *linear expression* is a polynomial in which there is no variable with an exponent greater than 1. In fact, the exponents can be only 1 or 0. And there must be at least one variable term with an exponent of 1.

  Ex: $5y - 3$

- A *quadratic expression* is a polynomial in which there is no variable with an exponent greater than 2. In fact, the exponents can be only 2, 1, or 0. And there must be at least one variable term with an exponent of 2.

  Ex. $\frac{1}{2}x^2 - 13x$

Relating operations with symbols

The basics of algebra involve symbols. Algebra uses symbols for quantities, operations, relations, or grouping. The symbols are shorthand and are much more efficient than writing out the words or meanings. But you need to know what the symbols represent, and the following list shares some of that information. The operations are covered thoroughly in Chapter 6.

- + means *add* or *find the sum* or *more than* or *increased by;* the result of addition is the *sum.* It also is used to indicate a *positive number.*

- – means *subtract* or *minus* or *decreased by* or *less;* the result is the *difference.* It's also used to indicate a *negative number.*

- × means *multiply* or *times.* The values being multiplied together are the *multipliers* or *factors;* the result is the *product.* Some other symbols meaning *multiply* can be grouping symbols: ( ), [ ], { }, ·, *. In algebra, the × symbol is used infrequently because it can be confused with the variable *x*. The · symbol is popular because it's easy to write. The grouping symbols are used when you need to contain many terms or a messy expression. By themselves, the grouping symbols don't mean to multiply, but if you put a value in front of or behind a grouping symbol, it means to multiply.

- ÷ means *divide*. The number that's going into the *dividend* is the *divisor*. The result is the *quotient*. Other signs that indicate division are the fraction line and slash, /.
- $\sqrt{\ }$ means to take the *square root* of something — to find the number, which, multiplied by itself, gives you the number under the sign. (See Chapter 6 for more on square roots.)
- $|\ |$ means to find the *absolute value* of a number, which is the number itself or its distance from 0 on the number line. (For more on absolute value, turn to Chapter 2.)
- π is the Greek letter pi that refers to the irrational number: 3.14159. It represents the relationship between the diameter and circumference of a circle.
- [ ] is the *greatest integer* operation. It tells you to evaluate what's in the brackets and replace it with the biggest integer that is not larger than what's in them.

EXAMPLE

Q. Use mathematical symbols to write the expression: "The product of 6 and *a* is divided by the difference between the square of *a* and 1 and added to the square root of the difference between pi and *r* cubed."

A. $\frac{6a}{a^2+1}+\sqrt{\pi-r^3}$. You don't need a dot between the 6 and the *a*. Writing the two factors together indicates a product. Putting the binomial a^2+1 in the denominator indicates that you're dividing the $6a$ by that expression. The exponent of 3 indicates *r* is being cubed. The two terms in the difference both appear under the radical.

Q. Use mathematical symbols to write the expression: "The absolute value of the sum of *x* and 8 times the greatest integer value of the quotient of *x* and 3."

A. $|x+8|\cdot\left[\frac{x}{3}\right]$. The dot between the absolute value and greatest integer operations isn't really necessary, but it helps define better what you're expressing.

YOUR TURN

13 Use mathematical symbols to write the expression: "Four times *z* plus the square root of 11."

14 Use mathematical symbols to write the expression: "The difference between *x* and two is divided by pi."

15 Use mathematical symbols to write the expression: "The product of six and the absolute value of the sum of eight and y is divided by the square root of the difference between 9 and x."

Taking Aim at Algebra-Speak

Everything you study requires some understanding of the vocabulary and any special notation. When you can use one word like "introductory" instead of "all that good stuff that comes before the meat of the matter," then you've saved time and space and gotten to the point quickly.

Algebra is full of good words and symbols, as you see in the previous section. And now you find how specific symbols and wording gets right to the point (and, yes, a point in algebra can mean multiply). You're "equal" to the challenge!

Herding numbers with grouping symbols

Before a car manufacturer puts together a car, several different things have to be done first. The engine experts have to construct the engine with all its parts. The body of the car then has to be mounted onto the chassis and also secured. Other car assemblers have to perform the tasks that they specialize in as well. When these tasks are all accomplished in order, the car can be put together. The same is true with algebra. You have to do what's inside the *grouping* symbol before you can use the result in the rest of the equation.

Grouping symbols tell you that you have to deal with the *terms* inside the grouping symbols *before* you deal with the larger problem. If the problem contains grouped items, do what's inside a grouping symbol first, and then follow the order of operations. The grouping symbols are as follows.

- **Parentheses ( ):** Parentheses are the most commonly used symbols for grouping.
- **Brackets [ ] and braces { }:** Brackets and braces are also used frequently for grouping and have the same effect as parentheses. Using the different types of symbols helps when there's more than one grouping in a problem. It's easier to tell where a group starts and ends.

» **Radical** $\sqrt{\ }$: This is used for finding roots.

» **Fraction line (called the *vinculum*):** The fraction line also acts as a grouping symbol — everything above the line (in the *numerator*) is grouped together, and everything below the line (in the *denominator*) is grouped together.

Even though the order of operations and grouping-symbol rules are fairly straightforward, it's hard to describe, in words, all the situations that can come up in these problems. The explanations and examples in Chapters 3 and 7 should clear up any questions you may have.

EXAMPLE

Q. What are the operations found in the expression: $3y-\frac{y}{4}+\sqrt{2y}$?

A. The operations, in order from left to right, are multiplication, subtraction, division, addition, multiplication, and square root. The term $3y$ means to multiply 3 times y. The subtraction symbol separates the first and second terms. Writing y over 4 in a fraction means to divide. Then that term has the radical added to it. The 2 and y are multiplied under the radical, and then the square root is taken.

Q. Identify the grouping symbols shown in $\frac{4[6-2(x+1)]}{x^2+11}$.

A. The first grouping symbol to recognize is the fraction line. It separates the term in the numerator, $4[6-2(x+1)]$, from the terms in the denominator, x^2+11. Next, you see the brackets, which contain the two terms forming a subtraction, $[6-2(x+1)]$. The last grouping symbols are the parentheses, which have a multiplier of 2 and, inside, the sum of two terms.

YOUR TURN

16 Write the expression using the correct symbols: "The square root of x is subtracted from 3 times y."

17 Write the expression using the correct symbols: "Add 2 and y; then divide that sum by 11."

18 Identify the grouping symbols in $5\left\{16+\frac{4(11-z)}{12}\right\}$.

Defining relationships

Algebra is all about relationships — not the he-loves-me-he-loves-me-not kind of relationship, but the relationships between numbers or among the terms of an expression. Although algebraic relationships can be just as complicated as romantic ones, you have a better chance of understanding an algebraic relationship. The symbols for the relationships are given here. The equations are found in Chapters 14 through 18, and inequalities are found in Chapter 19.

- = means that the first value *is equal to* or the same as the value that follows.
- ≠ means that the first value *is not equal to* the value that follows.
- ≈ means that one value is *approximately the same* or *about the same* as the value that follows; this is used when rounding numbers.
- ≤ means that the first value is *less than or equal to* the value that follows.
- < means that the first value is *less than* the value that follows.
- ≥ means that the first value is *greater than or equal to* the value that follows.
- > means that the first value is *greater than* the value that follows.

EXAMPLE

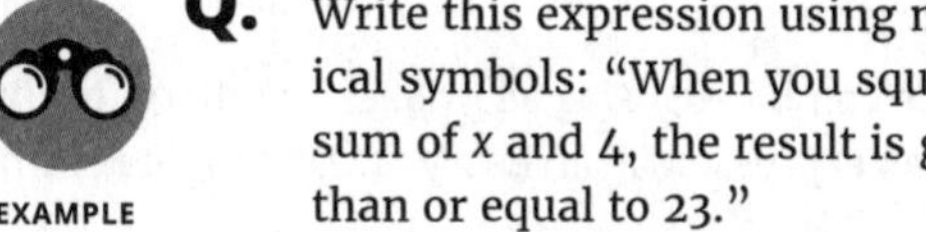

Q. Write this expression using mathematical symbols: "When you square the sum of x and 4, the result is greater than or equal to 23."

A. $(x+4)^2 \geq 23$. The point of the inequality symbol always faces the smaller value.

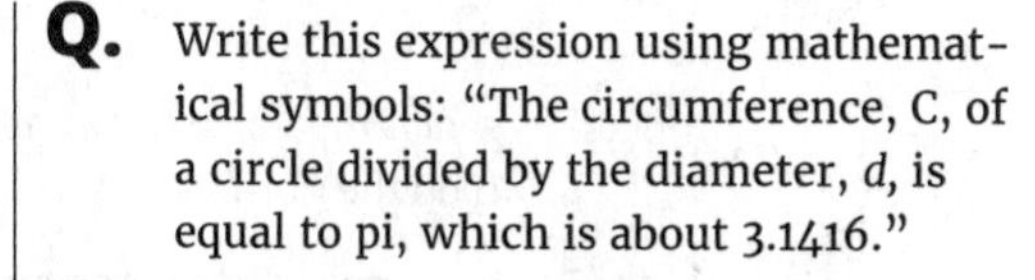

Q. Write this expression using mathematical symbols: "The circumference, C, of a circle divided by the diameter, *d*, is equal to pi, which is about 3.1416."

A. $\frac{C}{d} = \pi \approx 3.1416$. The wavy equal sign means "approximately" or "about."

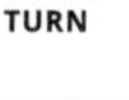

YOUR TURN

Write the expression using the correct symbols.

19 When you multiply the difference between *z* and 3 by 9, the product is equal to 13.

20 Dividing 12 by *x* is approximately the cube of 4.

21 The sum of y and 6 is less than the product of x and -2.

22 The square of m is greater than or equal to the square root of n.

Taking on algebraic tasks

Algebra involves symbols, such as variables and operation signs, which are the tools that you can use to make algebraic expressions more usable and readable. These things go hand in hand with simplifying, factoring, and solving problems, which are easier to solve if broken down into basic parts. Using symbols is actually much easier than wading through a bunch of words.

- To *simplify* means to combine all that can be combined, using allowable operations, to cut down on the number of terms, and to put an expression in an easily understandable form.
- To *factor* means to change two or more terms to just one term using multiplication. (See Chapters 11 through 13 for more on factoring.)
- To *solve* means to find the answer. In algebra, it means to figure out what the variable stands for. You solve for the variable to create a statement that is true. (You see solving equations and inequalities in Chapters 14 through 19.)
- To *check* your answer means to replace the variable with the number or numbers you have found when solving an equation or inequality, and show that the statement is true.

EXAMPLE

Q. *Simplify* the expression: $4+9-x$

A. You can add the two numbers. Simplified, you get $13-x$. You see all the types of simplifying methods in Chapter 7.

Q. *Factor* the expression: $5\cdot 6+5\cdot 11$

A. The two terms have a common factor of 5. Make this one term by taking out the common factor and writing it times the sum of the remaining factors. This can be $5(6+11)$ or, simplifying, it's $5(17)=85$. Factoring methods are found in Chapters 11 through 13.

Q. *Solve* the equation for the value of x: $x - 3 = 0$.

A. The only number that will make this equation a true statement is the number 3. So $x = 3$. Solving equations is covered in Chapters 14 through 18.

Q. *Check* to see if it's true that $x = -2$ in the equation $5 + x = 3$.

A. Replace the x in the equation and simplify on the left. You have $5 + (-2) = 3$, which is the same as $5 - 2 = 3$ or $3 = 3$. Yes, this is the solution to the equation.

YOUR TURN

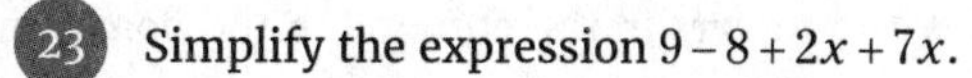

23 Simplify the expression $9 - 8 + 2x + 7x$.

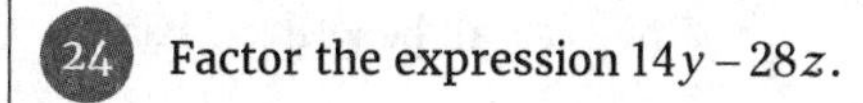

24 Factor the expression $14y - 28z$.

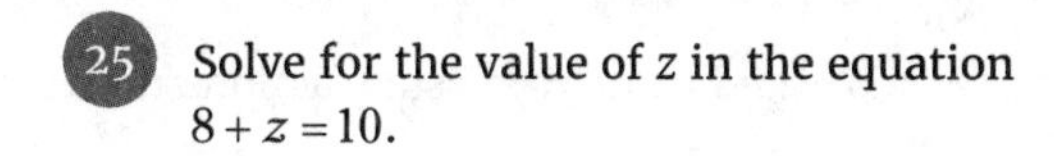

25 Solve for the value of z in the equation $8 + z = 10$.

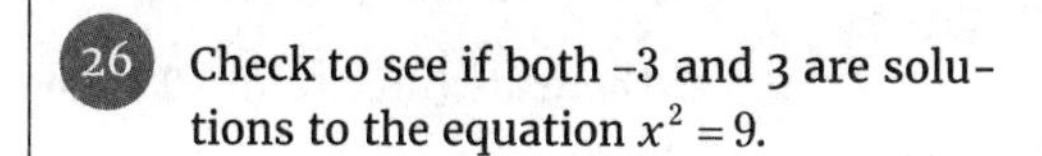

26 Check to see if both –3 and 3 are solutions to the equation $x^2 = 9$.

Practice Questions Answers and Explanations

1. **The numbers 15, 11, and $\sqrt{9}$ are natural numbers.** The number $\sqrt{9}$ qualifies, because it simplifies to the number 3.

2. **The numbers $-41, 15, 11, -\frac{12}{3}$, and $\sqrt{9}$ are integers.** The fraction $-\frac{12}{3}$ simplifies to -4, and $\sqrt{9}$ is equal to 3.

3. **The numbers -41, 15, -5.2, 11, 3.2121..., $-\frac{12}{3}$, $\frac{14}{11}$, and $\sqrt{9}$ are all rational numbers.** They can all be written as a fraction with integers in the numerator and denominator.

4. **The number $\sqrt{10}$ is the only irrational number listed here.** Technically, $\sqrt{-5}$ can be written as $\sqrt{5}i$, but the i makes the number imaginary. And the $\sqrt{9}$ can be simplified to create a rational number.

5. **The numbers 11 and 3 are prime.** Yes, $\sqrt{9}$ appears again!

6. **The only imaginary number is $\sqrt{-5}$ or $\sqrt{5}i$.**

7. 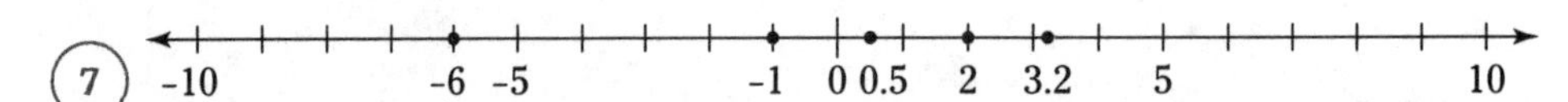

The points 0.5 and 3.2 are approximated, because the tick marks aren't on the number line.

8. 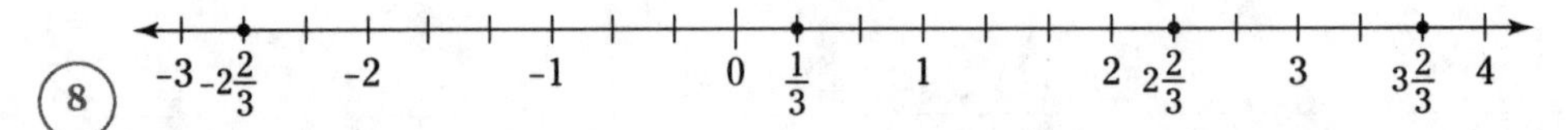

9. **3.** There are three terms. The term $4x$ is separated from $3x^3$ by subtraction and from 11 by addition.

10. **5.** There are two terms, and each has a different number of factors. The first term, $3xy$, has three factors: the 3 and the x and the y are multiplied together. The second term, $2z$, has two factors, the 2 and the z. So there are a total of five factors.

11. **x and y; h, k, a, b, 1.** The two variables are x and y. The constants are h, k, a, b, and 1.

12. **2, $\frac{1}{2}$, and 1.** The exponent in the term z^2 is the 2; the z is the base. The exponent in $z^{1/2}$ is $\frac{1}{2}$. And, even though it isn't showing, there's an implied exponent in the term z; it's assumed to be a 1, and the term can be written as z^1.

13. $4z+\sqrt{11}$. You could put the dot between the 4 and the z to indicate multiplication, but writing them together assumes they're being multiplied.

14. $\frac{x-2}{\pi}$. Use a fraction to indicate multiplication, rather than the slash.

15. $\frac{6|8+y|}{\sqrt{9-x}}$. Writing the 6 next to the absolute value indicates multiplication. The subtraction is written completely under the radical.

16. $3y-\sqrt{x}$.

17. $\frac{2+y}{11}$ **or** $(2+y)/11$.

18. The braces contain two terms: the 16 and the fraction. The 5 in front of the braces indicates that multiplication will be performed. The fraction line has the term $4(11-z)$ in the numerator and the number 12 in the denominator. And the parentheses have a multiplier of 4 in front and the two terms 11 and *z* inside, which need to be subtracted.

19. $\mathbf{(z-3)9=13}$ **or** $\mathbf{9(z-3)=13}$**.** The 9 can be written behind or in front of the parentheses.

20. $\frac{12}{x} \approx 4^3$. The *x* goes in the denominator.

21. $\mathbf{y+6<-2x}$ **or** $\mathbf{y+6<x(-2)}$**.** Use parentheses if the –2 follows the *x*.

22. $m^2 \geq \sqrt{n}$. Use the greater-than-or-equal-to symbol.

23. $\mathbf{1+9x}$**.** You subtract the 8 from the 9, and you add the 2*x* and 7*x* to get 9*x*. (It's like adding 2 apples to 7 apples!)

24. $\mathbf{14(y-2z)}$**.** Both 14*y* and –28 are divisible by 14. So you write the 14 outside the parentheses and put the division results inside the parentheses.

25. $\mathbf{z=2}$**.** The only number you can add to 8 to get a result of 10 is 2.

26. **Yes, they are both solutions.** $(3)^2=9$ and $(-3)^2=9$

If you're ready to test your skills a bit more, take the following chapter quiz that incorporates all the chapter topics.

Whaddya Know? Chapter 1 Quiz

Complete each problem. You can find the solutions and explanations in the next section.

1. Which of the following numbers can be termed *rational*?

   $-3,\ \frac{13}{11},\ \sqrt{-25},\ 4.431321,\ 102$

2. Which of the following is a solution of the equation $\frac{1+x}{5}=3$?

   A) 5 B) –4 C) 14 D) 15 E) 16

3. How many terms are there in $2x(x-4)$?

4. What is the exponent in $\frac{(x-h)^2}{3}-\frac{(y-k)^2}{4}=1$?

5. Which of the following is: "the product of four times the difference between *x* and eight divided by the sum of nine and x^2"?

   A) $\frac{4x-8}{9+x^2}$ B) $\frac{4(x-8)}{9+x^2}$ C) $\frac{4(x+8)}{9+x^2}$ D) $\frac{x-8}{4(9-x^2)}$ E) $\frac{9-x^2}{4(x-8)}$

6. Write the following using the corresponding mathematical symbols: "The quotient of *x* and 11 is greater than or equal to the sum of 9 and *y*."

7. Write the following using the corresponding mathematical symbols: "The difference between 9 and z is less than the product of 9 and z."

8. How many terms are there in $3x^2 - 5x - 2$?

9. What is a common factor of the terms $3a + 3a^2 + 3a^3 + 3$?

10. Which of the following numbers can be termed real?

$$-3.\overline{42},\ \frac{13}{111},\ 4+3i,\ 53^2,\ -\frac{1}{1001}$$

11. What is the constant term in $\frac{x}{4} + \frac{y}{9} = 1$?

12. What is the coefficient in $x^2 + 5x - 11$?

13. Place the following numbers on the number line: $-1.5,\ -0.6,\ 0.8$

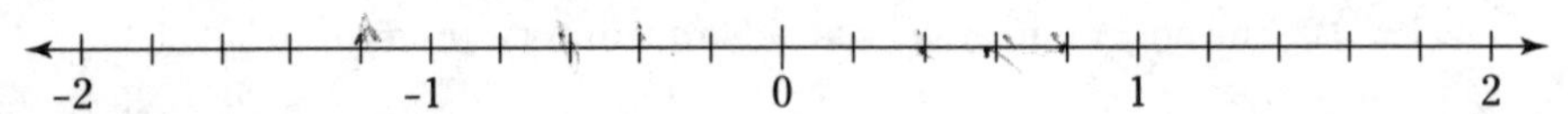

14. Write the following using the corresponding mathematical symbols: "The square root of 20 is about 4.5."

15. Simplify: $4x^2 + 3 - x^2 + 7$.

Answers to Chapter 1 Quiz

1. $-3, \frac{13}{11}, 4.431321, 102$. All of the terms are rational except for $\sqrt{-25}$. The number -3 is an integer and is rational, because it can be written as a fraction such as $-\frac{3}{1}$. The number $\frac{13}{11}$ is already written as a fraction and is rational. The number 4.431321 is a terminating decimal and can be written as $4\frac{431321}{1{,}000{,}000}$ or $\frac{4{,}431{,}321}{1{,}000{,}000}$. The number 102 is a whole number and can be written as a fraction over 1. The number $\sqrt{-25}$ is imaginary, so it can't be a rational number.

2. **C. 14.** Replacing the *x* with 5, you get $\frac{6}{5}$, which is not equal to 3. Replacing the *x* with –4, you get $\frac{-3}{5}$, which is not equal to 3. Replacing the *x* with 15, you get $\frac{16}{5}$, which is not equal to 3. Replacing the *x* with 16, you get $\frac{17}{5}$, which is not equal to 3. Replacing the *x* with 14, you get $\frac{15}{5}$, which IS equal to 3.

3. **1.** The 2*x* multiplies a binomial, making the expression all one term. There are two terms in the parentheses, but the expression is still just one term.

4. **2.** There are two exponents, each on the terms in the parentheses. They are both 2.

5. **B.** The answer $\frac{4x-8}{9+x^2}$ only multiplies the *x* and not the 8. The answer $\frac{4(x+8)}{9+x^2}$ finds the sum of *x* and 8, not the difference. The answer $\frac{x-8}{4(9-x^2)}$ has the 4 multiplier in the denominator instead of the numerator. And the answer $\frac{9-x^2}{4(x-8)}$ has reversed the numerator and denominator.

6. $\frac{x}{11} \geq 9+y$. The quotient refers to division, and the sum refers to addition. The point of the symbol faces the smaller side.

7. $9-z<9z$. The difference refers to subtraction, and the product refers to multiplication. The point of the symbol faces the smaller result.

8. **3.** The three terms are separated by the two subtraction symbols.

9. **3.** The number 3 divides each of the terms evenly (leaving no remainder).

10. $-3.\overline{42}, \frac{13}{111}, 53^2, -\frac{1}{1001}$ All of the terms are real except for $4+3i$. The number $-3.\overline{42}$ is a repeating decimal and is a rational number; it can be written as $-3\frac{14}{33}$ or $-\frac{113}{33}$. See Chapter 4 for more on repeating decimals. The number $\frac{13}{111}$ is a rational number and so is real. The number 53^2 is equal to 2,809 and is a whole number. And $-\frac{1}{1001}$ is a rational number, already written as a fraction. The number $4+3i$ has the imaginary factor of *i*, so it is imaginary and not real.

(11) **1.** The number 1 is a term that stands alone and isn't multiplying or dividing any other number. The 4 and 9 are both part of the coefficients of their respective terms.

(12) **5.** The 5 multiplies the variable x.

(13) 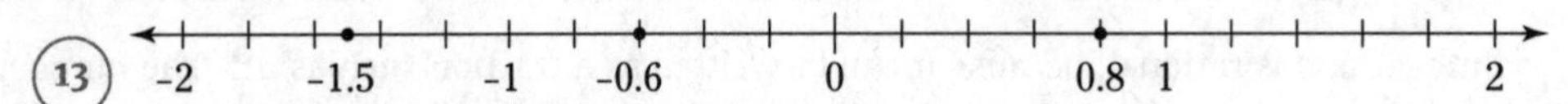

The number line is broken up into units of 0.2 in length. The number –1.5 has to be estimated, as it's halfway between –1.6 and –1.4. The other two numbers have tick marks to rest on.

(14) $\sqrt{20} \approx 4.5$. The "wavy equal sign" symbol means the answer is approximate and has been rounded.

(15) $3x^2 + 10$. The two x^2 terms are combined by subtracting 1 from 4. The two constants are added together.

» Recognizing operations

» Operating on signed numbers: adding, subtracting, multiplying, and dividing

Chapter 2
Deciphering Signs in Expressions

Numbers have many characteristics: They can be big, little, even, odd, whole, fractions, positive, negative, and sometimes cold and indifferent. (I'm kidding about that last one.) Chapter 1 describes numbers' different names and categories. But this chapter concentrates mainly on how numbers compare to one another, what their comparison looks like on the number line, the positive and negative characteristics of numbers, and how a number's sign reacts to different manipulations. This chapter tells you how to add, subtract, multiply, and divide signed numbers, no matter whether all the numbers are all the same sign or a combination of positive and negative.

Assigning Numbers Their Place

Positive numbers are greater than 0. They're on the opposite side of 0 from the negative numbers. If you were to arrange a tug-of-war between positive and negative numbers, the positive numbers would line up on the right side of 0. Negative numbers get smaller and smaller, the farther they are from 0. This situation can get confusing because you may think that −400 is *bigger* than −12. But just think of −400°F and −12°F. Neither is anything pleasant to think about, but −400°F is definitely less pleasant — colder, lower, smaller.

Using the number line

When comparing *negative* numbers, the number closer to 0 is the *bigger* or *greater* number. You may think that recognizing that 16 is bigger than 10 is an easy concept. But what about −1.6 and −1.04? Which of these numbers is bigger?

REMEMBER

The easiest way to compare numbers and to tell which is bigger or has a greater value is to find each number's position on the number line. The number line goes from negatives on the left to positives on the right (see Figure 2-1). Whichever number is farther to the right has the greater value, meaning it's bigger.

FIGURE 2-1: A number line.

EXAMPLE

Q. Using the number line in Figure 2-1, determine which is larger, −16 or −10.

A. The number −10 is to the right of −16, so it's the bigger of the two numbers.

Q. Which is larger, −1.6 or −1.04?

A. The number −1.04 is to the right of −1.6, so it's larger. A nice way to compare decimals is to write them with the same number of decimal places. So rewrite −1.6 as −1.60; it's easier to compare to −1.04 in this format.

Now that you've seen some examples of using a number line to compare numbers, try the following problems for practice. Use the number line found in Figure 2-2.

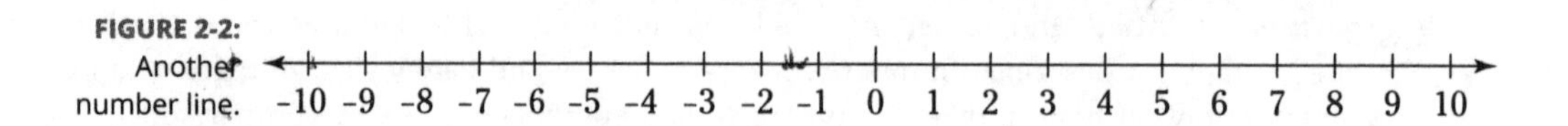

FIGURE 2-2: Another number line.

YOUR TURN

1 Which number is larger, 4.6 or −9.2?

2 Which number is larger, −1 or 0?

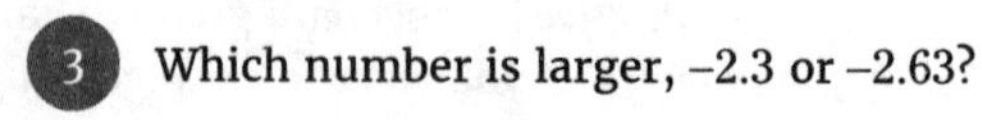

3 Which number is larger, −2.3 or −2.63?

Put the numbers in order from smallest to largest: 4, 0, –5, –3.2.

Put the numbers in order from smallest to largest: $-\frac{1}{3}, \frac{1}{6}, -2\frac{1}{4}, -3\frac{1}{8}$.

Comparing positives and negatives with symbols

Although my mom always told me not to compare myself to other people, comparing numbers to other numbers is often useful. And, when you compare numbers, the greater-than sign (>) and less-than sign (<) come in handy, which is why I use them in Table 2-1, where I put some positive- and negative-signed numbers in perspective.

TABLE 2-1 Comparing Positive and Negative Numbers

Comparison	What It Means
$6 > 2$	6 is greater than 2; 6 is farther from 0 than 2 is.
$10 > 0$	10 is greater than 0; 10 is positive and is bigger than 0.
$-5 > -8$	-5 is greater than -8; -5 is closer to 0 than -8 is.
$-300 > -400$	-300 is greater than -400; -300 is closer to 0 than -400 is.
$0 > -6$	0 is greater than -6; -6 is negative and is smaller than 0.
$7 > -80$	7 is greater than -80; -80 is negative and is smaller than 0.

REMEMBER

Positive numbers are always bigger than negative numbers.

Two other signs related to the greater-than and less-than signs are the greater-than-or-equal-to sign (≥) and the less-than-or-equal-to sign (≤).

So, putting the numbers 6, −2, −18, 3, 16, and −11 in order from smallest to largest gives you −18, −11, −2, 3, 6, and 16, which are shown as dots on a number line in Figure 2-3.

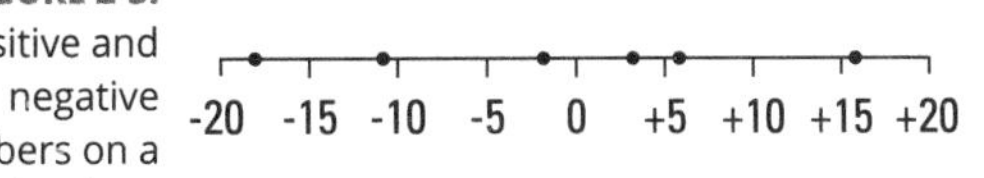

FIGURE 2-3: Positive and negative numbers on a number line.

EXAMPLE

Q. Write the description using math notation: –14 is greater than –20.

A. $-14 > -20$. The point of the arrow is always in the direction of the smaller of the two numbers.

Q. Write the description using math notation: $3\frac{1}{3}$ is less than or equal to 0.

A. $3\frac{1}{3} \leq 0$. You could also write the inequality as $0 \geq 3\frac{1}{3}$, which is read, "0 is greater than or equal to $3\frac{1}{3}$."

YOUR TURN

6 Write the description using math notation: –8 is less than 1.

7 Write the description using math notation: –3 is greater than –30.

8 Write the description using math notation: –11 is less than or equal to 11.

9 Write the description using math notation: –10 is greater than or equal to –16.

10 Write the description using math notation: –1 is greater than or equal to –1.

Zeroing in on Zero

But what about 0? I keep comparing numbers to see how far they are from 0. Is 0 positive or negative? The answer is that it's neither. Zero has the unique distinction of being neither positive nor negative. Zero separates the positive numbers from the negative ones — what a job! When using the number line to determine the order of numbers — which one is larger — you look at how far the number is from 0. You already know that a positive number is going to be larger than a negative number, but comparing two negative numbers can be a bit more challenging. You put the two negative numbers on the number line. The negative number that's farthest from 0 is the smaller of the two numbers.

EXAMPLE

Q. Which number is larger, -3 or -13?

A. -3. Look at the following number line. You see that -3 is to the right of -13. In terms of distance from 0, -13 is much farther away, so it is smaller than -3.

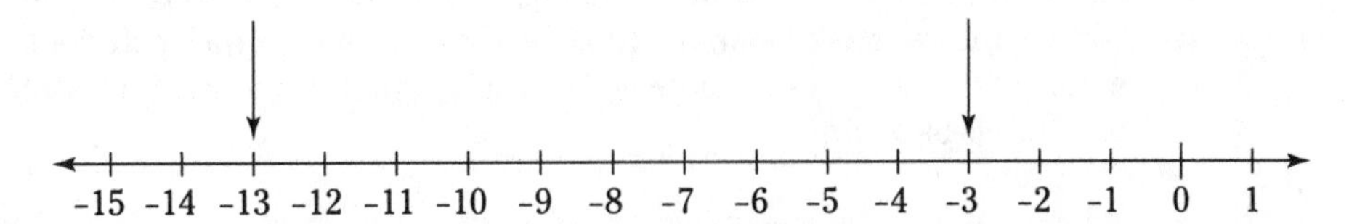

YOUR TURN

11 Which is larger, -2 or -8?

Which has the greater value, -13 or -1?

Which is bigger, -0.003 or -0.03?

Which is larger, $-\frac{1}{6}$ or $-\frac{2}{3}$?

Going in for Operations

Operations in algebra are nothing like operations in hospitals. Well, you get to dissect things in both, but dissecting numbers is a whole lot easier (and a lot less messy) than dissecting things in a hospital.

Algebra is just a way of generalizing arithmetic, so the operations and rules used in arithmetic work the same for algebra. Some new operations do crop up in algebra, though, just to make things more interesting than adding, subtracting, multiplying, and dividing. I introduce three of those new operations after explaining the difference between a binary operation and a non-binary operation.

Sorting out types of operations

Operations in mathematics come in all shapes and sizes. There are the basic operations that you first ran into when you started school, and then you have the operations that are special to one branch of mathematics or another. The operations are universal; they work in all languages and at all levels of math.

Breaking into binary operations

Bi means two. A *bi*cycle has two wheels. A *bi*gamist has two spouses. A *bi*nary operation involves two numbers. Addition, subtraction, multiplication, and division are all *binary operations* because you need two numbers to perform them. You can add $3+4$, but you can't add $3+$ if there's nothing after the plus sign. You need another number.

Introducing nonbinary operations

A *nonbinary operation* needs just one number to accomplish what it does. A nonbinary operation performs a task and spits out the answer. *Square roots* are nonbinary operations. You find $\sqrt{4}$ by performing this operation on just one number (see Chapter 6 for more on square roots). Another important nonbinary operation is *absolute value*. It will be used in the upcoming sections, where you subtract numbers. And two other important nonbinary operations are *factorial* and *greatest integer*. It gets better and better!

Getting it absolutely right with absolute value

The absolute value operation, indicated by two vertical bars around a number, $|\ |$, is greatly related to the number line, because it tells you how far a number is from 0 without any regard to the sign of the number. The absolute value of a number is its value without a sign. The *absolute value* doesn't pay any attention to whether the number is less than or greater than 0; it just determines how *far* it is from 0.

The formal definition of the absolute value operation is:

$$|a| = \begin{cases} a \text{ if } a \geq 0 \\ -a \text{ if } a < 0 \end{cases}$$

So, essentially, if a number is positive or 0, then its absolute value is exactly that number. If the number you're evaluating is negative, then you find its opposite — or you make it a positive number.

Getting the facts straight with factorial

The *factorial* operation looks like someone took you by surprise. You indicate that you want to perform the operation by putting an exclamation point after a number. If you want 6 factorial, you write "6!". Okay, I've given you the symbol, but you need to know what to do with it.

To find the value of $n!$, you multiply that number by every positive integer smaller than n.

$$n! = n(n-1)(n-2)(n-3)\cdots 3\cdot 2\cdot 1$$

There's one special rule when using factorial: $0! = 1$. This is by definition. The value of 0! is designated as being 1. This result doesn't really fit the rule for computing the factorial, but the mathematicians who first described the factorial operation designated that 0! is equal to 1 so that it worked with their formulas involving permutations, combinations, and probability.

Deciphering Signs in Expressions

Getting the most for your math with the greatest integer

You may have never used the *greatest integer* function before, but you've certainly been its victim. Utility and phone companies and sales tax schedules use this function to get rid of fractional values. Do the fractions get dropped off? Why, of course not. The amount is rounded up to the next greatest integer.

REMEMBER

The greatest integer function takes any real number that isn't an integer and changes it to the greatest integer it exceeds. If the number is already an integer, then it stays the same.

The symbol for the greatest integer function is a set of brackets, [ ]. You put your number in question in the brackets, evaluate it, and out pops the answer.

$$[n] = \begin{cases} n \text{ if } n \text{ is an integer} \\ \text{"biggest integer not greater than } n\text{" if } n \text{ is not an integer} \end{cases}$$

EXAMPLE

Q. Find the absolute value: $|-4|$

A. $|-4| = 4$. The distance from −4 to 0 is 4 units.

Q. Evaluate: $|6-6|$

A. $|6-6| = |0| = 0$. You perform the operation inside the absolute value bars before evaluating.

Q. Evaluate 3!

A. $3! = 3\cdot 2\cdot 1 = 6$

Q. Evaluate 6!

A. $6! = 6\cdot 5\cdot 4\cdot 3\cdot 2\cdot 1 = 720$

Q. Evaluate: $\left[6\frac{1}{2}\right]$

A. $\left[6\frac{1}{2}\right]=6$. The number 6 is the biggest integer that is not larger than $6\frac{1}{2}$.

Q. Evaluate: $[-3.87]$

A. $[-3.87]=-4$. Just picture the number line. The number −3.87 is to the right of −4, so the greatest integer not exceeding −3.87 is −4. In fact, a good way to compute the greatest integer is to picture the value's position on the number line and slide back to the closest integer to the left — if the value isn't already an integer.

YOUR TURN

15 Determine which is greater: $\left[5\frac{1}{2}\right]$ or 3!

16 Determine which is greater: $\left[4\frac{1}{2}\right]$ or $|-4|$.

17 Determine which is greater: 5! or $|-100|$.

18 Determine which is greater: $[-9.99]$ or $|-9.9|$.

Tackling the Basic Binary Operations

What is a *binary operation*? A bicycle has two wheels. A biannual term lasts two years. And a binary operation requires two numbers. These operations are performed on two numbers — one written before the operation symbol and one after. Addition and subtraction are pretty familiar, but the multiplication and division symbols come in several varieties.

Adding signed numbers

If you're on an elevator in a building that has four floors above the ground floor and five floors below ground level, you can have a grand time riding the elevator all day, pushing buttons, and actually "operating" with signed numbers. If you want to go up five floors from the third sub-basement, you end up on the second floor above ground level.

You're probably too young to remember this, but people actually used to get paid to be elevator operators and push buttons all day. I wonder if these people had to understand algebra first.

Deciphering Signs in Expressions

Adding like to like: Same-signed numbers

When your first-grade teacher taught you that $1+1=2$, they probably didn't tell you that this was just one part of the whole big addition story. They didn't mention that adding one positive number to another positive number is really a special case. If they *had* told you this big-story stuff — that you can add positive and negative numbers together or add any combination of positive and negative numbers together — you might have packed up your little school bag and sack lunch and left the room right then and there.

Adding positive numbers to positive numbers is just a small part of the whole addition story, but it was enough to get you started at that time. This section gives you the big story — all the information you need to add numbers of any sign. The first thing to consider in adding signed numbers is to start with the easiest situation — when the numbers have the same sign. Look at what happens:

- You have three CDs and your friend gives you four new CDs:

  $(+3)+(+4)=+7$

  You now have seven CDs.

- You owed Jon $8 and had to borrow $2 more from him:

  $(-8)+(-2)=-10$

  Now you're $10 in debt.

TIP

There's a nice *S* rule for addition of positives to positives and negatives to negatives. See if you can say it quickly three times in a row: *When the signs are the same, you find the sum, and the sign of the sum is the same as the signs.* This rule holds when *a* and *b* represent any two real numbers:

$$(+a)+(+b)=+(|a|+|b|)$$
$$(-a)+(-b)=-(|a|+|b|)$$

I wish I had something as alliterative for all the rules, but this is math, not poetry!

Say you're adding −3 and −2. The signs are the same; so you find the sum of 3 and 2, which is 5. The sign of this sum is the same as the signs of −3 and −2, so the *sum* is also a negative.

Here are some examples of finding the sums of same-signed numbers:

- $(+8)+(+11)=+19$: The signs are both positive, and so is the sum.
- $(-14)+(-100)=-114$: The sign of the sum is the same as the signs.
- $(+4)+(+7)+(+2)=+13$: Because all the numbers are positive, add them and make the sum positive, too.
- $(-5)+(-2)+(-3)+(-1)=-11$: This time all the numbers are negative, so add them and give the sum a minus sign.

Adding same-signed numbers is a snap! (A little more alliteration for you.)

Adding different signs

Can a relationship between a Leo and a Gemini ever add up to anything? I don't know the answer to that question, but I do know that numbers with different signs add up very nicely. You just have to know how to do the computation, and, in this section, I tell you.

TIP

When the signs of two numbers are different, forget the signs for a while and find the *difference* between the numbers. This is the difference between their *absolute values* (see the "Getting it absolutely right with absolute value" section, earlier in this chapter). The number farther from 0 determines the sign of the answer.

$$(+a)+(-b)=+(|a|-|b|)$$ if the positive a is farther from 0.

$$(+a)+(-b)=-(|b|-|a|)$$ if the negative b is farther from 0.

Look what happens when you add numbers with different signs:

- You had $20 in your wallet and spent $12 for your movie ticket:

  $(+20)+(-12)=+8$

  After settling up, you have $8 left. You knew the answer would be positive, because +20 is farther from 0 than −12, and the difference between 20 and 12 is 8.

- I have $20, but it costs $32 to fill my car's gas tank:

  $(+20)+(-32)=-12$

  I'll have to borrow $12 to fill the tank. This time the answer will be negative, because −32 is farther from 0 than +20. The difference between the two numbers is 12.

Here's how to solve these two situations using the rules for adding signed numbers.

- **$(+20)+(-12)=+8$:** Find the difference between 20 and 12: $20-12=8$. Because 20 is farther from 0 than 12, the result is positive, so $(+20)+(-12)=+(20-12)=+8$.
- **$(+20)+(-32)=-12$:** Find the difference between 20 and 32: $32-20=12$. Because 32 is farther from 0 than 20 and is a negative number, the result is negative, so $(+20)+(-32)=-(32-20)=-12$.

Here are some more examples of finding the sums of numbers with different signs:

- **$(+6)+(-7)=-1$:** The difference between 6 and 7 is 1. Because 7 is farther from 0 than 6 is, and 7 is negative, the answer is –1.
- **$(-6)+(+7)=+1$:** This time the 7 is positive. It's still farther from 0 than 6 is, and so the answer is +1.
- **$(-4)+(+3)+(+7)+(-5)=+1$:** If you take these operations in order from left to right (although you can add in any order you like), you add the first two together to get –1. Add –1 to the +7 to get +6. Then add +6 to –5, the last number, to get +1.

EXAMPLE

Q. $(-6)+(-4)=-(6+4)=$

A. The signs are the same, so you find the sum and apply the common sign. The answer is -10.

Q. 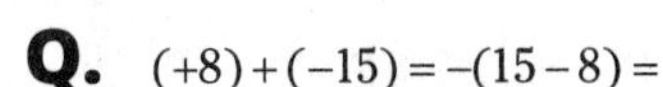
$(+8)+(-15)=-(15-8)=$

A. The signs are different, so you find the difference and use the sign of the number with the larger absolute value. The answer is -7.

YOUR TURN

19 $4+(-3)=$

20 $5+(-11)=$

21 $(-18)+(-5)=$

22 $47+(-33)=$

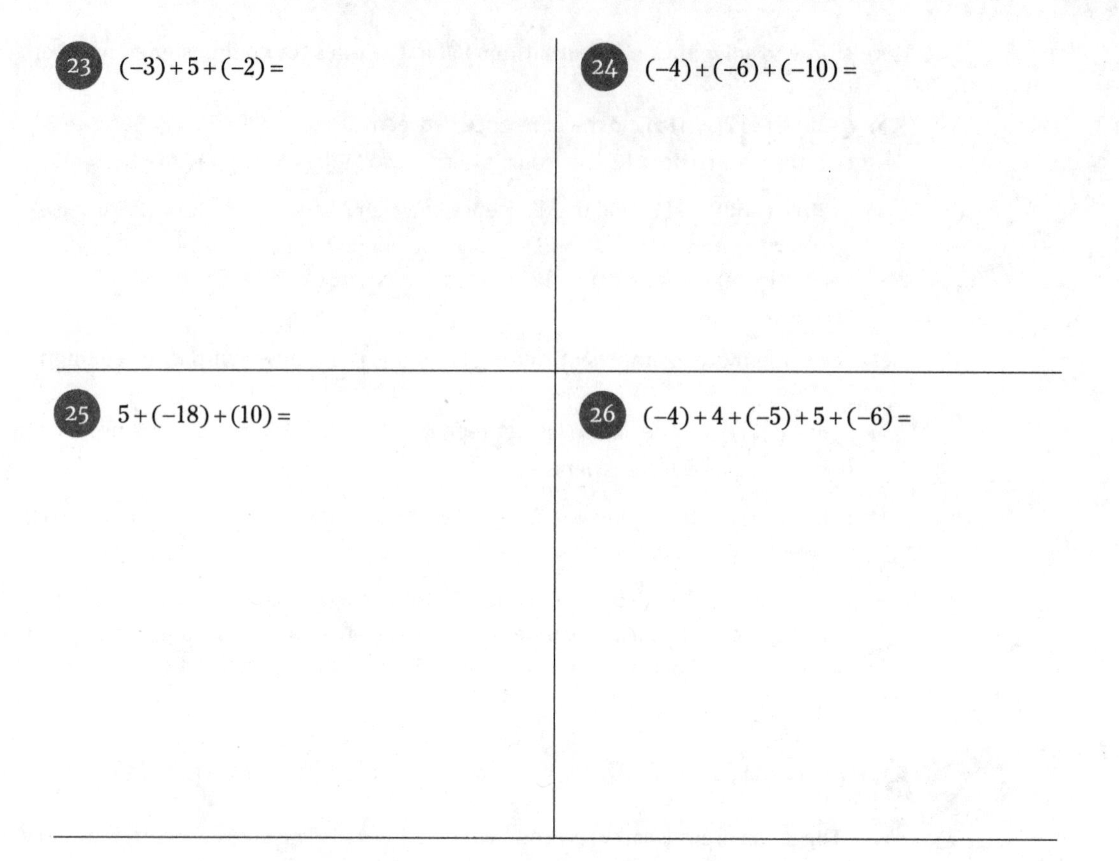

23. $(-3)+5+(-2)=$

24. $(-4)+(-6)+(-10)=$

25. $5+(-18)+(10)=$

26. $(-4)+4+(-5)+5+(-6)=$

Making a Difference with Signed Numbers

Subtracting signed numbers is really easy to do: You *don't!* Instead of inventing a new set of rules for subtracting signed numbers, mathematicians determined that it's easier to change the subtraction problems to addition problems and use the rules that you find in the previous section. Think of it as an original form of recycling.

Consider the method for subtracting signed numbers for a moment. Just change the subtraction problem into an addition problem? It doesn't make much sense, does it? Everybody knows that you can't just change an arithmetic operation and expect to get the same or right answer. You found out a long time ago that $10-4$ isn't the same as $10+4$. You can't just change the operation and expect it to come out correctly.

So, to make this work, you really change *two* things. (It almost seems to fly in the face of *two wrongs don't make a right,* doesn't it?)

TIP

When subtracting signed numbers, change the minus sign to a plus sign *and* change the number that the minus sign was in front of (the second number) to its opposite. Then just add the numbers using the rules for adding signed numbers.

- $(+a)-(+b)=(+a)+(-b)$
- $(+a)-(-b)=(+a)+(+b)$
- $(-a)-(+b)=(-a)+(-b)$
- $(-a)-(-b)=(-a)+(+b)$

EXAMPLE

These first examples put the process of subtracting signed numbers into real-life terms:

Q. The submarine was 60 feet below the surface when the skipper shouted, "Dive!" It went down another 40 feet. What is the submarine's depth now?

A. $-60-(+40)=-60+(-40)=-100$. Change from subtraction to addition. Change the 40 to its opposite, −40. Then use the addition rule. The submarine is now 100 feet below the surface.

Q. Some kids are pretending that they're on a reality-TV program and clinging to some footholds on a climbing wall. One team challenges the position of the opposing team's player. "You were supposed to go down 3 feet, then up 8 feet, then down 4 feet. You shouldn't be 1 foot higher than where you started!" The referee decides to check by having the player go backward, by making the player do the opposite, or subtracting the moves. What was the result?

A. Putting a negative sign in front of each assigned move, you have: $-(-3)-(+8)-(-4)=+(+3)+(-8)+(+4)=-5+(+4)=-1$. The player ended up 1 foot lower than where they started, so they had moved correctly in the first place.

And now here are some examples of subtracting signed numbers:

Q. Solve: $-16-4$

A. $-16-4=-16+(-4)=-20$. The subtraction becomes addition, and the +4 becomes negative. Then, because you're adding two signed numbers with the same sign, you find the sum and attach their common negative sign.

Q. Solve: $-3-(-5)$

A. $-3-(-5)=-3+(+5)=2$. The subtraction becomes addition, and the −5 becomes positive. When adding numbers with opposite signs, you find their difference. The 2 is positive because the +5 is farther from 0.

REMEMBER

To subtract two signed numbers:

$$a-(+b)=a+(-b) \quad and \quad a-(-b)=a+(+b)$$

YOUR TURN

27 $5-(-2)=$

28 $-6-(-8)=$

29 $4-87=$

30 $0-(-15)=$

31 $2.4-(-6.8)=$

32 $-15-(-11)=$

Multiplying Signed Numbers

When you multiply two or more signed numbers, you just multiply them without worrying about the sign of the answer until the end. Then, to assign the sign, just count the number of negative signs in the problem. If the number of negative signs is an even number, the answer is positive. If the number of negative signs is odd, the answer is negative.

REMEMBER

The product of two signed numbers:

$$(+)(+)=+ \quad \text{and} \quad (-)(-)=+$$
$$(+)(-)=- \quad \text{and} \quad (-)(+)=-$$

The product of more than two signed numbers:

$(+)(+)(+)(-)(-)(-)(-)$ has a *positive* answer because there are an *even* number of negative factors.

$(+)(+)(+)(-)(-)(-)$ has a *negative* answer because there are an *odd* number of negative factors.

EXAMPLE

Q. (−2)(−3) =

A. Multiply the two factors without their signs, and you get 6. There are two negative signs in the problem, so the result is positive. The answer is +6.

Q. (−2)(+3)(−1)(+1)(−4) =

A. There are three negative signs in the problem, so the result is negative. The product of the numbers (without their signs) is 24. The answer is −24.

YOUR TURN

33 (−6)(3) =

34 (14)(−1) =

35 (−6)(−3) =

36 (6)(−3)(4)(−2) =

37 (−1)(−1)(−1)(−1)(−1)(2) =

38 (−10)(2)(3)(1)(−1) =

Dividing Signed Numbers

The rules for assigning the sign of the answer when dividing signed numbers are exactly the same as those for multiplying signed numbers (see "Multiplying Signed Numbers" earlier in this chapter.) The rules do differ, though, because you have to divide, not multiply.

REMEMBER

When you divide signed numbers, just count the number of negative signs in the problem — in the numerator, in the denominator, and perhaps in front of the problem. If you have an even number of negative signs, the answer is positive. If you have an odd number of negative signs, the answer is negative.

EXAMPLE

Q. $\frac{-36}{-9} =$

A. There are two negative signs in the problem, which is even, so the answer is positive. The answer is +4.

Q. $\frac{-(-3)(-12)}{4} =$

A. There are three negative signs in the problem, which is odd, so the answer is negative. The answer is −9.

Q. $-\frac{121}{-11} =$

A. There are two negative signs — one of them in front of the fraction — so the answer is positive. The answer is +11.

YOUR TURN

39 $\frac{-22}{-11} =$

40 $\frac{24}{-3} =$

41 $-\frac{3(-4)}{-2} =$

42 $\frac{(-5)(2)(3)}{-1} =$

43 $\frac{(-2)(-3)(-4)}{(-1)(-6)} =$

44 $\frac{-1{,}000{,}000}{1{,}000{,}000} =$

Working with Nothing: Zero and Signed Numbers

What role does 0 play in the signed-number show? What does 0 do to the signs of the answers? Well, when you're doing addition or subtraction, what 0 does depends on where it is in the problem. When you multiply or divide, 0 tends to just wipe out the numbers and leave you with nothing.

Here are some general guidelines about 0:

- **Adding zero:** $0 + a$ is just a. Zero doesn't change the value of a. (This is also true for $a + 0$.)
- **Subtracting zero:** When doing subtraction with 0, order matters. Is 0 being subtracted or are you subtracting from 0?
  - $a - 0 = a$. When you subtract 0 from a number, you don't change it.
  - $0 - a = -a$. Use the rule for subtracting signed numbers: Change the operation from subtraction to addition and change the sign of the second number, giving you $0 + (-a)$.
- **Multiplying by 0:** $a \times 0 = 0$. Twice nothing is nothing; three times nothing is nothing; multiply by nothing and you get nothing; likewise, $0 \times a = 0$.
- **Dividing 0 by a number:** $0 \div a = 0$. Take you and your friends: If none of you has anything, dividing that *nothing* into shares just means that each share has nothing.

  Note: You can't use 0 as a divisor. Numbers can't be divided by 0; not even 0 can be divided by 0. The answers just don't exist.

So, working with 0 isn't too tricky. You follow normal addition and subtraction rules, and just keep in mind that multiplying and dividing with 0 (0 being divided) leaves you with nothing — literally.

Q. Perform the operation: $-3 + 0 =$

A. –3. Adding 0 to any number doesn't change the number – even when the number is negative!

Q. Perform the operations: $67(611 + 4{,}231) \times 0 =$

A. 0. You want to pay attention to the whole problem. The first operations result in a huge number, but the final result is 0. That's what 0 does best!

YOUR TURN

45 $4 + 0 =$

46 $0 - 4 =$

47 $4 \times 0 =$

48 $\frac{0}{4} =$

Practice Questions Answers and Explanations

1. **4.6.** The number 4.6 is to the right of 0 on the number line, and –9.2 is to the left of 0. And, of course, a positive number will be larger than a negative number.

2. **0.** Even though 0 isn't positive (or negative), it's to the right of –1, so 0 is larger.

3. **–2.3.** To make the comparison easier, rewrite –2.3 as –2.30, so it has the same number of decimal places as –2.63. The number –2.30 is closer to 0 (on the left of 0), so it is larger.

4. **–5, –3.2, 0, 4.** The two negative numbers are the smallest, with the –5 being farthest left and the smallest of the two. The number 4 is, of course, larger than 0.

5. **$-3\frac{1}{8}$, $-2\frac{1}{4}$, $-\frac{1}{3}$, $\frac{1}{6}$.** The only positive number is $\frac{1}{6}$, so it is largest. The number $-\frac{1}{3}$ is closest to 0, so it's the largest of the negative numbers. The number $-3\frac{1}{8}$ is the farthest from 0, so it's the smallest.

6. **$-8 < 1$.** The point of the inequality goes toward the smaller number.

7. **$-3 > -30$.** The point of the inequality goes toward the smaller number, the –30.

8. **$-11 \leq 11$.** The number –11 is smaller than 11, so the point of the inequality points in that direction.

9. **$-10 \geq -16$.** The point of the inequality always goes toward the smaller of the two numbers.

10. **$-1 \geq -1$.** The number –1 is equal to itself, so the "greater than" part is just the other choice.

11. **–2.** The following number line shows that the number –2 is to the right of –8. The number –8 is farther from 0, so it is the smaller of the two numbers.

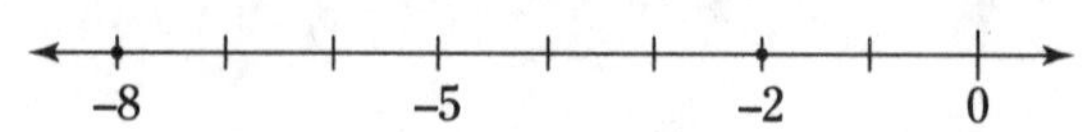

12. **–1.** The number –1 is to the right of –13. The number –13 is farther from 0.

13. **–0.003.** The following number line shows that the number –0.003 is to the right of –0.03, which means –0.003 is bigger than –0.03. You can also rewrite –0.03 as –0.030 for easier comparison.

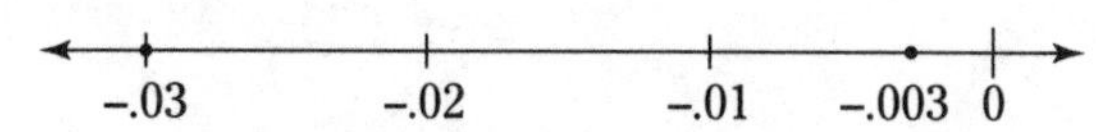

14. **$-\frac{1}{6}$.** The number $-\frac{2}{3} = -\frac{4}{6}$, and $-\frac{4}{6}$ is to the left of $-\frac{1}{6}$ on the following number line. So $-\frac{1}{6}$ is larger than $-\frac{2}{3}$.

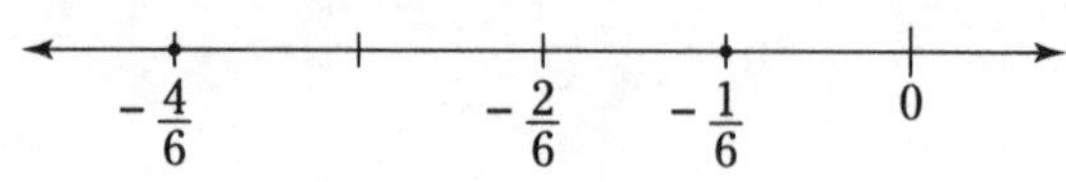

15. **3!** $\left[5\frac{1}{2}\right] = 5$ and $3! = 3 \cdot 2 \cdot 1 = 6$. $6 > 5$.

16. **Neither.** $\left[4\frac{1}{2}\right]=4$ and $|-4|=4$. They are equal.

17. **5!** $5!=5\cdot 4\cdot 3\cdot 2\cdot 1=120$ and $|-100|=100$. Thus $120>100$.

18. $|-9.9|$. $[-9.99]=-10$ and $|-9.9|=9.9$. Thus $9.9>-10$.

19. **+1.** The number 4 has the greater absolute value, so the answer is positive.

$$4+(-3)=+(4-3)=1$$

20. **−6**. The number −11 has the greater absolute value, so the answer is negative.

$$5+(-11)=-(11-5)=-6$$

21. **−23.** Both of the numbers have negative signs; when the signs are the same, find the sum of their absolute values.

$$(-18)+(-5)=-(18+5)=-23$$

22. **14.** Number 47 has the greater absolute value.

$$47+(-33)=+(47-33)=14$$

23. **0.**

$$(-3)+5+(-2)=[(-3)+5]+(-2)=(2)+(-2)=0$$

24. **−20.**

$$(-4)+(-6)+(-10)=-(4+6)+(-10)=(-10)+(-10)=-(10+10)=-20$$

25. **−3.**

$$5+(-18)+(10)=-(18-5)+10=-(13)+10=-(13-10)=-3$$

Or you may prefer to add the two numbers with the same sign first, like this:

$$5+(-18)+(10)=(5+10)+(-18)=15+(-18)=-(18-15)=-3$$

26. **−6**. You can do this because order and grouping (association) don't matter in addition.

$$(-4)+4+(-5)+5+(-6)=[(-4)+4]+[(-5)+5]+(-6)=0+0+(-6)=-6$$

You see more on the associative property in Chapter 3.

27. **7**. Change the −2 to +2 and the subtraction to addition. The signs are the same now, so find the sum.

$$5-(-2)=5+(+2)=7$$

28. **2**. Change the −8 to +8 and the subtraction to addition. The signs are different now, so find the difference.

$$-6-(-8)=-6+(+8)=8-6=2$$

(29) **−83**. Change the +87 to −87 and the subtraction to addition. Yes, it looks like the 87 is negative, but that is the operation of subtraction, not the sign of the number.

$4-87=4+(-87)=-(87-4)=-83$

(30) **15.** Change the −15 to +15 and the subtraction to addition.

$0-(-15)=0+15=15$

(31) **9.2.** Change the −6.8 to +6.8 and the subtraction to addition.

$2.4-(-6.8)=2.4+6.8=9.2$

(32) **−4.** Change the −11 to +11 and the subtraction to addition.

$-15-(-11)=-15+11=-(15-11)=-4$

(33) **−18.** The multiplication problem has one negative, and 1 is an odd number.

(34) **−14.** The multiplication problem has one negative, and 1 is an odd number.

(35) **18.** The multiplication problem has two negatives, and 2 is an even number.

(36) **144.** The multiplication problem has two negatives.

(37) **−2.** The multiplication problem has five negatives.

(38) **60.** The multiplication problem has two negatives.

(39) **2.** The division problem has two negatives.

(40) **−8.** The division problem has one negative.

(41) **−6.** Three negatives result in a negative.

(42) **30.** The division problem has two negatives.

(43) **−4.** The division problem has five negatives.

(44) **−1.** The division problem has one negative.

(45) **4.** Adding 0 to a number doesn't change the number.

(46) **−4.** Change the problem to $0+(-4)$ and add.

(47) **0.** Multiplying by 0 always gives you 0 as a result.

(48) **0.** Dividing 0 by a nonzero number always gives you 0.

If you're ready to test your skills a bit more, take the following chapter quiz that incorporates all the chapter topics.

Whaddya Know? Chapter 2 Quiz

Quiz time! Complete each problem to test your knowledge on the various topics covered in this chapter. You can then find the solutions and explanations in the next section.

1. $(-4)(3)(-1)(-5)(0)(6)(-2) =$
2. Use the inequality symbols to write the statement: "14 is greater than or equal to 12.9."
3. $[-11.3] =$
4. $0 \div 8 =$
5. $-15 + (-2) =$
6. $\frac{-16}{8} =$
7. $-10 + 4 =$
8. $-1 - (-5) =$
9. Put the numbers in order from smallest to largest: $0,\ -1\frac{1}{3},\ -5.7,\ \frac{2}{7},\ 4$
10. $|-11.3| =$
11. $(-6)(-3)(-1) =$
12. $-\frac{-14}{-2} =$
13. Which number is larger, –8 or –6?
14. $\left[2\frac{1}{3}\right] \cdot |-1| =$ (And, yes, that's the greatest integer function.)
15. $4! =$
16. $-12 - 4 =$

Deciphering Signs in Expressions

Answers to Chapter 2 Quiz

1. **0.** Yes, there are four negative signs, but 0 has no sign.

2. $14 \geq 12.9$. The point of the inequality always faces the smaller of the two numbers.

3. **−12.** The number −12 is to the left of −11.3, so it is smaller. You want the largest negative integer that is smaller than the −11.3.

4. **0.** Any number dividing into 0 (except 0 itself) results in 0.

5. **−17.** When the signs are the same, you find the sum: $-15+(-2)=-(15+2)=-17$.

6. **−2.** There is one negative sign.

7. **−6.** When the signs are different, you find the difference: $-10+4=-(10-4)=-6$. And the sign of that difference is the sign of the number with the greater absolute value.

8. **4.** Change the subtraction to addition, and change the second number to the opposite sign: $-1-(-5)=-1+(+5)=+(5-1)=+4$.

9. $-5.7,\ -1\frac{1}{3},\ 0,\ \frac{2}{7},\ 4$. The two negative numbers come first, with −5.7 being the smaller of the two. Then comes 0, which separates the negative numbers from the positive numbers. The fraction is smaller than 1, so it's definitely smaller than 4.

10. **11.3.** This is the absolute value function. When applied, it produces positive numbers or 0.

11. **−18.** There are three negative signs.

12. **−7.** There are three negative signs, counting the sign in front of the fraction.

13. **−6.** The number −6 is closer to 0 than −8, so −8 is the smaller value.

14. **2.** $\left[2\frac{1}{3}\right]=2$ and $|-1|=1$. The product of the two results is 2.

15. **24.** $4!=4\cdot 3\cdot 2\cdot 1=24$.

16. **−16.** Change the subtraction to addition, and change the second number to the opposite sign: $-12-4=-12+(-4)=-(12+4)=-16$.

» **Distributing over addition and subtraction**

» **Incorporating inverses and identities**

» **Utilizing the associative and commutative rules**

Chapter 3
Incorporating Algebraic Properties

Algebra has rules for everything, including a sort of shorthand notation to save time and space. The notation that comes with any particular property cuts down on misinterpretation because it's very specific and universally known. (I give the guidelines for doing operations like addition, subtraction, multiplication, and division in Chapter 2.) In this chapter, you see the specific rules that apply when you use grouping symbols and rearrange terms. You also find how opposites attract — or not — in the form of inverses and identities.

Getting a Grip on Grouping Symbols

The most commonly used *grouping symbols* in algebra are (in order from most to least common):

- Parentheses ( )
- Brackets [ ]
- Braces { }
- Fraction lines / or —
- Radicals $\sqrt{}$
- Absolute value symbols | |

Here's what you need to know about grouping symbols: You must compute whatever is inside them (or under or over, in the case of the fraction line) first, before you can use that result to solve the rest of the problem. If what's inside isn't or can't be simplified into one term, then anything outside the grouping symbol that multiplies one of the terms has to multiply them all — that's the *distributive property,* which I cover in the next section.

EXAMPLE

Q. $16-(4+2)=$

A. Add the 4 and 2; then subtract the result from the 16: $16-(4+2)=16-6=10$.

Q. Simplify $2[6-(3-7)]$.

A. Work from the inside out. First subtract the 7 from the 3; then subtract the −4 from the 6 by changing it to an addition problem. You can then multiply the 2 by the 10:

$$2[6-(3-7)]=2[6-(-4)]=2[6+4]=2[10]=20.$$

Q. $1-|-8+19|+3(4+2)=$

A. Combine what's in the absolute value and parentheses first, before combining the results:

$$\begin{aligned}1-|-8+19|+3(4+2)&=1-|11|+3(6)\\&=1-11+18=-10+18=8\end{aligned}$$

When you get to the three terms with subtraction and addition signs, $1-11+18$, you always perform the operations in order, reading from left to right. (See Chapter 7 for more on this process, called the order of operations.)

Q. $\frac{32}{30-2(3+4)}=$

A. You have to complete the work in the denominator first before dividing the 32 by that result:

$$\frac{32}{30-2(3+4)}=\frac{32}{30-2(7)}=\frac{32}{30-14}=\frac{32}{16}=2$$

As you're working through the problems, just remember to:

- Work from the inside out when there are several grouping symbols.
- Move from left to right when performing addition and/or subtraction on several different terms.

1. $3(2-5)+14=$

2. $4[3(6-8)+2(5+9)]-11=$

3. $5\{8[2+(6-3)]-4\}=$

4. $\dfrac{\sqrt{19-3(6-8)}}{6[8-4(5-2)]-1}=$

5. $4-5|6-3(8-2)|=$

6. $\dfrac{(9-1)5-4(6)}{\sqrt{11-2}-\sqrt{4-\sqrt{2+7}}}=$

Incorporating Algebraic Properties

Spreading, Grouping, and Changing the Order

Three important processes in algebra are the distributive property, the associative property, and the commutative property. Because some operations have these properties, it makes working with the variables and numbers so much easier and nicer.

Distributing the wealth

When an estate is "distributed," everyone hopes to get an equal share. The distributive property works the same way. The *distributive property* is used when you perform an operation on each of the terms within a grouping symbol. The following rules show distributing multiplication over addition and distributing multiplication over subtraction:

$$a(b+c)=a\cdot b+a\cdot c \quad \text{and} \quad a(b-c)=a\cdot b-a\cdot c$$

The distributive property is frequently employed when terms within parentheses or other grouping symbols cannot be combined. This allows for each term to interact with the multiplier.

EXAMPLE

Q. $3(6+x)=$

A. Distribute the 3 over the $6+x$ by multiplying each term by 3: $3(6+x)=3\cdot 6+3\cdot x$. Then simplify the terms: $3\cdot 6+3\cdot x=18+3x$.

Q. $5\left(a-\frac{1}{5}\right)=$

A. $5\times a-5\times\frac{1}{5}=5a-1$

YOUR TURN

7. $4(7+y)=$

8. $-3(x-11)=$

9. $\frac{2}{3}(6+15y)=$

10. $-8\left(\frac{1}{2}-\frac{1}{4}+\frac{3}{8}\right)=$

11 $4a(\pi - 2) =$

12 $5\left(z + \frac{4}{5} - 2\right) =$

Making Associations Work

The *associative property* has to do with grouping — that is, how you deal with two or more terms when you perform operations on them. Think about what the word *associate* means. When you associate with someone, you're close to the person, or you're in the same group with them. Say that Anika, Becky, and Cora associate. Whether Anika drives over to pick up Becky and the two of them go to Cora's house and pick her up, or Cora is at Becky's house and Anika picks up both of them at the same time, the same result occurs: all three are in the car at the end.

The *associative property* means that even if a particular grouping of the operation changes, the result remains the same. (If you need a reminder about grouping, refer to the section, "Getting a Grip on Grouping Symbols," earlier in this chapter.) Addition and multiplication are associative operations. Subtraction and division are *not* associative. So,

$$a + (b + c) = (a + b) + c$$
$$a \cdot (b \cdot c) = (a \cdot b) \cdot c$$
$$a - (b - c) \neq (a - b) - c \text{ (except in a few special cases)}$$
$$a \div (b \div c) \neq (a \div b) \div c \text{ (except in a few special cases)}$$

You can always find a few cases where the associative property works even though it isn't supposed to. For example, in the subtraction problem $5 - (4 - 0) = (5 - 4) - 0$, the property seems to work. Also, in the division problem $6 \div (3 \div 1) = (6 \div 3) \div 1$, it seems to work. I just picked numbers very carefully that would make it seem like you could associate using subtraction and division. Although there are exceptions, a rule must work *all* the time, not just in special cases.

Here's how the associative property works:

$4 + (5 + 8) = 4 + 13 = 17$ and $(4 + 5) + 8 = 9 + 8 = 17$, so $4 + (5 + 8) = (4 + 5) + 8$
$3 \times (2 \times 5) = 3 \times 10 = 30$ and $(3 \times 2) \times 5 = 6 \times 5 = 30$, so $3 \times (2 \times 5) = (3 \times 2) \times 5$

This rule is special to addition and multiplication. It doesn't work for subtraction or division. You're probably wondering why you would even use this rule. That's because it can sometimes make the computation easier.

EXAMPLE

Q. Use the associative property to create an easier problem: $14+(-14+111)=$

A. With the current grouping, you need to add –14 to 111 and then add the result to 14. Instead, take advantage of the associative property and regroup: $(14-14)+111$. Now you have a resulting problem of $0+111$ — much easier!

Q. Use the associate property to create an easier problem: $16\left(\frac{7}{8}\times 9\right)=$

A. Regroup, putting the first two numbers together:

$$16\left(\frac{7}{8}\times 9\right)=\left(16\times\frac{7}{8}\right)\cdot 9=\left(\cancel{16}^{2}\times\frac{7}{\cancel{8}}\right)\cdot 9=(14)\cdot 9=126$$

YOUR TURN

13 $16+(-16+47)=$

14 $(5-13)+13=$

15 $18\left(\frac{5}{9}\times 7\right)=$

16 $(110\times 8)\frac{1}{8}=$

Computing by Commuting

Before discussing the commutative property, take a look at the word *commute*. You probably commute to work or school and know that whether you're traveling from home to work or from work to home, the distance is the same: The distance doesn't change because you change directions (although getting home during rush hour may make that distance *seem* longer).

The same principle is true of *some* algebraic operations: It doesn't matter whether you add $1+2$ or $2+1$, the answer is still 3. Likewise, multiplying 2×3 or 3×2 yields 6.

TIP

The *commutative property* means that you can change the order of the numbers in an operation without affecting the result. Addition and multiplication are commutative. Subtraction and division are not. So,

$$a+b=b+a$$
$$a\cdot b=b\cdot a$$
$$a-b\neq b-a\ (\text{except in a few special cases})$$
$$a\div b\neq b\div a\ (\text{except in a few special cases})$$

In general, subtraction and division are *not* commutative. The special cases occur when you choose the numbers carefully. For example, if a and b are the same number, then the subtraction appears to be commutative because switching the order doesn't change the answer. In the case of division, if a and b are opposites, then you get -1 no matter which order you divide them in. By the way, this is why, in mathematics, big deals are made about proofs. A few special cases of something may work, but a real rule or theorem has to work *all* the time.

You can use this rule to your advantage when doing math computations. Sometimes, changing the order in addition or multiplication situations can make the work much easier.

Incorporating Algebraic Properties

EXAMPLE

Q. $5\times\frac{4}{7}\times\frac{1}{5}=$

A. You don't really want to multiply fractions unless it's necessary. Notice that the first and last factors are multiplicative inverses of one another:

$$5\times\frac{4}{7}\times\frac{1}{5}=$$

Just switch the order of the last two numbers:

$$5\times\frac{1}{5}\times\frac{4}{7}=\left(5\times\frac{1}{5}\right)\times\frac{4}{7}=(1)\frac{4}{7}=\frac{4}{7}$$

Q. $-3+16+303=$

A. The second and last terms are reversed, and then the first two terms are grouped. $-3+16+303=-3+303+16=(-3+303)+16=300+16=316$.

YOUR TURN

17 $8+5+(-8)=$

18 $5\times 47\times 2=$

19 $\frac{3}{5}\times 13\times 10=$

20 $-23+47+23-47+8=$

21 $16\times 18\times 25\times\frac{4}{5}\times\frac{7}{9}\times\frac{1}{8}=$

Relating Inverses and Identities

In mathematics, inverses and identities are closely related. The definition of an inverse includes references to an identity. And when describing the identity of an operation, you call up the inverses.

Investigating Inverses

In mathematics, the *inverse* of a number is tied to a specific operation.

The *additive inverse* of the number 5 is −5; the *additive inverse* of the number $-\frac{1}{3}$ is $\frac{1}{3}$. When you add a number and its additive inverse together, you always get 0, the *additive identity.* Every real number has an additive inverse, even the number 0. The number 0 is its own additive inverse. And all real numbers (except 0) and their inverses have opposite signs; the number 0 is neither positive nor negative, so there is no sign.

The *multiplicative inverse* of the number 5 is $\frac{1}{5}$; the *multiplicative inverse* of the number $-\frac{1}{3}$ is -3. When you multiply a number and its multiplicative inverse together, you always get 1, the *multiplicative identity.* Every real number except the number 0 has a multiplicative inverse. A number and its multiplicative inverse are always the same sign.

EXAMPLE

Q. Find the additive and multiplicative inverses of the number -14.

A. The additive inverse is 14, because $-14+14=0$. The multiplicative inverse of -14 is $-\frac{1}{14}$, because $-14\left(-\frac{1}{14}\right)=1$.

Q. Find the additive and multiplicative inverses of the number $\frac{3}{4}$.

A. The additive inverse is $-\frac{3}{4}$, because $\frac{3}{4}+\left(-\frac{3}{4}\right)=0$. The multiplicative inverse of $\frac{3}{4}$ is $\frac{4}{3}$, because $\frac{3}{4}\cdot\frac{4}{3}=\frac{\cancel{3}^1}{\cancel{4}^1}\cdot\frac{\cancel{4}^1}{\cancel{3}^1}=\frac{1}{1}=1$.

YOUR TURN

Find the additive and multiplicative inverses of the given number.

22 11

23 $-\frac{1}{3}$

24 $\frac{7}{3}$

25 -1

26 4.5

27 $-3\frac{3}{4}$

Identifying Identities

The term *identity* in mathematics is most frequently used in terms of a specific operation. When using addition, the *additive identity* is the number 0. You can think of it as allowing another number to keep its identity when 0 is added. If you add $7+0$, the result is 7. The number 7 doesn't change. When using multiplication, the *multiplicative identity* is the number 1. When you multiply 7×1, the result is 7. Again, the number 7 doesn't change.

When adding a number and its additive inverse together, you get the additive identity. So $-5+5=0$. And when multiplying a number and its multiplicative inverse together, you get the multiplicative identity. Multiplying, $6\left(\frac{1}{6}\right)=1$.

EXAMPLE

Q. Use an additive identity to change the expression $4x+5$ to an expression with only the variable term.

A. The additive inverse of 5 is −5. If you add −5 to the expression, you have $4x+5+(-5)$. Use the associative property to group the 5 and −5 together: $4x+(5+(-5))$. The sum of a number and its additive inverse is 0, so the expression becomes $4x+0$. Because 0 is the additive identity, $4x+0=4x$.

Q. Use a multiplicative identity to change the expression $\frac{x}{9}+1$ into an expression without a fraction.

A. Because the term $\frac{x}{9}$ can be written $\frac{1}{9}x$, you can identify the coefficient and work on it to find a multiplicative inverse. The multiplicative inverse of $\frac{1}{9}$ is 9. So multiply both terms by 9.

$$9\left(\frac{x}{9}+1\right)=9\cdot\frac{x}{9}+9\cdot1=\cancel{9}\cdot\frac{x}{\cancel{9}}+9\cdot1=x+9$$

Notice how the distributive function works for you here!

YOUR TURN

28 Use an additive identity to change the expression $9x-8$ to one with only the variable term.

29 Use an additive identity to change the expression $6-3x$ to one with only the variable term.

30 Use a multiplicative identity to change the term $-7x$ to one with only the variable factor.

31 Use a multiplicative identity to change the expression $\frac{x}{4}+\frac{3}{4}$ to one with the variable factor having an integer for a coefficient.

Working with Factorial

The nonbinary operation called *factorial* is important in problems involving probability, counting items, and, of course, is a basic function used in algebra. When you see $n!$, you know to take the number n and multiply it by every natural number smaller than it is:

$$n! = n(n-1)(n-2)\cdots 3\cdot 2\cdot 1$$

The number n must be a whole number. This means that 1! is rather redundant. There's no natural number smaller, so $1! = 1$. And then there's 0!. There's no natural number smaller than 0; in fact, 0 isn't a natural number itself! By definition, $0! = 1$. How can that be? Well, mathematicians decided it is so. And it wasn't arbitrary or a flip-of-the-coin. By assigning 0! the value 1, it makes all sorts of formulas for counting and other applications work consistently.

A particular challenge when working with factorials is to reduce fractions containing those functions. Basically, you find the common factors in the factorials in the numerator and denominator, eliminate them, and determine what's left.

EXAMPLE

Q. What is $\frac{4!}{3!} = ?$

A. If these were just the two numbers 4 and 3, you would either leave it as is, because 4 and 3 don't have any common factors other than 1, or you would write this as a mixed number. It's different with factorials. Rewrite the fraction after expanding the factorial values.

$$\frac{4!}{3!} = \frac{4\cdot 3\cdot 2\cdot 1}{3\cdot 2\cdot 1} = \frac{4\cdot \cancel{3}\cdot \cancel{2}\cdot \cancel{1}}{\cancel{3}\cdot \cancel{2}\cdot \cancel{1}} = 4$$

Q. What is $\frac{10!}{8!} = ?$

A. Write out the factorials and reduce the fraction.

$$\frac{10!}{8!} = \frac{10\cdot 9\cdot 8\cdot 7\cdot 6\cdot 5\cdot 4\cdot 3\cdot 2\cdot 1}{8\cdot 7\cdot 6\cdot 5\cdot 4\cdot 3\cdot 2\cdot 1} = \frac{10\cdot 9\cdot \cancel{8}\cdot \cancel{7}\cdot \cancel{6}\cdot \cancel{5}\cdot \cancel{4}\cdot \cancel{3}\cdot \cancel{2}\cdot \cancel{1}}{\cancel{8}\cdot \cancel{7}\cdot \cancel{6}\cdot \cancel{5}\cdot \cancel{4}\cdot \cancel{3}\cdot \cancel{2}\cdot \cancel{1}}$$
$$= \frac{10\cdot 9}{1} = 90$$

Q. What is $3!0! = ?$

A. The value of 0! is defined to be 1. This is by design/definition.

$$3!0! = 3\cdot 2\cdot 1\cdot 1 = 6$$

Q. What is $\frac{100!}{97!} = ?$

A. "Yikes!", you say. I have to write out the product of all the numbers from 100 down to 1 and then from 97 down to 1? "No," is the answer. You can take a shortcut. In the numerator, instead of writing from 97 down to 1, just use 97!. Here's how it works:

$$\frac{100!}{97!} = \frac{100\cdot 99\cdot 98\cdot 97!}{97!} = \frac{100\cdot 99\cdot 98\cdot \cancel{97!}}{\cancel{97!}} = \frac{100\cdot 99\cdot 98}{1} = 970{,}200$$

YOUR TURN

32 $\frac{6!}{5!} =$

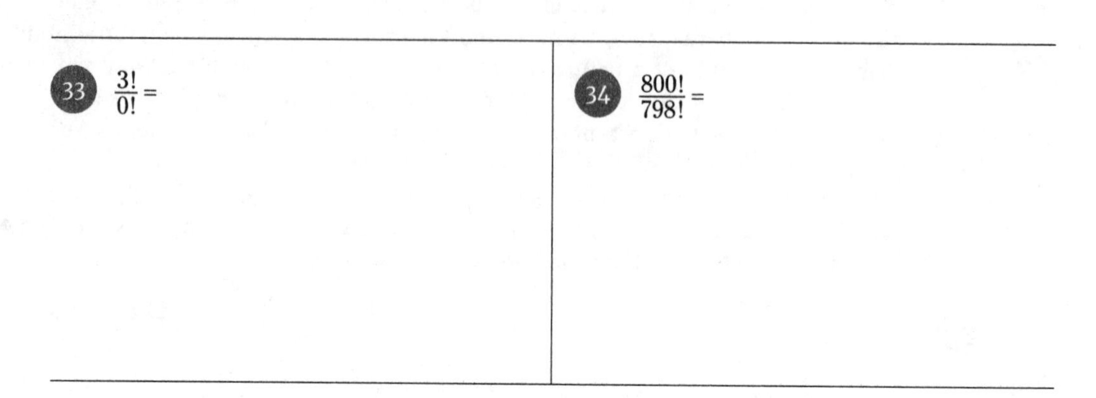

33 $\frac{3!}{0!} =$

34 $\frac{800!}{798!} =$

Applying the Greatest Integer Function

The *greatest integer function* is one of the nonbinary functions that is frequently used in algebra and its applications. This function is a method of rounding numbers. When you "round" a number to its nearest integer or tenth or thousandth or thousand, and so on, you move up or down to get to the closer value. With the greatest integer function, it doesn't matter how close, it matters in which direction.

When rounding numbers, you determine what digits need to be dropped and whether the target place value or number goes up by 1 or stays the same.

Suppose you want to round 3.667 to the nearest integer. The target number is the 3. Because the next digit, the 6, is greater than 5, you round the 3 up to 4 and eliminate the rest of the digits. You can replace them with 0 or, in this case, just keep the number 4. So, in rounded form, $3.667 \approx 4.000$ or $3.667 \approx 4$.

What if you're asked to round 1234.567 to the nearest hundred? The target number is the 2. The next digit down is 3, which is smaller than 5, so you leave the 2 as is and replace the 3 and the rest of the smaller digits with 0s. $1234.567 \approx 1200.000$ or $1234.567 \approx 1200$.

Note: When the digit below the target digit is exactly 5, and nothing else, you round up.

The greatest integer function acts a bit differently than rounding. It's similar in that the greatest integer function eliminates unwanted fractions or decimals, but the greatest integer function only goes in one direction: down. Or, if you already have an integer, the greatest integer leaves the number alone.

EXAMPLE

Q. Which is larger: 5.7 rounded to the nearest integer or $[5.7]$?

A. Rounding 5.7 to the nearest integer, you home in on the 7, which will be eliminated. Because 7 is larger than 5, you round up: $5.7 \approx 6.0$ or $5.7 \approx 6$. Using the greatest integer function, $[5.7] = 5$. So, rounding gives you the larger value.

Q. Compare $13\frac{1}{3}$ rounded to the nearest integer and $\left[13\frac{1}{3}\right]$.

A. Rounding $13\frac{1}{3}$ to the nearest integer, you drop off the $\frac{1}{3}$, because its decimal value is about 0.333, or less than half; so, $13\frac{1}{3}$ rounds to 13. Applying the greatest integer function, $\left[13\frac{1}{3}\right] \approx 13$. So the values are the same.

YOUR TURN

35 Round 146.95 to the nearest integer.

$[146.95] =$

Practice Question Answers and Explanations

1. **5.** First do the subtraction in the parentheses. Then multiply the result by 3. Finally, add 14.

$$3(2-5)+14=3(-3)+14=(-9)+14=5$$

2. **77.** Work inside the brackets first. Perform the two operations in the parentheses. Then perform the multiplications on the two results. Add the products. Then multiply that result by 4. Finally, subtract 11.

$$\begin{aligned}4\left[3(6-8)+2(5+9)\right]-11&=4\left[3(-2)+2(14)\right]-11\\&=4\left[-6+28\right]-11\\&=4\left[22\right]-11=88-11=77\end{aligned}$$

3. **180.** Multiplying by the 5 outside the braces will come last. First, do the subtraction in the parentheses. Then add inside the brackets. Multiply that result by 8, and then subtract 4. Finally, multiply by 5.

$$\begin{aligned}&5\{8[2+(6-3)]-4\}=5\{8[2+3]-4\}\\&=5\{8[5]-4\}=5\{40-4\}=5\{36\}=180\end{aligned}$$

4. $-\frac{1}{5}$**.** Your last step will be to divide the result in the numerator by the result in the denominator. In the numerator, first do the subtraction in the parentheses, multiply that result by 3, and then subtract that product from 19. Finally, find the square root of the difference. In the denominator, do the subtraction in the parentheses, multiply the result by 4, subtract that product from 8, and then multiply the difference by 6. Finally, subtract 1 from the product. Then you can do the division indicated by the fraction line.

$$\frac{\sqrt{19-3(6-8)}}{6\left[8-4(5-2)\right]-1}=\frac{\sqrt{19-3(-2)}}{6\left[8-4(3)\right]-1}=\frac{\sqrt{19+6}}{6[8-12]-1}=\frac{\sqrt{25}}{6[-4]-1}=\frac{5}{-24-1}=\frac{5}{-25}=-\frac{1}{5}$$

5. **−56.** Working inside the absolute value bars, first do the subtraction in the parentheses, multiply the difference by 3, and then subtract the product from 6. Find the absolute value of the result before multiplying by 5. Then subtract the product from 4.

$$\begin{aligned}4-5|6-3(8-2)|&=4-5|6-3(6)|=4-5|6-18|\\&=4-5|-12|=4-5(12)\\&=4-60=-56\end{aligned}$$

6. **8.** Work on the numerator and denominator separately — leaving the division for the last step. In the numerator, do the subtraction in the first parentheses and do the multiplication in the last term. Multiply the remainder by 5. Then subtract the second term from the first. In the denominator, do the subtraction under the first radical. Then look at the radical-within-the-radical and do the addition. Evaluate the two radical values you've formed. Next, find the difference under the remaining radical and evaluate it. Perform the subtraction in the denominator — and, finally, divide.

$$\frac{(9-1)5-4(6)}{\sqrt{11-2-\sqrt{4-\sqrt{2+7}}}}=\frac{(8)5-24}{\sqrt{9}-\sqrt{4-\sqrt{9}}}=\frac{40-24}{3-\sqrt{4-3}}=\frac{16}{3-\sqrt{1}}=\frac{16}{3-1}=\frac{16}{2}=8$$

If you're ready to test your skills a bit more, take the following chapter quiz that incorporates all the chapter topics.

7. $28+4y$. Multiply each term in the parentheses by 4.

$$4(7+y)=4\times 7+4\times y=28+4y$$

8. $-3x+33$. Multiply each term in the parentheses by −3.

$$-3(x-11)=(-3)x-(-3)(11)=-3x+33$$

9. $4+10y$. This problem involves operations with fractions. You will see more on fractions in Chapter 4.

$$\frac{2}{3}(6+15y)=\frac{2}{3}\times 6+\frac{2}{3}(15y)=\frac{2\times 6}{3}+\frac{2\times 15}{3}y=4+10y$$

10. **−5.**

$$-8\left(\frac{1}{2}-\frac{1}{4}+\frac{3}{8}\right)=(-8)\left(\frac{1}{2}\right)-(-8)\left(\frac{1}{4}\right)+(-8)\left(\frac{3}{8}\right)=-\frac{8}{2}+\frac{8}{4}-\frac{8\times 3}{8}$$
$$=-4+2-3$$
$$=(-4+2)-3=-2-3=-5$$

11. $4a\pi-8a$.

$$4a(\pi-2)=(4a\times\pi)-(4a\times 2)=4a\pi-8a$$

12. $5z-6$.

$$5\left(z+\frac{4}{5}-2\right)=5\times z+5\left(\frac{4}{5}\right)-5(2)=5z+\frac{5\times 4}{5}-10$$
$$=5z+4-10=5z-6$$

13. **47.** Regroup so that the first two numbers are together. Their sum is 0!

$$16+(-16+47)=[16+(-16)]+47=0+47=47$$

14. **5.** Regroup so that the second two numbers are together.

$$(5-13)+13=[5+(-13)]+13=5+[(-13)+13]=5+0=5$$

15. **70.** Regroup so that the first two numbers are together. The fraction reduces to make the multiplication easier.

$$18\left(\frac{5}{9}\times 7\right)=\left(18\times\frac{5}{9}\right)7=\left(\frac{18\times 5}{9}\right)7=(10)7=70$$

(16) **110.** Regroup so that the second two numbers are together. The multiplication is simple!

$$(110\times 8)\frac{1}{8}=110\left(8\times\frac{1}{8}\right)=110(1)=110$$

(17) **5.** Switch the order of the second and third numbers.

$$8+5+(-8)=5+8+(-8)=5+[8+(-8)]=5+0=5$$

(18) **470.** Switch the order of the second and third numbers.

$$5\times 47\times 2=5\times 2\times 47=(5\times 2)47=10\times 47=470$$

(19) **78.** Switch the order of the second and third numbers.

$$\frac{3}{5}\times 13\times 10=\frac{3}{5}\times 10\times 13=\left(\frac{3}{5}\times 10\right)13=\left(\frac{3\times 10}{5}\right)13=6\times 13=78$$

(20) **8.** Switch the order of the second and third numbers.

$$\begin{aligned}-23+47+23-47+8&=-23+23+47-47+8\\&=(-23+23)+(47-47)+8\\&=0+0+8=8\end{aligned}$$

(21) **560.** There's a lot of switching that goes on here to make the problem easier. Your goal is to be able to multiply fractions times numbers that will eliminate their denominators. In this particular problem, I'm bringing the $\frac{1}{8}$ up next to the 16 and the $\frac{7}{9}$ up next to the 18.

$$16\times 18\times 25\times\frac{4}{5}\times\frac{7}{9}\times\frac{1}{8}=16\times\frac{1}{8}\times 18\times\frac{7}{9}\times 25\times\frac{4}{5}$$

Then I group, reduce, multiply, and find the product of the three "nice" numbers!

$$\begin{aligned}\left(16\times\frac{1}{8}\right)\times\left(18\times\frac{7}{9}\right)\times\left(25\times\frac{4}{5}\right)&=\left(\cancel{16}^{2}\times\frac{1}{\cancel{8}}\right)\times\left(\cancel{18}^{2}\times\frac{7}{\cancel{9}}\right)\times\left(\cancel{25}^{5}\times\frac{4}{\cancel{5}}\right)\\&=(2\times 1)\times(2\times 7)\times(5\times 4)\\&=2\times 14\times 20=560\end{aligned}$$

(22) **-11 and $\frac{1}{11}$.** The sum of 11 and −11 is 0; the product of 11 and $\frac{1}{11}$ is 1.

(23) **$\frac{1}{3}$ and −3.** The sum of $-\frac{1}{3}$ and $\frac{1}{3}$ is 0; the product of $-\frac{1}{3}$ and −3 is 1.

(24) **$-\frac{7}{3}$ and $\frac{3}{7}$.** The sum of $\frac{7}{3}$ and $-\frac{7}{3}$ is 0; the product of $\frac{7}{3}$ and $\frac{3}{7}$ is 1.

(25) **1 and −1.** The sum of −1 and 1 is 0; the product of −1 and −1 is 1. The number is its own multiplicative inverse.

(26) -4.5 **and** $\frac{1}{4.5}$ **or** $\frac{2}{9}$**.** The sum of -4.5 and 4.5 is 0. The product of 4.5 and $\frac{1}{4.5}$ is 1. You usually don't want to write decimals in fractions, so you can change $\frac{1}{4.5}$ by writing it with a fractional denominator and simplifying:

$$\frac{1}{4.5}=\frac{1}{4\frac{1}{2}}=\frac{1}{9/2}=\frac{2}{9}$$

You'll see more on working with fractions in Chapter 4.

(27) $3\frac{3}{4}$ **and** $-\frac{4}{15}$**.** The additive inverse is just the positive version of the same number. To write the multiplicative inverse, you change the mixed number to an improper fraction and just flip it:

$$-3\frac{3}{4}=-\frac{15}{4}$$

There's a full coverage of fractions in Chapter 4.

(28) **Use 8.** The additive inverse of -8 is 8. If you add 8 to the expression, you have $9x-8+8$. Use the associative property to group the -8 and 8 together: $9x+(-8+8)$. The sum of a number and its additive inverse is 0, so the expression becomes $9x+0$. Because 0 is the additive identity, $9x+0=9x$.

(29) **Use −6.** The additive inverse of 6 is -6. If you add -6 to the expression, you have $6+(-6)-3x$. Use the associative property to group the 6 and -6 together: $(6+(-6))-3x$. The sum of a number and its additive inverse is 0, so the expression becomes $0-3x$. Because 0 is the additive identity, $0-3x=-3x$.

(30) **Use** $-\frac{1}{7}$**.** The multiplicative inverse of -7 is $-\frac{1}{7}$. If you multiply the expression by $-\frac{1}{7}$, you have $-7x\left(-\frac{1}{7}\right)$. Use the commutative property to rearrange the factors and the associative property to group the -7 and $-\frac{1}{7}$ together: $-7\cdot\left(-\frac{1}{7}\right)x=\left(-\frac{7}{1}\cdot-\frac{1}{7}\right)x$. The product of a number and its multiplicative inverse is 1, so the expression becomes $1x$. Because 1 is the multiplicative identity, $1x=x$.

(31) **Use 4.** The expression $\frac{x}{4}$ can be written as $\frac{1}{4}x$. The multiplicative inverse of $\frac{1}{4}$ is 4. If you multiply the expression by 4, you have

$$4\left(\frac{x}{4}+\frac{3}{4}\right)=4\cdot\frac{x}{4}+4\cdot\frac{3}{4}=\cancel{4}\cdot\frac{x}{\cancel{4}}+\cancel{4}\cdot\frac{3}{\cancel{4}}=x+3$$

An added bonus here is that the constant also becomes an integer.

(32) **6.** $\frac{6!}{5!}=\frac{6\cdot\cancel{5!}}{\cancel{5!}}=6$

(33) **6.** $\frac{3!}{0!}=\frac{3\cdot 2\cdot 1}{1}=6$ By definition, the value of 0! is 1.

(34) **639,200.** $\frac{800!}{798!}=\frac{800\cdot 799\cdot \cancel{798!}}{\cancel{798!}}=800\cdot 799=639{,}200$

(35) **147.** The 9, which will be dropped with the 5 following it, is greater than 5. Round the 6 in the ones place to a 7.

(36) **146.** The greatest integer smaller than 146.95 is 146.

If you're ready to test your skills a bit more, take the following chapter quiz that incorporates all the chapter topics.

Whaddya Know? Chapter 3 Quiz

Quiz time! Complete each problem to test your knowledge on the various topics covered in this chapter. You can then find the solutions and explanations in the next section.

1. $\{6-3[3+(7-4)]\}+(7.54-6.54)=$
2. Which is larger: 5.932 rounded to the nearest integer or $[5.932]$?
3. $\frac{82!}{80!}=$
4. Distribute: $-3(7+2a)=$
5. $23\times 47\times\frac{1}{23}\times\frac{2}{47}=$
6. What should you add to the expression $4y-5$ to get the variable term alone?
7. $\frac{7!}{3!}=$
8. Distribute: $5\left(\frac{1}{5}-\frac{x}{10}\right)=$
9. Apply the greatest integer function: $\left[6\frac{7}{8}\right]=$
10. $\frac{-5+\sqrt{25-4(3)(-2)}}{2(-2)}=$
11. $-60+4+60=$
12. Round to the nearest integer: $6\frac{7}{8}$
13. $2[4-(3-6)]=$
14. What should you multiply the expression $\frac{x}{3}+2$ by to get rid of the fraction?

Answers to Chapter 3 Quiz

1. **−11.** Work on the braces first. Do the subtraction in the parentheses and add the result to 3. Multiply that sum by 3 and subtract the product from 6. In the last parentheses, find the difference. Then add the result from the braces to this last number.

$$\begin{aligned}\{6-3[3+(7-4)]\}+(7.54-6.54) &= \{6-3[3+(3)]\}+(7.54-6.54)\\ &= \{6-3[6]\}+(7.54-6.54)\\ &= \{6-18\}+(7.54-6.54)\\ &= \{-12\}+(7.54-6.54)\\ &= -12+(1)\\ &= -11\end{aligned}$$

2. **5.932.** 5.932 rounds up to 6, and $[5.932]=5$.

3. **6,642.** $\frac{82!}{80!}=\frac{82\cdot 81\cdot \cancel{80!}}{\cancel{80!}}=82\cdot 81=6{,}642$

4. $-21-6a$. Multiply each term in the parentheses by -3.

$$-3(7+2a)=-3\cdot 7+(-3)\cdot 2a=-21-6a$$

5. **2.** Use the commutative property to reverse the order of the middle two factors.

$$\begin{aligned}23\times 47\times\frac{1}{23}\times\frac{2}{47} &= 23\times\frac{1}{23}\times 47\times\frac{2}{47}=\left(23\times\frac{1}{23}\right)\times\left(47\times\frac{2}{47}\right)=\\ &= \left(\cancel{23}\times\frac{1}{\cancel{23}}\right)\times\left(\cancel{47}\times\frac{2}{\cancel{47}}\right)=1\times 2=2\end{aligned}$$

You will find more on operations involving fractions in Chapter 4.

6. **5.** The additive inverse of −5 is 5. $4y-5+5=4y$

7. **840.** $\frac{7!}{3!}=\frac{7\cdot 6\cdot 5\cdot 4\cdot 3!}{3!}=\frac{7\cdot 6\cdot 5\cdot 4\cdot \cancel{3!}}{\cancel{3!}}=7\cdot 6\cdot 5\cdot 4=840$

8. $1-\frac{x}{2}$. Multiply each term in the parentheses by 5.

$$5\left(\frac{1}{5}-\frac{x}{10}\right)=5\cdot\frac{1}{5}-5\cdot\frac{x}{10}=\cancel{5}\cdot\frac{1}{\cancel{5}}-\cancel{5}\cdot\frac{x}{\cancel{10}^{2}}=1-\frac{x}{2}$$

9. **6.** The largest integer that is not bigger than $6\frac{7}{8}$ is 6.

10 $-\frac{1}{2}$. Work under the radical first. Multiply those last three factors together. Then subtract the result from 25. Find the square root of the difference. Then add -5 to the root. Multiply the two numbers in the denominator. And, finally, divide the numerator by the denominator.

$$\frac{-5+\sqrt{25-4(3)(-2)}}{2(-2)}=\frac{-5+\sqrt{25-(-24)}}{2(-2)}=\frac{-5+\sqrt{25+24}}{2(-2)}$$
$$=\frac{-5+\sqrt{49}}{2(-2)}=\frac{-5+7}{2(-2)}=\frac{2}{2(-2)}$$
$$=\frac{2}{-4}=-\frac{1}{2}$$

11 **4.** Use the commutative property to reverse the order of the last two numbers.

$$-60+4+60=-60+60+4=(-60+60)+4=0+4=4$$

12 **7.** $6\frac{7}{8}$ is equal to 6.875. Because the 8 is greater than 5, you round up to the next-larger integer.

13 **14.** First, do the subtraction in the parentheses. Then subtract that result from 4. Finally, multiply by 2.

$$2[4-(3-6)]=2[4-(-3)]=2[7]=14$$

14 **3.** The multiplicative inverse of $\frac{1}{3}$ is 3. Multiply both terms by this number:

$$3\left(\frac{x}{3}+2\right)=\cancel{3}\cdot\frac{x}{\cancel{3}}+3\cdot 2=x+6$$

» Making proportions work for you

» Operating on fractions and decimals

» Linking fractions and decimals

Chapter 4

Coordinating Fractions and Decimals

At one time or another, most math students wish that the world were made up of whole numbers only. But those non-whole numbers called *fractions* really make the world a wonderful place. (Well, that may be stretching it a bit.) In any case, fractions are here to stay, and this chapter helps you delve into them in all their wondrous workings. Compare developing an appreciation for fractions with watching or playing a sport: If you want to enjoy and appreciate a game, you have to understand the rules. You know that this is true if you watch soccer games. That offside rule is hard to understand at first. But, finally, you figure it out, discover the basics of the game, and love the sport. This chapter gets down to basics with the rules involving fractions so you can "play the game."

You may not think that decimals belong in a chapter on fractions, but there's no better place for them. Decimals are just a shorthand notation for the most favorite fractions. Think about the words that are often used and abbreviated, such as Mister (Mr.), Doctor (Dr.), Tuesday (Tue.), October (Oct.), and so on! Decimals are just fractions with denominators of 10, 100, 1,000, and so on, and they're abbreviated with periods, or *decimal points.*

Understanding fractions, where they come from, and why they look the way they do helps when you're working with them. A fraction has two parts:

$$\frac{\text{top}}{\text{bottom}}$$

or

$$\frac{\text{numerator}}{\text{denominator}}$$

REMEMBER

The *denominator* of a fraction, or bottom number, tells you the total number of items. The *numerator*, or top number, tells you how many of that total (the bottom number) are being considered.

In all the cases using fractions, the denominator tells you how many *equal* portions or pieces there are. Without the equal rule, you could get different pieces in various sizes. For example, in a recipe calling for $\frac{1}{2}$ cup of flour, if you didn't know that the one part was one of two *equal* parts, then there could be two *unequal* parts — one big and one little. Should the big or the little part go into the cookies?

Along with terminology like *numerator* and *denominator*, fractions fall into one of three types — *proper, improper,* or *mixed* — which I cover in the following sections.

Converting Improper Fractions and Mixed Numbers

An *improper fraction* is one where the *numerator* (the number on the top of the fraction) has a value greater than or equal to the *denominator* (the number on the bottom of the fraction) — so the fraction is top heavy. Improper fractions can be written as *mixed numbers* or whole numbers, and vice versa. A *mixed number* contains both a whole number *and* a fraction.

To convert a mixed number to an improper fraction: Multiply the integer (A) times the denominator (D) and add the numerator (N). Put that result over the denominator (D).

$$A\frac{N}{D}=\frac{A\cdot D+N}{D}$$

To convert an improper fraction to a mixed number: Divide the numerator (N) by the denominator (D). Put the quotient (Q) in front as the integer, the remainder (R) as the numerator, and the denominator (D) in its usual place.

$$\frac{N}{D}\rightarrow D\overline{)N}\rightarrow \begin{array}{c}Q+R\\ D\overline{)N}\end{array}\rightarrow Q\frac{R}{D}$$

You sometimes want an improper fraction, and other times a mixed number; it just depends on what you're doing at the time. You can easily change from one form to the other.

EXAMPLE

Q. After the party, Maria puts all the leftover pieces of pizza together. There are 15 pieces, each $\frac{1}{8}$ of a pizza. How much pizza does Maria have left?

A. Maria has a whole pizza plus seven pieces more: $\frac{15}{8} = 1\frac{7}{8}$. (So she'll have to save two pizza boxes to put the leftovers in.)

Q. A recipe calls for $\frac{2}{3}$ cup of sugar, but you want to double the recipe (you have a hungry family). Doubling the sugar requires $\frac{4}{3}$ cups. If you're using a 1-cup measuring cup, your cup will runneth over. How much more than a cup of sugar is this?

A. $\frac{4}{3} = 1\frac{1}{3}$, so you'll need a full cup plus $\frac{1}{3}$ cup more.

YOUR TURN

1. Change the improper fraction $\frac{29}{8}$ to a mixed number.

2. Change the mixed number $4\frac{5}{9}$ to an improper fraction.

3. Change the mixed number $4\frac{7}{100}$ to an improper fraction.

4. Change the mixed number $-2\frac{1}{13}$ to an improper fraction.

5. Change the improper fraction $\frac{402}{11}$ to a mixed number.

6. Change the improper fraction $-\frac{19}{7}$ to a mixed number.

Finding Fraction Equivalences

In algebra, all sorts of computations and manipulations use fractions. In many problems, you have to change the fractions so that they have the same denominator or so that their form is compatible with what you need to solve the problem. Two fractions are *equivalent* if they have the same value, such as $\frac{1}{2}$ and $\frac{3}{6}$. To create an equivalent fraction from a given fraction, you multiply or divide both the numerator and denominator by the same number. This technique is basically the same one you use to reduce a fraction.

Rewriting fractions

When you multiply or divide the numerator and denominator of a fraction by the same number, you don't change the value of the fraction. In fact, you're basically multiplying or dividing by 1 because any time the numerator and denominator of a fraction are the same number, it equals 1.

For example, how much of a 32-ounce package are you using if your recipe calls for 12 ounces? You're using $\frac{12}{32}$ of the package. If you divide both the 12 and the 32 by 4, you're basically dividing $\frac{12}{32}$ by $\frac{4}{4}$, which equals 1. The same goes for multiplying the numerator and denominator by the same number. So, you're using $\frac{3}{8}$ of the package.

Here's another example: Create two new versions of the fraction $\frac{16}{20}$ — one with smaller numbers and one with larger numbers.

To come up with smaller numbers, you need to find something that divides both the 16 and the 20 evenly. There are two possibilities: 2 and 4.

$$\frac{16 \div 2}{20 \div 2} = \frac{8}{10} \text{ and } \frac{16 \div 4}{20 \div 4} = \frac{4}{5}$$

The fraction $\frac{4}{5}$ is said to be in "lowest possible terms" because there are no common factors available. No whole numbers can divide the two numbers evenly except for 1.

There is no end to the possibilities when creating multiples of a particular fraction:

$\frac{16 \times 3}{20 \times 3} = \frac{48}{60}$ and $\frac{16 \times 5}{20 \times 5} = \frac{80}{100}$ and $\frac{16 \times 4}{20 \times 4} = \frac{64}{80}$, and so on. You choose the numbers that work best for you in the situation.

However, not all fractions with large numbers can be changed to smaller numbers. Certain rules have to be followed so that the fraction maintains its integrity; the fraction has to have the same value as it did originally.

EXAMPLE

Q. Find a fraction equivalent to $\frac{7}{8}$ with a denominator of 40.

A. Because 5 times 8 is 40, you multiply both the numerator and denominator by 5. In reality, you're just multiplying by 1, which doesn't change the real value of anything.

$$\frac{7}{8}\times\frac{5}{5}=\frac{7\times5}{8\times5}=\frac{35}{40}$$

Q. Reduce $\frac{15}{36}$ by multiplying the numerator and denominator by $\frac{1}{3}$. The same thing is accomplished if you divide both the numerator and denominator by 3.

A. $\frac{15}{36}\times\frac{\frac{1}{3}}{\frac{1}{3}}=\frac{15\times\frac{1}{3}}{36\times\frac{1}{3}}=\frac{\frac{15}{3}}{\frac{36}{3}}=\frac{5}{12}$ or

$$\frac{15\div3}{36\div3}=\frac{5}{12}$$

YOUR TURN

7 Find an equivalent fraction with a denominator of 28 for $\frac{3}{7}$.

8 Find an equivalent fraction with a denominator of 30 for $\frac{x}{6}$.

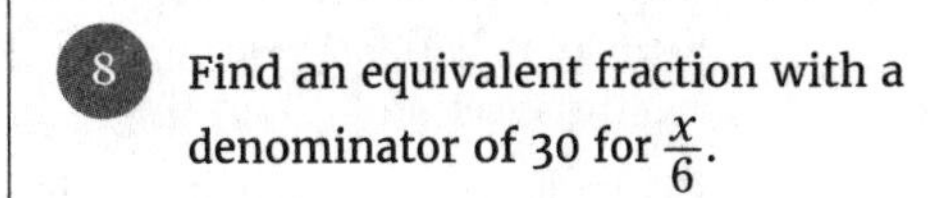

9 Reduce this fraction: $\frac{16}{60}$.

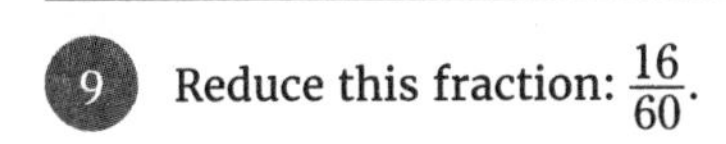

10 Reduce this fraction: $\frac{63}{84}$.

Determining lowest terms

A fraction is in its lowest common terms when there is no common factor of the numerator and denominator except the number 1. Nothing will divide both of them evenly.

REMEMBER

To reduce fractions to their lowest terms, follow these steps:

1. **Look for numbers that evenly divide both the numerator and the denominator.**

   If you find more than one number that divides both evenly, choose the largest.

2. **Divide both the numerator and the denominator by the number you chose, and put the results in their corresponding positions.**

When reducing fractions, your fraction isn't *wrong* if you don't choose the largest-possible divisor. It just means that you have to divide again to get to the lowest terms. When reducing the fraction $\frac{48}{60}$, you might have chosen to divide by 6 instead of 12. In that case, you'd get the fraction $\frac{8}{10}$, which can be reduced again by dividing the numerator and denominator by 2. Choosing the largest number possible just reduces the number of steps you have to take.

Q. Write the fraction $\frac{1001}{1540}$ in lowest terms.

A. The divisors of 1001 are 7, 11, and 13. Are any of these divisors of 1540? Yes, 7 divides 1540 and 11 divides 1540. $\frac{1001}{1540} = \frac{7 \times 11 \times 13}{2 \times 2 \times 5 \times 7 \times 11} = \frac{\cancel{7} \times \cancel{11} \times 13}{2 \times 2 \times 5 \times \cancel{7} \times \cancel{11}} = \frac{13}{20}$.

Q. Determine if the fraction $\frac{143}{333}$ is in lowest terms.

A. Yes. The numerator is divisible by 11 and 13. The denominator is divisible by 9 and 37. They don't share any common factors, so this fraction is in its lowest terms.

YOUR TURN

11 Determine if the fraction is in lowest terms: $\frac{14}{15}$

12 Determine if the fraction is in lowest terms: $\frac{39}{26}$

13 Determine if the fraction is in lowest terms: $\frac{105}{141}$

14 Determine if the fraction is in lowest terms: $\frac{484}{485}$

Making Proportional Statements

A *proportion* is an equation with two fractions equal to one another. Proportions have some wonderful properties that make them useful for solving problems — especially when you're comparing one quantity to another or one percentage to another.

Given the proportion $\frac{a}{b} = \frac{c}{d}$, the following are also true:

- $a \cdot d = c \cdot b$. (The cross-products form an equation.)
- $\frac{b}{a} = \frac{d}{c}$. (The "flip" is an equation.)
- $\frac{a}{b} = \frac{c \cdot \not{k}}{d \cdot \not{k}}$. (You can reduce either fraction vertically.)
- $\frac{a}{b \cdot \not{k}} = \frac{c}{d \cdot \not{k}}$. (You can reduce the numerator or denominator horizontally.)

EXAMPLE

Q. Find the missing value in the following proportion: $\frac{42}{66} = \frac{28}{d}$

A. The numerator and denominator in the fraction on the left have a common factor of 6. Multiply each by $\frac{1}{6}$. Flip the proportion to get the unknown in the numerator of the

Coordinating Fractions and Decimals

right-hand fraction. Then you see that the two bottom numbers both have a common factor of 7. Divide each by 7. Finally, cross-multiply to get your answer:

$$\frac{42}{66} \times \frac{\frac{1}{6}}{\frac{1}{6}} = \frac{7}{11} = \frac{28}{d}$$

$$\frac{11}{7} = \frac{d}{28}$$

$$\frac{11}{{}_1\cancel{7}} = \frac{d}{\cancel{28}_4}$$

$$\frac{11}{1} = \frac{d}{4}$$

$$11 \times 4 = 1 \times d$$

$$44 = d$$

Q. If Agnes can type 60 words per minute, how long will it take her to type a manuscript containing 4,020 words (if she can keep typing at the same rate)?

A. Set up a proportion with words in the two numerators and the corresponding number of minutes in the denominators:

$$\frac{60 \text{ words}}{1 \text{ minute}} = \frac{4{,}020 \text{ words}}{x \text{ minutes}}$$

Divide both numerators by 60 and then cross-multiply to solve for x.

$$\frac{{}^1\cancel{60}}{1} = \frac{\cancel{4{,}020}^{67}}{x}$$

$$1 \times x = 1 \times 67$$

$$x = 67$$

It will take her 67 minutes — just over an hour.

YOUR TURN

15 Solve for x: $\frac{7}{21} = \frac{x}{24}$

16 Solve for x: $\frac{45}{x} = \frac{60}{200}$

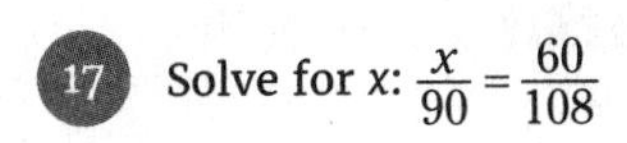

17 Solve for x: $\frac{x}{90} = \frac{60}{108}$

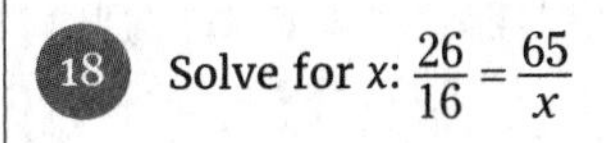

18 Solve for x: $\frac{26}{16} = \frac{65}{x}$

19 A recipe calls for 2 teaspoons of cinnamon and 4 cups of flour. You need to increase the flour to 6 cups. To keep the ingredients proportional, how many teaspoons of cinnamon should you use?

20 A factory produces two faulty tablets for every 500 tablets it produces. How many faulty tablets would you expect to find in a shipment of 1,250?

Finding Common Denominators

Before you can add or subtract fractions, you need to find a common denominator for them. Ideally, that common denominator is the *least common multiple* (the smallest number that each of the denominators can divide into without a remainder). A method of last resort, though, is to multiply the denominators together. Doing so gives you a number that the denominators divide evenly. You may have to work with larger numbers using this method, but you can always reduce the fractions at the end.

And then, there's the *box method*. This method is especially helpful when you have three or more fractions to deal with — and they have relatively large denominators.

Creating common denominators from multiples of factors

Common denominators (the *same numbers* in the denominators) are necessary for adding, subtracting, and comparing fractions. Carefully selected fractions that are equal to the number 1 are used to create common denominators because multiplying by 1 doesn't change a number's value.

Follow these steps to find a common denominator for two fractions and write the equivalent fractions:

1. **Find the least common multiple of the two denominators — the smallest number that both denominators divide evenly.**

   First, look to see if you can determine the common multiple by simple observation; you may know some multiples of the two numbers. If you find the common multiple by observation, go directly to Step 4. (Do not pass Go; do not collect $200.)

2. **If the common multiple isn't easily determined, start your search by choosing the larger denominator.**

   Check to see if the smaller denominator divides the larger one evenly. If it does, then you've found the common denominator. Go to Step 4.

3. **Check consecutive multiples of the larger denominator until you find one that the smaller one divides.**

   That's your common denominator.

4. **When you find a common denominator, rewrite both fractions as equivalent fractions with that denominator.**

   Multiply both numerator and denominator of each fraction by the equivalent of 1 that creates fractions with the common denominator.

EXAMPLE

Q. Find the least common denominator for the two fractions $\frac{7}{18}$ and $\frac{5}{24}$.

A. 72. Look at the multiples of 24: 24, 48, 72, 96. You can stop with the multiple 72, because that's also a multiple of 18. The LCD is 72.

Q. How would you write the fractions $\frac{1}{3}$ and $\frac{3}{4}$ with the same denominator?

A. The fractions $\frac{1}{3}$ and $\frac{3}{4}$ have denominators with no factors in common, so the least common denominator is 12, the product of the two numbers. Now you can write them both as fractions with a denominator of 12:

$$\frac{1}{3}\times\frac{4}{4}=\frac{4}{12} \text{ and } \frac{3}{4}\times\frac{3}{3}=\frac{9}{12}$$

YOUR TURN

21 Rewrite the fractions $\frac{2}{7}$ and $\frac{3}{8}$ with a common denominator.

22 Rewrite the fractions $\frac{5}{12}$ and $\frac{7}{18}$ with a common denominator.

23 Rewrite the fractions $\frac{9}{x}$ and $\frac{5}{6}$ with a common denominator.

24 Rewrite the fractions $\frac{5}{x}$ and $\frac{1}{x+6}$ with a common denominator.

25 Rewrite the fractions $\frac{1}{2}$, $\frac{1}{3}$, and $\frac{1}{5}$ with a common denominator.

26 Rewrite the fractions $\frac{2}{3}$, $\frac{5}{x}$, and $\frac{3}{2x}$ with a common denominator.

Using the box method

The *box method* is a very nicely structured process that has an added bonus. You can use it to find the least common denominator of two or more fractions, and you can also use it to find the greatest common factor of two or more numbers.

Consider the addition problem $\frac{7}{36}+\frac{49}{60}+\frac{95}{96}$. You don't want to multiply them all together. And, yes, you could look at multiples of the largest denominator. But, as an option, look at the box method.

1. **Write the three denominators in an "upside-down" division box.**

   $$\begin{array}{r|lll} & 36 & 60 & 96 \\ \hline \end{array}$$

2. **Outside the box, on the left, write a number that divides all of the denominators evenly.**

   For this first try, I'll use 2.

   $$\begin{array}{r|lll} 2 & 36 & 60 & 96 \\ \hline \end{array}$$

3. **Divide 2 into each denominator, putting the quotients under the respective denominators.**

   $$\begin{array}{r|lll} 2 & 36 & 60 & 96 \\ \hline & 18 & 30 & 48 \end{array}$$

4. **Now put a division box around these quotients and find another divisor. Repeat the process until there are no more common factors.**

   This time I chose 3. The order of choices doesn't really matter.

   $$\begin{array}{r|lll} 2 & 36 & 60 & 96 \\ \hline 3 & 18 & 30 & 48 \\ \hline & 6 & 10 & 16 \end{array}$$

   One more time:

   $$\begin{array}{r|lll} 2 & 36 & 60 & 96 \\ \hline 3 & 18 & 30 & 48 \\ \hline 2 & 6 & 10 & 16 \\ \hline & 3 & 5 & 8 \end{array}$$

5. **You're finished when the quotients don't have any common factors. All three have to have the same common factor. You can't divide just two of them.**

   So, what do you have here? First, you have the greatest common factor of the three numbers: $\text{GCF}(36,\ 60,\ 96)=2\cdot 3\cdot 2=12$. You multiply the divisors, the numbers down the left side, and get the biggest number that will divide all three numbers evenly.

But you're looking for a common denominator! To get that, find the product of those three numbers in the GCF times the product of the three numbers left on the bottom.

The least common denominator is: $2 \cdot 3 \cdot 2 \cdot 3 \cdot 5 \cdot 8 = 1{,}440$. This method is a bit nicer than looking at all the multiples of 96.

Applying Fractional Operations

Now that you have the tools necessary, you can investigate ways to perform binary operations on fractions. Addition and subtraction go together, because they both require common denominators. Multiplication and division are paired, because they can be performed without having to create the same denominator. And division is just "multiplication adjusted"!

Adding and subtracting fractions

You can add fractions together or subtract one from another if they have a common denominator. After you find the common denominator and change the fractions to their equivalents, you can add the numerators together or subtract them (keeping the denominators the same).

Adding and subtracting fractions takes a little special care. You can add quarts and gallons if you change them to the same unit. It's the same with fractions. You can add thirds and sixths if you find the common denominator first.

To add or subtract fractions:

1. **Convert the fractions so that they have the same value in the denominators.**

   Find out how to do this in the section, "Finding common denominators."

2. **Add or subtract the numerators.**

   Leave the denominators alone.

3. **Reduce the answer, if needed.**

EXAMPLE

Q. In her will, Jane gave $\frac{4}{7}$ of her money to the Humane Society and $\frac{1}{3}$ of her money to other charities. How much was left for her children's inheritance?

A. The fractions $\frac{4}{7}$ and $\frac{1}{3}$ aren't compatible because the denominators don't have any factors in common. The fraction $\frac{4}{7}$ has the larger denominator and can be written as $\frac{8}{14}$ or $\frac{12}{21}$ or $\frac{16}{28}$, and so on. You can stop at $\frac{12}{21}$ because 3 divides 21 evenly: $\frac{4}{7} = \frac{12}{21}$ and $\frac{1}{3} = \frac{7}{21}$. Add the numerators to get the total designation to charity in Jane's will: $\frac{12}{21} + \frac{7}{21} = \frac{19}{21}$. Subtract that total from the whole of Jane's proceeds to find what portion is allotted to her children: $\frac{21}{21} - \frac{19}{21} = \frac{2}{21}$. Jane's children will be awarded $\frac{2}{21}$ of her estate.

Q. $\frac{5}{6}+\frac{7}{8}=$

A. First find the common denominator, 24, and then complete the addition:

$$\left(\frac{5}{6}\times\frac{4}{4}\right)+\left(\frac{7}{8}\times\frac{3}{3}\right)=\frac{20}{24}+\frac{21}{24}$$
$$=\frac{41}{24}=1\frac{17}{24}$$

Q. $2\frac{1}{2}+\left(-1\frac{1}{3}\right)+5\frac{3}{10}=$

A. You need a common denominator of 30:

$$2\frac{1}{2}+\left(-1\frac{1}{3}\right)+5\frac{3}{10}$$
$$=2+\left(\frac{1}{2}\times\frac{15}{15}\right)-1-\left(\frac{1}{3}\times\frac{10}{10}\right)+5+\left(\frac{3}{10}\times\frac{3}{3}\right)$$

The whole number parts are separated from the fractional parts to keep the numbers in the computations smaller. Be sure to apply the subtraction to both the whole number and fraction when needed.

$$=2+\left(\frac{15}{30}\right)-1-\left(\frac{10}{30}\right)+5+\left(\frac{9}{30}\right)$$
$$=2-1+5+\left(\frac{15-10+9}{30}\right)=6+\frac{14}{30}=6\frac{7}{15}$$

Q. $2\frac{1}{8}-1\frac{1}{7}=$

A. In this problem, you see another option: you can change both mixed numbers to improper fractions. The common denominator is 56:

$$\frac{17}{8}-\frac{8}{7}=\left(\frac{17}{8}\times\frac{7}{7}\right)-\left(\frac{8}{7}\times\frac{8}{8}\right)=\frac{119}{56}-\frac{64}{56}=\frac{55}{56}$$

YOUR TURN

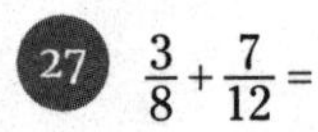

27 $\frac{3}{8}+\frac{7}{12}=$

28 $3\frac{1}{3}+4\frac{3}{5}+\frac{7}{15}=$

29 $1\frac{5}{12} - \frac{7}{9} =$

30 $3\frac{2}{3} - \left(-6\frac{1}{2}\right) =$

31 $\frac{1}{72} + \frac{1}{108} - \frac{1}{180} =$

Multiplying and dividing fractions

Multiplying fractions is really a much easier process than adding or subtracting fractions, because you don't have to find a common denominator. Furthermore, you can take some creative steps and reduce the fractions before you even multiply them.

When multiplying fractions, you can pair up the numerator of any fraction in the problem with the denominator of any other fraction; then divide each by the same number (reduce). Doing so saves you from having large numbers to multiply and then to reduce later.

Yes, multiplying fractions is a tad easier than adding or subtracting them. Multiplying is easier because you don't need to find a common denominator first. The only catch is that you have to change any mixed numbers to improper fractions. Then, at the end, you may have to change the fraction back again to a mixed number. Small price to pay.

When multiplying fractions, follow these steps:

1. **Change all mixed numbers to improper fractions.**
2. **Reduce any numerator-denominator combinations, if possible.**

3. **Multiply the numerators together and the denominators together.**
4. **Reduce the answer if necessary.**

Here's an example: Suppose Sadie worked $10\frac{2}{3}$ hours at time-and-a-half. How many hours will she get paid for?

Write the problem as $10\frac{2}{3}\times 1\frac{1}{2}$ and then rewrite the mixed numbers as improper fractions. $10\frac{2}{3}\times 1\frac{1}{2}=\frac{32}{3}\times\frac{3}{2}$. Reducing the fractions *before* multiplying can make multiplying the fractions easier. Smaller numbers are more manageable, and if you reduce the fractions before you multiply, you don't have to reduce them afterward.

The product $\frac{32}{3}\times\frac{3}{2}=$ has a 32 in the first numerator and a 2 in the second denominator. Even though the 32 and 2 aren't in the same fraction, you can reduce them because this is a multiplication problem. Multiplication is *commutative*, meaning that it doesn't matter what order you multiply the numbers. You can pretend that the 32 and 2 are in the same fraction. So, dividing the first numerator by 2 and the second denominator by 2, you get

$$\frac{\cancel{32}^{16}}{3}\times\frac{3}{\cancel{2}^{1}}=$$

But $\frac{16}{3}\times\frac{3}{1}=$ has a 3 in the first denominator and a 3 in the second numerator, so you can divide by 3: $\frac{16}{{}_1\cancel{3}}\times\frac{\cancel{3}^1}{1}=\frac{16}{1}=16$.

Dividing fractions is as easy as (dividing) pie — that is, dividing the pie into enough pieces so that everybody at your table gets an equal share. Actually, dividing fractions uses the same techniques as multiplying fractions, except that there's an additional "first step": the numerator and the denominator of the second fraction first have to change places — the fraction does a "flip."

When dividing fractions:

1. **Change all mixed numbers to improper fractions.**
2. **Flip the second fraction, placing the bottom number on top and the top number on the bottom.**
3. **Change the division sign to multiplication.**
4. **Continue as with the multiplication of fractions.**

The *flip* of a fraction is called its *reciprocal.* All real numbers except 0 have a reciprocal. The product of a number and its reciprocal is equal to 1.

Consider this example: If you buy $6\frac{1}{2}$ pounds of sirloin steak and want to cut it into pieces that weigh $\frac{3}{4}$ pound each, how many pieces will you have?

First, change the mixed number to an improper fraction. Then flip the second fraction, and change the division to multiplication:

$$6\frac{1}{2}\div\frac{3}{4}=\frac{13}{2}\div\frac{3}{4}=\frac{13}{2}\times\frac{4}{3}$$

Now reduce the fraction and multiply. Change the answer to a mixed number:

$$\frac{13}{{}_1\cancel{2}}\times\frac{\cancel{4}^{2}}{3}=\frac{13}{1}\times\frac{2}{3}=\frac{26}{3}=8\frac{2}{3}$$

Having $8\frac{2}{3}$ pieces means that you'll have eight pieces weighing the full $\frac{3}{4}$ pound and one piece left over that's smaller. (That's the cook's bonus or mean Aunt Martha's piece.)

EXAMPLE

Q. Multiply the three fractions: $\frac{15}{16}\times\frac{21}{75}\times\frac{24}{49}$.

A. You can make the problem easier if you reduce the fractions first. The 15 and 75 are both divisible by 15, the 21 and 49 are both divisible by 7, and the 16 and 24 are both divisible by 8:

$$\frac{{}^{1}\cancel{15}}{16}\times\frac{21}{\cancel{75}_{5}}\times\frac{24}{49}=\frac{1}{16}\times\frac{\cancel{21}^{3}}{5}\times\frac{24}{\cancel{49}_{7}}$$
$$=\frac{1}{{}_2\cancel{16}}\times\frac{3}{5}\times\frac{\cancel{24}^{3}}{7}$$
$$=\frac{1}{2}\times\frac{3}{5}\times\frac{3}{7}=\frac{9}{70}$$

Q. Divide: $2\div\frac{3}{5}=$

A. First change the 2 to a fraction: $\frac{2}{1}$. Then change the divide to multiply and the second (right) fraction to its reciprocal. Then do the multiplication problem to get the answer $\frac{10}{3}$, which can be changed to the mixed number.

$$2\div\frac{3}{5}=\frac{2}{1}\times\frac{5}{3}=\frac{10}{3}=3\frac{1}{3}$$

YOUR TURN

32 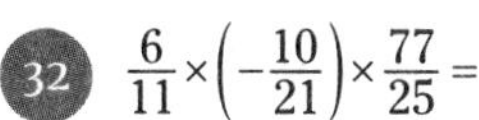 $\frac{6}{11}\times\left(-\frac{10}{21}\right)\times\frac{77}{25}=$

33 $4\frac{1}{5}\times\frac{25}{49}=$

Coordinating Fractions and Decimals

34. $\left(-\frac{7}{27}\right)\times\frac{18}{25}\times\left(-\frac{15}{28}\right)=$

35. $-\frac{15}{14}\div\left(-\frac{20}{21}\right)=$

36. $2\frac{1}{2}\div\frac{3}{4}=$

37. $7\frac{1}{7}\div 3\frac{3}{14}=$

Simplifying Complex Fractions

A *complex fraction* is a fraction within a fraction. If a fraction has another fraction in its numerator or denominator (or both), it's called *complex.* Fractions with this structure are awkward to deal with and need to be simplified. To simplify a complex fraction, you first work at creating improper fractions or integers in the numerator and denominator, independently, and then you divide the numerator by the denominator. You need to boil this down to one term in the numerator and one in the denominator.

EXAMPLE

Q. $\frac{4\frac{1}{2}}{\frac{6}{7}}=$

A. First, change the mixed number in the numerator to an improper fraction. Then divide the two fractions by multiplying the numerator by the reciprocal of the denominator.

$$\frac{4\frac{1}{2}}{\frac{6}{7}} = \frac{\frac{9}{2}}{\frac{6}{7}} = \frac{9}{2} \div \frac{6}{7} = \frac{9}{2} \times \frac{7}{6}$$

$$= \frac{{}^{3}\cancel{9}}{2} \times \frac{7}{\cancel{6}_{2}} = \frac{21}{4} = 5\frac{1}{4}$$

Q. $\dfrac{1\frac{9}{25} - \frac{4}{5}}{\frac{1}{2} + \frac{1}{3} + \frac{4}{5}} =$

A. First, find a common denominator for the fractions in the numerator and a separate one for those in the denominator. Then subtract the fractions in the numerator and add the fractions in the denominator. Finally, divide the two fractions by multiplying the numerator by the reciprocal of the denominator.

$$\frac{1\frac{9}{25} - \frac{4}{5}}{\frac{1}{2} + \frac{1}{3} + \frac{4}{5}} = \frac{\frac{34}{25} - \frac{4}{5}}{\frac{1}{2} + \frac{1}{3} + \frac{4}{5}} = \frac{\frac{34}{25} - \frac{20}{25}}{\frac{15}{30} + \frac{10}{30} + \frac{24}{30}} =$$

$$= \frac{\frac{14}{25}}{\frac{49}{30}} = \frac{14}{25} \div \frac{49}{30} = \frac{14}{25} \times \frac{30}{49}$$

$$= \frac{{}^{2}\cancel{14}}{{}_{5}\cancel{25}} \times \frac{\cancel{30}^{6}}{\cancel{49}_{7}} = \frac{12}{35}$$

YOUR TURN

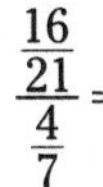

38 $\dfrac{\frac{16}{21}}{\frac{4}{7}} =$

39 $\dfrac{3\frac{1}{3}}{\frac{2}{5}} =$

40 $\dfrac{4\frac{2}{7}}{1\frac{1}{14}} =$

41 $\dfrac{2\frac{1}{3}+4\frac{1}{5}}{10-1\frac{5}{6}} =$

Performing Operations with Decimals

Decimals are essentially fractions whose denominators are powers of 10. This property makes for much easier work when adding, subtracting, multiplying, or dividing.

- When adding or subtracting decimal numbers, just line up the decimal points and fill in zeros, if necessary.
- When multiplying decimals, just ignore the decimal points until you're almost finished. Count the number of digits to the right of the decimal point in each multiplier, and the total number of digits is how many decimal places you should have in your answer.
- Dividing has you place the decimal point *first,* not last. Make your divisor a whole number by moving the decimal point to the right. Then adjust the number you're dividing into by moving the decimal point the same number of places. Put the decimal point in your answer directly above the decimal point in the number you're dividing into (the dividend).

EXAMPLE

Q. $14.536 + 0.000004 - 2.3 =$

A. Line up the decimal points in the first two numbers and add. Put in zeros to help you line up the digits. Then subtract the last number from the result.

$$\begin{array}{r} 14.536000 \\ \underline{+0.000004} \\ 14.536004 \end{array} \qquad \begin{array}{r} 14.536004 \\ \underline{-2.300000} \\ 12.236004 \end{array}$$

Q. $5.6 \times 0.123 \div 0.6 =$

A. Multiply the first two numbers together, creating an answer with four decimal places to the right of the decimal point. Then divide the result by 0.6, after moving the decimal point one place to the right in both divisor and dividend.

$$
\begin{array}{r}
0.123 \\
\times \quad 5.6 \\
\hline
738 \\
615 \\
\hline
0.6888
\end{array}
$$

$$
\begin{array}{r}
1.148 \\
0.6_{\wedge}\overline{)0.6_{\wedge}888} \\
\underline{6} \\
8 \\
\underline{6} \\
28 \\
\underline{24} \\
48 \\
\underline{48}
\end{array}
$$

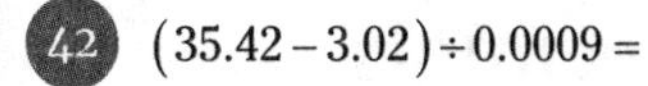

42 $(35.42 - 3.02) \div 0.0009 =$

43 $5.2 \times 0.00001 - 3 =$

Changing Fractions to Decimals and Vice Versa

Decimals are nothing more than glorified fractions. Decimals are special because, when written as fractions, their denominators are always powers of 10 — for example, 10, 100, 1,000, and so on. Because decimals are such special fractions, you don't even have to bother with the denominator part. Just write the numerator and use a decimal point to indicate that it's really a fraction with a denominator that's a power of 10.

The number of digits (decimal places) to the right of the decimal point in a number tells you the number of zeros in the power of 10 that is written in the denominator of the corresponding fraction.

Here are some examples of changing decimals to fractions:

EXAMPLE

Q. Change 0.408 to a fraction.

A. $0.408 = \frac{408}{1,000}$. The decimal has three digits, 408, to the right of the decimal point, so you use the power of 10 with three zeros.

Q. Change 60.00009 to a fraction.

A. $60.00009 = 60\frac{9}{100,000}$. The decimal has five digits, 00009, to the right of the decimal point, so you'll need 100,000 in the denominator. The 60 is written in front of the fraction and doesn't affect the decimal value. The *lead zeros* are not written in front of the 9 in the numerator of the fraction. You start by writing the first nonzero digit.

REMEMBER

A *digit* is any single number from 0 through 9. (But, when you count the ten *digits* at the end of your feet, you start with 1 and end with 10.)

Decimal fractions are great because you can add, subtract, multiply, and divide them so easily. The ease in computation (and typing) is why changing a fraction to a decimal is often desirable.

Making fractions become decimals

All fractions can be changed to decimals. In Chapter 1, you are told that rational numbers have decimals that can be written exactly as fractions. The decimal forms of rational numbers either end (terminate) or repeat in a pattern.

To change a fraction to a decimal, just divide the top by the bottom.

EXAMPLE

Q. Write $\frac{15}{8}$ as a decimal.

A. $\frac{15}{8}$ becomes $8\overline{)15.000}$ and $8\overline{)15.000} = 1.875$ so $\frac{15}{8} = 1.875$.

Q. Write $\frac{4}{11}$ as a decimal.

A. $\frac{4}{11}$ becomes $11\overline{)4.000000\ldots}$ and $11\overline{)4.000000\ldots} = 0.363636\ldots$ so $\frac{4}{11} = 0.363636\ldots$. The division never ends, so the three dots (ellipses) tell you that the pattern repeats forever.

If the division doesn't come out evenly, you can either show the repeating digits or stop after a certain number of decimal places and *round off*. Another way to show repeating digits is to draw a bar over the digits that repeat. So $0.363636\ldots$ can be written as $0.\overline{36}$.

Rounding decimals

To round a number means to create an approximate value. If you're measuring the distance from one side of the street to the other and have a measurement of 37 feet, $3\frac{3}{16}$ inches, you probably don't need a number that precise. Depending on what you're doing, you may be fine with 37 feet — which is the number rounded to the nearest foot. Or 37 feet 3 inches may do it for you. It just depends on the circumstances.

To round decimal numbers:

1. **Determine the number of places you want and look one further to the right.**
2. **Increase the last place you want by one number if the one further is 5 or bigger.**
3. **Leave the last place you want as it is, if the one further is less than 5.**

The symbol ≈ means *approximately equal* or *about equal.* This symbol is useful when you're rounding a number.

Here are some examples of rounding each decimal to the nearer thousandth (three decimal places):

EXAMPLE

Q. Round 0.363636 to the nearest thousandth.

A. 0.364. When rounded to three decimal places, you look at the fourth digit (one further). The fourth digit is 6, which is greater than 5, so you increase the third digit by 1, making the 3 a 4.

Q. Round 0.03125 to the nearest thousandth.

A. 0.031. When rounded to three decimal places, you look at the fourth digit. The fourth digit is 2, which is smaller than 5, so you leave the third digit as it is.

Writing decimals as equivalent fractions

Decimals representing rational numbers come in two varieties: terminating decimals and repeating decimals. When changing from decimals to fractions, you put the digits in the decimal over some other digits and reduce the fraction.

Getting terminal results with terminating decimals

To change a terminating decimal into a fraction, put the digits to the right of the decimal point in the numerator. Put the number 1 in the denominator, followed by as many zeros as the numerator has digits. Reduce the fraction if necessary.

EXAMPLE

Q. Change 0.36 into a fraction.

A. There are two digits in 36, so the 1 in the denominator is followed by two zeros. Both 36 and 100 are divisible by 4, so the fraction reduces.

$$0.36 = \frac{36}{100} = \frac{9}{25}$$

Q. Change 0.0005 into a fraction.

A. Don't forget to count the zeros in front of the 5 when counting the number of digits. The fraction reduces.

$$0.0005 = \frac{5}{10{,}000} = \frac{1}{2{,}000}$$

Repeating yourself with repeating decimals

When a decimal repeats itself, you can always find the fraction that corresponds to the decimal. In this chapter, I only cover the decimals that show every digit repeating.

To change a *repeating decimal* (in which every digit is part of the repeated pattern) into its corresponding fraction, write the repeating digits in the numerator of a fraction and, in the denominator, as many 9's as there are repeating digits. Reduce the fraction if necessary.

EXAMPLE

Q. Write the decimal 0.126126126... as a fraction in lowest terms.

A. $0.126126126\ldots = \frac{126}{999} = \frac{14}{111}$. The three repeating digits are 126. Placing the 126 over a number with three 9's, you reduce by dividing the numerator and denominator by 9.

Q. Write the decimal 0.857142857142857142... as a fraction in lowest terms.

A. The six repeating digits are put over six 9's. Reducing the fraction takes a few divisions. The common factors of the numerator and denominator are 11, 13, 27, and 37. When completely reduced, you have $\frac{857142}{999999} = \frac{6}{7}$.

YOUR TURN

44 Change $\frac{3}{5}$ to a decimal.

45 Change $\frac{40}{9}$ to a decimal.

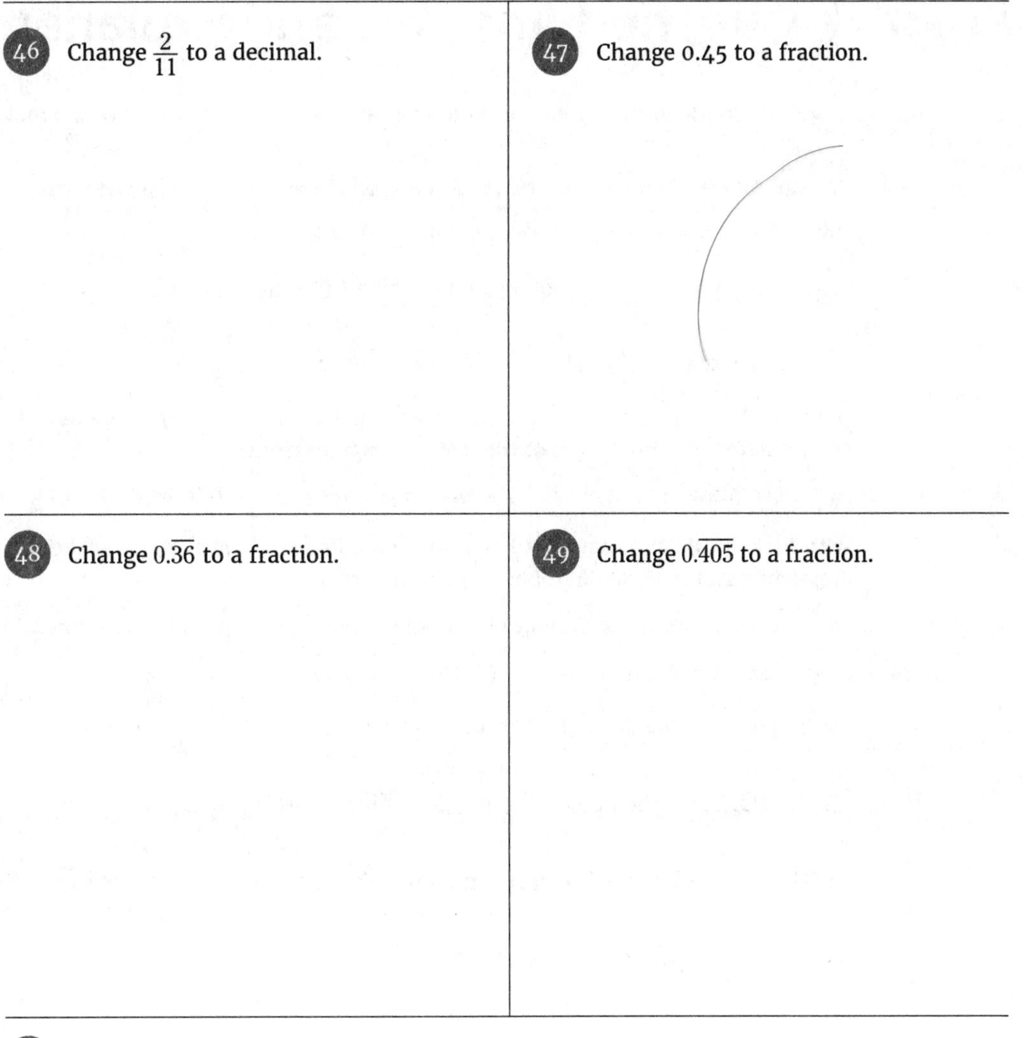

46 Change $\frac{2}{11}$ to a decimal.

47 Change 0.45 to a fraction.

48 Change $0.\overline{36}$ to a fraction.

49 Change $0.\overline{405}$ to a fraction.

50 Round the decimal 4.172797 to the nearest thousandth.

Practice Question Answers and Explanations

1. $3\frac{5}{8}$. First, divide the 29 by 8. The number 8 divides 29 three times with a remainder of 5.

2. $\frac{41}{9}$. Multiply the 4 and 9 and then add the 5, which equals 41. Then write the fraction with this result in the numerator and the 9 in the denominator.

3. $\frac{407}{100}$. Multiply the 4 times 100 and add the 7. Put the sum over 100.

4. $-\frac{27}{13}$. Ignore the negative sign at first; you don't want it involved in the computation. First multiply the 2 times 13 to get 26. Add the 1 to get 27. You have 27 in the numerator, 13 in the denominator, and now you put the negative sign in front.

5. $36\frac{6}{11}$. The number 11 divides 402 a total of 36 times with 6 left over. The 36 goes in front, with the 6 in the numerator and 11 in the denominator. This example makes it especially apparent that the mixed number is more understandable.

6. $-2\frac{5}{7}$. Divide 7 into 19, for a quotient of 2. The remainder 5 goes in the numerator. Put the negative sign in front of the 2.

7. $\frac{12}{28}$. To get 28 in the denominator, multiply 7 by 4: $\frac{3}{7}=\frac{3}{7}\times\frac{4}{4}=\frac{12}{28}$

8. $\frac{5x}{30}$. Multiply both the numerator and denominator by 5: $\frac{x}{6}=\frac{x}{6}\times\frac{5}{5}=\frac{5x}{30}$

9. $\frac{4}{15}$. The number 4 is the greatest common divisor of 16 and 60 because $16=4\times 4$ and $60=15\times 4$. So multiply $\frac{16}{60}$ by $\frac{\frac{1}{4}}{\frac{1}{4}}$ to get

$$\frac{16}{60}=\frac{16}{60}\times\frac{\frac{1}{4}}{\frac{1}{4}}=\frac{16\times\frac{1}{4}}{60\times\frac{1}{4}}=\frac{4}{15}$$

Or, if you prefer, divide both the numerator and denominator by 4:

$$\frac{16}{60}=\frac{16\div 4}{60\div 4}=\frac{4}{15}$$

10. $\frac{3}{4}$. The largest common divisor of 63 and 84 is 21, because $63=3\times 21$ and $84=4\times 21$. So

$$\frac{63}{84}=\frac{63}{84}\times\frac{\frac{1}{21}}{\frac{1}{21}}=\frac{63\times\frac{1}{21}}{84\times\frac{1}{21}}=\frac{3}{4}$$

But let's say you do this in two steps — both dividing by a common factor.

You see that both 63 and 84 are divisible by 7. So divide both the numerator and denominator by 7.

$$\frac{63}{84}=\frac{63\div 7}{84\div 7}=\frac{9}{12}$$

Now you see that the new version has a numerator and denominator divisible by 3.

$$\frac{9\div 3}{12\div 3}=\frac{3}{4}$$

It took two steps instead of one, but you have the same answer.

11 $\frac{14}{15}$. Yes, this is completely reduced. You can write 14 as 2×7 and 15 as 3×5, but they don't share any common factors.

12 $\frac{3}{2}$. No. Both 39 and 26 are divisible by 13. You may not be completely familiar with the multiples of 13, but if you just note that 39 is divisible by 3 and 26 is divisible by 2, you come up with $39=3\times 13$ and $26=2\times 13$ and you can reduce the fraction: $\frac{39\div 13}{26\div 13}=\frac{3}{2}$

13 $\frac{35}{47}$. No. Here's another situation where you look for common factors. Both of these numbers are divisible by 3. (Chapter 8 contains all the rules of divisibility you'll be needing.)

$$\frac{105}{141}=\frac{105\div 3}{141\div 3}=\frac{35}{47}$$

You can stop right there, because 47 is a prime number.

14 $\frac{484}{485}$. Yes. This is completely reduced.

Just to check the factors: $\frac{484}{485}=\frac{2\cdot 2\cdot 11\cdot 11}{5\cdot 97}$. The numerator shows a prime factorization.

You see more of this in Chapter 8.

15 $x=8$. The left fraction can be reduced by dividing by 7. Then the two denominators can be reduced by dividing by 3. Then find the cross-product.

$$\frac{7}{21}=\frac{x}{24}\rightarrow\frac{7\div 7}{21\div 7}=\frac{x}{24}$$
$$\frac{1}{3}=\frac{x}{24}\rightarrow\frac{1}{{}_{1}\not{3}}=\frac{x}{{}_{8}\not{24}}$$
$$\frac{1}{1}=\frac{x}{8}\rightarrow 1\times 8=x\times 1$$
$$x=8$$

Coordinating Fractions and Decimals

(16) $x = 150$. The right fraction can be reduced by dividing by 20. Then the two numerators can be reduced by dividing by 3. Then find the cross-product.

$$\frac{45}{x} = \frac{60}{200} \rightarrow \frac{45}{x} = \frac{60 \div 20}{200 \div 20}$$
$$\frac{45}{x} = \frac{3}{100} \rightarrow \frac{{}^{15}\cancel{45}}{x} = \frac{{}^{1}\cancel{3}}{10}$$
$$\frac{15}{x} = \frac{1}{10} \rightarrow 15 \times 10 = 1 \times x$$
$$x = 150$$

(17) $x = 50$. The right fraction can be reduced by dividing by 12. Then the two denominators can be reduced by dividing by 9. Then find the cross-product.

$$\frac{x}{90} = \frac{60}{108} \rightarrow \frac{x}{90} = \frac{60 \div 12}{108 \div 12}$$
$$\frac{x}{90} = \frac{5}{9} \rightarrow \frac{x}{{}^{10}\cancel{90}} = \frac{5}{{}^{1}\cancel{9}}$$
$$\frac{x}{10} = \frac{5}{1} \rightarrow 1 \times x = 5 \times 10$$
$$x = 50$$

(18) $x = 40$. The left fraction can be reduced by dividing by 2. Then the two numerators can be reduced by dividing by 13. Then find the cross-product.

$$\frac{26}{16} = \frac{65}{x} \rightarrow \frac{26 \div 2}{16 \div 2} = \frac{65}{x}$$
$$\frac{13}{8} = \frac{65}{x} \rightarrow \frac{{}^{1}\cancel{13}}{8} = \frac{{}^{5}\cancel{65}}{x}$$
$$\frac{1}{8} = \frac{5}{x} \rightarrow 1 \times x = 5 \times 8$$
$$x = 40$$

(19) **$x = 3$ (3 teaspoons).** Fill in the proportion: $\frac{\text{original cinnamon}}{\text{original flour}} = \frac{\text{new cinnamon}}{\text{new flour}}$. Then let x represent the new cinnamon:

$$\frac{2}{4} = \frac{x}{6} \rightarrow \frac{2 \div 2}{4 \div 2} = \frac{x}{6}$$
$$\frac{1}{2} = \frac{x}{6} \rightarrow \frac{1}{{}^{1}\cancel{2}} = \frac{x}{{}^{3}\cancel{6}}$$
$$\frac{1}{1} = \frac{x}{3} \rightarrow 1 \times x = 1 \times 3$$
$$x = 3$$

(20) **$x = 5$ (5 faulty tablets).** Set up the proportion: $\frac{2\text{ faulty tablets}}{500\text{ tablets}} = \frac{x\text{ faulty tablets}}{1{,}250\text{ tablets}}$

$$\frac{2}{500} = \frac{x}{1{,}250} \rightarrow \frac{\cancel{2}^{1}}{\cancel{500}_{250}} = \frac{x}{1{,}250}$$
$$\frac{1}{250} = \frac{x}{1{,}250} \rightarrow \frac{1}{{}_{1}\cancel{250}} = \frac{x}{{}_{5}\cancel{1{,}250}}$$
$$\frac{1}{250} = \frac{x}{5} \rightarrow 1 \times x = 1 \times 5$$
$$x = 5$$

21. $\frac{2}{7}=\frac{16}{56}$ **and** $\frac{3}{8}=\frac{21}{56}$. The largest common factor of 7 and 8 is 1. So the least common denominator is 56, the product of the denominators: $\left(\frac{7\times8}{1}=56\right)$. Here are the details:

$$\frac{2}{7}=\frac{2}{7}\times\frac{8}{8}=\frac{16}{56} \text{ and } \frac{3}{8}=\frac{3}{8}\times\frac{7}{7}=\frac{21}{56}$$

22. $\frac{5}{12}=\frac{15}{36}$ **and** $\frac{7}{18}=\frac{14}{36}$. The largest common factor of 12 and 18 is 6. The least common denominator is 36. $\left(\frac{12\times18}{6}=36\right)$. Here are the details:

$$\frac{5}{12}=\frac{5}{12}\times\frac{3}{3}=\frac{15}{36} \text{ and } \frac{7}{18}=\frac{7}{18}\times\frac{2}{2}=\frac{14}{36}$$

23. $\frac{9}{x}=\frac{54}{6x}$ **and** $\frac{5}{6}=\frac{5x}{6x}$. The largest common factor of x and 6 is 1. The least common denominator is their product: 6x. Break it down: $\frac{9}{x}=\frac{9}{x}\times\frac{6}{6}=\frac{54}{6x}$ and $\frac{5}{6}=\frac{5}{6}\times\frac{x}{x}=\frac{5x}{6x}$.

24. $\frac{5}{x}=\frac{5x+30}{x(x+6)}$ **and** $\frac{1}{x+6}=\frac{x}{x(x+6)}$. The largest common factor of x and $x+6$ is 1. Their least common denominator is their product: $x(x+6)$. Here's the long of it: $\frac{5}{x}=\frac{5}{x}\times\frac{x+6}{x+6}=\frac{5x+30}{x(x+6)}$ and $\frac{1}{x+6}=\frac{1}{x+6}\times\frac{x}{x}=\frac{x}{x(x+6)}$.

25. $\frac{1}{2}=\frac{15}{30}$, $\frac{1}{3}=\frac{10}{30}$, **and** $\frac{1}{5}=\frac{6}{30}$. The least common denominator of fractions with denominators of 2, 3, and 5 is 30. Write it out:

$$\frac{1}{2}=\frac{1}{2}\times\frac{15}{15}=\frac{15}{30},\ \frac{1}{3}=\frac{1}{3}\times\frac{10}{10}=\frac{10}{30},\text{ and }\frac{1}{5}=\frac{1}{5}\times\frac{6}{6}=\frac{6}{30}$$

26. $\frac{2}{3}=\frac{4x}{6x}$, $\frac{5}{x}=\frac{30}{6x}$, **and** $\frac{3}{2x}=\frac{9}{6x}$. The last two denominators, $\frac{5}{x}$ and $\frac{3}{2x}$, have a common factor of x. And the product of all three denominators is $6x^2$. Divide the product by x and you get 6x. In long hand:

$$\frac{2}{3}=\frac{2}{3}\times\frac{2x}{2x}=\frac{4x}{6x},\ \frac{5}{x}=\frac{5}{x}\times\frac{6}{6}=\frac{30}{6x},\text{ and }\frac{3}{2x}=\frac{3}{2x}\times\frac{3}{3}=\frac{9}{6x}$$

27. $\frac{23}{24}$. The least common denominator is 24.

$$\frac{3}{8}+\frac{7}{12}=\left(\frac{3}{8}\times\frac{3}{3}\right)+\left(\frac{7}{12}\times\frac{2}{2}\right)=\frac{9}{24}+\frac{14}{24}=\frac{23}{24}$$

(28) $8\frac{2}{5}$. The least common denominator is 15.

$$3\frac{1}{3}+4\frac{3}{5}+\frac{7}{15}=\frac{10}{3}+\frac{23}{5}+\frac{7}{15}=\left(\frac{10}{3}\times\frac{5}{5}\right)+\left(\frac{23}{5}\times\frac{3}{3}\right)+\frac{7}{15}$$

$$=\frac{50}{15}+\frac{69}{15}+\frac{7}{15}=\frac{126}{15}\times\frac{\frac{1}{3}}{\frac{1}{3}}=\frac{42}{5}=8\frac{2}{5}$$

Or, leaving the whole number parts separate:

$$3\frac{1}{3}+4\frac{3}{5}+\frac{7}{15}=3+\left(\frac{1}{3}\times\frac{5}{5}\right)+4+\left(\frac{3}{5}\times\frac{3}{3}\right)+\frac{7}{15}$$

$$=3+4+\left(\frac{5}{15}+\frac{9}{15}+\frac{7}{15}\right)=7+\frac{21}{15}=7+1+\frac{6}{15}=8\frac{6}{15}=8\frac{2}{5}$$

(29) $\frac{23}{36}$.

$$1\frac{5}{12}-\frac{7}{9}=\frac{17}{12}-\frac{7}{9}=\left(\frac{17}{12}\times\frac{3}{3}\right)-\left(\frac{7}{9}\times\frac{4}{4}\right)=\frac{51}{36}-\frac{28}{36}=\frac{23}{36}$$

(30) $10\frac{1}{6}$.

$$3\frac{2}{3}-\left(-6\frac{1}{2}\right)=3\frac{2}{3}+6\frac{1}{2}=\frac{11}{3}+\frac{13}{2}=\left(\frac{11}{3}\times\frac{2}{2}\right)+\left(\frac{13}{2}\times\frac{3}{3}\right)=\frac{22}{6}+\frac{39}{6}=\frac{61}{6}=10\frac{1}{6}$$

(31) $\frac{19}{1{,}080}$. Here's a chance to use the box method.

$$\begin{array}{r|rrr} 6 & 72 & 108 & 180 \\ \hline 6 & 12 & 18 & 30 \\ \hline & 2 & 3 & 5 \end{array}$$

The least common denominator is $6\cdot 6\cdot 2\cdot 3\cdot 5=1{,}080$.

$$\frac{1}{72}\times\frac{15}{15}+\frac{1}{108}\times\frac{10}{10}-\frac{1}{180}\times\frac{6}{6}=\frac{15}{1{,}080}+\frac{10}{1{,}080}-\frac{6}{1{,}080}=\frac{15+10-6}{1{,}080}=\frac{19}{1{,}080}$$

(32) $-\frac{4}{5}$. Go ahead and put the negative sign in front of the work in the first step. With one of the three fractions being negative, you know that the answer will be negative.

$$\frac{6}{11}\times\left(-\frac{10}{21}\right)\times\frac{77}{25}=-\frac{{}^{2}\cancel{6}\times 10\times 77}{11\times\cancel{21}_{7}\times 25}=-\frac{2\times\overset{2}{\cancel{10}}\times 77}{11\times 7\times\cancel{25}_{5}}$$

$$=-\frac{2\times 2\times\cancel{77}^{7}}{{}_{1}\cancel{11}\times 7\times 5}=-\frac{2\times 2\times\cancel{7}^{1}}{1\times\cancel{7}_{1}\times 5}=-\frac{2\times 2\times 1}{1\times 1\times 5}=-\frac{2\times 2}{5}=-\frac{4}{5}$$

(33) $2\frac{1}{7}$. Change the first fraction to a mixed number.

$$4\frac{1}{5}\times\frac{25}{49}=\frac{21}{5}\times\frac{25}{49}=\frac{{}^{3}\cancel{21}\times 25}{5\times\cancel{49}_{7}}=\frac{3\times\cancel{25}^{5}}{{}_{1}\cancel{5}\times 7}=\frac{3\times 5}{1\times 7}=\frac{15}{7}=2\frac{1}{7}$$

(34) $\frac{1}{10}$. This time the answer is positive, because there are two negative fractions in the problem.

$$\left(-\frac{7}{27}\right)\times\frac{18}{25}\times\left(-\frac{15}{28}\right)=+\frac{\overset{1}{\cancel{7}}\times18\times15}{27\times25\times\underset{4}{\cancel{28}}}=\frac{1\times\overset{2}{\cancel{18}}\times15}{\underset{3}{\cancel{27}}\times25\times4}$$

$$=\frac{1\times2\times\overset{3}{\cancel{15}}}{3\times\underset{5}{\cancel{25}}\times4}=\frac{1\times2\times3}{3\times5\times4}=\frac{1\times2\times\overset{1}{\cancel{3}}}{\underset{1}{\cancel{3}}\times5\times4}=\frac{1\times\overset{1}{\cancel{2}}\times1}{1\times5\times\underset{2}{\cancel{4}}}=\frac{1\times1\times1}{1\times5\times2}=\frac{1}{10}$$

(35) $1\frac{1}{8}$. The answer will be positive, because there are two negative fractions.

$$-\frac{15}{14}\div\left(-\frac{20}{21}\right)=-\frac{15}{14}\times\left(-\frac{21}{20}\right)=\frac{15}{14}\times\frac{21}{20}=\frac{\overset{3}{\cancel{15}}\times21}{14\times\underset{4}{\cancel{20}}}$$

$$=\frac{3\times\overset{3}{\cancel{21}}}{\underset{2}{\cancel{14}}\times4}=\frac{3\times3}{2\times4}=\frac{9}{8}=1\frac{1}{8}$$

(36) $3\frac{1}{3}$.

$$2\frac{1}{2}\div\frac{3}{4}=\frac{5}{2}\div\frac{3}{4}=\frac{5}{2}\times\frac{4}{3}=\frac{5\times4}{2\times3}=\frac{5\times\overset{2}{\cancel{4}}}{\underset{1}{\cancel{2}}\times3}=\frac{5\times2}{1\times3}=\frac{10}{3}=3\frac{1}{3}$$

(37) $2\frac{2}{9}$.

$$7\frac{1}{7}\div3\frac{3}{14}=\frac{50}{7}\div\frac{45}{14}=\frac{50}{7}\times\frac{14}{45}=\frac{\overset{10}{\cancel{50}}\times14}{7\times\underset{9}{\cancel{45}}}=\frac{10\times\overset{2}{\cancel{14}}}{\underset{1}{\cancel{7}}\times9}$$

$$=\frac{10\times2}{1\times9}=\frac{20}{9}=2\frac{2}{9}$$

(38) $1\frac{1}{3}$.

$$\frac{\frac{16}{21}}{\frac{4}{7}}=\frac{16}{21}\times\frac{7}{4}=\frac{\overset{4}{\cancel{16}}}{21}\times\frac{7}{\underset{1}{\cancel{4}}}=\frac{4}{\underset{3}{\cancel{21}}}\times\frac{\overset{1}{\cancel{7}}}{1}=\frac{4}{3}=1\frac{1}{3}$$

(39) $8\frac{1}{3}$.

$$\frac{3\frac{1}{3}}{\frac{2}{5}}=\frac{\frac{10}{3}}{\frac{2}{5}}=\frac{10}{3}\times\frac{5}{2}=\frac{\overset{5}{\cancel{10}}}{3}\times\frac{5}{\underset{1}{\cancel{2}}}=\frac{25}{3}=8\frac{1}{3}$$

(40) 4.

$$\frac{4\frac{2}{7}}{1\frac{1}{14}}=\frac{\frac{30}{7}}{\frac{15}{14}}=\frac{30}{7}\times\frac{14}{15}=\frac{\overset{2}{\cancel{30}}}{7}\times\frac{14}{\underset{1}{\cancel{15}}}=\frac{2}{\underset{1}{\cancel{7}}}\times\frac{\overset{2}{\cancel{14}}}{1}=\frac{4}{1}=4$$

(41) $\frac{4}{5}$.

$$\frac{2\frac{1}{3}+4\frac{1}{5}}{10-1\frac{5}{6}}=\frac{\frac{7}{3}+\frac{21}{5}}{10-\frac{11}{6}}=\frac{\frac{35}{15}+\frac{63}{15}}{\frac{60}{6}-\frac{11}{6}}=\frac{\frac{98}{15}}{\frac{49}{6}}=\frac{98}{15}\times\frac{6}{49}=\frac{{}^{2}\cancel{98}}{{}_{5}\cancel{15}}\times\frac{\cancel{6}^{2}}{\cancel{49}_{1}}=\frac{4}{5}$$

(42) **36,000.** Simplify inside the parentheses first.

$$\begin{array}{r} 35.42 \\ \underline{-3.02} \\ 32.4 \end{array} \qquad \begin{array}{r} 3\,6000 \\ 0.0009_{\wedge}\overline{)32.4000_{\wedge}} \\ \underline{27} \\ 5\,4 \\ \underline{5\,4} \\ 0 \\ \underline{0} \\ 0 \\ \underline{0} \\ 0 \\ \underline{0} \end{array}$$

(43) **−2.999948.** By the order of operations, you multiply first and then subtract 3 from the result.

$$\begin{array}{r} 000005.2 \\ \underline{\times 0.00001} \\ 0.000052 \end{array} \qquad \begin{array}{r} 0.000052-3= \\ -3.000000 \\ \underline{0.000052} \\ -2.999948 \end{array}$$

(44) **0.6.** Dividing, you have $5\overline{)3.0}$ with quotient 0.6.

(45) **4.$\bar{4}$.** $\frac{40}{9}=4\frac{4}{9}=4.\bar{4}$ because

$$\begin{array}{r} 0.44 \\ 9\overline{)4.00} \\ \underline{36} \\ 40 \\ \underline{36} \\ 4 \\ \vdots \end{array}$$

46 $0.\overline{18}$.

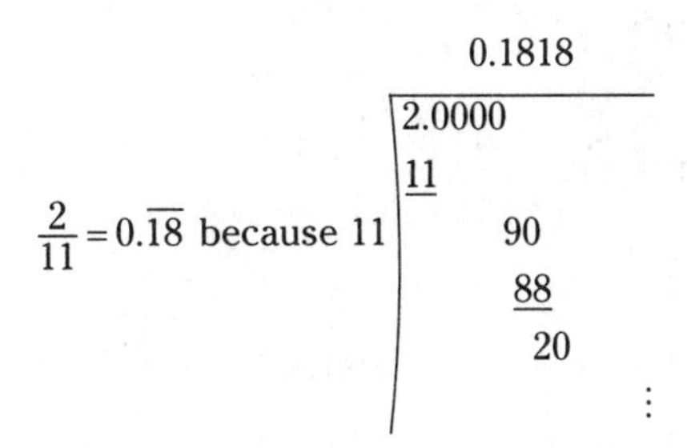

47 $\frac{9}{20}$. $0.45 = \frac{45}{100} = \frac{\cancel{45}^{9}}{\cancel{100}_{20}} = \frac{9}{20}$.

48 $\frac{4}{11}$.

$0.\overline{36} = \frac{36}{99}$ because two digits repeat $= \frac{\cancel{36}^{4}}{\cancel{99}_{11}} = \frac{4}{11}$.

49 $\frac{15}{37}$.

$0.\overline{405} = \frac{405}{999}$ because three digits repeat: $\frac{\cancel{405}^{45}}{\cancel{999}_{111}} = \frac{\cancel{45}^{15}}{\cancel{111}_{37}} = \frac{15}{37}$.

50 **4.173.** The number 7 follows the 2 in the thousandths place. Because 7 is greater than 5, you round up.

If you're ready to test your skills a bit more, take the following chapter quiz that incorporates all the chapter topics.

Whaddya Know? Chapter 4 Quiz

1. $7.7 \times 0.013 \div 1.001 =$
2. Write the fraction $\frac{8}{7}$ as an equivalent fraction with a denominator of 42.
3. Rewrite the fractions with the least common denominator: $\frac{8}{x+1} - \frac{5}{6}$
4. Write the fraction as a decimal: $\frac{8}{11}$
5. Solve the proportion for x: $\frac{x}{9} = \frac{16}{12}$
6. Change the decimal to a fraction in lowest terms: $1.2\overline{727}$.
7. Rewrite the fractions with the least common denominator: $\frac{4}{15} + \frac{7}{20}$
8. $\frac{5}{16} \times \frac{8}{9} \div \frac{11}{12} =$

9. Change the mixed number $-3\frac{3}{4}$ to an improper fraction.
10. Simplify the complex fraction: $\dfrac{8\frac{1}{4}}{4\frac{2}{5}}$
11. Write the fraction $\frac{90}{300}$ in lowest terms.
12. $5.32 + 0.006 - 0.000049 =$
13. Change the decimal to a fraction in lowest terms: 0.125
14. Change the improper fraction $\frac{25}{7}$ to a mixed number.
15. $3\frac{1}{3} + \frac{3}{4} - \frac{1}{6} =$

Answers to Chapter 4 Quiz

1. **0.1.** Multiply the first two numbers; then divide the result by the third.

$$\begin{array}{r} 0.013 \\ \times\ 7.7 \\ \hline .1001 \end{array} \qquad 1.001_{\wedge}\overline{)0.100_{\wedge}1}^{\;\;.1}$$

2. $\frac{48}{42}$**.** Multiply both the numerator and denominator by 6.

3. $\frac{48}{6(x+1)} - \frac{5(x+1)}{6(x+1)}$**.** The fractions have no common denominator, so you multiply the first fraction by $\frac{6}{6}$ and the second fraction by $\frac{x+1}{x+1}$.

4. $0.72\overline{72}$**.** Dividing the denominator into the numerator, you get a repeating decimal.

5. $\boldsymbol{x = 12}$**.** Both denominators are divisible by 3. Divide to get $\frac{x}{\cancel{9}^3} = \frac{16}{\cancel{12}^4} \rightarrow \frac{x}{3} = \frac{16}{4}$. The fraction on the right can be reduced by dividing each number by 4. $\frac{x}{3} = \frac{\cancel{16}^4}{\cancel{4}^1} \rightarrow \frac{x}{3} = \frac{4}{1}$. Now, when you cross-multiply, you have $x = 12$.

6. $1\frac{3}{11}$**.** Put the repeating digits, 27, over 99 in a fraction and reduce. $\frac{\cancel{27}^3}{\cancel{99}^{11}} = \frac{3}{11}$.

7. $\frac{16}{60} + \frac{21}{60}$**.** Multiply the first fraction by $\frac{4}{4}$ and the second fraction by $\frac{3}{3}$.

8. $\frac{10}{33}$**.** Multiply the first two fractions. Then divide the result by the third fraction.

$$\begin{aligned} \frac{5}{16} \times \frac{8}{9} \div \frac{11}{12} &= \left(\frac{5}{\cancel{16}^2} \times \frac{\cancel{8}^1}{9}\right) \div \frac{11}{12} \\ &= \frac{5}{18} \div \frac{11}{12} \\ &= \frac{5}{18} \times \frac{12}{11} \\ &= \frac{5}{\cancel{18}^3} \times \frac{\cancel{12}^2}{11} = \frac{10}{33} \end{aligned}$$

9. $-\frac{15}{4}$**.** Multiply the whole number 3 times the denominator 4 and add the numerator.

10. $1\frac{7}{8}$**.** $\frac{8\frac{1}{4}}{4\frac{2}{5}} = \frac{\frac{33}{4}}{\frac{22}{5}} = \frac{33}{4} \times \frac{5}{22} = \frac{\cancel{33}^3}{4} \times \frac{5}{\cancel{22}^2} = \frac{15}{8} = 1\frac{7}{8}$

(11) $\frac{3}{10}$**.** Both the numerator and denominator are divisible by 30. Divide and write the results in the corresponding positions.

(12) **5.325951.** Add the first two numbers and then subtract the third from the result.

$$\begin{array}{r} 5.32 \\ +0.006 \\ \hline 5.326 \end{array} \qquad \begin{array}{r} 5.326000 \\ -.000049 \\ \hline 5.325951 \end{array}$$

(13) $\frac{1}{8}$**.** Put 125 over 1,000 in a fraction and reduce.

(14) $3\frac{4}{7}$**.** Divide 7 into 25. The remainder goes in the numerator.

(15) $3\frac{11}{12}$**.** Write the three fractions with their least common denominator, 12.

$3\frac{1}{3}+\frac{3}{4}-\frac{5}{6}=3\frac{4}{12}+\frac{9}{12}-\frac{2}{12}$. Now, add the first two fractions and subtract the third.

$$\begin{aligned} 3\frac{4}{12}+\frac{9}{12}-\frac{2}{12} &= 3+\frac{4}{12}+\frac{9}{12}-\frac{2}{12} \\ &= 3+\frac{13}{12}-\frac{2}{12} \\ &= 3+\frac{11}{12} \end{aligned}$$

2
Operating on Operations

Contents at a Glance

- » Trimming down radical fractions
- » Working in fractional exponents
- » Performing operations using fractional exponents

Chapter 5
Taming Rampaging Radicals

The operation of taking a square root, cube root, or any other root is an important one in algebra (as well as in science and other areas of mathematics). The radical symbol ($\sqrt{\ }$) indicates that you want to take a *root* (what multiplies by itself to give you the number or value) of an expression. A more convenient notation, though, is to use an exponent, or power. This exponent is easily incorporated into algebraic work and makes computations easier to perform and results easier to report.

When you do square roots, the symbol for that operation is a radical, $\sqrt{\ }$. A cube root has a small 3 in front of the radical, $\sqrt[3]{\ }$; a fourth root has a small 4, $\sqrt[4]{\ }$, and so on.

The radical is a *nonbinary* operation (involving just one number) that asks you, "What number times itself gives you this number under the radical?" Another way of saying this is: "If $\sqrt{a}=b$, then $b^2=a$, or if $\sqrt[3]{c}=d$, then $d^3=c$," and so on.

When working through these examples and doing the problems, you will see the associative and commutative properties of addition and multiplication in action. Being able to change the order and regroup is essential in these processes. Refer to Chapter 3 for more on those properties.

Simplifying Radical Terms

Radical expressions such as $\sqrt{40}$ and $\sqrt[3]{54}$ may look harmless enough, and in fact, they are just asking to be simplified. When the number under a radical has a factor that is a perfect square or perfect cube and so on, you can pull that wonderful factor out by performing the root operation.

The rules you want to use when doing these simplifications are:

- $\sqrt{a^2 \cdot b} = \sqrt{a^2} \cdot \sqrt{b} = a\sqrt{b}$
- $\sqrt[3]{a^3 \cdot b} = \sqrt[3]{a^3} \cdot \sqrt[3]{b} = a\sqrt[3]{b}$

To help you with these simplifications, here are the first perfect squares and perfect cubes.

n	1	2	3	4	5	6	7	8	9	10	11	12
n^2	1	4	9	16	25	36	49	64	81	100	121	144
n^3	1	8	27	64	125	216	343	512	729	1,000	1,331	1,728

Sometimes recognizing factors of numbers can be a challenge. The Rules of Divisibility are found in Chapter 8, if you want to take a peek.

EXAMPLE

Q. Simplify $\sqrt{720}$.

A. You may not always recognize the largest perfect square factor. Let's assume you recognize that 9 is a factor of 720. Then you have $\sqrt{720} = \sqrt{9 \cdot 80} = \sqrt{9} \cdot \sqrt{80} = 3\sqrt{80}$. That's fine, but it's not completely simplified. You may notice that 4 is a factor of 80, but, even better, 16 is a factor of 80: $3\sqrt{80} = 3\sqrt{16 \cdot 5} = 3\sqrt{16} \cdot \sqrt{5} = 3 \cdot 4\sqrt{5} = 12\sqrt{5}$.

Q. Simplify $\sqrt[3]{54}$.

A. The perfect cube factor you're looking for is 27: $\sqrt[3]{54} = \sqrt[3]{27 \cdot 2} = \sqrt[3]{27} \cdot \sqrt[3]{2} = 3\sqrt[3]{2}$.

YOUR TURN

1 Simplify $\sqrt{162}$.

2 Simplify $\sqrt{175}$.

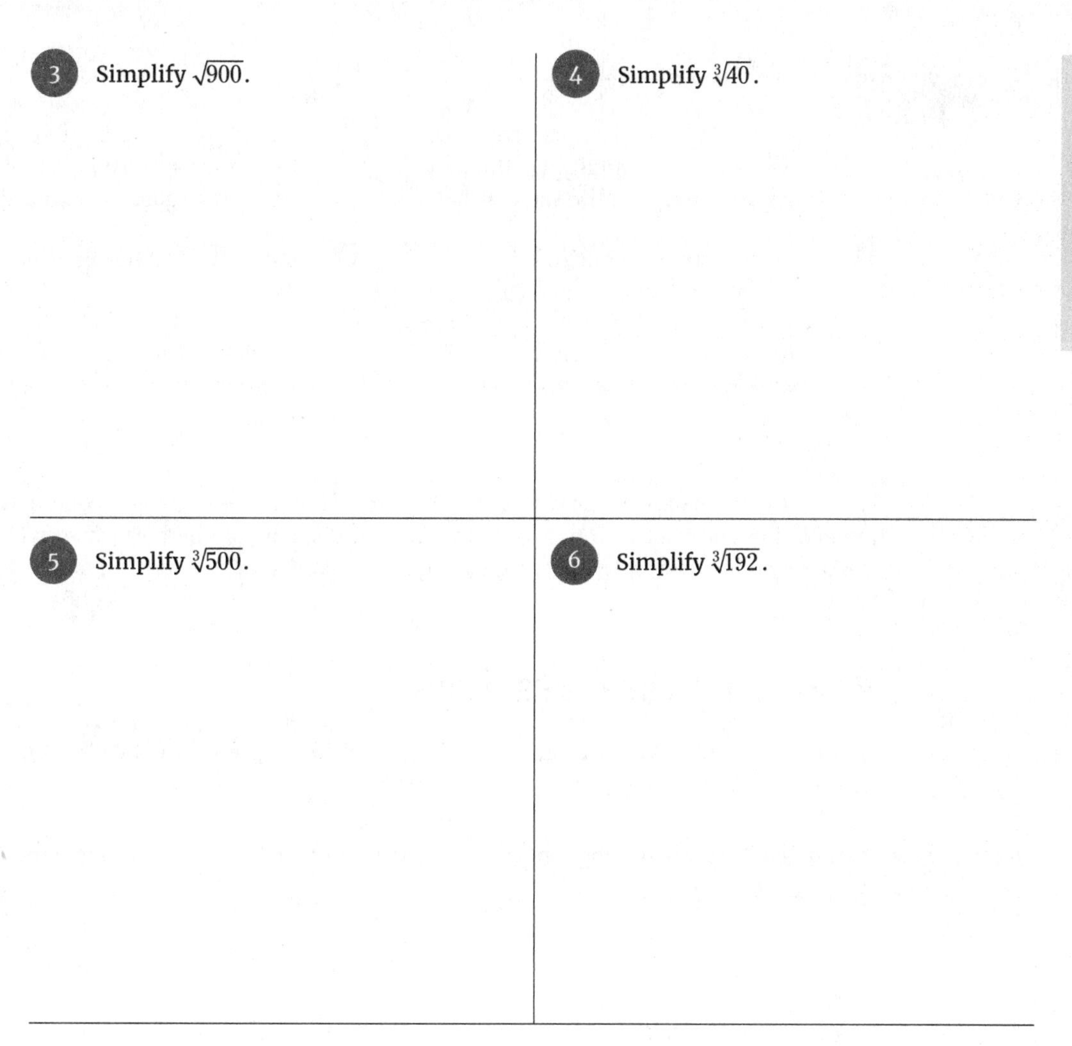

Taming Rampaging Radicals

Working through Radical Expressions

Simplifying and working through a radical expression means rewriting it as something equivalent using the smallest possible numbers under the radical symbol. If the number under the radical isn't a perfect square or cube or whichever power for the particular root, then you want to see whether that number has a factor that's a perfect square or cube (and so on) and factor it out.

Recognizing perfect square terms

Finding square roots is a relatively common operation in algebra, but working with and combining the roots isn't always so clear.

Expressions with radicals can be multiplied or divided as long as the root power *or* the value under the radical is the same. Expressions with radicals cannot be added or subtracted unless *both* the root power *and* the value under the radical are the same. Here are some examples of simplifying the radical expressions when possible.

EXAMPLE

Q. Can $\sqrt{2}\cdot\sqrt{3}$ be simplified?

A. Yes. $\sqrt{2}\cdot\sqrt{3}=\sqrt{6}$. These can be combined because it's multiplication, and the root powers are the same.

Q. Can $\sqrt{8}\div\sqrt{4}$ be simplified?

A. Yes. $\sqrt{8}\div\sqrt{4}=\sqrt{2}$. These can be combined because it's division, and the root powers are the same.

Q. Can $\sqrt{2}+\sqrt{3}$ be simplified?

A. No. $\sqrt{2}+\sqrt{3}$. These cannot be combined because it's addition, and the values under the radicals are not the same.

Q. Can $4\sqrt{3}+2\sqrt{3}$ be simplified?

A. Yes. $4\sqrt{3}+2\sqrt{3}=6\sqrt{3}$. These can be combined because the root powers and the numbers under the radicals are the same.

When the numbers inside the radical are the same, you can see some nice combinations involving addition and subtraction. Multiplication and division can be performed whether they're the same or not. The *root power* refers to square root ($\sqrt{\ }$), cube root ($\sqrt[3]{\ }$), fourth root ($\sqrt[4]{\ }$), and so on.

Rewriting radical terms

Here are the rules for adding, subtracting, multiplying, and dividing radical expressions. Assume that *a* and *b* are positive values.

- Addition and subtraction can be performed if the root power and value under the radical are the same:
  - $m\sqrt{a}+n\sqrt{a}=(m+n)\sqrt{a}$
  - $m\sqrt{a}-n\sqrt{a}=(m-n)\sqrt{a}$
- Multiplication and division can be performed if the root powers are the same:
  - $\sqrt{a}\sqrt{b}=\sqrt{ab}$
  - $\dfrac{\sqrt{a}}{\sqrt{b}}=\sqrt{\dfrac{a}{b}}$

Here are some of the more frequently used square roots:

$\sqrt{1}=1$ $\sqrt{4}=2$ $\sqrt{9}=3$

$\sqrt{16}=4$ $\sqrt{25}=5$ $\sqrt{36}=6$

$\sqrt{49}=7$ $\sqrt{64}=8$ $\sqrt{81}=9$

$\sqrt{100}=10$ $\sqrt{10{,}000}=100$ $\sqrt{1{,}000{,}000}=1{,}000$

Notice that the square root of a 1 followed by an even number of zeros is always a 1 followed by half that many zeros.

EXAMPLE

Q. Simplify the radical expression: $\sqrt{3}\left(\sqrt{6}\right)+\sqrt{10}\left(\sqrt{5}\right)$.

A. First, multiply: $\sqrt{3}\left(\sqrt{6}\right)+\sqrt{10}\left(\sqrt{5}\right)=\sqrt{18}+\sqrt{50}$.

Now factor each product under the radicals into a perfect square times another number: $\sqrt{18}+\sqrt{50}=\sqrt{9\cdot 2}+\sqrt{25\cdot 2}$.

Rewrite each radical as a product and simplify the roots of perfect squares: $\sqrt{9\cdot 2}+\sqrt{25\cdot 2}=\sqrt{9}\sqrt{2}+\sqrt{25}\sqrt{2}=3\sqrt{2}+5\sqrt{2}$.

Now the two terms can be added: $3\sqrt{2}+5\sqrt{2}=8\sqrt{2}$.

In the following two examples, the numbers under the radicals aren't perfect squares, so the numbers are written as the product of two factors — one of them is a perfect-square factor. You also see here how to apply the rule for roots of products and write the expression in simplified form.

EXAMPLE

Q. $\sqrt{60}=$

A. The number 60 can be written as the product of 3 and 20 or 5 and 12, but none of those numbers is a perfect square. Instead, you use 4 and 15 because 4 is a perfect square.

$$\sqrt{60}=\sqrt{4\times 15}=\sqrt{4}\times\sqrt{15}=2\sqrt{15}$$

Q. $\sqrt{8}+\sqrt{18}=$

A. You can't add the two radicals together the way they are, but after you simplify them, the two terms have the same radical factor, and you can add them together.

$$\begin{aligned}\sqrt{8}+\sqrt{18}&=\sqrt{4\times 2}+\sqrt{9\times 2}\\&=\sqrt{4}\times\sqrt{2}+\sqrt{9}\times\sqrt{2}\\&=2\sqrt{2}+3\sqrt{2}=5\sqrt{2}\end{aligned}$$

YOUR TURN

7 $\left(\sqrt{12}\right)\left(\sqrt{6}\right)=$

8 $\left(\sqrt{200}\right)\left(\sqrt{200}\right)=$

9. $(\sqrt{63})(\sqrt{21}) =$

10. $(\sqrt{15})(\sqrt{10})(\sqrt{6}) =$

11. $\sqrt{24} + \sqrt{54} =$

12. $\sqrt{72} - \sqrt{50} =$

13. $6\sqrt{3} + 4\sqrt{2} - 5\sqrt{3} + 7\sqrt{2} =$

14. $\frac{\sqrt{40}}{\sqrt{10}} + \frac{\sqrt{60}}{\sqrt{10}} - \frac{3\sqrt{12}}{\sqrt{2}} =$

15. $(10\sqrt{6})(4\sqrt{12}) + (3\sqrt{10})(6\sqrt{5}) =$

16. $\frac{\sqrt{40} + \sqrt{90}}{\sqrt{5}} =$

Rationalizing Fractions

You rationalize a fraction with a radical in its denominator (bottom) by changing the original fraction to an equivalent fraction that has a multiple of that radical in the numerator (top). Usually, you want to remove radicals from the denominator. The square root of a number that isn't a perfect square is *irrational.* Dividing with an irrational number is difficult because, when expressed as decimals, those numbers never end and never have a repeating pattern.

To rationalize a fraction with a square root in the denominator, multiply both the numerator and denominator by that square root.

EXAMPLE

Q. Rationalize $\frac{10}{\sqrt{5}}$.

A. Recall the property that $\sqrt{a \times b} = \sqrt{a} \times \sqrt{b}$. It works both ways: $\sqrt{a} \times \sqrt{b} = \sqrt{a \times b}$.

Multiplying the denominator by itself creates a perfect square (so there'll be no radical). Simplify and reduce the fraction.

$$\frac{10}{\sqrt{5}} = \frac{10}{\sqrt{5}} \times \frac{\sqrt{5}}{\sqrt{5}} = \frac{10\sqrt{5}}{\sqrt{25}}$$
$$= \frac{10\sqrt{5}}{5} = \frac{{}^{2}\cancel{10}\sqrt{5}}{\cancel{5}_{1}} = 2\sqrt{5}$$

Q. Rationalize $\frac{\sqrt{6}}{\sqrt{10}}$.

A. You multiply both of the radicals by the radical in the denominator. The products lead to results that can be simplified nicely.

$$\frac{\sqrt{6}}{\sqrt{10}} = \frac{\sqrt{6}}{\sqrt{10}} \times \frac{\sqrt{10}}{\sqrt{10}} = \frac{\sqrt{60}}{\sqrt{100}}$$
$$= \frac{\sqrt{4}\sqrt{15}}{10} = \frac{2\sqrt{15}}{10} = \frac{{}^{1}\cancel{2}\sqrt{15}}{\cancel{10}_{5}} = \frac{\sqrt{15}}{5}$$

YOUR TURN

17 Rationalize $\frac{1}{\sqrt{2}}$.

18 Rationalize $\frac{4}{\sqrt{3}}$.

19 Rationalize $\frac{3}{\sqrt{6}}$.

20 Rationalize $\frac{\sqrt{21}}{\sqrt{15}}$.

Managing Radicals as Exponential Terms

The radical symbol indicates that you're to do the operation of *taking a root* — or figuring out what number is multiplied by itself to give you the value under the radical. An alternate notation, a *fractional exponent,* also indicates that you're to take a root, but fractional exponents are much more efficient when you perform operations involving powers of the same number.

The equivalence between the square root of a and the fractional power notation is $\sqrt{a} = a^{1/2}$. The 2 in the bottom of the fractional exponent indicates a square root. The general equivalence between all roots, powers, and fractional exponents is $\sqrt[n]{a^m} = a^{m/n}$.

EXAMPLE

Q. $\sqrt[3]{x^7} =$

A. The root is 3; you're taking a cube root. When creating the exponent, the 3 goes in the fraction's denominator. The 7 goes in the fraction's numerator. The answer is $x^{7/3}$.

Q. $\frac{1}{\sqrt{x^3}} =$

A. The exponent becomes negative when you bring up the factor from the fraction's denominator. (Refer to Chapter 6 for more on negative exponents.) Also, when no root is showing on the radical, it's assumed that a 2 goes there because it's a square root. The answer is $x^{-3/2}$.

YOUR TURN

21 Write the radical form in exponential form: $\sqrt{6}$.

22 Write the radical form in exponential form: $\sqrt[3]{x}$.

23 Write the radical form in exponential form: $\sqrt{7^5}$.

24 Write the radical form in exponential form (assume that y is positive): $\sqrt[4]{y^3}$.

25 Write the radical form in exponential form (assume that x is positive): $\frac{1}{\sqrt{x}}$.

26 Write the radical form in exponential form: $\frac{3}{\sqrt[5]{2^2}}$.

Using Fractional Exponents

Fractional exponents by themselves are fine and dandy. They're a nice, compact way of writing an operation to be performed on the power of a number. What's even nicer is when you can simplify or evaluate an expression, and its result is an integer. You want to take advantage of these simplification situations.

If a value is written $a^{m/n}$, the easiest way to evaluate it is to take the root first and then raise the result to the power. Doing so keeps the numbers relatively small — or at least smaller than the power might become. The answer comes out the same either way. Being able to compute these problems in your head saves time.

EXAMPLE

Q. $8^{4/3} =$

A. Finding the cube root first is easier than raising 8 to the fourth power, which is 4,096, and then taking the cube root of that big number. By finding the cube root first, you can do all the math in your head. If you write out the solution, here's what it looks like: $8^{4/3} = \left(8^{1/3}\right)^4 = \left(\sqrt[3]{8}\right)^4 = (2)^4 = 16$

Q. $\left(\frac{1}{9}\right)^{3/2} =$

A. See Chapter 6 for the rule on raising a fraction to a power. It says that when a fraction is to be raised to a particular power, you raise both the numerator and denominator to that power. When the number 1 is raised to any power, the result is always 1. The rest involves the denominator.

$$\left(\frac{1}{9}\right)^{3/2} = \frac{1^{3/2}}{9^{3/2}} = \frac{1}{\left(9^{1/2}\right)^3} = \frac{1}{3^3} = \frac{1}{27}$$

YOUR TURN

27 Compute the value of $4^{5/2}$.

Compute the value of $27^{2/3}$.

Compute the value of $\left(\frac{1}{4}\right)^{3/2}$.

Compute the value of $\left(\frac{8}{27}\right)^{4/3}$.

Making the switch to fractional exponents

The convention that mathematicians have adopted is to use fractions in the powers to indicate that this stands for a root or a radical. The fractional exponents are easier to use when combining factors, and they're easier to type — for example, $\sqrt{x} = x^{\frac{1}{2}}$, $\sqrt[3]{x} = x^{\frac{1}{3}}$, and $\sqrt[4]{x} = x^{\frac{1}{4}}$.

Notice that when there's no number outside and to the upper left of the radical, you assume that it's a 2, for a square root. Also, recall that when raising a power to a power, you multiply the exponents.

When changing from radical form to fractional exponents:

- $\sqrt[n]{a} = a^{\frac{1}{n}}$. The *n*th root of *a* can be written as a fractional exponent with *a* raised to the reciprocal of that power.
- $\sqrt[n]{a^m} = a^{\frac{m}{n}}$. When the *n*th root of a^m is taken, it's raised to the $\frac{1}{n}$th power. Using the "Powers of Powers" rule, the *m* and the $\frac{1}{n}$ are multiplied together.

This rule involving changing radicals to fraction exponents allows you to simplify the following expressions. Note that when using the "Powers of Powers" rule, the bases still have to be the same.

EXAMPLE

Q. Simplify $6x^2 \cdot \sqrt[3]{x}$.

A. Change the factor with the radical into a factor with a fractional exponent.

$$6x^2 \cdot \sqrt[3]{x} = 6x^2 \cdot x^{\frac{1}{3}} = 6x^{2+\frac{1}{3}} = 6x^{\frac{7}{3}}$$

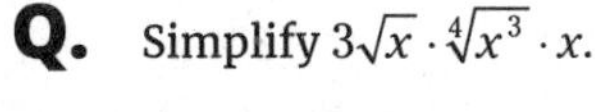

Q. Simplify $3\sqrt{x} \cdot \sqrt[4]{x^3} \cdot x$.

A. Change the two factors with radicals into factors with fractional exponents.

$$3\sqrt{x} \cdot \sqrt[4]{x^3} \cdot x = 3x^{\frac{1}{2}} \cdot x^{\frac{3}{4}} \cdot x^1 = 3x^{\frac{1}{2}+\frac{3}{4}+1} = 3x^{\frac{9}{4}}$$

Don't write the exponent as a mixed number; leave it as $\frac{9}{4}$.

YOUR TURN

Perform the operations using fractional exponents.

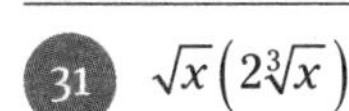

31 $\sqrt{x}\left(2\sqrt[3]{x}\right)$

32 $\dfrac{12\sqrt[4]{y}}{3\sqrt[6]{y}}$

Simplifying expressions with exponents

Writing expressions using fractional exponents is better than writing them as radicals because fractional exponents are easier to work with in situations where something complicated or messy needs to be simplified into something neater. The simplifying is done when you multiply and/or divide factors with the same base. When the bases are the same, you use the rules for multiplying (add exponents), dividing (subtract exponents), and raising to powers (multiply exponents). Refer to Chapter 6 if you need a reminder on these concepts.

EXAMPLE

Q. $\left(2^{4/3}\right)\left(2^{5/3}\right) =$

A. When numbers with the same base are multiplied together, you add the exponents:

$$\left(2^{4/3}\right)\left(2^{5/3}\right) = 2^{4/3+5/3} = 2^{9/3} = 2^3 = 8$$

Q. $\dfrac{5^{9/2}}{25^{1/4}} =$

A. Notice that the numbers don't have the same base! But 25 is a power of 5, so you can rewrite it and then apply the fourth root:

$$\frac{5^{9/2}}{25^{1/4}} = \frac{5^{9/2}}{\left(5^2\right)^{1/4}} = \frac{5^{9/2}}{5^{1/2}}$$

Now do the division by subtracting the exponents:

$$\frac{5^{9/2}}{5^{1/2}} = 5^{9/2-1/2} = 5^4 = 625$$

YOUR TURN

33 Simplify $\left(3^{1/4}\right)\left(3^{3/4}\right)$.

 Simplify $\dfrac{6^{14/5}}{6^{4/5}}$.

 Simplify $\dfrac{4^{7/4}}{8^{1/6}}$.

 Simplify $\dfrac{9^{3/4} \times 3^{7/2}}{27^{3/2}}$.

Estimating Answers

Radicals appear in many mathematical applications. You often need to simplify radical expressions, but it's also important to have an approximate answer in mind before you start. Doing so lets you evaluate whether the answer makes sense, based on your estimate. If you just keep in mind that $\sqrt{2}$ is about 1.4, $\sqrt{3}$ is about 1.7, and $\sqrt{5}$ is about 2.2, you can estimate many radical values.

REMEMBER

Estimating isn't always going to get you close. What you're really looking for is an error that's "way off." If you estimate the answer to be about 140 and you come up with 14, then you know there's something wrong. Always keep your eye on the reasonable answer.

EXAMPLE

Q. Estimate the value of $\sqrt{200}$.

A. Simplifying the radical, you get $\sqrt{200} = \sqrt{100 \times 2} = \sqrt{100} \times \sqrt{2} = 10\sqrt{2}$. If $\sqrt{2}$ is about 1.4, then 10(1.4) is 14. And, by the way, when rounded to three decimal places, $\sqrt{200}$ is 14.142.

Q. Estimate the value of $\sqrt{20} + \sqrt{27}$.

A. Simplifying the radicals, you get $\sqrt{20} + \sqrt{27} = \sqrt{4 \times 5} + \sqrt{9 \times 3} = 2\sqrt{5} + 3\sqrt{3}$. If $\sqrt{5}$ is about 2.2, then 2(2.2) is 4.4. Multiplying 3(1.7) for 3 times root three, you get 5.1. The sum of 4.4 and 5.1 is 9.5. How close it this? Rounded to three decimal places, the answer is 9.668.

YOUR TURN

37 Estimate $\sqrt{32}$.

Estimate $\sqrt{125}$.

Estimate $\sqrt{12} + \sqrt{18}$.

Estimate $\sqrt{160}$.

Practice Questions Answers and Explanations

1. **$9\sqrt{2}$.** Factor under the radical: $\sqrt{162}=\sqrt{81\cdot 2}$. Now write the product and simplify: $\sqrt{81\cdot 2}=\sqrt{81}\cdot\sqrt{2}=9\sqrt{2}$.

2. **$5\sqrt{7}$.** The perfect square factor of 175 is 25: $\sqrt{175}=\sqrt{25\cdot 7}=\sqrt{25}\cdot\sqrt{7}=5\sqrt{7}$.

3. **30.** If you didn't recognize that 900 is a perfect square, you may have written the number under the radical as a product involving 100: $\sqrt{900}=\sqrt{100\cdot 9}=\sqrt{100}\cdot\sqrt{9}=10\cdot 3=30$.

4. **$2\sqrt[3]{5}$.** Factor under the radical: $\sqrt[3]{40}=\sqrt[3]{8\cdot 5}$. The 8 is a perfect cube, so $\sqrt[3]{8\cdot 5}=\sqrt[3]{8}\cdot\sqrt[3]{5}=2\sqrt[3]{5}$.

5. **$5\sqrt[3]{4}$.** The perfect cube of 125 is a factor of 500: $\sqrt[3]{500}=\sqrt[3]{125\cdot 4}=\sqrt[3]{125}\cdot\sqrt[3]{4}=5\sqrt[3]{4}$.

6. **$4\sqrt[3]{3}$.** The greatest perfect cube factor is 64: $\sqrt[3]{192}=\sqrt[3]{64\cdot 3}=\sqrt[3]{64}\cdot\sqrt[3]{3}=4\sqrt[3]{3}$.

7. **$6\sqrt{2}$.** Multiply the two numbers together. Then factor using the perfect square 36.

$$\left(\sqrt{12}\right)\left(\sqrt{6}\right)=\sqrt{72}=\sqrt{36\cdot 2}=\sqrt{36}\cdot\sqrt{2}=6\sqrt{2}$$

8. **200.** You can multiply the two numbers, giving you the huge result 400,000. Yes, that will break down into two perfect squares. But consider this alternative: Factor both and simplify. Then multiply the results and simplify.

$$\begin{aligned}\left(\sqrt{200}\right)\left(\sqrt{200}\right)&=\left(\sqrt{100}\cdot\sqrt{2}\right)\left(\sqrt{100}\cdot\sqrt{2}\right)=\left(10\sqrt{2}\right)\left(10\sqrt{2}\right)\\&=100\sqrt{4}=100\cdot 2=200\end{aligned}$$

9. **$21\sqrt{3}$.** Factor the four numbers before multiplying. You'll see the pairing of numbers that will create a perfect square.

$$\begin{aligned}\left(\sqrt{63}\right)\left(\sqrt{21}\right)&=\left(\sqrt{9\cdot 7}\right)\left(\sqrt{3\cdot 7}\right)=\left(\sqrt{9}\sqrt{7}\right)\left(\sqrt{3}\sqrt{7}\right)=3\sqrt{7}\sqrt{3}\sqrt{7}\\&=3\sqrt{3}\left(\sqrt{7}\sqrt{7}\right)=3\sqrt{3}(7)=21\sqrt{3}\end{aligned}$$

10. **30.** Factor the numbers under the radicals. Then group the common factors to create perfect squares.

$$\begin{aligned}\left(\sqrt{15}\right)\left(\sqrt{10}\right)\left(\sqrt{6}\right)&=\left(\sqrt{3}\sqrt{5}\right)\left(\sqrt{2}\sqrt{5}\right)\left(\sqrt{2}\sqrt{3}\right)\\&=\sqrt{3}\sqrt{5}\sqrt{2}\sqrt{5}\sqrt{2}\sqrt{3}\\&=\sqrt{3}\sqrt{3}\sqrt{5}\sqrt{5}\sqrt{2}\sqrt{2}\\&=3\cdot 5\cdot 2=30\end{aligned}$$

11. **$5\sqrt{6}$.** First factor the numbers under the radicals to identify perfect-square factors.

$$\sqrt{24}+\sqrt{54}=\sqrt{4\times 6}+\sqrt{9\times 6}=\sqrt{4}\sqrt{6}+\sqrt{9}\sqrt{6}=2\sqrt{6}+3\sqrt{6}=5\sqrt{6}$$

12. **$\sqrt{2}$.** First factor the numbers under the radicals to identify perfect-square factors.

$$\sqrt{72}-\sqrt{50}=\sqrt{36\times 2}-\sqrt{25\times 2}=\sqrt{36}\sqrt{2}-\sqrt{25}\sqrt{2}=6\sqrt{2}-5\sqrt{2}=\sqrt{2}$$

13. $\mathbf{\sqrt{3}+11\sqrt{2}}$. Combine the like terms: $6\sqrt{3}-5\sqrt{3}+4\sqrt{2}+7\sqrt{2}=1\sqrt{3}+11\sqrt{2}$.

14. $\mathbf{2-2\sqrt{6}}$. First divide each term: $\frac{\sqrt{40}}{\sqrt{10}}+\frac{\sqrt{60}}{\sqrt{10}}-\frac{3\sqrt{12}}{\sqrt{2}}=\sqrt{4}+\sqrt{6}-3\sqrt{6}$. Now simplify and combine like terms: $\sqrt{4}+\sqrt{6}-3\sqrt{6}=2+1\sqrt{6}-3\sqrt{6}=2-2\sqrt{6}$.

15. $\mathbf{330\sqrt{2}}$. Multiply the coefficients and the numbers under the radicals separately: $40\sqrt{72}+18\sqrt{50}$. Now simplify the radicals by finding perfect-square factors: $40\sqrt{72}+18\sqrt{50}=40\sqrt{36\cdot 2}+18\sqrt{25\cdot 2}=40\sqrt{36}\sqrt{2}+18\sqrt{25}\sqrt{2}$. Find the square roots and simplify: $40\cdot 6\cdot\sqrt{2}+18\cdot 5\cdot\sqrt{2}=240\sqrt{2}+90\sqrt{2}=330\sqrt{2}$.

16. $\mathbf{5\sqrt{2}}$. First perform the divisions: $\frac{\sqrt{40}+\sqrt{90}}{\sqrt{5}}=\frac{\sqrt{40}}{\sqrt{5}}+\frac{\sqrt{90}}{\sqrt{5}}=\sqrt{8}+\sqrt{18}$. Now find the perfect-square factors under the radicals and simplify: $\sqrt{8}+\sqrt{18}=\sqrt{4\cdot 2}+\sqrt{9\cdot 2}=\sqrt{4}\sqrt{2}+\sqrt{9}\sqrt{2}=2\sqrt{2}+3\sqrt{2}=5\sqrt{2}$.

17. $\mathbf{\frac{\sqrt{2}}{2}}$. Multiply both the numerator and denominator by $\sqrt{2}$.

$$\frac{1}{\sqrt{2}}=\frac{1}{\sqrt{2}}\times\frac{\sqrt{2}}{\sqrt{2}}=\frac{\sqrt{2}}{\sqrt{4}}=\frac{\sqrt{2}}{2}$$

18. $\mathbf{\frac{4\sqrt{3}}{3}}$. Multiply both the numerator and denominator by $\sqrt{3}$.

$$\frac{4}{\sqrt{3}}=\frac{4}{\sqrt{3}}\times\frac{\sqrt{3}}{\sqrt{3}}=\frac{4\sqrt{3}}{\sqrt{9}}=\frac{4\sqrt{3}}{3}$$

19. $\mathbf{\frac{\sqrt{6}}{2}}$. Multiply both the numerator and denominator by $\sqrt{6}$.

$$\frac{3}{\sqrt{6}}=\frac{3}{\sqrt{6}}\times\frac{\sqrt{6}}{\sqrt{6}}=\frac{3\sqrt{6}}{\sqrt{36}}=\frac{{}^{1}\not{3}\sqrt{6}}{\not{6}_{2}}=\frac{\sqrt{6}}{2}$$

20. $\mathbf{\frac{\sqrt{35}}{5}}$. Multiply both the numerator and denominator by $\sqrt{15}$.

$$\frac{\sqrt{21}}{\sqrt{15}}=\frac{\sqrt{21}}{\sqrt{15}}\times\frac{\sqrt{15}}{\sqrt{15}}=\frac{\sqrt{21\times 15}}{\sqrt{15\times 15}}=\frac{\sqrt{7\times 3\times 3\times 5}}{15}=\frac{\sqrt{9}\sqrt{7\times 5}}{15}=\frac{{}^{1}\not{3}\sqrt{35}}{\not{15}_{5}}=\frac{\sqrt{35}}{5}$$

21. $\mathbf{6^{1/2}}$. The root is 2, so the exponent is $\frac{1}{2}$.

22. $\mathbf{x^{1/3}}$. The root is 3, so the exponent is $\frac{1}{3}$.

23. $\mathbf{7^{5/2}}$. The root is 2 and the power is 5, so the exponent is $\frac{5}{2}$.

24. $\mathbf{y^{3/4}}$. The root is 4 and the power is 3, so the exponent is $\frac{3}{4}$.

25. $\mathbf{x^{-1/2}}$. The root is 2 and the factor is in the denominator, so the exponent is negative.

26. $\mathbf{3\times 2^{-2/5}}$. The root is 5 and the power is 2. The factor is in the denominator, so the exponent is negative. The numerator multiplies the factor in the denominator.

(27) **32.** Find the root first.

$$4^{5/2} = \left(4^{1/2}\right)^5 = \left(\sqrt{4}\right)^5 = 2^5 = 32$$

(28) **9.** Find the root first.

$$27^{2/3} = \left(27^{1/3}\right)^2 = \left(\sqrt[3]{27}\right)^2 = 3^2 = 9$$

(29) $\mathbf{\frac{1}{8}}$. Both the numerator and denominator are raised to the power.

$$\left(\frac{1}{4}\right)^{3/2} = \frac{1^{3/2}}{4^{3/2}} = \frac{1}{\left(4^{1/2}\right)^3} = \frac{1}{2^3} = \frac{1}{8}$$

(30) $\mathbf{\frac{16}{81}}$. Both the numerator and denominator are raised to the power.

$$\left(\frac{8}{27}\right)^{4/3} = \frac{8^{4/3}}{27^{4/3}} = \frac{\left(8^{1/3}\right)^4}{\left(27^{1/3}\right)^4} = \frac{2^4}{3^4} = \frac{16}{81}$$

(31) $\mathbf{2x^{5/6}}$. First, change the radical expressions to those with fractional exponents: $\sqrt{x}\left(2\sqrt[3]{x}\right) = x^{1/2}\left(2x^{1/3}\right)$.

Multiply the variables by adding the exponents: $x^{1/2}\left(2x^{1/3}\right) = 2x^{1/2}x^{1/3} = 2x^{1/2+1/3} = 2x^{5/6}$.

(32) $\mathbf{4y^{1/12}}$. First, change the radical expressions to those with fractional exponents: $\frac{12\sqrt[4]{y}}{3\sqrt[6]{y}} = \frac{12y^{1/4}}{3y^{1/6}}$.

Divide the coefficients; then divide the variable by subtracting the exponents: $\frac{12y^{1/4}}{3y^{1/6}} = 4y^{1/4-1/6} = 4y^{1/12}$.

(33) **3.** Add the exponents:

$$3^{1/4} \times 3^{3/4} = 3^{1/4+3/4} = 3^{4/4} = 3^1 = 3$$

(34) **36.** Subtract the exponents:

$$\frac{6^{14/5}}{6^{4/5}} = 6^{14/5-4/5} = 6^{10/5} = 6^2 = 36$$

(35) **8.** First, change the two bases to a base of 2:

$$\frac{4^{7/4}}{8^{1/6}} = \frac{\left(2^2\right)^{7/4}}{\left(2^3\right)^{1/6}} = \frac{2^{7/2}}{2^{1/2}}$$

Now subtract the exponents:

$$\frac{2^{7/2}}{2^{1/2}} = 2^{7/2-1/2} = 2^{6/2} = 2^3 = 8$$

36) $\sqrt{3}$**.** Change two of the factors to a base of 3:

$$\frac{9^{3/4} \times 3^{7/2}}{27^{3/2}} = \frac{\left(3^2\right)^{3/4} \times 3^{7/2}}{\left(3^3\right)^{3/2}} = \frac{3^{3/2} \times 3^{7/2}}{3^{9/2}}$$

Add the exponents in the numerator to perform the multiplication. Then subtract the exponent in the denominator:

$$\frac{3^{3/2} \times 3^{7/2}}{3^{9/2}} = \frac{3^{3/2+7/2}}{3^{9/2}} = \frac{3^{10/2}}{3^{9/2}} = 3^{10/2-9/2} = 3^{1/2} = \sqrt{3}$$

37) **About 5.6.**

$$\sqrt{32} = \sqrt{16} \times \sqrt{2} = 4\sqrt{2} \approx 4(1.4) = 5.6$$

38) **About 11.**

$$\sqrt{125} = \sqrt{25} \times \sqrt{5} = 5\sqrt{5} \approx 5(2.2) = 11$$

39) **About 7.6.**

$$\sqrt{12} + \sqrt{18} = \left(\sqrt{4} \times \sqrt{3}\right) + \left(\sqrt{9} \times \sqrt{2}\right) = 2\sqrt{3} + 3\sqrt{2} \approx 2(1.7) + 3(1.4) = 7.6$$

40) **About 12.32.**

$$\sqrt{160} = \sqrt{16} \times \sqrt{10} = \sqrt{16} \times \sqrt{2} \times \sqrt{5} = 4\sqrt{2}\sqrt{5} \approx 4(1.4)(2.2) = 12.32$$

If you're ready to test your skills a bit more, take the following chapter quiz that incorporates all the chapter topics.

Whaddya Know? Chapter 5 Quiz

Quiz time! Complete each problem to test your knowledge on the various topics covered in this chapter. You can then find the solutions and explanations in the next section.

1. Write in radical form: $x^{-3/2}$.
2. $(5\sqrt{11})(\sqrt{22}) =$
3. Write in exponential form: $\sqrt[3]{x^4}$.
4. $(\sqrt{8})(\sqrt{6}) + (\sqrt{15})(\sqrt{5}) =$
5. Simplify: $\sqrt{250}$.
6. Estimate to one decimal place: $\sqrt{80}$.
7. $\sqrt{27} + \sqrt{75} =$
8. Simplify: $(200x^4)^{1/2} + (125x^3)^{2/3}$.
9. $\frac{\sqrt{60}}{\sqrt{15}} + \frac{2\sqrt{300}}{\sqrt{12}} =$
10. Compute: $\left(\frac{25}{36}\right)^{3/2}$.
11. Simplify: $\sqrt[3]{80}$.
12. Simplify: $2\sqrt{z}\left(3\sqrt[3]{z^2}\right)$.
13. $(\sqrt{15})(\sqrt{33}) =$
14. Rationalize the denominator: $\frac{12\sqrt{5}}{\sqrt{10}}$.

Answers to Chapter 5 Quiz

1. $\frac{1}{\sqrt[2]{x^3}}$. The negative sign puts the radical in the denominator.

2. $55\sqrt{2}$. Before multiplying the two radicals together, write the second radical as a product: $(5\sqrt{11})(\sqrt{22})=(5\sqrt{11})(\sqrt{2}\sqrt{11})$. The multiplication is now much simpler: $(5\sqrt{11})(\sqrt{2}\sqrt{11})=5\sqrt{11}\sqrt{11}\sqrt{2}=5\cdot 11\sqrt{2}=55\sqrt{2}$.

3. $x^{4/3}$. The 3 indicates the root.

4. $9\sqrt{3}$. Multiply the two products, and then simplify the results to see if they can be added: $(\sqrt{8})(\sqrt{6})+(\sqrt{15})(\sqrt{5})=(\sqrt{48})+(\sqrt{75})=(\sqrt{16\cdot 3})+(\sqrt{25\cdot 3})=(4\sqrt{3})+(5\sqrt{3})$. The numbers under the radicals are the same, so they can be added: $(4\sqrt{3})+(5\sqrt{3})=9\sqrt{3}$.

5. $5\sqrt{10}$. First, write 250 as the product of a perfect square and another factor. Then write the factors under radicals and evaluate: $\sqrt{250}=\sqrt{25\cdot 10}=\sqrt{25}\sqrt{10}=5\sqrt{10}$.

6. **8.8.** Simplify the radical: $\sqrt{80}=\sqrt{16\cdot 5}=4\sqrt{5}$. Use the estimate of 2.2 for $\sqrt{5}$, and you get $4\sqrt{5}\approx 4\cdot 2.2=8.8$.

7. $8\sqrt{3}$. Simplify the radicals: $\sqrt{27}+\sqrt{75}=\sqrt{9\cdot 3}+\sqrt{25\cdot 3}=3\sqrt{3}+5\sqrt{3}$. Because the numbers under the radicals are the same, they can be added: $3\sqrt{3}+5\sqrt{3}=8\sqrt{3}$.

8. $(10\sqrt{2}+25)x^2$. Apply the exponents to both factors in each term.

$$\begin{aligned}(200x^4)^{1/2}+(125x^3)^{2/3}&=200^{1/2}(x^4)^{1/2}+125^{2/3}(x^3)^{2/3}\\&=\sqrt{200}x^2+(\sqrt[3]{125})^2x^2\\&=\sqrt{100}\sqrt{2}x^2+5^2x^2\\&=10\sqrt{2}x^2+25x^2\end{aligned}$$

Now you can add the terms, although the coefficients don't add well together.

$$10\sqrt{2}x^2+25x^2=(10\sqrt{2}+25)x^2$$

9. **12.** First do the division, simplify, and then add.

$$\frac{\sqrt{60}}{\sqrt{15}}+\frac{2\sqrt{300}}{\sqrt{12}}=\sqrt{4}+2\sqrt{25}=2+2\cdot 5=2+10=12$$

10. $\frac{125}{216}$. Apply the exponent to both the numerator and denominator: $\left(\frac{25}{36}\right)^{3/2}=\frac{25^{3/2}}{36^{3/2}}$. Evaluate by taking the roots first: $\frac{25^{3/2}}{36^{3/2}}=\frac{(\sqrt{25})^3}{(\sqrt{36})^3}=\frac{5^3}{6^3}=\frac{125}{216}$.

11. $2\sqrt[3]{10}$. First, write 80 as the product of a perfect cube and another factor. Then write the factors under radicals and evaluate: $\sqrt[3]{80}=\sqrt[3]{8\cdot 10}=\sqrt[3]{8}\sqrt[3]{10}=2\sqrt[3]{10}$.

(12) $6z^{7/6}$. Change the radicals to exponential form: $2\sqrt{z}\left(3\sqrt[3]{z^2}\right)=2z^{1/2}\left(3z^{2/3}\right)$. Now multiply the two integers and the two variables: $2z^{1/2}\left(3z^{2/3}\right)=2\cdot3\cdot z^{1/2}\cdot z^{2/3}=6z^{7/6}$. Adding the two exponents involved finding a common denominator. Refer to Chapter 4 if you need a review of adding exponents.

(13) $3\sqrt{55}$. Write the numbers as products before multiplying. Then pair up factors that create a perfect square.

$$\begin{aligned}\left(\sqrt{15}\right)\left(\sqrt{33}\right)&=\left(\sqrt{3\cdot5}\right)\left(\sqrt{3\cdot11}\right)=\left(\sqrt{3}\sqrt{5}\right)\left(\sqrt{3}\sqrt{11}\right)\\&=\sqrt{3}\sqrt{3}\sqrt{5}\sqrt{11}=3\sqrt{5}\sqrt{11}=3\sqrt{55}\end{aligned}$$

(14) $6\sqrt{2}$. Multiply both the numerator and denominator by $\sqrt{10}$: $\frac{12\sqrt{5}}{\sqrt{10}}\cdot\frac{\sqrt{10}}{\sqrt{10}}=\frac{12\sqrt{50}}{10}$.

Now simplify the radical and reduce the fraction.

$$\frac{12\sqrt{50}}{10}=\frac{12\sqrt{25}\sqrt{2}}{10}=\frac{12\cdot5\sqrt{2}}{10}=\frac{12\cdot\cancel{5}\sqrt{2}}{\cancel{10}^{2}}=\frac{\cancel{12}^{6}\sqrt{2}}{\cancel{2}}=6\sqrt{2}$$

» Recognizing the power of powers

» Operating on exponents

» Investigating scientific notation

Chapter 6
Exploring Exponents

In the big picture of mathematics, exponents are a fairly new development. The principle behind exponents has always been there, but mathematicians had to first agree to use algebraic symbols such as x and y for values, before they could agree to the added shorthand of superscripts to indicate how many times the values were to be used. As a result, instead of writing $x \cdot x \cdot x \cdot x \cdot x$, you get to write the x with a superscript of $5: x^5$. In any case, be grateful. Exponents make life a lot easier.

This chapter introduces to you how exponents can be used (and abused), how to recognize scientific notation on a calculator, and how eeeeeasy it is to use *e*. What's this *e* business? The letter *e* was named for the mathematician Leonhard Euler; the Euler number, *e*, is approximately 2.71828 and is used in business and scientific calculations.

Powering up with Exponential Notation

Writing numbers with exponents is one thing; knowing what these exponents mean and what you can do with them is another thing altogether. Using exponents is so convenient that it's worth the time and trouble to find out the rules for using them correctly.

The base of an exponential expression can be any *real number.* Real numbers are the rational and irrational numbers combined. (Chapter 1 has a full explanation of what real numbers are.) The exponent (the power) can be any real number, too. An exponent can be positive, negative, fractional, or even a radical. What power!

When a number x is involved in repeated multiplication of x times itself, the number n can be used to describe how many multiplications are involved: $x^n = x \cdot x \cdot x \cdot x \cdot x$ a total of n times.

Even though the x in the expression x^n can be any real number and the n can be any real number, they can't both be 0 at the same time. For example, 0^0 really has no meaning in algebra. It takes a calculus course to prove why this restriction is so. Also, if x is equal to 0, then n can't be negative.

Here are the details about 0 in the term x^n:

- In x^n, the x can be 0 and the n can be any positive number. In all these cases, $0^n = 0, n > 0$.
- In x^n, the n can be 0 and the x can be any number except 0. In all these cases, $x^0 = 1, x \neq 0$.

There are two special types of exponents: negative and fractional. A negative exponent indicates that the factor belongs in the denominator of a fraction. And a fractional exponent indicates that you're working with a root and a power.

PAYING OFF A ROYAL DEBT EXPONENTIALLY

There's an old story about a king who backed out on his promise to the knight who saved his castle from a fire-breathing dragon. The king was supposed to pay the knight two bags of gold for his bravery and for the successful endeavor.

After the knight had slain the dragon, the king was reluctant to pay up — after all, no more fire breathing in the neighborhood! So, the frustrated knight, wanting to get his just reward, struck a bargain with the king: On January 1, the king would pay him 1 pence, and he would double the amount every day until the end of April. So, on January 2, the king would pay him 2 pence. On January 3, the king would pay him 4 pence. On January 4, the king would pay him 8 pence. On January 5, the king would pay him 16 pence. And this would continue through April 30.

The king thought that this was a pretty good deal. After all, the knight was just asking for some of the smallest coins that the king had. So, he agreed and started paying off the knight. It went pretty well until the end of January. On January 20, he had to pay 524,288 pence. Then, on February 20, he had to pay 1,125,899,906,842,624 pence. On the last day, April 30, he had to pay over 664,613, 998,000,000,000,000,000,000,000,000,000 pence. Add up all the pence on all the days, and the total amount was more than 1,329,227,000,000,000,000,000,000,000,000,000 pence. If a *pence* is close to a penny, then this is way over a trillion trillion dollars! Guess who was king then?

EXAMPLE

Q. $\left(\frac{3}{5}\right)^3 =$

A. $\left(\frac{3}{5}\right)^3 = \frac{3^3}{5^3} = \frac{3\cdot 3\cdot 3}{5\cdot 5\cdot 5} = \frac{27}{125}$

Q. $(-4)^5 =$

A. $(-4)^5 = (-4)\cdot(-4)\cdot(-4)\cdot(-4)\cdot(-4) = -1{,}024$

Q. $16^{1/2} =$

A. $16^{1/2} = \sqrt{16} = 4$

Q. Write the expression $3^3x^2y^4z^6(w-2)^2$ without exponents.

A.
$$\begin{aligned}3^3x^2y^4z^6 &= 3\cdot 3\cdot 3\cdot x\cdot x\cdot y\cdot y\cdot y\cdot y\cdot z\cdot z\cdot z\cdot z\cdot z\cdot z\cdot(w-2)(w-2)\\ &= 27x\cdot x\cdot y\cdot y\cdot y\cdot y\cdot z\cdot z\cdot z\cdot z\cdot z\cdot z\cdot(w-2)(w-2)\end{aligned}$$

YOUR TURN

1. $3^5 =$

2. $(-1)^{10} =$

3. $\left(\frac{2}{3}\right)^4 =$

4. $25^{1/2} =$

5. Write $4^2x^3(y+z)^5$ without any exponents.

Exploring Exponents

Using Negative Exponents

Negative exponents are very useful in algebra because they allow you to do computations on numbers with the same base without having to deal with pesky fractions.

When you use the negative exponent b^{-n}, you're saying $b^{-n} = \frac{1}{b^n}$ and also $\frac{1}{b^{-n}} = b^n$. And, of course, b cannot be 0.

Another nice feature of negative exponents is how they affect fractions. Look at this rule:

$$\left(\frac{a^p}{b^q}\right)^{-n} = \left(\frac{b^q}{a^p}\right)^n = \frac{b^{qn}}{a^{pn}}$$

A quick, easy way of explaining this rule is to just say that a negative exponent flips the fraction and then applies a positive power to the factors.

Negative exponents are a neat little creation. They mean something very specific and have to be handled with care, but they are oh, so convenient to have. You can use a negative exponent to write a fraction without writing a fraction! Using negative exponents is a way to combine expressions with the same base, whether the different factors are in the numerator or denominator. It's a way to change division problems into multiplication problems.

Negative exponents are a way of writing powers of fractions or decimals without using the fraction or decimal. For example, instead of writing $\left(\frac{1}{10}\right)^{14}$, you can write 10^{-14}.

The following examples involve changing numbers with negative exponents to fractions with positive exponents.

$z^{-4} =$

The reciprocal of z^4 is $\frac{1}{z^4} = z^{-4}$. In this case, z cannot be 0.

$6^{-1} =$

The reciprocal of 6 is $\frac{1}{6} = 6^{-1}$.

But what if you start out with a negative exponent in the denominator? What happens then? Look at the fraction $\frac{1}{3^{-4}}$. If you write the denominator as a fraction, you get $\frac{1}{\frac{1}{3^4}}$. Then, you change the *complex fraction* (a fraction with a fraction in it) to a division problem: $\frac{1}{\frac{1}{3^4}} = 1 \div \frac{1}{3^4} = 1 \cdot \frac{3^4}{1} = 3^4$. (Refer to division of fractions in Chapter 4, if you need a refresher.) So, to simplify a fraction with a negative exponent in the denominator, you can do a switcheroo: $\frac{1}{3^{-4}} = 3^4$.

EXAMPLE

Q. $\frac{x^2y^3}{3z^{-4}} =$

A. $\frac{x^2y^3}{3z^{-4}} = \frac{x^2y^3z^4}{3}$. Just bring the *z* and its negative exponent up to the numerator and change it to a positive exponent.

Q. $\frac{3a^{-2}}{4b^{-3}} =$

A. Yes, the negative exponent flips the factor to the denominator: $\frac{3a^{-2}}{4b^{-3}} = \frac{3b^3}{4a^2}$

Q. $\frac{4a^3b^5c^6d}{a^{-1}b^{-2}} =$

A. $\frac{4a^3b^5c^6d}{a^{-1}b^{-2}} = 4a^3a^1b^5b^2c^6d = 4a^4b^7c^6d.$ This time, you get to multiply like-factors after bringing the factors with negative exponents up to the numerator.

Q. $\left(\frac{3^4 \times 2^3}{3^7 \times 2^2}\right)^{-2} =$

A. First flip and then simplify the common bases before raising each factor to the second power.

$$\left(\frac{3^4 \times 2^3}{3^7 \times 2^2}\right)^{-2} = \left(\frac{3^7 \times 2^2}{3^4 \times 2^3}\right)^{2}$$

$$= \left(\frac{3^{\cancel{7}3} \times \cancel{2^2}}{\cancel{3^4} \times 2^{\cancel{3}1}}\right)^{2} = \left(\frac{3^3}{2^1}\right)^{2} = \frac{3^6}{2^2} = \frac{729}{4}$$

YOUR TURN

6 Rewrite $\frac{1}{3^6}$, using a negative exponent.

7 Rewrite $\frac{1}{5^{-5}}$, getting rid of the negative exponent.

8 Simplify $\left(\frac{3^{-4}}{2^3}\right)^{-2}$, leaving no negative exponent.

9. Simplify $\frac{\left(2^3 \times 3^2\right)^4}{\left(2^5 \times 3^{-1}\right)^5}$, leaving no negative exponent.

10. Simplify $\frac{z^{-1}}{5xy}$ =, leaving no negative exponent.

Multiplying and Dividing Exponentials

The number 16 can be written as 2^4, and the number 64 can be written as 2^6. When multiplying these two numbers together, you can either write $16 \times 64 = 1{,}024$ or multiply their two exponential forms together to get $2^4 \times 2^6 = 2^{10}$, which is equal to 1,024. The computation is easier — the numbers are smaller — when you use the exponential forms. Exponential forms are also better for writing very large or very small numbers. They make it easier to compare numbers and usually don't take up as much room.

Multiplying the same base

To multiply numbers with the same base (b), you add their exponents. The bases must be the same, or this rule doesn't work.

$$b^m \cdot b^n = b^{m+n}$$

You can multiply many exponential expressions together without having to change their form into the big or small numbers they represent. The only requirement is that the bases of the exponential expressions that you're multiplying have to be the same. The answer is then a nice, neat exponential expression.

You *can* multiply $2^4 \cdot 2^6$ and $a^5 \cdot a^8$, but you *cannot* multiply $3^6 \cdot 4^9$ because the bases are not the same.

EXAMPLE

Q. $2^4 \cdot 2^9 =$

A. $2^4 \cdot 2^9 = 2^{4+9} = 2^{13}$

Q. $a^5 \cdot a^8 =$

A. $a^5 \cdot a^8 = a^{5+8} = a^{13}$

Often, you find algebraic expressions with a whole string of factors; you want to simplify the expression, if possible. When there's more than one base in a term with powers of the bases, you combine the numbers with the same bases, compute the values, and then rewrite the single term.

EXAMPLE

Q. $3^2 \cdot 2^2 \cdot 3^3 \cdot 2^4 =$

A. Combine the two factors with base 3 and the two factors with base 2.

$$3^2 \cdot 2^2 \cdot 3^3 \cdot 2^4 = 3^{2+3} \cdot 2^{2+4} = 3^5 \cdot 2^6$$

Q. $4x^6y^5x^4y =$

A. The number 4 is a coefficient, which is written before the rest of the factors.

$$4x^6y^5x^4y = 4x^{6+4}y^{5+1} = 4x^{10}y^6$$

When there's no exponent showing, such as with *y*, you assume that the exponent is 1. In the preceding example, you see that the factor *y* was written as y^1, so its exponent could be added to that in the other *y* factor.

Multiplying the same power

You can add exponents when multiplying numbers with the same base. And you can *multiply* numbers that have the same *power* (in a multiplication problem) but different bases. This is the only exception to the rule, that the bases have to be the same when multiplying numbers with exponents.

The rule is that $a^n \cdot b^n = (a \cdot b)^n$.

EXAMPLE

Q. $4^8 \times 7^8 =$

A. $4^8 \times 7^8 = (4 \times 7)^8 = 28^8$. You'd rather leave the simplified expression as the power of 28, because the actual number is huge!

Q. $a^6 \cdot 3^4 \cdot b^4 \cdot c^6 =$

A. Rearrange the factors and multiply the factors with the same exponents.

$$a^6 \cdot 3^4 \cdot b^4 \cdot c^6 = a^6 \cdot c^6 \cdot 3^4 \cdot b^4 = (a \cdot c)^6 \cdot (3 \cdot b)^4 = (ac)^6(3b)^4$$

It's usually preferred to have the numerical factor in front. Because raising 3 to the fourth isn't too bad, you can write $(ac)^6(3b)^4 = (3b)^4(ac)^6 = 3^4b^4(ac)^6 = 81b^4(ac)^6$.

YOUR TURN

11 $3 \times 3^5 =$

12 $2^8 \times 2^{-3} =$

13 $2^3 \times 3^4 \times 2^6 \times 3^2 =$

14 $5^{-4} \times 6^2 \times 5^5 =$

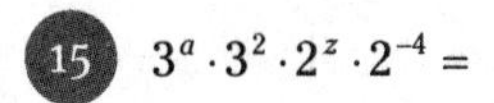

15 $3^a \cdot 3^2 \cdot 2^z \cdot 2^{-4} =$

16 $2^5 \cdot 9^5 + 3^4 b =$

Dividing with exponents

When numbers appear in exponential form, you can divide them by simply subtracting their exponents. As with multiplication, the bases have to be the same in order to perform this operation.

When the bases are the same and two factors are divided, subtract their exponents:

$$\frac{b^m}{b^n} = b^{m-n}$$

Remember that b cannot be 0. You can divide exponential expressions, leaving the answers as exponential expressions, as long as the bases are the same. Division is the reverse of multiplication, so it makes sense that, because you add exponents when multiplying numbers with the same base, you *subtract* the exponents when dividing numbers with the same base. Easy enough?

EXAMPLE

Q. $2^{10} \div 2^4 =$

A. $2^{10} \div 2^4 = 2^{10-4} = 2^6$. These exponentials represent the problem $1{,}024 \div 16$, which equals 64. It's much easier to leave the numbers as bases with exponents.

Q. $\frac{3^4}{3^3} =$

A. $\frac{3^4}{3^3} = 3^{4-3} = 3^1 = 3$

Q. $\frac{8^2 \times 3^5}{8^{-1} \times 3} =$

A. The bases of 8 and 3 are different, so you have to simplify the separate bases before multiplying the results together.

$$\frac{8^2 \times 3^5}{8^{-1} \times 3} = \frac{8^2}{8^{-1}} \times \frac{3^5}{3^1} = 8^{2-(-1)} \times 3^{5-1}$$
$$= 8^{2+1} \times 3^4 = 8^3 \times 3^4$$
$$= 512 \times 81 = 41{,}472$$

Q. $\dfrac{4x^6y^3z^2}{2x^4y^3z} =$

A. 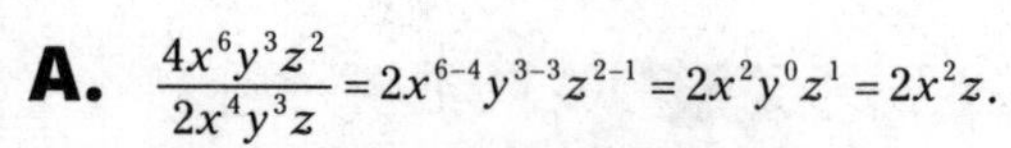

$\dfrac{4x^6y^3z^2}{2x^4y^3z} = 2x^{6-4}y^{3-3}z^{2-1} = 2x^2y^0z^1 = 2x^2z.$

The variables represent numbers, so writing this out the long way would look like this:

$$\frac{2\cdot 2\cdot x\cdot x\cdot x\cdot x\cdot x\cdot x\cdot y\cdot y\cdot y\cdot z\cdot z}{2\cdot x\cdot x\cdot x\cdot x\cdot y\cdot y\cdot y\cdot z}$$

$$= \frac{2\cdot \cancel{2}\cdot \cancel{x}\cdot \cancel{x}\cdot \cancel{x}\cdot \cancel{x}\cdot x\cdot x\cdot \cancel{y}\cdot \cancel{y}\cdot \cancel{y}\cdot \cancel{z}\cdot z}{\cancel{2}\cdot \cancel{x}\cdot \cancel{x}\cdot \cancel{x}\cdot \cancel{x}\cdot \cancel{y}\cdot \cancel{y}\cdot \cancel{y}\cdot \cancel{z}}$$

By crossing out the common factors, all that's left is $2x^2z$.

YOUR TURN

17 $3^{11} \div 3^6 =$

18 $\dfrac{7^9}{7} =$

19 $\dfrac{a^5}{a^{10}} =$

20 $\dfrac{3^2 \times 2^{-1}}{3 \times 2^{-5}} =$

21 $\dfrac{7^{-3} \times 2^4 \times 5}{7^{-7} \times 2^4 \times 5^{-1}} =$

Raising Powers to Powers

Raising a power to a power means that you take a number in exponential form and raise it to some power. For instance, raising 3^6 to the fourth power means to multiply the sixth power of 3 by itself four times: $3^6 \times 3^6 \times 3^6 \times 3^6$. As a power of a power, it looks like this: $\left(3^6\right)^4$. Raising something to a power tells you how many times it's multiplied by itself. To perform this operation, you use simple multiplication.

Here are the rules for raising a power to a power:

- $\left(b^m\right)^n = b^{m\times n}$. So to raise 3^6 to the fourth power, you write $\left(3^6\right)^4 = 3^{6\times 4} = 3^{24}$.
- $(a \times b)^m = a^m \times b^m$ and $\left(a^p \times b^q\right)^m = a^{p\times m} \times b^{q\times m}$.
- $\left(\frac{a}{b}\right)^m = \frac{a^m}{b^m}$ and $\left(\frac{a^p}{b^q}\right)^m = \frac{a^{p\times m}}{b^{q\times m}}$.

These rules say that if you multiply or divide two numbers and are raising the product or quotient to a power, then each factor gets raised to that power. (***Remember:*** A *product* is the result of multiplying, and a *quotient* is the result of dividing.)

EXAMPLE

Q. $\left(z^6\right)^8 =$

A. $\left(z^6\right)^8 = z^{6\times 8} = z^{48}$

Q. $\left(6^{-3}\right)^4 =$

A. $\left(6^{-3}\right)^4 = 6^{-3\cdot 4} = 6^{-12} = \frac{1}{6^{12}}$. You first multiply the exponents, then rewrite the product to create a positive exponent.

Q. $\left(3^{-4} \times 5^6\right)^7 =$

A. $\left(3^{-4} \times 5^6\right)^7 = 3^{-4\times 7} \times 5^{6\times 7} = 3^{-28} \times 5^{42}$

$$= \frac{1}{3^{28}} \times 5^{42}$$

Q. $\left(\frac{2^5}{5^2}\right)^3 =$

A. $\left(\frac{2^5}{5^2}\right)^3 = \frac{2^{5\times 3}}{5^{2\times 3}} = \frac{2^{15}}{5^6}$

Q. $\left(3x^2y^3\right)^2 =$

A. $\left(3x^2y^3\right)^2 = 3^2x^{2\cdot 2}y^{3\cdot 2} = 9x^4y^6$. Each factor in the parentheses is raised to the power outside the parentheses.

Q. $\left(3x^{-2}y\right)^2\left(2xy^{-3}\right)^4 =$

A.

$$\left(3x^{-2}y\right)^2\left(2xy^{-3}\right)^4$$
$$= \left(3^2x^{-2\cdot 2}y^{1\cdot 2}\right)\left(2^4x^{1\cdot 4}y^{-3\cdot 4}\right)$$
$$= \left(9x^{-4}y^2\right)\left(16x^4y^{-12}\right) = 144x^0y^{-10} = \frac{144}{y^{10}}$$

YOUR TURN

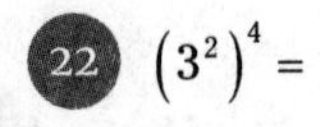

22 $\left(3^2\right)^4 =$

23 $\left(2^{-6}\right)^{-8} =$

24 $\left(2^3 \times 3^2\right)^4 =$

25 $\left(\left(3^5\right)^2\right)^6 =$

26 $\left(\dfrac{2^2 \times 3^4}{5^2 \times 3}\right)^3 =$

27 $\left(\dfrac{2^3}{e^5}\right)^2 =$

Testing the Power of Zero

If x^3 means $x \cdot x \cdot x$, what does x^0 mean? Well, it doesn't mean x times 0, so the answer isn't 0. The letter x represents some unknown real number; real numbers can be raised to the 0 power — except that the base just can't be 0. To understand how this works, use the following rule for division of exponential expressions involving 0.

Any number to the power of 0 equals 1 as long as the base number is not 0. In other words, $a^0 = 1$ as long as $a \neq 0$.

Consider the situation where you divide 2^4 by 2^4 by using the rule for dividing exponential expressions, which says that if the base is the same, you subtract the two exponents in the order that they're given. Doing this, you find that the answer is $2^{4-4} = 2^0$. But $2^4 = 16$, so $2^4 \div 2^4 = 16 \div 16 = 1$. That means that $2^0 = 1$. This is true of all numbers that can be written as a division problem, which means that it's true for all numbers except those with a base of 0.

Here are some examples of simplifying, using the rule that when you raise a real number a to the 0 power, you get 1.

EXAMPLE

Q. $4x^3y^4z^7 \div 2x^3y^3z^7 =$

A. $4x^3y^4z^7 \div 2x^3y^3z^7 = 2x^{3-3}y^{4-3}z^{7-7} = 2x^0y^1z^0 = 2y$. Both x and z end up with exponents of 0, so those factors become 1. Neither x nor z may be equal to 0.

Q. $\dfrac{(2x^2+3x)^4}{(2x^2+3x)^4} =$

A. $\dfrac{(2x^2+3x)^4}{(2x^2+3x)^4} = (2x^2+3x)^{4-4} = (2x^2+3x)^0 = 1$

YOUR TURN

28 $\dfrac{4x^2yzw^3}{4x^2yzw} =$

29 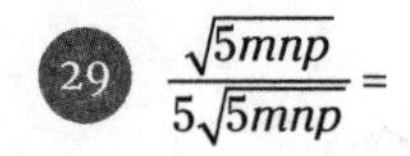

$$\frac{\sqrt{5mnp}}{5\sqrt{5mnp}} =$$

30 $$\frac{(6x + y^6)^4}{(6x + y^6)^4} =$$

Writing Numbers with Scientific Notation

Scientific notation is a standard way of writing in a more compact and useful way for numbers that are very small or very large. When a scientist wants to talk about the distance to a star being 45,600,000,000,000,000,000,000,000 light-years away, having it written as 4.56×10^{25} makes any comparisons or computations easier.

A number written in scientific notation is the product of a number between 1 and 10 and a power of 10. The power tells how many decimal places the original decimal point was moved in order to make that first number be between 1 and 10. The power is negative when you're writing a very small number and positive when writing a very large number with lots of zeros in front of the decimal point.

To write a number in scientific notation:

1. **Determine where the decimal point is in the number and move it left or right until you have exactly one digit to the left of the decimal point.**

   This gives you a number between 1 and 10.

2. **Count how many places (digits) you had to move the decimal point from its original position.**

   This is the absolute value of your exponent.

3. **If you moved the original decimal point to the left, your exponent is positive. If you moved the original decimal point to the right, your exponent is negative.**

4. **Rewrite the number in scientific notation by making a product of your new number that's between 1 and 10 times a 10 raised to the power of your exponent.**

EXAMPLE

Q. Write 41,000 in scientific notation.

A. A decimal point is *implied* (assumed there) after the last 0 in 41,000. Move the decimal place four spaces to the left, creating the number 4.1. The exponent is +4. $41{,}000 = 4.1 \times 10^4$

Q. Write 312,000,000,000 in scientific notation.

A. The decimal place is moved 11 spaces to the left. $312{,}000{,}000{,}000 = 3.12 \times 10^{11}$

Q. Write 0.00000031 in scientific notation.

A. The decimal place is moved seven spaces to the *right* this time. This is a very *small* number, and the exponent is negative. $0.00000031 = 3.1 \times 10^{-7}$

Q. Write 0.2 in scientific notation.

A. The decimal place is moved one space to the right. $0.2 = 2 \times 10^{-1}$

In modern scientific calculators, you see a different way of expressing scientific notation. It's indicated with the letter E. So, if you are doing some computation and see 3.2 E 10 for your answer, you translate it into scientific notation. 3.2 E 10 is short for 3.2×10^{10}, which is scientific notation for 32,000,000,000.

When writing numbers that are currently in scientific notation back in their full form, you just reverse the process with the decimal point. When you see a positive exponent on the 10, you move the decimal point that many places to the right (make it a big number). If the exponent is negative, then move the decimal point that many places to the left. You will probably have to add some zeros.

EXAMPLE

Q. Write 7.13×10^8 without scientific notation.

A. Move the decimal point eight places to the right. You'll have to add six zeros.

$$7.13 \times 10^8 = 713{,}000{,}000$$

Q. Write 4.7×10^{-6} without scientific notation.

A. Move the decimal point six places to the left. You'll have to add five zeros.

$$4.7 \times 10^{-6} = 0.0000047$$

Q. Write the calculator result 4.17 E 7 without scientific notation.

A. Move the decimal point seven places to the left. You'll have to add five zeros.

$$4.17 \text{ E } 7 = 4.17 \times 10^7 = 41{,}700{,}000$$

Q. Write the calculator result 1.01 E −8 without scientific notation.

A. Move the decimal point eight places to the left. You'll have to add seven zeros.

$$1.01 \text{ E} - 8 = 1.01 \times 10^{-8} = 0.0000000101$$

YOUR TURN

31 Write 4.03×10^{14} without scientific notation.

32 Write 3.71×10^{-13} without scientific notation.

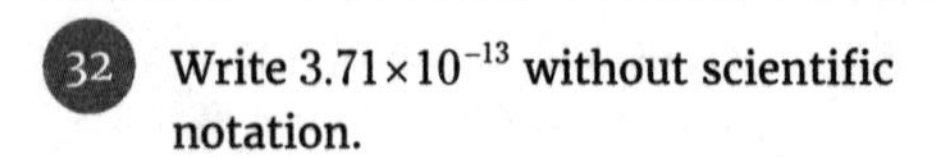

33 Write 4,500,000,000,000,000,000 using scientific notation.

34 Write 0.00000000000000003267 using scientific notation.

35 Write the calculator output 1.133 E 11 without scientific notation.

Practice Questions Answers and Explanations

1. **243.** $3^5 = 3\cdot3\cdot3\cdot3\cdot3 = 243$

2. **1.** $(-1)^{10} = 1$. There are ten negative signs, so the final answer is positive. See Chapter 2 for information on multiplying signed numbers.

3. $\mathbf{\frac{16}{81}}$. Raise both the numerator and denominator to the fourth power.

$$\left(\frac{2}{3}\right)^4 = \frac{2^4}{3^4} = \frac{2\cdot2\cdot2\cdot2}{3\cdot3\cdot3\cdot3} = \frac{16}{81}$$

4. **5.** Write the expression as a radical and evaluate. $25^{1/2} = \sqrt{25} = 5$

5. $\mathbf{16x \cdot x \cdot x(y+z)(y+z)(y+z)(y+z)(y+z)}$.

$$\begin{aligned}4^2x^3(y+z)^5 &= 4\cdot4x\cdot x\cdot x(y+z)(y+z)(y+z)(y+z)(y+z)\\ &= 16x\cdot x\cdot x(y+z)(y+z)(y+z)(y+z)(y+z)\end{aligned}$$

6. $\mathbf{3^{-6}}$. Just change the sign of the exponent when you move the base up.

7. $\mathbf{5^5}$. Just change the sign of the exponent when you move the base up.

8. $\mathbf{2^6 \times 3^8}$.

$$\left(\frac{3^{-4}}{2^3}\right)^{-2} = \left(\frac{2^3}{3^{-4}}\right)^2 = \frac{2^{3\times2}}{3^{-4\times2}} = \frac{2^6}{3^{-8}} = 2^6\times3^8 \text{ because } \frac{1}{3^{-8}} = 3^8$$

9. $\mathbf{\frac{3^{13}}{2^{13}}}$.

$$\frac{\left(2^3\times3^2\right)^4}{\left(2^5\times3^{-1}\right)^5} = \frac{2^{3\times4}\times3^{2\times4}}{2^{5\times5}\times3^{(-1)\times5}} = \frac{2^{12}\times3^8}{2^{25}\times3^{-5}} = 2^{12-25}\times3^{8-(-5)} = 2^{-13}\times3^{13} = \frac{3^{13}}{2^{13}}$$

10. $\mathbf{\frac{1}{5xyz}}$. Put a 1 in the numerator and move the z factor down: $\frac{z^{-1}}{5xy} = \frac{1}{5xyz}$

11. **729.** The exponent on the first factor is 1. $3\times3^5 = 3^1\times3^5 = 3^{1+5} = 3^6 = 729$

12. **32.** Add the exponents. $2^8\times2^{-3} = 2^{8+(-3)} = 2^5 = 32$

13. $\mathbf{2^9 \times 3^6}$. Regroup the factors and multiply.

$$2^3\times3^4\times2^6\times3^2 = 2^3\times2^6\times3^4\times3^2 = 2^{3+6}\times3^{4+2} = 2^9\times3^6$$

14. **180.** $5^{-4}\times6^2\times5^5 = 5^{-4}\times5^5\times6^2 = 5^{-4+5}\times6^2 = 5\times36 = 180$

15. $\mathbf{3^{a+2}\cdot2^{z-4}}$. Add the exponents. $3^a\cdot3^2\cdot2^z\cdot2^{-4} = 3^{a+2}\cdot2^{z-4}$

16. $\mathbf{18^5 + 81b}$. You can multiply the first two factors. But don't try to find that huge number. The second term is just simplified. $2^5\cdot9^5 + 3^4b = (2\cdot9)^5 + 81b = 18^5 + 81b$

17) **243.** $3^{11} \div 3^6 = 3^{11-6} = 3^5 = 243$

18) **7^8.** The exponent on the 7 in the denominator is 1. $\frac{7^9}{7} = \frac{7^9}{7^1} = 7^{9-1} = 7^8$

19) **$\frac{1}{a^5}$.** $\frac{a^5}{a^{10}} = a^{5-10} = a^{-5} = \frac{1}{a^5}$

20) **48.** Divide the factors with the same bases.

$$\frac{3^2 \times 2^{-1}}{3 \times 2^{-5}} = 3^{2-1} \times 2^{-1-(-5)} = 3^1 \times 2^4 = 3 \times 2^4 = 48$$

21) **$7^4 \times 5^2$.** Divide the factors with the same bases.

$$\frac{7^{-3} \times 2^4 \times 5}{7^{-7} \times 2^4 \times 5^{-1}} = 7^{-3-(-7)} \times 2^{4-4} \times 5^{1-(-1)} = 7^4 \times 2^0 \times 5^2 = 7^4 \times 1 \times 5^2 = 7^4 \times 5^2$$

22) **3^8.** $\left(3^2\right)^4 = 3^{2\times4} = 3^8$

23) **2^{48}.** $\left(2^{-6}\right)^{-8} = 2^{(-6)\times(-8)} = 2^{48}$

24) **$2^{12} \times 3^8$.** $\left(2^3 \times 3^2\right)^4 = 2^{3\times4} \times 3^{2\times4} = 2^{12} \times 3^8$

25) **3^{60}.** $\left[\left(3^5\right)^2\right]^6 = \left(3^{5\times2}\right)^6 = 3^{(5\times2)6} = 3^{60}$

26) **$\frac{2^6 \times 3^9}{5^6}$.** $\left(\frac{2^2 \times 3^4}{5^2 \times 3}\right)^3 = \left(\frac{2^2 \times 3^{4-1}}{5^2}\right)^3 = \left(\frac{2^2 \times 3^3}{5^2}\right)^3 = \frac{2^{2\times3} \times 3^{3\times3}}{5^{2\times3}} = \frac{2^6 \times 3^9}{5^6}$

27) **$\frac{2^6}{e^{10}}$.** $\left(\frac{2^3}{e^5}\right)^2 = \frac{2^{3(2)}}{e^{5(2)}} = \frac{2^6}{e^{10}}$

28) **w^2.** All the factors except w have an exponent of 0 after the division.

$$\frac{4x^2yzw^3}{4x^2yzw} = 4^{1-1}x^{2-2}y^{1-1}z^{1-1}w^{3-1} = 4^0x^0y^0z^0w^2 = w^2$$

29) **$\frac{1}{5}$.** $\frac{\sqrt{5mnp}}{5\sqrt{5mnp}} = \frac{1}{5} \cdot \frac{\sqrt{5mnp}}{\sqrt{5mnp}} = \frac{1}{5}(5mnp)^{1/2-1/2} = \frac{1}{5}(5mnp)^0 = \frac{1}{5}$

30) **1.** $\frac{(6x+y^6)^4}{(6x+y^6)^4} = (6x+y^6)^{4-4} = (6x+y^6)^0 = 1$

31) **403,000,000,000,000.** Move the decimal point 14 places to the right.

32) **0.000000000000371.** Move the decimal point 13 places to the left.

33) **4.5×10^{18}.** The decimal point was moved 18 places. This is a very large number.

34) **3.267×10^{-16}.** The decimal point was moved 16 places. This is a very small number.

35) **113,300,000,000.** The calculator result 1.133 E 11 is 1.133×10^{11}. Move the decimal point 11 places to the right.

If you're ready to test your skills a bit more, take the following chapter quiz that incorporates all the chapter topics.

Whaddya Know? Chapter 6 Quiz

Quiz time! Complete each problem to test your knowledge on the various topics covered in this chapter. You can then find the solutions and explanations in the next section.

1. Write without any negative exponents: $\frac{5ab^{-2}}{6bx^{-3}}$
2. Simplify: 2^6
3. Write without any negative exponents: 3^{-2}
4. $\frac{25x^2y^4}{\left(5xy^2\right)^2} =$
5. Simplify, leaving no negative exponents: $x^3 \cdot y^2 \cdot x^{-4} \cdot y^{-6} \cdot x \cdot y^5$
6. Write without any negative exponents: $\frac{1}{x^{-2}}$
7. Simplify: $4^{3/2}$
8. Write the number 0.00007 in scientific notation.
9. $\left[\frac{3^a}{x^2}\right]^4 =$
10. Simplify, leaving no negative exponents: $2^3 \times 3^{-3} \times 4^{-3} \times 6$
11. $\frac{a^6b^7c^{-4}}{a^{-1}bc^{-2}} =$
12. $\frac{3^{11}}{3^8} =$
13. Simplify: $(-2)^3$
14. $\frac{(4+a)^{16}}{4+a} =$
15. Simplify, leaving no negative exponents: $2^{-2} \times 3^2 \times 4^3 \times 9^{-1}$
16. Write the number 5.316×10^5 without scientific notation.
17. Simplify, leaving no negative exponents: $\frac{\left(3x^2\right)^{-2}}{(2x)^3}$

Answers to Chapter 6 Quiz

1. $\frac{5ax^3}{6b^3}$**.** Move the negative exponents to the opposite part of the fraction. Then simplify.

$$\frac{5ab^{-2}}{6bx^{-3}} = \frac{5ax^3}{6bb^2} = \frac{5ax^3}{6b^3}$$

2. **64.** $2^6 = 2\times2\times2\times2\times2\times2 = 64$

3. $\frac{1}{9}$**.** $3^{-2} = \frac{1}{3^2} = \frac{1}{9}$

4. **1.** Raise the factors in the denominator to the second power. Then reduce the fraction.

$$\frac{25x^2y^4}{\left(5xy^2\right)^2} = \frac{25x^2y^4}{5^2x^2y^{2\cdot2}} = \frac{25x^2y^4}{25x^2y^4} = \frac{\cancel{25}\cancel{x^2}\cancel{y^4}}{\cancel{25}\cancel{x^2}\cancel{y^4}} = 1$$

5. **y.** First, rearrange the factors and then perform the operations.

$$x^3\cdot x^{-4}\cdot x\cdot y^{-6}\cdot y^2\cdot y^5 = x^{3-4+1}y^{-6+2+5} = x^0y^1 = y$$

6. x^2**.** Move the x with its negative exponent up to the numerator and change the negative to positive.

7. **8.** $4^{3/2} = \left(\sqrt[2]{4}\right)^3 = (2)^3 = 8$

8. 7×10^{-5}**.** Move the decimal point five places to the left. This indicates a negative exponent is needed.

9. $\frac{3^{4a}}{x^8}$**.** Raise both the numerator and denominator to the fourth power.

10. $\frac{1}{36}$**.** Change the 4 and 6 to a power of 2 and a multiple of 2 and 3. This gets all the factors alike. Raise the power of 2, and then rearrange the factors to get the like numbers together. Perform the multiplications. You can now see that 2 and 3 have the same power. Multiply and simplify.

$$\begin{aligned}2^3\times3^{-3}\times4^{-3}\times6 &= 2^3\times3^{-3}\times(2^2)^{-3}\times(2\times3)\\ &= 2^3\times3^{-3}\times2^{-6}\times2\times3\\ &= 2^3\times2^{-6}\times2\times3^{-3}\times3\\ &= 2^{3-6+1}\times3^{-3+1}\\ &= 2^{-2}\times3^{-2}\\ &= (2\times3)^{-2}\\ &= 6^{-2} = \frac{1}{6^2} = \frac{1}{36}\end{aligned}$$

11 $\frac{a^7b^6}{c^2}$**.** Subtract the respective exponents: $\frac{a^6b^7c^{-4}}{a^{-1}bc^{-2}} = a^{6-(-1)}b^{7-1}c^{-4-(-2)} = a^7b^6c^{-2}$. Then rewrite as a fraction to put the factor with the negative exponent in the denominator.

$$a^7b^6c^{-2} = \frac{a^7b^6}{c^2}$$

12 **27.** Subtract the exponents. $\frac{3^{11}}{3^8} = 3^{11-8} = 3^3 = 27$

13 **−8.** $(-2)^3 = (-2)(-2)(-2) = -8$. Three negative signs means the answer is negative.

14 $(4+a)^{15}$**.** Write the binomial in the denominator as a power; then subtract the exponents.

$$\frac{(4+a)^{16}}{4+a} = \frac{(4+a)^{16}}{(4+a)^1} = (4+a)^{16-1} = (4+a)^{15}$$

15 **16.** First, rewrite the 4 and 9 as powers of 2 and 3, respectively. Then apply the powers and combine the like factors.

$$\begin{aligned} 2^{-2} \times 3^2 \times 4^3 \times 9^{-1} &= 2^{-2} \times 3^2 \times (2^2)^3 \times (3^2)^{-1} \\ &= 2^{-2} \times 2^6 \times 3^2 \times 3^{-2} \\ &= 2^4 \times 3^0 = 2^4 = 16 \end{aligned}$$

16 **531,600.** Move the decimal place five places to the right. You'll need to add two zeros.

17 $\frac{1}{72x^7}$**.** Move the factors in the parentheses to the denominator and perform the indicated operations.

$$\frac{(3x^2)^{-2}}{(2x)^3} = \frac{1}{(2x)^3(3x^2)^2} = \frac{1}{2^3x^33^2x^4} = \frac{1}{8x^39x^4} = \frac{1}{72x^7}$$

Exploring Exponents

3 Making Things Simple by Simplifying

Contents at a Glance

IN THIS CHAPTER

- Making algebraic expressions more user friendly
- Maintaining order with the order of operations
- Evaluating expressions and checking your work
- Combining all the rules for simpler computing

Chapter 7
Simplifying Algebraic Expressions

Algebra had its start as expressions that were all words. Everything was literally spelled out. As symbols and letters were introduced, algebraic manipulations became easier. But, as more symbols and notations were added, the rules that went along with the symbols also became a part of algebra. All this shorthand is wonderful, as long as you know the rules and follow the steps that go along with them. The *order of operations* is a biggie that you use frequently when working in algebra. It tells you what to do first, next, and last in a problem, whether terms are in grouping symbols or raised to a power.

And, because you may not always remember the order of operations correctly, checking your work is very important. Making sure that the answer you get makes sense, and that it actually solves the problem, is the next-to-last step of working every problem. The very final step is writing the solution in a way that other folks can understand easily.

This chapter walks you through the order of operations, checking your answers, and writing them correctly. And remember: The most commonly used variable in algebra is x. Because the variable x looks so much like the times sign, $\times$, other multiplication symbols are used in algebra problems. The following are equivalent multiplications; you see two variables being multiplied and a constant and variable being multiplied.

$$\begin{aligned} x\times y &= x\cdot y = (x)(y) = x(y) = (x)y = xy \\ 2\times y &= 2\cdot y = (2)(y) = 2(y) = (2)y = 2y \end{aligned}$$

In spreadsheets and calculators, the asterisk (*) sign indicates multiplication.

Addressing the Order of Operations

When does it matter in what *order* you do things? Or does it matter at all? Well, take a look at a couple of real-world situations:

- When you're cleaning the house, it *doesn't* matter whether you clean the kitchen or the living room first.
- When you're getting dressed, it *does* matter whether you put on your shoes first or your socks first.

Sometimes the order matters; other times it doesn't. In algebra, the order depends on which mathematical operations are performed. If you're doing only addition or only multiplication, you can use any order you want. But as soon as you mix things up with addition and multiplication in the same expression, you have to pay close attention to the correct order. You can't just pick and choose what to do first, next, and last according to what you feel like doing.

For example, look at the different ways this problem could be done, if there were no rules. Notice that all four operations are represented here.

$$8-3\times 4+6\div 2=$$

One way to do the problem is to just go from left to right:

1. $\mathbf{8-3=5}$
2. $\mathbf{5\times 4=20}$
3. $\mathbf{20+6=26}$
4. $\mathbf{26\div 2=13}$

This gives you a final answer of 13.

Another approach is to group the 3×4 together in parentheses. Grouped terms tell you that you have to do the operation inside the grouping symbol first.

1. $\mathbf{8-(3\times 4)=8-12=-4}$
2. $\mathbf{-4+6=2}$
3. $\mathbf{2\div 2=1}$

This gives you a final answer of 1.

Using other groupings, I can make the answer come out to be 25, 60, or even 0. I won't go into how these answers are obtained because they're all wrong anyway.

What is the correct answer? It's -1. This is because the correct way to do the problem is to multiply the 3 and 4 and divide the 6 and 2 first. Then you have $8-12+3=-4+3=-1$. The *order of operations* tells you how this is obtained.

Mathematicians designed rules so that anyone reading a mathematical expression would do it the same way as everyone else and get the same *correct* answer. In the case of multiple signs and operations, working out the problems needs to be done in a specified *order*, from the first to the last. This is the *order of operations.*

According to the order of operations, you perform all powers (exponents) and roots first, then you do all multiplication and division, and finally, perform the addition and subtraction. But this is just the order when performing the particular operations. If you have any grouping symbols, they go to the head if the line. You first need to perform operations in grouping symbols, such as ( ), { }, [ ], above and below fraction lines, and inside radicals. And if there are grouping symbols within grouping symbols, you work from the inside out. If you have more than two operations of the same level, do them in order from left to right, following the order of operations.

The order of operations (at each level, working from left to right) is as follows:

1. **Grouping symbols.**

   Work from inside out.

   Powers and roots, then multiplication and division, then addition and subtraction.

2. **Powers and roots.**
3. **Multiplication and division.**
4. **Addition and subtraction.**

When the expression is written in fraction form, you perform all the operations in the numerator and denominator separately. Then, finally, divide.

EXAMPLE

Q. $2\times4-10\div5=$

A. First, do the multiplication and division and then subtract the results:

$$2\times4-10\div5=8-2=6$$

Q. $\frac{8+2^2\times5}{\sqrt{64}-1}=$

A. First, find the values of the power and root ($2^2=4$ and $\sqrt{64}=8$). Then multiply in the numerator. Next, add the two terms in the numerator and subtract in the denominator. Then you can perform the final division. Here's how it breaks down:

$$\frac{8+2^2\times5}{\sqrt{64}-1}=\frac{8+4\times5}{8-1}=\frac{8+20}{8-1}=\frac{28}{7}=4$$

Simplifying Algebraic Expressions

1 $5+3\times 4^2+6\div 2-5\sqrt{9}=$

2 $\dfrac{6\times 8-4^2}{2^3+8\left(3^2-1\right)}=$

3 $2+3^3+3\left(2^2+\sqrt{81}\right)=$

4 $\dfrac{4^2+3^2}{9(4)-11}=$

Adding and Subtracting Like Terms

In algebra, the expression *like terms* refers to a common structure for the terms under consideration. *Like terms* have exactly the same variables in them, and each variable is "powered" the same (if *x* is squared and *y* cubed in one term, then *x* squared and *y* cubed occur in a *like term*). When adding and subtracting algebraic terms, the terms must be *alike,* with the same variables raised to exactly the same power, but the numerical coefficients can be different. For example, two terms that are *alike* are $2a^3b$ and $5a^3b$. Two terms that aren't *alike* are $3xyz$ and $4x^2yz$, where the power on the x term is different in the two terms.

When adding or subtracting terms that have *exactly* the same variables, perform the operations on the coefficients. When adding $2a+5a+4a$, what is the result?

$$2a+5a+4=(2+5+4)a=11a$$

Why does this work? Just look at the three terms in another way:

$$2a=a+a \quad 5a=a+a+a+a+a \quad 4a=a+a+a+a$$

So, $2a+5a+4a=a+a+a+a+a+a+a+a+a+a+a$. That's a total of 11 a variables altogether. Notice that the numbers in front — the coefficients 2, 5, and 4 — add up to 11.

REMEMBER

When there is no number in front of the variable, assume that the coefficient is a 1:

$$a=1a \qquad x=1x$$

An expression can have two or more different variables. You perform the operations on just the like variables. To simplify the expression $a+3a+x+2x$, combine the a's and the x's.

$$a+3a+x+2x=$$
$$1a+3a+1x+2x=$$
$$(1+3)a+(1+2)x=$$
$$4a+3x$$

Notice that you add terms that have the same variables because they represent the same amounts. You don't try to add the terms with different variables. And the same goes for subtraction.

To simplify the following expression $3x+4y-2x-8y+x$:

$$3x+4y-2x-8y+x=$$
$$(3-2+1)x+(4-8)y=$$
$$2x-4y$$

EXAMPLE

Q. Simplify: $5az+4az-2a+6-3b-2b$

A. Notice that the 6 doesn't have a variable. It stands by itself; it isn't multiplying anything.

Also, a term with az is different from a term with just a, so they don't combine.

$$5az+4az-2a+6-3b-2b=$$
$$(5+4)az-2a+(-3-2)b+6=$$
$$9az-2a-5b+6$$

Q. $6a+2b-4ab+7b+5ab-a+7=$

A. First, change the order and group the like terms together; then compute:

$$6a+2b-4ab+7b+5ab-a+7=(6a-a)+(2b+7b)+(-4ab+5ab)+7=$$
$$5a+9b+ab+7$$

The parentheses aren't necessary, but they help to keep track of what you can combine.

Q. $8x^2-3x+4xy-9x^2-5x-20xy=$

A. Again, combine like terms and compute:

$$8x^2-3x+4xy-9x^2-5x-20xy=(8x^2-9x^2)+(-3x-5x)+(4xy-20xy)=-x^2-8x-16xy$$

Q. Simplify: $x+3x+4x^2+5x^2+6x^3$

A. $4x+9x^2+6x^3$

Notice that the terms that combine *always* have exactly the same variables with exactly the same powers. (For more on powers, or exponents, see Chapter 6.) In order to add or subtract terms with the same variable, the exponents of the variable must be the same. Perform the required operations on the coefficients, leaving the variable and exponent as they are. Because x and x² don't represent the same amount, they can't be added together.

YOUR TURN

5 Combine the like terms in $4a+3ab-2ab+6a$.

6 Combine the like terms in $3x^2y-2xy^2+4x^3-8x^2y$.

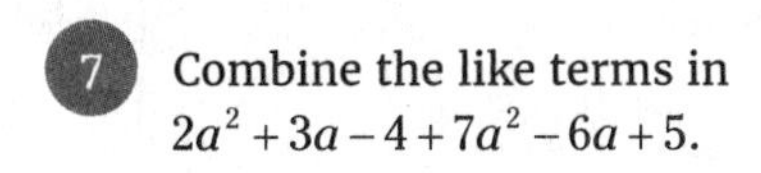

7 Combine the like terms in $2a^2 + 3a - 4 + 7a^2 - 6a + 5$.

8 Combine the like terms in $ab + bc + cd + de - ab + 2bc + e$.

Multiplying and Dividing Algebraically

Multiplying and dividing algebraic expressions is somewhat different from adding and subtracting them. When multiplying and dividing, the terms don't have to be exactly alike. You can multiply or divide all variables with the same base — using the laws of exponents (check out Chapter 6 for more information) — and you multiply or divide the number factors.

Dealing with factors

When multiplying factors containing variables, multiply the coefficients and variables as usual. If the bases are the same, you can multiply the bases by merely adding their exponents. (See more on the multiplication of exponents in Chapter 6.)

EXAMPLE

Q. $(4x^2y^2z^3)(3xy^4z^3) =$

A. The product of 4 and 3 is 12. Multiply the x's to get $x^2(x) = x^3$. Multiply the y's and then the z's and you get $y^2(y^4) = y^6$ and $z^3(z^3) = z^6$. Each variable has its own power determined by the factors multiplied together to get it. The answer is $12x^3y^6z^6$.

Q. $2 \cdot a \cdot a^2 \cdot a^3 \cdot a^4 \cdot 3 \cdot b \cdot b^5 \cdot 4 \cdot c =$

A. The three numbers have a product of 24. Multiplication is commutative, so you can multiply them in any order. Add the exponents on the like factors.
$2 \cdot a \cdot a^2 \cdot a^3 \cdot a^4 \cdot 3 \cdot b \cdot b^5 \cdot 4 \cdot c = 24a^{10}b^6c$

Q. $(2a^2b^2c^3)(4a^3b^2c^4) =$

A. $(2a^2b^2c^3)(4a^3b^2c^4) = 2(4)a^{2+3}b^{2+2}c^{3+4} = 8a^5b^4c^7$

Q. $(3x^2yz^{-2})(4x^{-2}y^2z^4)(3xyz) =$

A. $(3x^2yz^{-2})(4x^{-2}y^2z^4)(3xyz) = 3(4)(3)x^{2-2+1}y^{1+2+1}z^{-2+4+1} = 36x^1y^4z^3 = 36xy^4z^3$

In division of whole numbers, such as $27 \div 5$, the answers don't have to come out even. There can be a *remainder* (a value left over when one number is divided by another). But you usually don't want remainders when dividing algebraic expressions — the remainders would be new terms. So, be sure you don't leave any remainders lying around.

Diving into dividing

When dividing variables, write the problem as a fraction. Using the greatest common factor (GCF), divide the numbers and reduce. Use the rules of exponents (see Chapter 6) to divide variables that are the same. Dividing variables is fairly straightforward. Each variable is considered separately.

First, let me illustrate this rule with aluminum cans. Four friends decided to collect aluminum cans for recycling (and money). They collected $12x^3$ cans, and they're going to get y^2 cents per can. The total amount of money collected is then $12x^3y^2$ cents. How will they divvy this up?

Divide the total amount by 4 to get the individual amount that each of the four friends will receive:

$$\frac{12x^3y^2}{4} = 3x^3y^2 \text{ cents each}$$

The only thing that divides here is the coefficient. If you want the number of cans each will get paid for, divide by $4y^2$ instead of just 4:

$$\frac{12x^3y^2}{4y^2} = \frac{12x^3\cancel{y^2}}{4\cancel{y^2}} = 3x^3 \text{ cans}$$

Why is using variables better than using just numbers in this aluminum-can story? Because if the number of cans or the value per can changes, then you still have all the shares worked out. Just let the *x* and *y* change in value.

EXAMPLE

Q. Simplify the expression $\frac{6a^2}{3a}$.

A. Three divides 6 twice. Using the rules of exponents, $a^2 \div a = a$. $\frac{6a^2}{3a} = 2a^{2-1} = 2a^1 = 2a$

Q. Simplify $\frac{8x^2y^3}{2x^4y^2}$.

A. $\frac{8x^2y^3}{2x^4y^2} = 4x^{2-4}y^{3-2} = 4x^{-2}y^1 = \frac{4y}{x^2}$. It is customary to write the answer putting *x* in the denominator with a positive exponent rather than in the numerator with a negative exponent.

YOUR TURN

9 Multiply $(3x)(2x^2)$.

10 Multiply $(4y^2)(-x^4y)$.

11 Multiply $(6x^3y^2z^2)(8x^3y^4z)$.

12 Divide (write all exponents as positive numbers): $\frac{10x^2y^3}{-5xy^2}$.

13 Divide (write all exponents as positive numbers): $\frac{24x}{3x^2}$.

14 Divide (write all exponents as positive numbers): $\frac{13x^3y^4}{26x^8y^3}$.

Gathering Terms with Grouping Symbols

In algebra problems, parentheses, brackets, and braces are all used for grouping. Terms inside the grouping symbols have to be operated upon before they can be applied to anything outside the grouping symbol. All the grouping types have equal weight; none is more powerful or acts differently from the others.

If the problem contains grouped items, do what's inside a grouping symbol first, and then follow the order of operations. The grouping symbols are listed here.

- **Parentheses ( ):** Parentheses are the most commonly used symbols for grouping.
- **Brackets [ ] and braces { }:** Brackets and braces are also used frequently for grouping and have the same effect as parentheses. Using the different types of symbols helps when there's more than one grouping in a problem. It's easier to tell where a group starts and ends.
- **Radical $\sqrt{\ }$:** This represents an operation used for finding roots.
- **Fraction line (called the *vinculum*):** The fraction line also acts as a grouping symbol; everything above the line in the numerator is grouped together, and everything below the line in the denominator is grouped together.
- **Absolute value $|\ |$:** This represents an operation used to find the unsigned value of a number.

Even though the order of operations and grouping-symbol rules are fairly straightforward, it's hard to describe, in words, all the situations that can come up in these problems. The examples I show here should clear up many questions you may have.

Simplify: $2+3^2(5-1)$.

Use both the order of operations and grouping symbols:

1. **Subtract the 1 from the 5 in the parentheses to get 4.**

   $2+3^2(4)$

2. **Raise the 3 to the second power to get 9.**

   $2+9(4)$

3. **Multiply the 9 and 4 to get 36.**

   $2+36$

4. **Add to get the final answer.**

   38

Simplify: $\frac{5\left[3+\left(12-2^2\right)\right]}{|8-23|}+\frac{\sqrt{16-7}}{(-3)^2}$.

1. **Working from the inside out, first square the 2 before subtracting it from the 12.** You can also subtract the numbers in the absolute value and the numbers under the radical. Go ahead and square the −3.

   You can do all these steps at once because none of the results interacts with the others yet.

$$\frac{5\left[3+\left(12-2^2\right)\right]}{|8-23|}+\frac{\sqrt{16-7}}{(-3)^2}=\frac{5\left[3+(12-4)\right]}{|-15|}+\frac{\sqrt{9}}{9}$$
$$=\frac{5\left[3+(8)\right]}{|-15|}+\frac{\sqrt{9}}{9}$$

2. **Add the numbers in the brackets, find the absolute value of the −15, and find the square root.**

$$\frac{5\left[3+(8)\right]}{|-15|}+\frac{\sqrt{9}}{9}=\frac{5[11]}{15}+\frac{3}{9}$$

3. **Multiply the 5 and 11. Then simplify the two fractions by reducing them.**

$$\frac{5[11]}{15}+\frac{3}{9}=\frac{55}{15}+\frac{3}{9}=\frac{\cancel{55}^{11}}{\cancel{15}^{3}}+\frac{\cancel{3}^{1}}{\cancel{9}^{3}}=\frac{11}{3}+\frac{1}{3}$$

4. **You can now add the fractions quite nicely.**

$$\frac{11}{3}+\frac{1}{3}=\frac{12}{3}=4$$

WARNING

Be sure to catch the subtle difference between the two expressions: -2^4 and $(-2)^4$. Simplifying the expression -2^4, you get −16 because the order of operations says to first raise to the fourth power and then apply the negative sign. The expression $(-2)^4=16$ because the entire expression in parentheses is raised to the fourth power. This is equivalent to multiplying −2 by itself four times. The multiplication involves an even number of negative signs, so the result is positive.

TIP

In general, if you want a negative number raised to a power, you have to put it in parentheses with the power outside.

EXAMPLE

Q. Use grouping to simplify $\left[8 \div (5 - 3)\right] \times 5$.

A. The two grouping symbols here are brackets and parentheses. Work from the inside out. First, perform the subtraction in the parentheses. Then divide 8 by the result. Finally, multiply by 5.

$$\left[8 \div (5 - 3)\right] \times 5 = \left[8 \div (2)\right] \times 5 = \left[4\right] \times 5 = 20$$

Q. Simplify: $\frac{4(7+5)}{2+1}$.

A. Simplify the numerator and denominator separately. The final step is to perform the division.

$$\frac{4(7+5)}{2+1} = \frac{4(12)}{3} = \frac{48}{3} = 16$$

YOUR TURN

Simplify the following.

15 $[5(6+2)-7] \div 3$

16 $\frac{\sqrt{6^2+13}}{5^2-4}$

17 $\frac{-(-7)+\sqrt{(-7)^2-4(-2)(-3)}}{2(-2)}$

18 $\frac{\left|2^2-6^3\right|}{5 \cdot 3^2-(-2)^3}$

Evaluating Expressions

Evaluating an expression means that you want to change it from a bunch of letters and numbers to a specific value — some number. After you solve an equation or inequality, you want to go back and check to see whether your solution really works — so you evaluate the expression with that answer. For example, if you let $x = 2$ in the expression $3x^2 - 2x + 1$, you replace all the x's with 2's and apply the order of operations when doing the calculations. In this case, you get $3(2)^2 - 2(2) + 1 = 3(4) - 4 + 1 = 12 - 4 + 1 = 9$. Can you see why knowing that you square the 2 before multiplying by the 3 is so important? If you multiply by the 3 first, you end up with that first term being 36 instead of 12. It makes a big difference.

EXAMPLE

Q. Evaluate $\frac{5y - y^2}{2x}$ when $y = -4$ and $x = -3$.

A. $\frac{5y - y^2}{2x} = \frac{5(-4) - (-4)^2}{2(-3)} = \frac{5(-4) - 16}{2(-3)} = \frac{-20 - 16}{-6} = \frac{-36}{-6} = 6$

Q. Evaluate $\frac{n!}{r!(n-r)!}$ when $n = 8$ and $r = 3$.

A. What's with this exclamation (the $n!$)? The exclamation indicates an operation called *factorial*. This operation has you multiply the number in front of the ! by every positive whole number smaller than it. You see a lot of factorials in statistics and higher mathematics. The order of operations is important here, too.

$$\frac{n!}{r!(n-r)!} = \frac{8!}{3!(8-3)!} = \frac{8!}{3!5!}$$
$$= \frac{8 \cdot 7 \cdot 6 \cdot \cancel{5 \cdot 4 \cdot 3 \cdot 2 \cdot 1}}{3 \cdot 2 \cdot 1 \cdot \cancel{5 \cdot 4 \cdot 3 \cdot 2 \cdot 1}}$$
$$= \frac{8 \cdot 7 \cdot \cancel{6}}{\cancel{3 \cdot 2} \cdot 1} = 56$$

YOUR TURN

19 Evaluate $3x^2$ if $x = -2$.

20 Evaluate $9y - y^2$ if $y = -1$.

21 Evaluate $-(3x-2y)$ if $x=4$ and $y=3$.

22 Evaluate $6x^2-xy$ if $x=2$ and $y=-3$.

23 Evaluate $\frac{2x+y}{x-y}$ if $x=4$ and $y=1$.

24 Evaluate $\frac{x^2-2x}{y^2+2y}$ if $x=3$ and $y=-1$.

25 Evaluate $\frac{-b+\sqrt{b^2-4ac}}{2a}$ if $a=3$, $b=-2$, and $c=-1$.

26 Evaluate $\frac{n!}{r!}+\frac{n!}{r!(n-r)!}$ if $n=5$ and $r=2$.

Checking Your Answers

Every once in a while, I make a math mistake. Yes! Even me! That's why I'm such a big fan of checking answers before broadcasting my results. For example, pretend that I solve the equation $3x-2=2x+1$ and say that the answer is −1. (This is the result of my adding −2 instead of +2). If I take the time to check my answer, I see my error. Checking means to put the result back into the original statement. So, if $x=-1$ in the equation $3x-2=2x+1$, then I have $3(-1)-2=2(-1)+1$, giving me $-3-2=-2+1$ or that $-5=-1$. Oops! That is not a correct statement. Time to go back and redo the problem!

Another common error occurs when working with decimals or lots of zeros in numbers. Consider a problem where you're figuring out how much flour to purchase to make 10 loaves of bread. You come out with the answer 2,000 cups of flour. Oops! Even if you aren't much of a baker, you can certainly recognize that 2,000 is much too large. The decimal point must be in the wrong place. Go back and check your work. It's probably 20 cups that you need, not 2,000.

Checking your answers when doing algebra is always a good idea, just like reconciling your checkbook with your bank statement is a good idea. Actually, checking answers in algebra is easier and more fun than reconciling a checking account. Or maybe your checking account is more fun than mine.

Check your answers in algebra on two levels.

- **Level 1: Does the answer make any sense?** If your checkbook balance shows $40 million, does that make any sense? Sure, we'd all *like* it to be that, but for most of us, this would be a red flag that something is wrong with our computations.
- **Level 2: Does actually putting the answer back into the problem give you a true statement? Does it *work?*** This is the more critical check because it gives you more exact information about your answer. The first level helps weed out the obvious errors. This is the final check.

The next sections help you make even more sense of these checks.

Seeing if it makes sense

To check whether an answer makes any sense, you have to know something about the topic. A problem will be meaningful if it's about a situation you're familiar with. Just use your common sense. You'll have a good feeling as to whether the money amount in an answer is reasonable.

For example, your answer to an algebra problem is $x=5$. If you're solving for Jon's weight in pounds, unless Jon is a guinea pig instead of a person, you probably want to go back and redo the work. Five pounds or 5 ounces or 5 tons doesn't make any sense as an answer in this context.

On the other hand, if the problem involves a number of pennies in a person's pocket, then five pennies seems reasonable. Getting five as the number of home runs a player hit in one ballgame may at first seem quite possible, but if you think about it, five home runs in one game is a lot — even for Ryan Howard or Albert Pujols. You may want to double-check.

Plugging in values

Actually plugging in your answer requires you to go through the algebra and arithmetic manipulations in the problem. You add, subtract, multiply, and divide to see if you get a true statement using your answer.

Suppose Jack's cellular plan has 400 more minutes than Jill's. If the two of them have a total of 1,400 minutes altogether, then how many minutes does Jill have? Does $x = 500$ work for an answer?

1. **Write the problem.**

   Let x represent the number of minutes that Jill has. Jack has $x + 400$ minutes. That means $x + (x + 400) = 1{,}400$. The number of minutes Jill has plus the number of minutes Jack has equals 1,400.

2. **Insert the answer into the equation.**

   Replace the variable, x, with your answer of 500 to get $500 + (500 + 400) = 1{,}400$.

3. **Do the operations and check to see if the answer works.**

   $500 + 900 = 1{,}400$ is a true statement, so the problem checks. Jill has 500 minutes; Jack has 400 more than that, or 900 minutes; together, they have 1,400 minutes.

You can apply a variation of the preceding steps to check whether $x = 2$ works in the equation $5x[x + 3(x^2 - 3)] + 1 = 0$.

1. **Write out the equation.**

   $5x[x + 3(x^2 - 3)] + 1 = 0$

2. **Replace the variable with 2.**

   $5 \times 2[2 + 3(2^2 - 3)] + 1 = 0$

3. **Do the operations and simplify.**

   Square the 2 to get $5 \times 2[2 + 3(4 - 3)] + 1 = 0$.

   Subtract in the parentheses to get $5 \times 2[2 + 3(1)] + 1 = 0$.

   Add in the brackets to get $5 \times 2[5] + 1 = 0$.

   Multiply the 5, 2, and 5 to get $50 + 1 \neq 0$.

This time the work does *not* check. You should go back and try again to find a value for x that works.

EXAMPLE

Q. Given $\frac{x^2-4}{x+3}=\frac{x+2}{2}$, which works: $x=7$ or $x=-7$?

A. Substituting 7 for x, you have $\frac{7^2-4}{7+3}=\frac{7+2}{2}$, which simplifies to $\frac{45}{10}=\frac{9}{2}$. Reducing the fraction on the left, you find that $\frac{9}{2}=\frac{9}{2}$. So 7 works. Substituting –7 for x, you have $\frac{(-7)^2-4}{-7+3}=\frac{-7+2}{2}$, which simplifies to $\frac{45}{-4}=\frac{-5}{2}$. Neither fraction reduces, and, if you cross-multiply, you get $90=20$, which is false. The –7 doesn't work.

Q. Given $x^4+x^2-2=0$, which works: $x=1$ or $x=-1$?

A. Substituting 1 for x, you have $1^4+1^2-2=1+1-2=0$. This is true. And for $x=-1$, $(-1)^4+(-1)^2-2=1+1-2=0$. Both are solutions.

Check to see which answer works in the given equation.

YOUR TURN

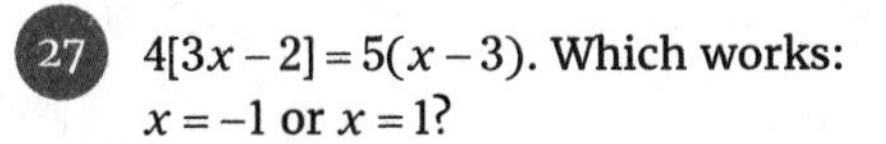

27 $4[3x-2]=5(x-3)$. Which works: $x=-1$ or $x=1$?

28 $z^3+2z^2-z-2=0$. Which works: $z=1$ or $z=-1$ or $z=-2$?

Practice Questions Answers and Explanations

1. **41.**

   Powers and roots: $5+3\times4^2+6\div2-5\sqrt{9}=5+3\times16+6\div2-5\cdot3$

   Multiply and divide: $5+3\times16+6\div2-5\cdot3=5+48+3-15$

   Add and subtract: $5+48+3-15=53+3-15=56-15=41$

2. $\mathbf{\frac{4}{9}}$**.**

   Parentheses (power, subtract): $\frac{6\times8-4^2}{2^3+8(3^2-1)}=\frac{6\times8-4^2}{2^3+8(9-1)}=\frac{6\times8-4^2}{2^3+8(8)}$

   Powers and roots: $\frac{6\times8-4^2}{2^3+8(8)}=\frac{6\times8-16}{8+8(8)}$

   Multiply and divide: $\frac{6\times8-16}{8+8(8)}=\frac{48-16}{8+64}$

   Add and subtract: $\frac{48-16}{8+64}=\frac{32}{72}$

   Reduce the fraction: $\frac{32}{72}=\frac{4}{9}$

3. **68.**

   Parentheses (power, root, add): $2+3^3+3(2^2+\sqrt{81})=2+3^3+3(4+9)=2+3^3+3(13)$

   Power and roots: $2+3^3+3(13)=2+27+3(13)$

   Multiply and divide: $2+27+3(13)=2+27+39$

   Add and subtract: $2+27+39=29+39=68$

4. **1.**

   Powers and roots: $\frac{4^2+3^2}{9(4)-11}=\frac{16+9}{9(4)-11}$

   Multiply and divide: $\frac{16+9}{9(4)-11}=\frac{16+9}{36-11}$

   Add and subtract: $\frac{16+9}{36-11}=\frac{25}{25}$

   Reduce the fraction: $\frac{25}{25}=1$

5. $\mathbf{10a+ab}$**.**

   $$\begin{aligned}4a+3ab-2ab+6a&=4a+6a+3ab-2ab\\&=(4+6)a+(3-2)ab\\&=10a+1ab=10a+ab\end{aligned}$$

6. $\mathbf{-5x^2y-2xy^2+4x^3}$**.**

   $$\begin{aligned}3x^2y-2xy^2+4x^3-8x^2y&=3x^2y-8x^2y-2xy^2+4x^3\\&=3x^2y-8x^2y-2xy^2+4x^3\\&=(3-8)x^2y-2xy^2+4x^3\\&=-5x^2y-2xy^2+4x^3\end{aligned}$$

7. $\mathbf{9a^2 - 3a + 1}$.

8. $\mathbf{3bc + cd + de + e}$.

9. $\mathbf{6x^3}$.

10. $\mathbf{-4x^4y^3}$. The coefficient of the second factor is –1.

    $(4y^2)(-x^4y) = (4y^2)(-1x^4y^1)$.

11. $\mathbf{48x^6y^6z^3}$.

12. $\mathbf{-2xy}$. $\frac{10x^2y^3}{-5xy^2} = -2x^{2-1}y^{3-2} = -2x^1y^1$.

13. $\mathbf{\frac{8}{x}}$. $\frac{24x}{3x^2} = 8x^{1-2} = 8x^{-1} = \frac{8}{x}$.

14. $\mathbf{\frac{y}{2x^5}}$. Reducing the fraction formed by the coefficients and dividing the variable factors,

    $\frac{13x^3y^4}{26x^8y^3} = \frac{1}{2}x^{3-8}y^{4-3} = \frac{1}{2}x^{-5}y^1 = \frac{y}{2x^5}$.

15. **11.** Working from the inside out, first add the 6 and 2, and then multiply the sum by 5. Subtract the 7.

    $$[5(6+2)-7] \div 3 = [5(8)-7] \div 3 = [40-7] \div 3 = [33] \div 3$$

    Now, dividing, $[33] \div 3 = 11$.

16. $\mathbf{\frac{1}{3}}$. Working under the radical, you first square the 6 and then add 13. In the denominator, you first square the 5 and then subtract 4.

    $$\frac{\sqrt{6^2+13}}{5^2-4} = \frac{\sqrt{36+13}}{25-4} = \frac{\sqrt{49}}{21}$$

    Now find the square root of 49. Then reduce the fraction.

    $$\frac{\sqrt{49}}{21} = \frac{7}{21} = \frac{1}{3}$$

17. **−3.** Work under the radical first. Square the −7 and multiply the 4, −2, and −3.

    $$\frac{-(-7)+\sqrt{(-7)^2-4(-2)(-3)}}{2(-2)} = \frac{-(-7)+\sqrt{49-24}}{2(-2)}$$

    Now simplify the first term in the numerator, subtract under the radical, and simplify the denominator.

    $$\frac{-(-7)+\sqrt{49-24}}{2(-2)} = \frac{7+\sqrt{25}}{-4}$$

    Find the square root of 25, and add that to 7. Then divide the sum by −4.

    $$\frac{7+\sqrt{25}}{-4} = \frac{7+5}{-4} = \frac{12}{-4} = -3$$

(18) **4.** Working in the numerator, first square the 2 and cube the 6; then find the difference between the results.

$$\frac{|2^2-6^3|}{5\cdot3^2-(-2)^3}=\frac{|4-216|}{5\cdot3^2-(-2)^3}=\frac{|-212|}{5\cdot3^2-(-2)^3}$$

In the denominator, first square the 3, and then multiply it by 5. Cube the −2.

$$\frac{|-212|}{5\cdot3^2-(-2)^3}=\frac{|-212|}{5\cdot9-(-2)^3}=\frac{|-212|}{45-(-8)}$$

The absolute value of −212 is +212. Find the difference of the terms in the denominator. Then simplify the fraction.

$$\frac{|-212|}{45-(-8)}=\frac{212}{53}=\frac{4\cdot\cancel{53}}{\cancel{53}}=4$$

(19) **12.**

$$3x^2=3(-2)^2=3(4)=12$$

(20) **−10.**

$$9y-y^2=9(-1)-(-1)^2=9(-1)-1=-9-1=-10$$

(21) **−6.**

$$-(3x-2y)=-(3\cdot4-2\cdot3)=-(12-6)=-(6)=-6$$

(22) **30.**

$$6x^2-xy=6(2)^2-2(-3)=6(4)-2(-3)=24-(-6)=30$$

(23) **3.**

$$\frac{2x+y}{x-y}=\frac{2(4)+1}{4-1}=\frac{8+1}{4-1}=\frac{9}{3}=3$$

(24) **−3.**

$$\frac{x^2-2x}{y^2+2y}=\frac{3^2-2\cdot3}{(-1)^2+2(-1)}=\frac{9-2\cdot3}{1+2(-1)}=\frac{9-6}{1+(-2)}=\frac{3}{-1}=-3$$

(25) **1.**

$$\frac{-b+\sqrt{b^2-4ac}}{2a}=\frac{-(-2)+\sqrt{(-2)^2-4(3)(-1)}}{2(3)}=\frac{-(-2)+\sqrt{4-4(3)(-1)}}{2(3)}=\frac{2+\sqrt{4-(-12)}}{2(3)}=$$

$$=\frac{2+\sqrt{16}}{6}=\frac{2+4}{6}=\frac{6}{6}=1$$

(26) **70.**

$$\frac{n!}{r!}+\frac{n!}{r!(n-r)!}=\frac{5!}{2!}+\frac{5!}{2!(5-2)!}=\frac{5!}{2!}+\frac{5!}{2!3!}$$
$$=\frac{5\cdot4\cdot3\cdot2\cdot1}{2\cdot1}+\frac{5\cdot4\cdot3\cdot2\cdot1}{2\cdot1\cdot3\cdot2\cdot1}$$
$$=\frac{5\cdot4\cdot3\cdot\cancel{2\cdot1}}{\cancel{2\cdot1}}+\frac{5\cdot4\cdot\cancel{3\cdot2\cdot1}}{2\cdot1\cdot\cancel{3\cdot2\cdot1}}$$
$$=60+\frac{20}{2}=60+10=70$$

(27) $\boldsymbol{x=-1}$**.** Substituting –1 for x, you have $4[3(-1)-2]=5(-1-3)$, which simplifies to $4[-3-2]=5(-4)$ or $-20=-20$. The value works. Substituting 1 for x, you have $4[3(1)-2]=5(1-3)$, which simplifies to $4[1]=5(-2)$. But 4 does not equal –10, so it doesn't work.

(28) $\boldsymbol{z=1,-1,}$ **and** $\boldsymbol{-2}$**.** Substituting 1 for z, you have $(1)^3+2(1)^2-1-2=1+2-1-2=0$. So $z=1$. Now, trying –1 for z, you have $(-1)^3+2(-1)^2-(-1)-2=-1+2+1-2=0$, which also works. Finally, letting $z=-2$, you have $(-2)^3+2(-2)^2-(-2)-2=-8+8+2-2=0$. All three numbers work.

If you're ready to test your skills a bit more, take the following chapter quiz that incorporates all the chapter topics.

Whaddya Know? Chapter 7 Quiz

Simplifying Algebraic Expressions

1. Which value of x makes the statement $x^3+x^2-12x=\left(x^2+4x\right)\left(x-3\right)$ true: $x=4$ or $x=3$?
2. $4-3\times2^2+10\div5+2\sqrt{25}=$
3. $\left(5a^2b\right)\left(6ab^3\right)=$
4. $\frac{-2-\sqrt{2^2-4(1)(-15)}}{2(1)}=$
5. $\left(-2xyz\right)\left(3x^2y\right)\left(-4x^3y^2z^4\right)=$
6. $\frac{4\left(3^2+4^2\right)}{8^2+6^2}=$
7. $\frac{16a^2b}{8ab^2}=$
8. Simplify: $3x^2-4xy+5y^2+2x^2+4xy$
9. $\frac{36xyz^2}{20x^2yz}=$
10. $6+\left[9-(3^2+1)\right]-5(2+7)=$
11. Evaluate $\frac{3x^2-2x+1}{y^2+3y-2}$ when $x=2$ and $y=-1$.
12. Simplify: $2a+3b-4c+5a-6b+11$
13. $\frac{6^2+\sqrt{64}}{|4-15|}=$

Answers to Chapter 7 Quiz

1. **Both.** When $x = 4$,

$$\begin{aligned} x^3 + x^2 - 12x &= \left(x^2 + 4x\right)\left(x - 3\right) \\ 4^3 + 4^2 - 12(4) &\stackrel{?}{=} \left(4^2 + 4(4)\right)(4 - 3) \\ 64 + 16 - 48 &\stackrel{?}{=} (16 + 16)(1) \\ 32 &= 32 \end{aligned}$$

When $x = 3$,

$$\begin{aligned} x^3 + x^2 - 12x &= \left(x^2 + 4x\right)\left(x - 3\right) \\ 3^3 + 3^2 - 12(3) &\stackrel{?}{=} \left(3^2 + 4(3)\right)(3 - 3) \\ 27 + 9 - 36 &\stackrel{?}{=} (9 + 12)(0) \\ 0 &= 0 \end{aligned}$$

2. **4.** Perform the power and root, then multiply, divide, and multiply. Finally, subtract and add.

$$\begin{aligned} 4 - 3 \times 2^2 + 10 \div 5 + 2\sqrt{25} &= 4 - 3 \times 4 + 10 \div 5 + 2(5) \\ &= 4 - 12 + 2 + 10 \\ &= -8 + 2 + 10 \\ &= -6 + 10 = 4 \end{aligned}$$

3. $\boldsymbol{30a^3b^4}$**.** Add the exponents of the variables. $\left(5a^2b\right)\left(6ab^3\right) = (5)(6)a^{2+1}b^{1+3} = 30a^3b^4$

4. **–5.** Simplify under the radical first.

$$\begin{aligned} \frac{-2 - \sqrt{2^2 - 4(1)(-15)}}{2(1)} &= \frac{-2 - \sqrt{4 - 4(1)(-15)}}{2(1)} = \frac{-2 - \sqrt{4 + 60}}{2(1)} \\ &= \frac{-2 - \sqrt{64}}{2(1)} = \frac{-2 - 8}{2(1)} = \frac{-10}{2} = -5 \end{aligned}$$

5. $\boldsymbol{24x^6y^4z^5}$**.** Add the exponents of the variables.

$$\left(-2xyz\right)\left(3x^2y\right)\left(-4x^3y^2z^4\right) = (-2)(3)(-4)x^{1+2+3}y^{+1+1+2}z^{1+4} = 24x^6y^4z^5$$

6. **1.** First, square the numbers, and then add them together. Multiply the sum in the numerator by 4, and then reduce the fraction. $\dfrac{4\left(3^2 + 4^2\right)}{8^2 + 6^2} = \dfrac{4(9 + 16)}{64 + 36} = \dfrac{4(25)}{100} = \dfrac{100}{100} = 1$

7. $\boldsymbol{\frac{2a}{b}}$**.** Subtract the exponents of the variables. $\dfrac{16a^2b}{8ab^2} = 2a^{2-1}b^{1-2} = 2a^1b^{-1} = \dfrac{2a}{b}$

8. $\boldsymbol{5x^2 + 5y^2}$**.** Combine the like terms.

$$3x^2 - 4xy + 5y^2 + 2x^2 + 4xy = 3x^2 + 2x^2 - 4xy + 4xy + 5y^2 = 5x^2 + 5y^2$$

9. $\mathbf{\frac{9z}{5x}}$. Subtract the exponents of the variables. $\frac{36xyz^2}{20x^2yz} = \frac{36}{20}x^{1-2}y^{1-1}z^{2-1} = \frac{9}{5}x^{-1}y^0z^1 = \frac{9z}{5x}$

10. **–40.** Working in the bracket, first simplify in the parentheses and subtract the result from 9. Add the two numbers in the last parentheses and multiply the result by 5. Finally, add and subtract the numbers.

$$\begin{aligned}6+\left[9-(3^2+1)\right]-5(2+7) &= 6+[9-(9+1)]-5(2+7)\\ &= 6+[9-(10)]-5(2+7)\\ &= 6+[-1]-5(2+7)\\ &= 6+[-1]-5(9)\\ &= 6+[-1]-45\\ &= 5-45=-40\end{aligned}$$

11. $\mathbf{-\frac{9}{4}}$.

$$\frac{3x^2-2x+1}{y^2+3y-2} = \frac{3(2)^2-2(2)+1}{(-1)^2+3(-1)-2} = \frac{3(4)-2(2)+1}{1+3(-1)-2} = \frac{12-4+1}{1-3-2} = \frac{9}{-4} = -\frac{9}{4}$$

12. $\mathbf{7a-3b-4c+11}$. Combine the like terms.

$$2a+3b-4c+5a-6b+11 = 2a+5a+3b-6b-4c+11 = 7a-3b-4c+11$$

13. **4.** Simplify the numerator and denominator before dividing.

$$\frac{6^2+\sqrt{64}}{|4-15|} = \frac{36+8}{|-11|} = \frac{44}{11} = 4$$

» Bringing big numbers down to size: Divisibility rules!

» Investigating composite numbers with prime factorizations

» Finding and using factoring methods

Chapter 8

Working with Numbers in Their Prime

Prime numbers (whole numbers evenly divisible only by themselves and 1) have been the subject of discussions between mathematicians and nonmathematicians for centuries. Prime numbers and their mysteries have intrigued philosophers, engineers, and astronomers. These folks and others have discovered plenty of information about prime numbers, but many unproven conjectures remain. So why are prime numbers important? Why study them? Prime numbers play an important role in coding (encrypting passwords and protecting information). And who knows what else we'll be able to do with them in the future!

Probably the biggest mystery is determining what prime number will be discovered next. Computers have aided the search for a comprehensive list of prime numbers, but because numbers go on forever without end, and because no one has yet found a pattern or method for listing prime numbers, the question involving the *next big one* remains.

Beginning with the Basics

Prime numbers are important in algebra because they help you work with the smallest-possible numbers. Big numbers are often unwieldy and can produce more computation errors when you perform operations and solve equations. So, reducing fractions to their lowest terms and factoring expressions to make problems more manageable are basic and very desirable tasks.

The first and smallest prime number is the number 2. It's the only *even* prime number. All primes after 2 are odd because all even numbers can be divided evenly by 1, themselves, and 2. So even numbers greater than 2 don't fit the definition of a prime number.

Here are the first 46 prime numbers:

2	3	5	7	11	13	17	19
23	29	31	37	41	43	47	53
59	61	67	71	73	79	83	89
97	101	103	107	109	113	127	131
137	139	149	151	157	163	167	173
179	181	191	193	197	199		

TIP

When you already recognize that a number is prime, you don't waste time trying to find things to divide into it when you're reducing a fraction or factoring an expression. There are so many primes that you can't memorize or recognize them all, but just knowing or memorizing the primes smaller than 100 is a big help, and memorizing the first 46 (all the primes smaller than 200) would be a bonus.

YOUR TURN

Identify which number is prime.

WHY ISN'T THE NUMBER 1 PRIME?

By tradition and definition, the number 1 is not prime. The definition of a prime number is that it can be divided evenly only by itself and 1. In this case, there would be a double hit, because 1 is itself.

Many theorems and conjectures involving primes don't work if 1 is included. Mathematicians around the time of Pythagoras sometimes even excluded the number 2 from the list of primes because they didn't consider 1 or 2 to be *true numbers* — they were just generators of all other even and odd numbers. Sometimes it seems that mathematical rules are a bit arbitrary. But in this case, it just makes everything else work better if 1 isn't a prime.

1. 3, 15, 27, or 39

2. 21, 33, 37, or 49

3. 35, 43, 51, or 81

4. 77, 85, 97, or 111

Working with Numbers in Their Prime

MERSENNE PRIMES

Mersenne primes are special prime numbers that can be written as 1 less than a power of 2. The numbers 3 and 7 are Mersenne primes, because $2^2 - 1 = 3$ and $2^3 - 1 = 7$. But, if you try $2^4 - 1 = 15$, you see that 15 isn't a prime. So this formula doesn't always give you a prime; it's just that there are many primes that can be written this way, as 1 less than a power of 2.

In 1996, the Great Internet Mersenne Prime Search was launched. This involved a contest to find large Mersenne primes. A gentleman, on his home computer, found a Mersenne prime of sufficient size, and the Electronic Frontier Foundation awarded him $50,000. In 2020, the 51st Mersenne prime was found to be $2^{82,589,933} - 1$. For more information on the Great Internet Mersenne Prime Search, go to `www.mersenne.org`.

Composing Composite Numbers

Prime numbers are interesting to think about, but they can also be a dead end in terms of factoring algebraic expressions or reducing fractions. The opposite of prime numbers, *composite numbers*, can be broken down into factorable, reducible pieces. In this section, you see how every composite number is the product of prime numbers, in a process known as prime factorization. Every number's prime factorization is unique.

The *prime factorization* of a number is the unique product of prime numbers that results in the given number. A prime number's prime factorization consists of just that prime number, by itself.

Here are some examples of prime factorization:

- $6 = 2 \times 3$
- $12 = 2 \times 2 \times 3 = 2^2 \times 3$
- $16 = 2 \times 2 \times 2 \times 2 = 2^4$
- $250 = 2 \times 5 \times 5 \times 5 = 2 \times 5^3$
- $510{,}510 = 2 \times 3 \times 5 \times 7 \times 11 \times 13 \times 17$
- $42{,}059 = 137 \times 307$

Okay, so that last one is a doozy. Finding that prime factorization without a calculator, computer, or list of primes is difficult.

The factors of some numbers aren't always obvious, but I do have some techniques to help you write prime factorizations, so check out the next section.

Writing Prime Factorizations

Writing the prime factorization of a composite number is one way to be absolutely sure you've left no stone unturned when reducing fractions or factoring algebraic expressions. These factorizations show you the one and only way a number can be factored. Two favorite ways of creating prime factorizations are upside-down division and trees.

Dividing while standing on your head

A slick way of writing out prime factorizations is to do an upside-down division. You put a *prime factor* (a prime number that evenly divides the number you're working on) on the outside left and the result or *quotient* (the number of times it divides evenly) underneath. You divide the quotient (the number underneath) by another prime number and keep doing this until the bottom number is a prime. Then you can stop. The order you do the divisions in doesn't matter. You get the same result or list of prime factors no matter what order you use. So, if you like to get all the even factors out first, just divide by 2 until you can't any longer.

Say you want to find the prime factorization of 120 using upside-down division. You start with the only even prime, 2. Put the quotient under the number you're dividing by. Then continue dividing by prime numbers until you end up with a prime at the bottom.

$$\begin{array}{r|l} 2 & 120 \\ 2 & 60 \\ 2 & 30 \\ 3 & 15 \\ & 5 \end{array}$$

Starting with the only even prime, I divided by 2, first, and got 60. Because 60 isn't prime, I divided by 2 again and got 30. Then I divided by 2 again and got 15. The number 15 is divisible by 3, and the result of the division is 5. Because the number 5 is prime, I stopped dividing and used the results (all the numbers running down the left side and the bottom) to write the prime factorization. You don't have to divide in this order; you'll get the same prime numbers regardless.

Looking at the numbers found for the prime factorization, you see that they act the same as the divisors in a division problem — only, in this case, they're all prime numbers. Although many composite numbers could have played the role of divisor for the number 120, the numbers for the prime factorization of 120 must be prime-number divisors.

When using this process, you usually do all the 2's first, then all the 3's, then all the 5's, and so on to make the prime factorization process easier, but you can do this in any order: $120 = 2 \times 2 \times 2 \times 3 \times 5 = 2^3 \times 3 \times 5$. In the next example, start with 13 because it seems obvious that it's a factor. The rest are all in a mixed-up order.

EXAMPLE

Q. Find the prime factorization of 13,000.

A. Sometimes you want to get rid of the larger primes first.

$$\begin{array}{r|l} 13 & 13{,}000 \\ 5 & 1{,}000 \\ 2 & 200 \\ 2 & 100 \\ 5 & 50 \\ 2 & 10 \\ & 5 \end{array}$$

So $13{,}000 = 13 \times 5 \times 2 \times 2 \times 5 \times 2 \times 5 = 2^3 \times 5^3 \times 13$. Even though the numbers started out in a mixed-up order, it's standard procedure to write the prime factorization going from the smallest prime up through the largest, each with its corresponding exponent (if it's bigger than 1).

YOUR TURN

Use upside-down division to write the prime factorization of the number.

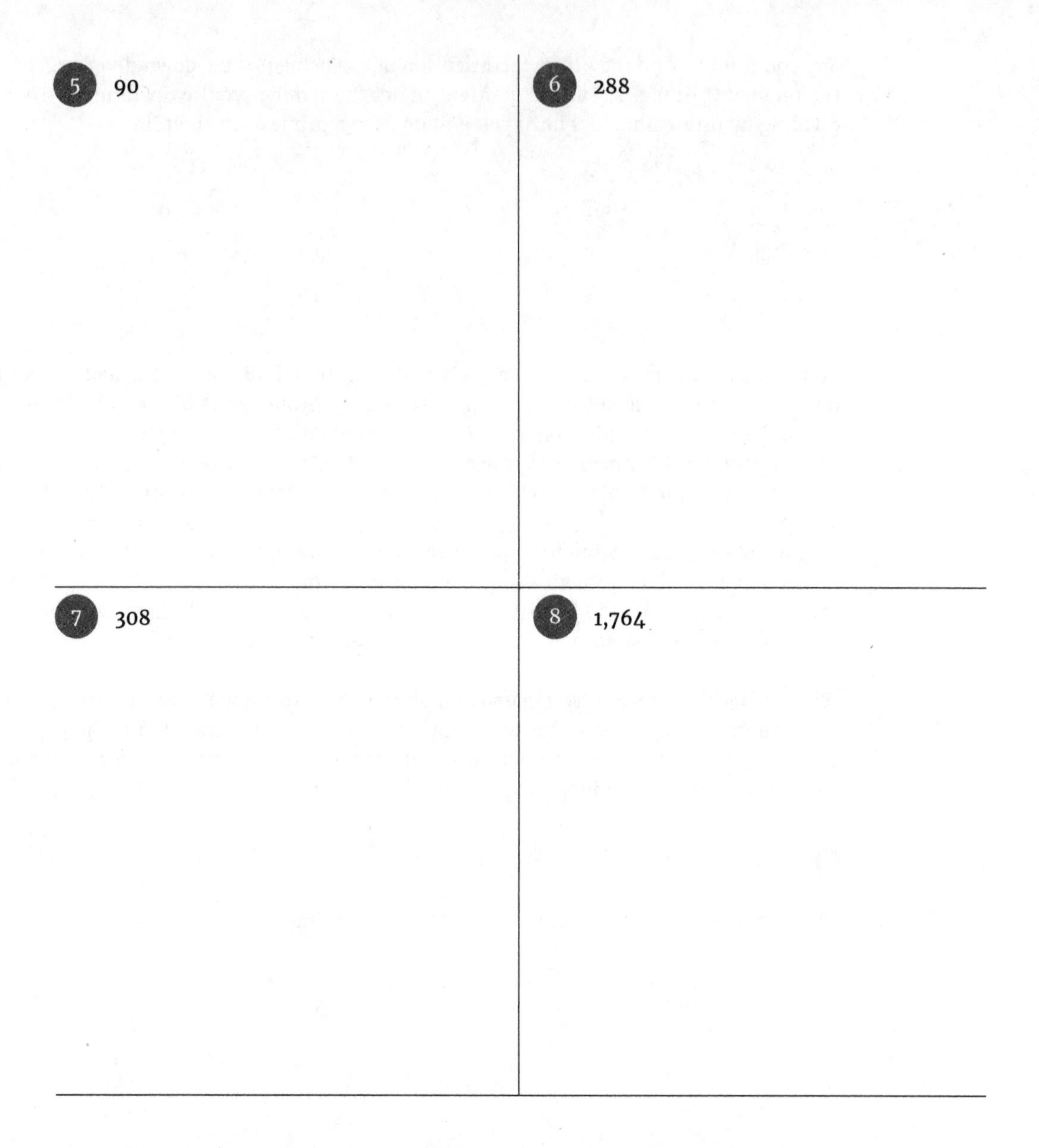

Getting to the root of primes with a tree

Another popular method for finding prime factorizations is to use a tree. Think of the number you start with as being the trunk of the tree and the prime factors as being at the ends of the roots.

To use the tree method, you write down your number and find two factors whose product is that number. The factors don't have to be prime; choose your favorite! Then you find factors for the two factors, and factors for the factors of the factors, and so on. You're finished when the lowest part of any root system is a prime number. Then you collect all those prime numbers for the factorization.

Figure 8-1 shows an example of finding the prime factorization of 6,350,400 using a factor tree.

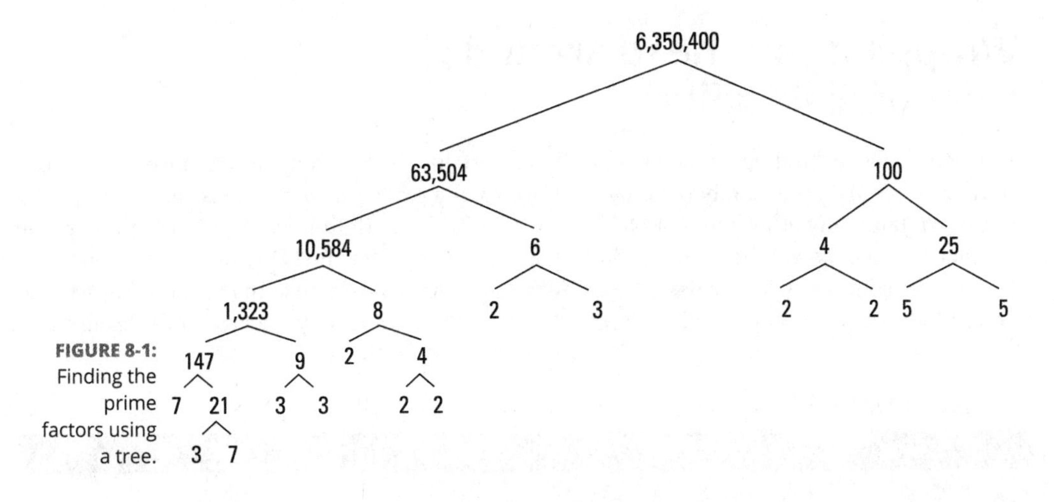

FIGURE 8-1: Finding the prime factors using a tree.

Now you collect all the prime numbers at the ends of the roots. I see the prime factors as 7, 3, 7, 3, 3, 2, 2, 2, 2, 3, 2, 2, 5, 5. Putting them in order, I get:

$$\begin{aligned} 6{,}350{,}400 &= 2\cdot 2\cdot 2\cdot 2\cdot 2\cdot 2\cdot 3\cdot 3\cdot 3\cdot 3\cdot 5\cdot 5\cdot 7\cdot 7 \\ &= 2^6\cdot 3^4\cdot 5^2\cdot 7^2 \end{aligned}$$

REMEMBER

You may not have created a tree the same way I did. Everyone sees different multiples and factors and has their favorites as far as dividing. I like to stick to numbers I can divide in my head. But you may be a calculator person. The great thing is that every way works and gives you the same final answer.

YOUR TURN

Use a tree to write the prime factorization of the number.

9 160	**10** 3,025
11 2,079	**12** 32,670

Wrapping your head around the rules of divisibility

The techniques for finding prime factorizations work just fine, as long as you have a good head start on what divides a number evenly. You probably already know the rules for dividing by 2 or 5 or 10. But many other numbers have very helpful rules or gimmicks for just looking at the number and seeing whether it's divisible by a particular factor. In Table 8-1, I give you many of the more commonly used rules of divisibility. Some are easier to use than others. Notice that I don't have the numbers in order; I prefer to group the numbers by the types of rules used.

TABLE 8-1 Rules of Divisibility

Number	Rule
2	The number ends in 0, 2, 4, 6, or 8.
5	The number ends in 0 or 5.
10	The number ends in 0.
4	The last two digits form a number divisible by 4.
8	The last three digits form a number divisible by 8.
3	The sum of the digits is a number divisible by 3.
9	The sum of the digits is a number divisible by 9.
11	The difference between the sums of the alternating digits is divisible by 11.
6	The number is divisible by both 2 and 3 (use both rules).
12	The number is divisible by both 3 and 4 (use both rules).
7	Double the last digit, subtract from remaining digits. Repeat until divisibility is recognizable (refer to the example below).

EXAMPLE

Q. Use the rules of divisibility to determine what divides 360 evenly.

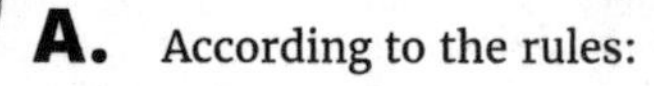

A. According to the rules:

- The number 360 ends in 0, so it's divisible by 2, 5, and 10.
- The last two digits of 360 form the number 60, which is divisible by 4, so the whole number is divisible by 4.
- The last three digits of 360 (okay, so it's all of them) form a number divisible by 8, so the whole number is divisible by 8.
- The sum of the digits in 360 is 9, so the number is divisible by both 3 and 9.
- The difference between the sums of the alternating digits is 3, so the number is not divisible by 11. (To get that difference, I added the $3+0$ to get 3, and the 6 had nothing to add to it. The difference between 6 and 3 is 3.)
- The number 360 is divisible by both 2 and 3, so it's divisible by 6.
- The number 360 is divisible by both 3 and 4, so it's divisible by 12.

Q. Use the rules of divisibility to determine what divides 1,056.

A. According to the rules:

- The number 1,056 ends in 6, so it's divisible by 2.
- The last two digits of 1,056 form the number 56, which is divisible by 4, so the whole number is divisible by 4.
- The last three digits of 1,056 form the number 56 (the 0 in front is ignored), which is divisible by 8, so the whole number is divisible by 8.
- The sum of the digits is 12, which is divisible by 3, so the whole number is divisible by 3.
- The difference between the sums of the alternating digits is 0, which is divisible by 11, so the whole number is divisible by 11.
- The number is divisible by both 2 and 3, so it's divisible by 6.
- The number is divisible by both 3 and 4, so it's divisible by 12.

Q. Show that the number 86,492 is divisible by 7.

A. Follow these steps:

- Drop the 2 on the end; double it and subtract the 4 from 8,649: $8{,}649 - 4 = 8{,}645$.
- Drop the 5 on the end; double it and subtract the 10 from 864: $864 - 10 = 854$.
- Drop the 4 on the end; double it and subtract the 8 from 85: $85 - 8 = 77$.
- You can probably stop here, if you remember that 77 is divisible by 7. If not, then drop the 7 on the end; double it and subtract the 14 from 7: $7 - 14 = -7$, which is divisible by 7.

Q. Use the rules of divisibility to determine what divides 77,077.

A. Using some good judgments:

- Even though you could use the rule for divisibility by 7, it's quicker just to divide. $77{,}077 \div 7 = 11{,}011$.
- The difference between the sums of the alternating digits is 0, which is divisible by 11, so the whole number is divisible by 11. $11{,}011 \div 11 = 1001$.
- And 11 divides again! $1001 \div 11 = 91$.
- If you don't recognize 91 as being a multiple of 7, just use the rule: double the 1 and subtract 2 from 9 to get 7! $91 \div 7 = 13$.

Determine what the number can be divided by using the rules of divisibility.

13 1,260	**14** 9,625
15 6,534	**16** 22,176

Making Use of a Prime Factor

Doing the actual factoring in algebra is easier when you can recognize which numbers are composite and which are prime. If you know in which category a number belongs, then you know what to do with it. When reducing fractions or factoring out many-termed expressions, you look for what the numbers have in common. If a number is prime, you stop looking. Now, try putting all this knowledge to work!

Taking primes into account

Prime factorizations are useful when you reduce fractions. Sure, you can do repeated reductions — first divide the numerator and denominator by 5 and then divide them both by 3 and so on. But a much more efficient use of your time is to write the prime factorizations of the numerator and denominator and then have an easy task of finding the common factors all at once.

EXAMPLE

Q. Reduce the fraction $\frac{120}{165}$.

A. Follow these steps:

1. **Find the prime factorization of the numerator.**

   120 is $2^3 \times 3 \times 5$.

2. **Find the prime factorization of the denominator.**

   165 is $3 \times 5 \times 11$.

3. **Next, write the fraction with the prime factorizations in it.**

   $\frac{120}{165} = \frac{2^3 \cdot 3 \cdot 5}{3 \cdot 5 \cdot 11}$

4. **Cross out the factors the numerator shares with the denominator to see what's left — the reduced form.**

   $\frac{120}{165} = \frac{2^3 \cdot 3 \cdot 5}{3 \cdot 5 \cdot 11} = \frac{2^3 \cdot \cancel{3} \cdot \cancel{5}}{\cancel{3} \cdot \cancel{5} \cdot 11} = \frac{2^3}{11} = \frac{8}{11}$

Q. Reduce the fraction $\frac{48x^3y^2z}{84xy^2z^3}$.

A. Use these steps. Note the addition of variables.

1. **Find the prime factorization of the numerator.**

   $48x^3y^2z = 2^4 \times 3 \times x^3y^2z$.

2. **Find the prime factorization of the denominator.**

   $84xy^2z^3 = 2^2 \times 3 \times 7 \times xy^2z^3$.

Working with Numbers in Their Prime

3. **Write the fraction with the prime factorization.**

$$\frac{48x^3y^2z}{84xy^2z^3} = \frac{2^4 \cdot 3 \cdot x^3y^2z}{2^2 \cdot 3 \cdot 7 \cdot xy^2z^3}$$

4. **Cross out the factors in common.**

$$\frac{2^4 \cdot 3 \cdot x^3 \cdot y^2 \cdot z}{2^2 \cdot 3 \cdot 7 \cdot x \cdot y^2 \cdot z^3} = \frac{\cancel{2^4}^{2^2} \cdot \cancel{3} \cdot \cancel{x^3}^{x^2} \cdot \cancel{y^2} \cdot \cancel{z}}{\cancel{2^2} \cdot \cancel{3} \cdot 7 \cdot \cancel{x} \cdot \cancel{y^2} \cdot \cancel{z^3}^{z^2}} = \frac{2^2 \cdot x^2}{7 \cdot z^2} = \frac{4x^2}{7z^2}$$

By writing the prime factorizations, you can be certain that you haven't missed any factors that the numerator and denominator may have in common.

YOUR TURN

Reduce the fraction by writing the prime factorizations of the numerator and denominator to determine common factors.

17 $\frac{48}{80}$

18 $\frac{600}{475}$

19 $\frac{1,764}{1,694}$

20 $\frac{10,560}{20,250}$

Pulling out factors and leaving the rest

Pulling out common factors from lists of terms or the sums or differences of a bunch of terms is done for a good reason. It's a common task when you're simplifying expressions and solving equations. The common factor that makes the biggest difference in these problems is the greatest common factor (GCF). When you recognize the GCF and factor it out, it does the most good.

The *greatest common factor* is the largest-possible term that evenly divides each term of an expression containing two or more terms (or evenly divides the numerator and denominator of a fraction).

In any factoring discussion, the GCF, the most common and easiest factoring method, always comes up first. And it's helpful to know about the GCF when solving equations. In an expression with two or more terms, finding the greatest common factor can make the expression more understandable and manageable.

When simplifying expressions, the best-case scenario is to recognize and pull out the GCF from a list of terms. Sometimes, though, the GCF may not be so recognizable. It may have some strange factors, such as 7, 13, or 23. It isn't the end of the world if you don't recognize one of these numbers as being a multiplier; it's just nicer if you do.

When factoring an algebraic expression, use the following procedure.

1. **Determine any common numerical factors.**
2. **Determine any common variable factors.**
3. **Write the prime factorizations of each term.**
4. **Find the GCF.**
5. **Divide each term by the GCF.**
6. **Write the result as the product of the GCF and the results of the division.**

For example: Find the GCF of $12x^2y^4 + 16xy^3 - 20x^3y^2$ and write the factorization.

Use the following steps to solve:

1. **Determine any common numerical factors.**

   Each term has a coefficient that is divisible by a power of 2, which is $2^2 = 4$.

2. **Determine any common variable factors.**

   Each term has *x* and *y* factors.

Working with Numbers in Their Prime

3. **Write the prime factorizations of each term.**

   $12x^2y^4 = 2^2 \times 3 \times x^2y^4$

   $16xy^3 = 2^4 \times xy^3$

   $-20x^3y^2 = -2^2 \times 5 \times x^3y^2$

4. **Find the GCF.**

   The GCF is the product of all the factors that all three terms have in common. The GCF contains the *lowest* power of each variable and number that occurs in any of the terms. Each variable in the sample problem has a factor of 2. If the lowest power of 2 that shows in any of the factors is 2^2, then 2^2 is part of the GCF.

   Each factor has a power of x. Because the lowest power of x that shows up in any of the factors is 1, x^1 is part of the GCF.

   Each factor has a power of y. Because the lowest power of y that shows in any of the factors is 2, y^2 is part of the GCF.

   The GCF of $12x^2y^4 + 16xy^3 - 20x^3y^2$ is $2^2xy^2 = 4xy^2$.

5. **Divide each term by the GCF.**

   The respective terms are divided as shown:

   - $\frac{12x^2y^4}{4xy^2} = 3xy^2$
   - $\frac{16xy^3}{4xy^2} = 4y$
   - $\frac{-20x^3y^2}{4xy^2} = -5x^2$

   Notice that the three different results of the division have nothing in common. Each of the first two results has a y and the first and third both have an x, but nothing is shared by all the results. This is the best factoring situation, which is what you want.

6. **Write the result as the product of the GCF and the results of the division.**

   Rewriting the original expression with the GCF factored out and in parentheses, you get $12x^2y^4 + 16xy^3 - 20x^3y^2 = 4xy^2(3xy^2 + 4y - 5x^2)$.

In the following example problems, I show you the shortened version of these steps.

EXAMPLE

Q. Find the GCF and write the factorization of $40a^5x + 80a^5y - 120a^5z$.

A. The factorizations of the terms are: $2^3 \cdot 5 \cdot a^5x$, $2^4 \cdot 5 \cdot a^5y$, and $-2^3 \cdot 3 \cdot 5a^5z$. Each term has a factor of $2^3 \cdot 5$, and a^5, so the GCF is $40a^5$, and you can write the expression as the product of the GCF and the results of dividing each term by the GCF: $40a^5\,(x + 2y - 3z)$.

Q. Find the GCF and write the factorization of $18x^2y + 25z^3 + 49z^2$.

A. Even though none of these terms is prime, the three terms have nothing in common — nothing that *all three* share. The following prime factorizations demonstrate:

- $18x^2y = 2 \times 3^2x^2y$
- $25z^3 = 5^2\,z^3$
- $49z^2 = 7^2\,z^2$

The last two terms do have a factor of z in common, but the first term doesn't. This expression is said to be *prime* because it can't be factored.

YOUR TURN

Find the GCF of the terms and write the factorization.

21 $36x^2y^3 + 45xy^4$

22 $99a^5b^3 - 132a^2b^6$

23 $240xyz^5 + 672xy^3z^3$

24 $462m^3p^2 + 700m^3n - 1260m^3$

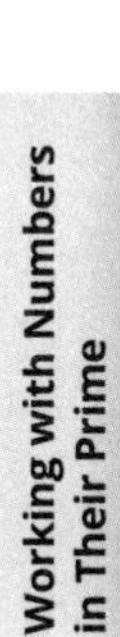

Practice Questions Answers and Explanations

1. **3.** The numbers 15, 27, and 39 are all multiples of 3.

2. **37.** The number 21 is a multiple of 3 and 7. The number 33 is a multiple of 3 and 11. The number 49 is the square of 7.

3. **43.** The number 35 is a multiple of 5 and 7. The number 51 is a multiple of 3 and 17. The number 81 is the square of 9 and is also 3^4.

4. **97.** The number 77 is a multiple of 7 and 11. The number 85 is a multiple of 5 and 17. The number 111 is a multiple of 3 and 37.

5. $\mathbf{2 \cdot 3^2 \cdot 5}$

$$\begin{array}{r|l} 2 & 90 \\ 3 & 45 \\ 3 & 15 \\ & 5 \end{array}$$

6. $\mathbf{2^5 \cdot 3^2}$

$$\begin{array}{r|l} 2 & 288 \\ 2 & 144 \\ 2 & 72 \\ 2 & 36 \\ 2 & 18 \\ 3 & 9 \\ & 3 \end{array}$$

7. $\mathbf{2^2 \cdot 7 \cdot 11}$

$$\begin{array}{r|l} 2 & 308 \\ 2 & 154 \\ 7 & 77 \\ & 11 \end{array}$$

8. $\mathbf{2^2 \cdot 3^2 \cdot 7^2}$

$$\begin{array}{r|l} 2 & 1{,}764 \\ 2 & 882 \\ 3 & 441 \\ 3 & 147 \\ 7 & 49 \\ & 7 \end{array}$$

(9) $\mathbf{2^5 \cdot 5.}$

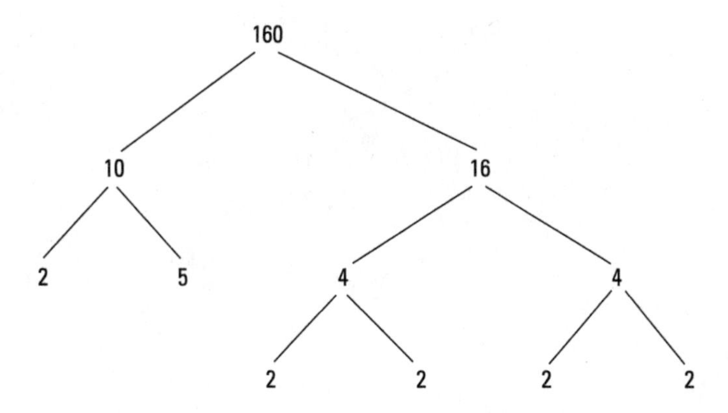

The bottom entries are all prime numbers: 2, 5, 2, 2, 2, 2. So the prime factorization is $2^5 \times 5$.

(10) $\mathbf{5^2 \cdot 11^2.}$

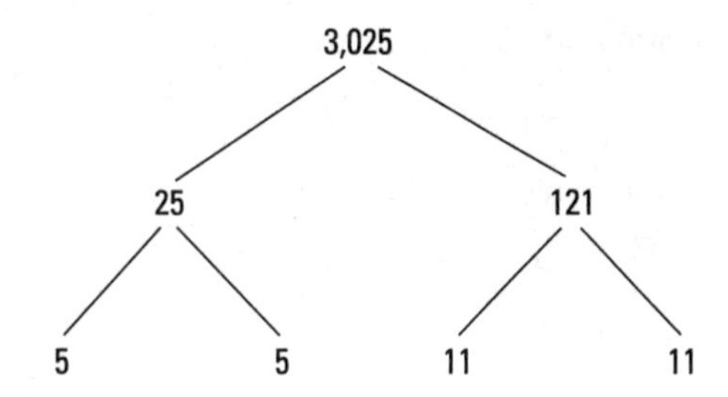

The bottom entries are all prime numbers: 5, 5, 11, 11. So the prime factorization is $5^2 \times 11^2$.

(11) $\mathbf{3^3 \cdot 7 \cdot 11.}$

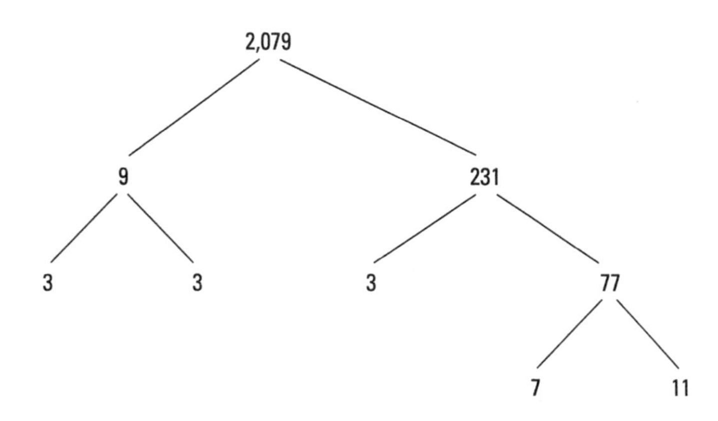

The bottom entries are all prime numbers: 3, 3, 3, 7, 11. So the prime factorization is $3^3 \times 7 \times 11$.

(12) **$2 \cdot 3^3 \cdot 5 \cdot 11^2$.**

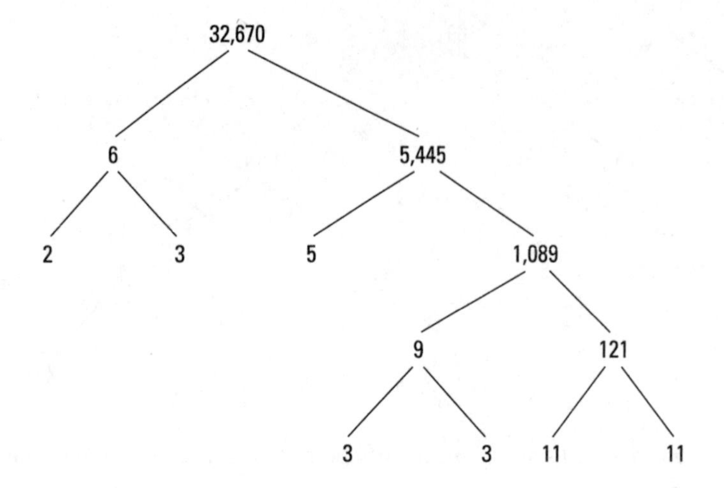

The bottom entries are all prime numbers: 2, 3, 5, 3, 3, 11, 11. So the prime factorization is $2 \times 3^3 \times 5 \times 11^2$.

(13) **2, 3, 4, 5, 6, 7, 9, 10, and 12.** It's divisible by:

- 2, 5 and 10, because it ends in 0.
- 3 and 9, because the sum of the digits is 9.
- 4, because the last two digits form the number 60, and 4 divides 60 evenly.
- 6, because it's divisible by both 2 and 3.
- 7, because doubling 6 and subtracting from 12 leaves a remainder of 0.
- 12, because it's divisible by both 3 and 4.

(14) **5, 7, and 11.** It's divisible by:

- 5, because it ends in 5.
- 7, because doubling 5 and subtracting from 962 leaves 952. Doubling 2 and subtracting from 95 leaves 91. Doubling 1 and subtracting from 9 leaves 7.
- 11, because the sum of the first and third digits is 11 and the sum of the second and fourth digits is 11; the difference of the sums is 0.

(15) **2, 3, 6, 9, and 11.** It's divisible by:

- 2, because it ends in 4.
- 3 and 9, because the sum of the digits is 18.
- 6, because it's divisible by both 2 and 3.
- 11, because the sum of the first and third digits is 9 and the sum of the second and fourth digits is 9; the difference of the sums is 0.

(16) **2, 3, 4, 6, 7, 8, 9, 11, and 12.** It's divisible by:

- 2, because it ends in 6.
- 3 and 9, because the sum of the digits is 18.
- 4, because the last two digits form the number 76, and 4 divides 76 evenly.
- 6, because it's divisible by both 2 and 3.
- 7, because doubling 6 and subtracting from 2,217 leaves 2,205. Doubling 5 and subtracting from 220 leaves 210. Doubling 0 and subtracting from 21 leaves 21. The number 7 divides 21 evenly.
- 8, because the last three digits form the number 176, and 8 divides 176 evenly.
- 11, because the sum of the first, third, and fifth digits is 9, and the sum of the second and fourth digits is 9. The difference between those sums is 0.
- 12, because it's divisible by both 3 and 4.

(17) $\mathbf{\frac{3}{5}}$. $\frac{48}{80}=\frac{2^4\cdot 3}{2^4\cdot 5}=\frac{\cancel{2^4}\cdot 3}{\cancel{2^4}\cdot 5}=\frac{3}{5}$

(18) $\mathbf{1\frac{5}{19}}$. $\frac{600}{475}=\frac{2^3\cdot 3\cdot 5^2}{5^2\cdot 19}=\frac{2^3\cdot 3\cdot \cancel{5^2}}{\cancel{5^2}\cdot 19}=\frac{2^3\cdot 3}{19}=\frac{24}{19}=1\frac{5}{19}$

(19) $\mathbf{1\frac{5}{121}}$. $\frac{1{,}764}{1{,}694}=\frac{2^2\cdot 3^2\cdot 7^2}{2\cdot 7\cdot 11^2}=\frac{\cancel{2^2}^1\cdot 3^2\cdot \cancel{7^2}^1}{\cancel{2}\cdot \cancel{7}\cdot 11^2}=\frac{2\cdot 3^2\cdot 7}{11^2}=\frac{126}{121}=1\frac{5}{121}$

(20) $\mathbf{\frac{352}{675}}$. $\frac{10{,}560}{20{,}250}=\frac{2^6\cdot 3\cdot 5\cdot 11}{2\cdot 3^4\cdot 5^3}=\frac{\cancel{2^6}^5\cdot \cancel{3}\cdot \cancel{5}\cdot 11}{\cancel{2}\cdot \cancel{3^4}^3\cdot \cancel{5^3}^2}=\frac{2^5\cdot 11}{3^3\cdot 5^2}=\frac{352}{675}$

(21) $\mathbf{9xy^3(4x+5y)}$. Rewriting the terms using their prime factorizations, you get $36x^2y^3+45xy^4=2^2\cdot 3^2\cdot x^2y^3+3^2\cdot 5\cdot xy^4$. The GCF of the two terms is $3^2\cdot xy^3$. Dividing each term by the GCF, $\frac{2^2\cdot \cancel{3^2}\cdot \cancel{x^2}^1\cancel{y^3}}{\cancel{3^2}\cancel{x}\cancel{y^3}}+\frac{\cancel{3^2}\cdot 5\cdot \cancel{x}\cancel{y^4}^1}{\cancel{3^2}\cancel{x}\cancel{y^3}}=2^2\cdot x^1+5\cdot y^1=4x+5y$. So the expression in factored form is $9xy^3(4x+5y)$.

(22) $\mathbf{33a^2b^3(3a^3-4b^3)}$. Rewriting the terms using their prime factorizations, you get $99a^5b^3-132a^2b^6=3^2\cdot 11\cdot a^5b^3-2^2\cdot 3\cdot 11\ a^2b^6$. The GCF of the two terms is $3\cdot 11\cdot a^2b^3$. Dividing each term by the GCF, $\frac{\cancel{3^2}^1\cdot \cancel{11}\cdot \cancel{a^5}^3\cancel{b^3}}{\cancel{3}\cdot \cancel{11}\cdot \cancel{a^2}\cancel{b^3}}-\frac{2^2\cdot \cancel{3}\cdot \cancel{11}\cdot \cancel{a^2}\cancel{b^6}^3}{\cancel{3}\cdot \cancel{11}\cdot \cancel{a^2}\cancel{b^3}}=3^1\cdot a^3-2^2\cdot b^3=3a^3-4b^3$. So the expression in factored form is $33a^2b^3(3a^3-4b^3)$.

(23) $\mathbf{48xyz^3(5z^2+14y^2)}$. Rewriting the terms using their prime factorizations, you get $240xyz^5+672xy^3z^3=2^4\cdot 3\cdot 5\cdot xyz^5+2^5\cdot 3\cdot 7\cdot xy^3z^3$. The GCF of the two terms is $2^4\cdot 3\cdot xyz^3$. Dividing each term by the GCF, $\frac{\cancel{2^4}\cdot \cancel{3}\cdot 5\cdot \cancel{x}\cancel{y}\cancel{z^5}^2}{\cancel{2^4}\cdot \cancel{3}\cdot \cancel{x}\cancel{y}\cancel{z^3}}+\frac{\cancel{2^5}^1\cdot \cancel{3}\cdot 7\cdot \cancel{x}\cancel{y^3}^2\cancel{z^3}}{\cancel{2^4}\cdot \cancel{3}\cdot \cancel{x}\cancel{y}\cancel{z^3}}=5\cdot z^2+2\cdot 7\cdot y^2=5z^2+14y^2$. So the expression in factored form is $48xyz^3(5z^2+14y^2)$.

24 **$14m^3(33p^2+50n-90)$.** Rewriting the terms using their prime factorizations, you get $462m^3p^2+700m^3n-1260m^3=2\cdot3\cdot7\cdot11\cdot m^3p^2+2^2\cdot5^2\cdot7\cdot m^3n-2^2\cdot3^2\cdot5\cdot7\cdot m^3$. The GCF of the three terms is $2\cdot7\cdot m^3$. Dividing each term by the GCF,

$$\frac{\cancel{2}\cdot3\cdot\cancel{7}\cdot11\cdot\cancel{m^3}p^2}{\cancel{2}\cdot\cancel{7}\cdot\cancel{m^3}}+\frac{2^{\cancel{2}^1}\cdot5^2\cdot\cancel{7}\cdot\cancel{m^3}n}{\cancel{2}\cdot\cancel{7}\cdot\cancel{m^3}}-\frac{2^{\cancel{2}^1}\cdot3^2\cdot5\cdot\cancel{7}\cdot\cancel{m^3}}{\cancel{2}\cdot\cancel{7}\cdot\cancel{m^3}}=3\cdot11\cdot p^2+2^1\cdot5^2\cdot n-2^1\cdot3^2\cdot5=33p^2+50n-90.$$

So the expression in factored form is $14m^3(33p^2+50n-90)$.

If you're ready to test your skills a bit more, take the following chapter quiz that incorporates all the chapter topics.

Whaddya Know? Chapter 8 Quiz

Quiz time! Complete each problem to test your knowledge on the various topics covered in this chapter. You can then find the solutions and explanations in the next section.

1. Which of these numbers is prime: 39, 43, 51, 91?
2. Which of the numbers 2 through 12 is the number 660 divisible by?
3. Reduce the fraction: $\frac{1,050}{25,380}$
4. Find the GCF of the expression: $75a^3b^2c-125a^4bc^3+500a^5b^6$
5. Write the prime factorization of the number 180.
6. Which of the numbers 2 through 12 is the number 9,849 divisible by?
7. Find the GCF of the expression: $6x^2y^3+8x^3y^4-10x^4y^4$
8. Write the prime factorization of the number 616.
9. Reduce the fraction: $\frac{792}{2,808}$
10. Write the prime factorization of the number 162,000.

Answers to Chapter 8 Quiz

1. **43.** The number 39 is divisible by 3 and 13, 51 is divisible by 3 and 17, 91 is divisible by 7 and 13.

2. **2, 3, 4, 5, 6, 10, 11.** The number 660 ends in 0, making it divisible by 2, 5 and 10. The last two digits form the number 60, making it divisible by 4. The sum of the digits is 12, making it divisible by 3. And, because it's divisible by both 2 and 3, it's divisible by 6. The difference between the sums of the alternate digits $(6-6)$ is 0, so it's divisible by 11.

3. $\mathbf{\frac{35}{846}}$. $\frac{1{,}050}{25{,}380}=\frac{2\cdot 3\cdot 5^2\cdot 7}{2^2\cdot 3^3\cdot 5\cdot 7}=\frac{\cancel{2}\cdot\cancel{3}\cdot 5^{\cancel{2}1}\cdot 7}{2^{\cancel{2}1}\cdot 3^{\cancel{3}2}\cdot\cancel{5}\cdot 47}=\frac{5\cdot 7}{2\cdot 3^2\cdot 47}=\frac{35}{846}$

4. $\mathbf{25a^3b}$. $25a^3b\left(3bc-5ac^3+20a^2b^5\right)=75a^3b^2c-125a^4bc^3+500a^5b^6$

5. $\mathbf{180=2^2\cdot 3^2\cdot 5}$. Either the upside-down division or tree diagram works well here.

6. **3 and 7.** The number 9,849 is odd, so that immediately throws out all the even divisors. The sum of the digits is 30, so it's divisible by 3. And, using the rule for divisibility by 7:

$$\begin{array}{rrrr} 9 & 8 & 4 & \cancel{9} \\ - & 1 & 8 & \\ \hline 9 & 6 & \cancel{6} & \\ -1 & 2 & & \\ \hline 8 & \cancel{4} & & \\ -8 & & & \\ \hline 0 & & & \end{array}$$

7. $\mathbf{2x^2y^3}$. $2x^2y^3\left(3+4xy-5x^2y\right)=6x^2y^3+8x^3y^4-10x^4y^4$

8. $\mathbf{616=2^3\cdot 7\cdot 11}$.

9. $\mathbf{\frac{11}{39}}$. $\frac{792}{2{,}808}=\frac{2^3\cdot 3^2\cdot 11}{2^3\cdot 3^3\cdot 13}=\frac{\cancel{2^3}\cdot\cancel{3^2}\cdot 11}{\cancel{2^3}\cdot 3^{\cancel{3}1}\cdot 13}=\frac{11}{3\cdot 13}=\frac{11}{39}$

10. $\mathbf{162{,}000=2^4\cdot 3^4\cdot 5^3}$. You should use the upside-down division with a number this large.

FOIL

- Squaring and cubing binomials
- Calling on Pascal to raise binomials to many powers
- Incorporating special rules when multiplying

Chapter 9

Specializing in Multiplication Matters

In Chapter 6, you see how to multiply algebraic expressions together using the rules involving exponents and unlike terms. When performing multiplication involving one or many terms, it's helpful to understand how to take advantage of special situations, like multiplying the sum and difference of the same two values, such as $(xy+16)(xy-16)$, and raising a binomial to a power, such as $(a+3)^4$, which seem to pop up frequently. Fortunately, various procedures have been developed to handle such situations with ease.

You also see the nitty-gritty of multiplying factor by factor and term by term (the long way) compared to the neat little pattern called FOIL. With this information, you'll be able to efficiently and correctly multiply expressions in many different types of situations. You'll also be able to expand expressions using multiplication so that you can later go backward and factor expressions back into their original multiplication forms.

Algebra is full of contradictory actions. In one instance, you're asked to factor (see Chapters 11, 12, and 13 for facts on factoring), and in the next, you're told to distribute or "unfactor." As another example, you're first asked to reduce fractions, and then you're supposed to multiply and create bigger numbers. In other words, you're first asked to compress the math expression, and then to spread it all out again. Make up your mind!

But rest assured that all these seemingly contradictory processes have good reasoning behind them. You carefully wrap a birthday gift so it can be unwrapped the next day. You water and fertilize your lawn to make it grow — just so you can cut it. See, contradictions are everywhere!

In this chapter, I tell you when, why, and how to "unfactor." After all, you want to make informed decisions on how to proceed and then have the skills to execute them correctly.

Distributing One Factor Over Many

When things are handed out equitably, everyone or everything involved gets an equal share — and just one of the shares, not twice as many as others get. When a child is distributing their birthday treats to classmates, it's: "One for you, and one for you" In the game Mancala, the stones in a cup are distributed one to each of the next cups until they're gone. Any other way is cheating! In algebra, distributing is much the same process — each term gets a share.

Distributing items is an act of spreading them out equally. Algebraic distribution means to multiply each of the terms within the parentheses by another term that is outside the parentheses. Each term gets multiplied by the same amount.

To distribute a particular term over several other terms, you multiply each of the target terms by the one that is getting distributed. *Distribution* involves multiplying each individual term in a grouped series of terms by a value outside of the grouping.

$$a(b+c+d+e+\ldots)=ab+ac+ad+ae+\ldots$$

The addition signs could just as well be subtraction, and a is any real number: positive, negative, integer, fraction.

REMEMBER

A *term* is made up of variable(s) and/or number(s) joined by multiplication and/or division. Terms are separated from one another by addition or subtraction.

EXAMPLE

Q. Distribute the number 2 over the terms $4x+3y-6$.

A. Use these steps:

1. **Multiply each term by the number(s) and/or variable(s) outside of the parentheses.**

   $2(4x+3y-6)=2(4x)+2(3y)-2(6)$

2. **Perform the multiplication operation in each term.**

   $=8x+6y-12$

Q. Distribute the number x over the terms $4x+3y-6$.

A. Use these steps:

1. **Multiply each term by the number(s) and/or variable(s) outside of the parentheses.**

   $x(4x+3y-6)=x(4x)+x(3y)-x(6)$

2. **Perform the multiplication operation in each term.**

   $=4x^2+3xy-6x$

When you distribute some factor over several terms, you don't change the value of the original expression. The value is the same, whether you distribute first or add up what's in the parentheses first. When performing algebraic manipulations, you often have to make a judgment call as to whether to combine what's in the parentheses first (if you can do that) or to distribute first.

TIP

Distributing first to get the answer is the better choice when the multiplication of each term gives you nicer numbers. Fractions or decimals in the parentheses are sometimes changed into nice, whole numbers when the distribution is done first. The other choice — adding up what's in the parentheses first — is preferred when distributing gives you too many big multiplication problems. Sometimes it's easy to tell which case you have; other times, you just have to guess and try it.

YOUR TURN

Distribute and simplify.

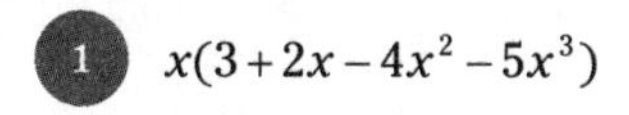

1 $x(3+2x-4x^2-5x^3)$

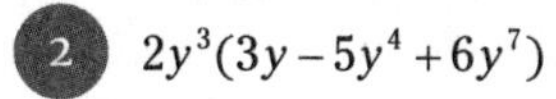

2 $2y^3(3y-5y^4+6y^7)$

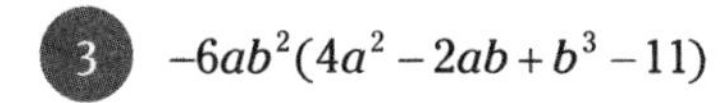

3 $-6ab^2(4a^2-2ab+b^3-11)$

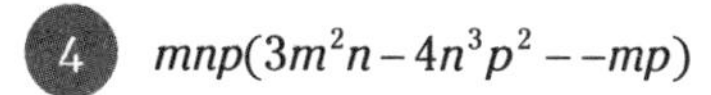

4 $mnp(3m^2n-4n^3p^2--mp)$

Distributing Signs

When a number is distributed over terms within parentheses, you multiply each term by that number. An even easier type of distribution is distributing a simple sign (no, you don't distribute a Leo or Libra — they'd object). But, what should be rather simple is often done in error. Hence, I'm devoting some time to signs.

Positive (+) and negative (−) signs are simple to distribute, but distributing a negative sign can cause errors. Distributing a positive sign makes no difference in the signs of the terms.

WARNING

One mistake to avoid when you're distributing a negative sign is not distributing over *all* the terms. This is especially the case when the process is *hidden*. By hidden, I mean that a negative sign may not be in front of the whole expression, where it sticks out. It can be between terms, showing a subtraction and not being recognized for what it is. Don't let the negative signs ambush you.

EXAMPLE

Q. Distribute the negative sign: $-(4x+2y-3z+7)$.

A. $\mathbf{-4x-2y+3z-7}$. Using a negative sign in the expression $-(4x+2y-3z+7)$ is the same as multiplying through by −1:

$$-1(4x+2y-3z+7)=$$
$$-1(4x)-1(2y)-1(-3z)-1(7)=$$
$$-4x-2y+3z-7$$

Each term was changed to a term with the opposite sign.

Q. Simplify the expression by distributing and combining like terms: $4x(x-2)-(5x+3)$.

A. $\mathbf{4x^2-13x-3}$. Follow these steps:

1. **Distribute the 4x over the x and the −2 by multiplying both terms by 4x:**

   $$4x(x-2)=4x(x)-4x(2)$$

2. **Distribute the negative sign over the 5x and the 3 by changing the sign of each term.** Be careful — you can easily make a mistake if you stop after only changing the 5x.

   $$-(5x+3)=-(+5x)-(+3)$$

3. **Multiply and combine the like terms:**

   $$4x(x)-4x(2)-(+5x)-(+3)=$$
   $$4x^2-8x-5x-3=$$
   $$4x^2-13x-3$$

YOUR TURN

Distribute and simplify the following terms.

5 $4(2x+3y-5z)$	**6** $-3(9-4a+2b-c)$
7 $\frac{1}{2}(6x-4y+22z-3)$	**8** $-12\left(\frac{a}{3}+\frac{5b}{6}-\frac{7c}{9}-\frac{d}{12}\right)$

Mixing It up with Numbers and Variables

Distributing variables over the terms in an algebraic expression involves multiplication rules and the rules for exponents. When different variables are multiplied together, they can be written side by side without using any multiplication symbols between them. If the same variable is multiplied as part of the distribution, then the exponents are added together.

REMEMBER

When multiplying factors with the same base, add the exponents:

$$a^x \cdot a^y = a^{x+y}$$

Let me show you a couple of distribution problems involving factors with exponents.

EXAMPLE

Q. Distribute the a through the terms in the parentheses: $a(a^4+2a^2+3)$.

A. $\boldsymbol{a^5+2a^3+3a}$. Follow these steps.

1. **Multiply a times each term:**

$$a(a^4+2a^2+3)=$$
$$a\cdot a^4+a\cdot 2a^2+a\cdot 3$$

2. **Use the rules of exponents to simplify:**

$$a^5+2a^3+3a$$

Q. Distribute z^4 over the terms $2z^2-3z^{-2}+z^{-4}+5z^{\frac{1}{3}}$.

A. $\boldsymbol{2z^6-3z^2+1+5z^{\frac{13}{3}}}$. Follow these steps.

1. **Distribute the z^4 by multiplying it times each term:**

$$z^4\left(2z^2-3z^{-2}+z^{-4}+5z^{\frac{1}{3}}\right)=$$
$$z^4\cdot 2z^2-z^4\cdot 3z^{-2}+z^4z^{-4}+z^4\cdot 5z^{\frac{1}{3}}$$

2. **Simplify by adding the exponents:**

$$2z^{4+2}-3z^{4-2}+z^{4-4}+5z^{4+\frac{1}{3}}=$$
$$2z^6-3z^2+z^0+5z^{\frac{13}{3}}=2z^6-3z^2+1+5z^{\frac{13}{3}}$$

REMEMBER

The exponent 0 means the value of the expression is 1. So, $x^0=1$ for any real number x except 0.

You combine exponents with different signs by using the rules for adding and subtracting signed numbers. Fractional exponents are combined after finding common denominators. Exponents that are improper fractions are left in that form.

These next examples show what happens when you have more than one variable — and how you have to use the rule of adding exponents very carefully.

EXAMPLE

Q. Simplify the expression by distributing: $5x^2y^3\left(16x^2-2x+3xy+4y^3-11y^5+z-1\right)$.

A. Follow these steps.

1. **Multiply each term by $5x^2y^3$:**

$$5x^2y^3\cdot 16x^2-5x^2y^3\cdot 2x+5x^2y^3\cdot 3xy+5x^2y^3\cdot 4y^3-5x^2y^3\cdot 11y^5+5x^2y^3z-5x^2y^3$$

2. **Complete the multiplication in each term. Add exponents where needed:**

$$80x^4y^3-10x^3y^3+15x^3y^4+20x^2y^6-55x^2y^8+5x^2y^3z-5x^2y^3$$

3. **You're finished! There are no like terms to be combined.**

Q. Simplify by distributing: $-4xyzw(4-x-y-z-w)$.

A. Follow these steps.

1. **Multiply each term by $-4xyzw$:**

$$-4xyzw(4)-4xyzw(-x)-4xyzw(-y)-4xyzw(-z)-4xyzw(-w)$$

2. **Complete the multiplication in each term:**

$$-16xyzw+4x^2yzw+4xy^2zw+4xyz^2w+4xyzw^2$$

YOUR TURN

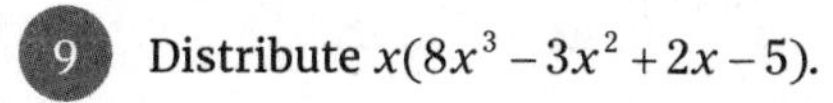

9 Distribute $x(8x^3-3x^2+2x-5)$.

10 Distribute $x^2y(2xy^2+3xyz+y^2z^3)$.

11 Distribute $-4y(3y^4-2y^2+5y-5)$.

12 Distribute $x^{1/2}y^{-1/2}\left(x^{3/2}y^{1/2}+x^{1/2}y^{3/2}\right)$.

Negative exponents yielding fractional answers

As the heading suggests, a base that has a negative exponent can be changed into a fractional form. The base and the exponent become part of the denominator of the fraction, but the exponent loses its negative sign in the process. Then you cap it all off with a 1 in the numerator.

REMEMBER

The format for changing negative exponents to fractions is $a^{-n}=\frac{1}{a^n}$. (See Chapter 6 for more details on negative exponents.)

In the following example, I show you how a negative exponent leads to a fractional answer.

PALINDROMES

The word *palindrome* comes from the Greek word *palindromos,* which means *running back again.* A palindrome is any word, sentence, or even a complete poem that reads the same backward as it does forward. For example, Leigh Mercer wrote, "A man, a plan, a canal — Panama" to honor the man responsible for building the Panama Canal. (Do you see that the letters in the sentence are the same — reading forward or backward?) Or, how about "Niagara, O roar again!" There are words that are palindromes: *rotator, Malayalam* (an East Indian language), and *redivider.*

Number palindromes have been of great interest to mathematicians over the years. Some perfect squares are palindromes — for example, 121 and 14,641. A palindromic date may be October 9, 1901 (1091901). Some couples choose their wedding dates by observing when a particular day is a palindrome.

You can create a palindrome by reversing the digits of almost any number and adding the reversal to the original number. For example, take 146, reverse the digits to get 641. Add them together: $146 + 641 = 787$. If you don't get a palindrome, just repeat the steps (and repeat) until you finally (and you will) get a palindrome.

EXAMPLE

Q. Distribute $5a^{-3}b^{-2}$ over each term in $\left(2ab^3 - 3a^2b^2 + 4a^4b - ab\right)$ and write the final answer without negative exponents.

A. Follow these steps.

1. **Multiplying the numbers and adding the exponents:**

$$5a^{-3}b^{-2}\left(2ab^3 - 3a^2b^2 + 4a^4b - ab\right)$$
$$= 5a^{-3}b^{-2}\left(2ab^3\right) - \left(5a^{-3}b^{-2}\right)\left(3a^2b^2\right) + \left(5a^{-3}b^{-2}\right)\left(4a^4b\right) - \left(5a^{-3}b^{-2}\right)\left(ab\right)$$
$$= 10a^{-3+1}b^{-2+3} - 15a^{-3+2}b^{-2+2} + 20a^{-3+4}b^{-2+1} - 5a^{-3+1}b^{-2+1}$$

2. **The factor of *b* with the 0 exponent becomes 1:**

$$10a^{-2}b^1 - 15a^{-1}b^0 + 20a^1b^{-1} - 5a^{-2}b^{-1}$$

3. **This step shows the final result without negative exponents — using the procedure for changing negative exponents to fractions (see earlier in this section):**

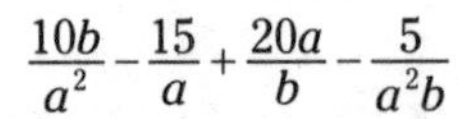

$$\frac{10b}{a^2} - \frac{15}{a} + \frac{20a}{b} - \frac{5}{a^2b}$$

YOUR TURN

Distribute and simplify. Leave no negative exponents in your answers.

13 $x^{-3}(3x^5+6x^4-2x^3)$

14 $ab^{-2}(4ab^2-a^2b+3b^3)$

15 $4y^{-1}z^{-4}(3y^3z^6+2y^2z^5-4yz^4)$

16 $m^{-3}n^{-2}(m^{-3}n^{-2}+m^{-2}n^{-3})$

Working with Fractional Powers

Exponents that are fractions work the same way as exponents that are integers. When multiplying factors with the same base, the exponents are added together. The only hitch is that the fractions must have the same denominator in order to be added. (The rules don't change just because the fractions are exponents.)

Distribute and simplify: $x^{\frac{1}{4}}y^{\frac{2}{3}}\left(x^{\frac{1}{2}}+x^{\frac{3}{4}}y^{\frac{1}{3}}-y^{-\frac{1}{3}}\right)$.

Follow these steps to solve:

1. **Multiply the factor times each term:**

$$x^{\frac{1}{4}}y^{\frac{2}{3}}\left(x^{\frac{1}{2}}+x^{\frac{3}{4}}y^{\frac{1}{3}}-y^{-\frac{1}{3}}\right)=x^{\frac{1}{4}}y^{\frac{2}{3}}\cdot x^{\frac{1}{2}}+x^{\frac{1}{4}}y^{\frac{2}{3}}\cdot x^{\frac{3}{4}}y^{\frac{1}{3}}-x^{\frac{1}{4}}y^{\frac{2}{3}}\cdot y^{-\frac{1}{3}}$$

2. **Rearrange the variables and add the exponents:**

$$=x^{\frac{1}{4}}x^{\frac{1}{2}}y^{\frac{2}{3}}+x^{\frac{1}{4}}x^{\frac{3}{4}}y^{\frac{2}{3}}y^{\frac{1}{3}}-x^{\frac{1}{4}}y^{\frac{2}{3}}y^{-\frac{1}{3}}=x^{\frac{1}{4}+\frac{1}{2}}y^{\frac{2}{3}}+x^{\frac{1}{4}+\frac{3}{4}}y^{\frac{2}{3}+\frac{1}{3}}-x^{\frac{1}{4}}y^{\frac{2}{3}-\frac{1}{3}}$$

3. **Finish up by adding the fractional exponents:**

$$x^{\frac{1}{4}+\frac{1}{2}}y^{\frac{2}{3}}+x^{\frac{1}{4}+\frac{3}{4}}y^{\frac{2}{3}+\frac{1}{3}}-x^{\frac{1}{4}}y^{\frac{2}{3}-\frac{1}{3}}$$
$$=x^{\frac{3}{4}}y^{\frac{2}{3}}+x^{1}y^{1}-x^{\frac{1}{4}}y^{\frac{1}{3}}$$
$$=x^{\frac{3}{4}}y^{\frac{2}{3}}+xy-x^{\frac{1}{4}}y^{\frac{1}{3}}$$

REMEMBER

Radicals can be changed to expressions with fractions as exponents. This is handy when you want to combine terms with the same bases and you have some of the bases under radicals:

- $\sqrt{x}=x^{\frac{1}{2}}$
- $\sqrt{xy}=\sqrt{x}\sqrt{y}=x^{\frac{1}{2}}y^{\frac{1}{2}}$
- $\sqrt{x^3}=\left(x^3\right)^{\frac{1}{2}}=x^{\frac{3}{2}}$
- $\sqrt[n]{a}=a^{\frac{1}{n}}$ and $\sqrt[n]{a^m}=a^{\frac{m}{n}}$

Distribution is easier when you have radicals in the problem if you first change everything to fractional exponents. (Turn to Chapter 5 for more on exponential operations within radicals.)

REMEMBER

The exponent rule for raising a product in parentheses to a power is to multiply each power in the parentheses by the outside power — for example, $(x^4y^3)^2=x^8y^6$.

EXAMPLE

Q. Simplify by distributing: $\sqrt{xy^3}\left(\sqrt{x^5y}-\sqrt{xy^7}\right)$.

A. Follow these steps.

1. **Change the radical notation to fractional exponents:**

$$\sqrt{xy^3}\left(\sqrt{x^5y}-\sqrt{xy^7}\right)=\left(xy^3\right)^{\frac{1}{2}}\left[\left(x^5y\right)^{\frac{1}{2}}-\left(xy^7\right)^{\frac{1}{2}}\right]$$

2. **Raise the powers of the factor and terms inside the parentheses:**

$$\left(xy^3\right)^{\frac{1}{2}}\left[\left(x^5y\right)^{\frac{1}{2}}-\left(xy^7\right)^{\frac{1}{2}}\right]=x^{\frac{1}{2}}y^{\frac{3}{2}}\left[x^{\frac{5}{2}}y^{\frac{1}{2}}-x^{\frac{1}{2}}y^{\frac{7}{2}}\right]$$

3. **Distribute the outside factor over each term within the parentheses:**

$$x^{\frac{1}{2}}y^{\frac{3}{2}}\left[x^{\frac{5}{2}}y^{\frac{1}{2}}-x^{\frac{1}{2}}y^{\frac{7}{2}}\right]=x^{\frac{1}{2}}y^{\frac{3}{2}}\left(x^{\frac{5}{2}}y^{\frac{1}{2}}\right)-x^{\frac{1}{2}}y^{\frac{3}{2}}\left(x^{\frac{1}{2}}y^{\frac{7}{2}}\right)$$

4. **Add the exponents of the variables:**

$$x^{\frac{1}{2}}y^{\frac{3}{2}}\left(x^{\frac{5}{2}}y^{\frac{1}{2}}\right)-x^{\frac{1}{2}}y^{\frac{3}{2}}\left(x^{\frac{1}{2}}y^{\frac{7}{2}}\right)=x^{\frac{1}{2}}x^{\frac{5}{2}}y^{\frac{3}{2}}y^{\frac{1}{2}}-x^{\frac{1}{2}}x^{\frac{1}{2}}y^{\frac{3}{2}}y^{\frac{7}{2}}$$
$$=x^{\frac{6}{2}}y^{\frac{4}{2}}-x^{\frac{2}{2}}y^{\frac{10}{2}}$$

5. **Simplify the fractional exponents:**

$$x^{\frac{6}{2}}y^{\frac{4}{2}}-x^{\frac{2}{2}}y^{\frac{10}{2}}=x^3y^2-x^1y^5=x^3y^2-xy^5$$

YOUR TURN

Distribute and simplify.

17 $x^{3/2}(x^2+x^{1/2}+3x^{-1/2}+2)$	18 $y^{4/3}(3y^{2/3}-2y^{-1/3})$
19 $z^{-1/4}(z^{1/4}+11z^{5/4}-6z^{9/4})$	20 $a^{5/6}(a^{-1/3}+a^{-1/2})$

Distributing More Than One Term

The preceding sections in this chapter describe how to distribute one term over several others. This section shows you how to distribute a *binomial* (a polynomial with two terms). You also discover how to distribute polynomials with three or more terms.

The word *polynomial* comes from *poly* meaning "many" and *nomen* meaning "name" or "designation." A polynomial is an algebraic expression with one or more terms in it. For example, a polynomial with one term is a *monomial;* a polynomial with two terms is a *binomial.* If there are three terms, it's a *trinomial.*

Distributing binomials

Distributing two terms (a *binomial*) over several terms amounts to just applying the distribution process twice. The following steps tell you how to distribute a binomial over some polynomial:

1. **Break the binomial into its two terms.**
2. **Distribute each term of the binomial over the other factor.**

3. **Perform the distributions you've created.**

4. **Simplify and combine any like terms.**

EXAMPLE

Q. Multiply using distribution: $(x^2+1)(y-2)$.

A. Follow these steps.

1. **Break the binomial into its two terms.**

   In this case, $(x^2+1)(y-2)$, break the first binomial into its two terms, x^2 and 1.

2. **Distribute each term over the other factor.**

   Multiply the first term, x^2, times the second binomial, and multiply the second term, 1, times the second binomial.

   $x^2(y-2)+1(y-2)$

3. **Perform the two distributions.**

   $x^2(y-2)+1(y-2)=x^2y-2x^2+y-2$

4. **Simplify and combine any like terms.**

   In this case, nothing can be combined; none of the terms are alike.

Now that you have the idea, try walking through a polynomial distribution that has variables in all the terms.

EXAMPLE

Q. Multiply using distribution: $(a^2+2b)(4a^2+3ab-2ab^2-b^2)$.

A. Follow these steps.

1. **Break the binomial into its two terms and multiply those terms times the second factor:**

   $a^2(4a^2+3ab-2ab^2-b^2)+2b(4a^2+3ab-2ab^2-b^2)$

2. **Perform the two distributions:**

   $a^2(4a^2)+a^2(3ab)-a^2(2ab^2)-a^2(b^2)+2b(4a^2)+2b(3ab)-2b(2ab^2)-2b(b^2)$

3. **Multiply and simplify:**

   $4a^4+3a^3b-2a^3b^2-a^2b^2+8a^2b+6ab^2-4ab^3-2b^3$

Again, there are no like terms to combine.

Distributing trinomials

A *trinomial* (a polynomial with three terms) can be distributed over another expression. Each term in the first factor is distributed separately over the second factor, and then the entire expression is simplified, combining anything that can be combined. This process can be extended to multiplying with any size polynomial. In this section, I show you trinomials and leave the extension to you when you run into anything larger.

The following problem introduces you to working through the distribution of trinomials.

EXAMPLE

Q. Multiply by distributing the first factor over the second: $(x+y+2)(x^2-2xy+y+1)$.

A. Distribute each term of the trinomial by multiplying them times the second factor:

$$x\left(x^2-2xy+y+1\right)+y\left(x^2-2xy+y+1\right)+2\left(x^2-2xy+y+1\right)$$

1. **Do the three distributions:**

$$x^3-2x^2y+xy+x+x^2y-2xy^2+y^2+y+2x^2-4xy+2y+2$$

2. **Simplify:**

$$\begin{aligned}&x^3-2x^2y+xy+x+x^2y-2xy^2+y^2+y+2x^2-4xy+2y+2\\&=x^3-2x^2y+x^2y+2x^2+x-2xy^2+y^2+xy-4xy+y+2y+2\\&=x^3-x^2y+2x^2+x-2xy^2+y^2-3xy+3y+2\end{aligned}$$

YOUR TURN

Distribute and simplify.

 21 $(x+3)(x^2-2x+1)$

 22 $(y^2-2)(1+2y-4y^2+7y^3)$

 23 $(z^4+1)(z^8-z^4+1)$

 24 $(x^2+2x-3)(x^2-4x+5)$

Curses, Foiled Again — Or Not

A common process found in algebra is that of multiplying two binomials together. A *binomial* is an expression with two terms, such as $x+7$. One possible way to multiply the binomials together is to distribute the two terms in the first binomial over the two terms in the second.

But some math whiz came up with a great acronym, *FOIL*, which translates to F for First, O for Outer, I for Inner, and L for Last. This acronym helps you save time and makes multiplying binomials easier. These letters refer to the terms' positions in the product of two binomials.

When you FOIL the product $(a+b)(c+d)$,

- The product of the **F**irst terms is *ac*.
- The product of the **O**uter terms is *ad*.
- The product of the **I**nner terms is *bc*.
- The product of the **L**ast terms is *bd*.

The result is then $ac+ad+bc+bd$. Usually *ad* and *bc* are like terms and can be combined.

EXAMPLE

Q. Use FOIL to multiply $(x-8)(x-9)$.

A. Using FOIL, multiply the First, the Outer, the Inner, and the Last, and then combine the like terms:

$$\begin{aligned}(x-8)(x-9) &= x\cdot x + x(-9) + (-8)x + (-8)(-9)\\ &= x^2 - 9x - 8x + 72\\ &= x^2 - 17x + 72\end{aligned}$$

Q. Use FOIL to multiply $(2y^2+3)(y^2-4)$.

A. Using FOIL, multiply the First, the Outer, the Inner, and the Last, and then combine the like terms:

$$\begin{aligned}(2y^2+3)(y^2-4) &= 2y^2\cdot y^2 + 2y^2(-4) + 3\cdot y^2 + 3(-4)\\ &= 2y^4 - 8y^2 + 3y^2 - 12\\ &= 2y^4 - 5y^2 - 12\end{aligned}$$

YOUR TURN

Use FOIL to multiply.

25 $(2x+1)(3x-2)$.

26 $(x-7)(3x+5)$.

27 $(x^2-2)(x^2-4)$.

28 $(3x+4y)(4x-3y)$.

Squaring Binomials

You can always use FOIL or the distributive law to square a binomial, but there's a helpful pattern that makes the work quicker and easier. You square the first and last terms, and then you put twice the product of the terms between the two squares.

REMEMBER

The squares of binomials are $(a+b)^2 = a^2+2ab+b^2$ and $(a-b)^2 = a^2-2ab+b^2$.

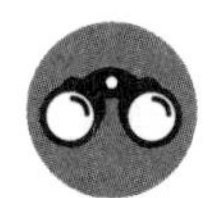

EXAMPLE

Q. $(x+5)^2 =$

A. The 10*x* is twice the product of the *x* and 5: $(x+5)^2 = (x)^2+2(x)(5)+(5)^2 = \mathbf{x^2+10x+25}$.

Q. $(3y-7)^2 =$

A. The 42*y* is twice the product of 3*y* and 7. And because the 7 is negative, the Inner and Outer products are, too: $(3y-7)^2 = (3y)^2-2(3y)(7)+(7)^2 = \mathbf{9y^2-42y+49}$.

YOUR TURN

29 $(x+3)^2 =$

30 $(2y-1)^2 =$

31 $(3a-2b)^2 =$

32 $(5xy+z)^2 =$

Multiplying the Sum and Difference of the Same Two Terms

When you multiply two binomials together, you can always just FOIL them. You save yourself some work, though, if you recognize when the terms in the two binomials are the same — except for the sign between them. If they're the sum and difference of the same two variables or numbers, then their product is just the difference between the squares of the two terms.

REMEMBER

The product of $(a+b)(a-b)$ is a^2-b^2.

This special product occurs because applying the FOIL method results in two opposite terms that cancel one another out: $(a+b)(a-b)=a^2-ab+ab-b^2=a^2-b^2$.

EXAMPLE

Q. $(x+5)(x-5)=$

A. x^2-25

Q. $(3ab^2-4)(3ab^2+4)=$

A. $9a^2b^4-16$. You square $3ab^2$, giving you $9a^2b^4$.

YOUR TURN

33 $(x+3)(x-3)=$

34 $(2x-7)(2x+7)=$

35 $(a^3-3)(a^3+3)=$

36 $(2x^2h+9)(2x^2h-9)=$

Specializing in Multiplication Matters

Powering Up Binomials

A binomial can be squared, cubed, and raised to any power. Yes, the results can get a bit lengthy, but there are several techniques that can help you. I describe the pattern in squaring binomials in an earlier section in this chapter. You'll now see how to cube binomials and make use of Pascal's Triangle to make life easier when doing even higher powers.

Cubing binomials

To *cube* something in algebra is to multiply it by itself and then multiply the result by itself again. When cubing a binomial, you have a couple of options. With the first option, you square the binomial and then multiply the original binomial times the square. This process involves distributing and then combining the like terms. Not a bad idea, but I have a better one.

When two binomials are cubed, two patterns occur.

- In the first pattern, the *coefficients* (numbers in front of and multiplying each term) in the answer start out as 1-3-3-1. The first coefficient is 1, the second is 3, the third is 3, and the last is 1.
- In the second pattern, the powers on the variables decrease and increase by ones. The powers of the first term in the binomial go down by one with each step, and the powers of the second term go up by one each time.

REMEMBER

To cube a binomial, follow this rule:

The cube of $(a+b)$, $(a+b)^3$, is $a^3+3a^2b+3ab^2+b^3$.

When one or more of the variables has a coefficient, the powers of the coefficient get incorporated into the pattern, and the 1-3-3-1 seems to disappear.

EXAMPLE

Q. $(y+4)^3=$

A. The answer is built by incorporating the two patterns — the 1-3-3-1 of the coefficients and powers of the two terms: $(y+4)^3=y^3+3y^2(4^1)+3y(4^2)+4^3=y^3+12y^2+48y+64$. Notice how the powers of *y* go down one step each term. Also, the powers of 4, starting with the second term, go up one step each time. The last part of using this method is to simplify each term. The 1-3-3-1 pattern gets lost when you do the simplification, but it's still part of the answer — just hidden. ***Note:*** In binomials containing subtraction, the terms in the answer will have alternating signs: +, −, +, −.

Q. $(2x-3)^3=$

A. Starting with the 1-3-3-1 pattern and adding in the multipliers:

$$(2x-3)^3=(2x)^3+3(2x)^2(-3)^1+3(2x)(-3)^2+(-3)^3=\mathbf{8x^3-36x^2+54x-27}$$

YOUR TURN

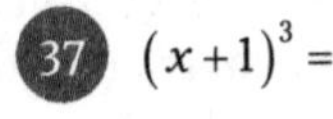

37 $(x+1)^3=$

38 $(y-2)^3=$

39 $(3z+1)^3 =$

40 $(5-2y)^3 =$

Raising Binomials to Higher Powers

The nice pattern for cubing binomials, 1-3-3-1, gives you a start on what the coefficients of the different terms are — at least what they start out to be before simplifying the terms. Similar patterns also exist for raising binomials to the fourth power, fifth power, and so on. They're all based on mathematical *combinations* and are easily pulled out with Pascal's Triangle. Check out Figure 9-1 to see a small piece of Pascal's Triangle with the powers of the binomial identified.

Pascal's Triangle	Power
1	$(a+b)^0$
1 1	$(a+b)^1$
1 2 1	$(a+b)^2$
1 3 3 1	$(a+b)^3$
1 4 6 4 1	$(a+b)^4$
1 5 10 10 5 1	$(a+b)^5$

FIGURE 9-1: Pascal's Triangle can help you find powers of binomials.

How is Pascal's Triangle created? How can you take it further — to higher powers? Note that the numbers in each row (after the first two) are the sum of the two numbers diagonally above it. The 1's on each end occur, because you assume there's a 0 diagonally to the left or right above that number.

EXAMPLE

Q. Refer to Figure 9-1 and use the coefficients from the row for the fourth power of the binomial to raise $(x-3y)$ to the fourth power, $(x-3y)^4$.

A. Follow these steps.

1. **Insert the coefficients 1-4-6-4-1 from the row in Pascal's Triangle.**

   1 4 6 4 1

2. **Place the decreasing powers of the first term, x, after the coefficients.**

   $1x^4 \quad 4x^3 \quad 6x^2 \quad 4x^1 \quad 1$

   Technically, there's an x^0 with the last 1, but because it's equal to 1, you don't need it.

3. **Place the decreasing powers of $-3y$, moving from right to left, starting with the 1 on the left.**

   $1x^4 \quad 4x^3(-3y)^1 \quad 6x^2(-3y)^2 \quad 4x^1(-3y)^3 \quad 1(-3y)^4$

4. **Raise the factors to their respective powers.**

   $1x^4 \quad 4x^3(-3y) \quad 6x^2(9y^2)^2 \quad 4x^1(-27y^3) \quad 1(81y^4)$

5. **Simplify each term and write the final answer.**

   $(x-3y)^4 = x^4 - 12x^3y + 54x^2y^2 - 108xy^3 + 81y^4$

Q. Raise $(2x-1)$ to the 6th power, $(2x-1)^6$.

A. Follow these steps.

1. **The 6th power isn't shown on the Pascal's Triangle in Figure 9-1.** So find the next row by adding adjacent numbers in the row for 5th power — beginning and ending with 1.

   1 6 15 20 15 6 1

2. **Fill in the decreasing powers of the first term from left to right, and decreasing powers of the second term from right to left.** Then simplify.

$$
\begin{array}{lllllll}
1(2x)^6 & 6(2x)^5 & 15(2x)^4 & 20(2x)^3 & 15(2x)^2 & 6(2x)^1 & 1 \\
1(2x)^6 & 6(2x)^5(-1)^1 & 15(2x)^4(-1)^2 & 20(2x)^3(-1)^3 & 15(2x)^2(-1)^4 & 6(2x)^1(-1)^5 & 1(-1)^6 \\
1(64x^6) & 6(32x^5)(-1) & 15(16x^4)(1) & 20(8x^3)(-1) & 15(4x^2)(1) & 6(2x)(-1) & 1(1)
\end{array}
$$

$(2x-1)^6 = 64x^6 - 192x^5 + 240x^4 - 160x^3 + 60x^2 - 12x + 1$

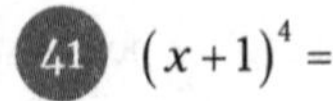

41 $(x+1)^4 =$

42 $(2y-1)^4 =$

43 $(z-1)^5 =$

44 $(3z+2)^5 =$

Creating the Sum and Difference of Cubes

A lot of what you do in algebra is to take advantage of patterns, rules, and quick tricks. Multiply the sum and difference of two values together, for example, and you get the difference of squares. Another pattern gives you the sum or difference of two cubes. These patterns are actually going to mean a lot more to you when you do the factoring of binomials, but for now, just practice with these patterns.

Here's the rule: If you multiply a binomial times a particular *trinomial* — one that has the squares of the two terms in the binomial, as well as the opposite of the product of them, such as $(y-5)(y^2+5y+25)$ — you get the sum or difference of two perfect cubes:

$$(a+b)(a^2-ab+b^2)=a^3+b^3 \quad \text{and} \quad (a-b)(a^2+ab+b^2)=a^3-b^3$$

So, using the second format, $(y-5)(y^2+5y+25)=y^3-125$, which is the difference of two cubes. In the trinomial, you see the square of y, the square of -5, and the opposite of the product of y and -5.

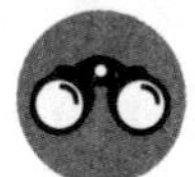

EXAMPLE

Q. $(y+3)(y^2-3y+9)=$

A. The product is y^3+27. If you don't believe me, multiply it out: Distribute the binomial over the trinomial and combine like terms. You find all the *middle* terms pairing up with their opposites and becoming 0, leaving just the two cubes.

Q. $(5y-1)(25y^2+5y+1)=$

A. The product is $125y^3-1$. The number 5 cubed is 125, and −1 cubed is −1. The other terms in the product drop out because of the opposites that appear there.

YOUR TURN

45 $(x-2)(x^2+2x+4)=$

46 $(y+1)(y^2-y+1)=$

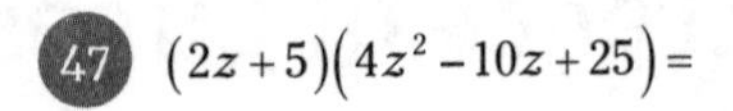

47 $(2z+5)(4z^2-10z+25)=$

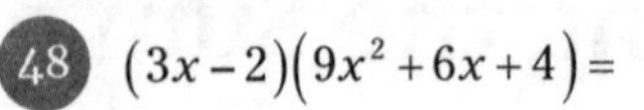

48 $(3x-2)(9x^2+6x+4)=$

Multiplying Conjugates

Two binomials are conjugates of one another when they are the sum and difference of the same two numbers. You see what happens in the earlier section of this chapter: Your result when multiplying by the sum and difference of the same two terms is the difference of two squares.

Identifying a conjugate and multiplying by it is important when you're dealing with radicals in the denominators of fractions. In Chapter 5, you see that by multiplying a square root times itself, the radical disappears! This is very important when dealing with fractions; you want your denominator to be a real number.

Conjugates are important not only when the denominator of your fraction has a radical, but also when that radical is part of a binomial. Multiply the denominator by its conjugate, and you get rid of the radical! The conjugate is just the same two terms but with a different sign.

EXAMPLE

Q. Multiply $x+\sqrt{3}$ by its conjugate.

A. x^2-3. The conjugate has subtraction instead of division, so you have

$$\left(x+\sqrt{3}\right)\left(x-\sqrt{3}\right)=x^2-\left(\sqrt{3}\right)^2=x^2-3$$

Q. Create an equivalent fraction by multiplying the denominator by its conjugate: $\frac{2}{y-\sqrt{6}}$.

A. The conjugate is $y+\sqrt{6}$. To create an equivalent fraction, you multiply both the numerator and denominator by that binomial.

$$\frac{2}{y-\sqrt{6}}\cdot\frac{y+\sqrt{6}}{y+\sqrt{6}}=\frac{2\left(y+\sqrt{6}\right)}{y^2-6}$$

Yes, you still have a radical in the numerator, but that's okay. You just don't want one in the denominator.

YOUR TURN

49 Multiply $\sqrt{5} - x$ by its conjugate.

50 Multiply $\sqrt{y} + \sqrt{2z}$ by its conjugate.

51 Create an equivalent fraction without a radical in the denominator: $\frac{x}{x + \sqrt{10}}$.

Practice Questions Answers and Explanations

1. $\mathbf{3x+2x^2-4x^3-5x^4}$. Multiply each term in the parentheses by x: $x(3)+x(2x)-x(4x^2)-x(5x^3)=3x+2x^2-4x^3-5x^4$.

2. $\mathbf{6y^4-10y^7+12y^{10}}$. Multiply each term in the parentheses by $2y^3$: $2y^3(3y)-2y^3(5y^4)+2y^3(6y^7)=6y^4-10y^7+12y^{10}$.

3. $\mathbf{-24a^3b^2+12a^2b^3-6ab^5+66ab^2}$. Multiply each term in the parentheses by $-6ab^2$: $-6ab^2(4a^2)-(-6ab^2)(2ab)+(-6ab^{2+)}(b^3)-(-6ab^2)(11)=-24a^3b^2+12a^2b^3-6ab^5+66ab^2$.

4. $\mathbf{3m^3n^2p-4mn^4p^3-m^2np^2}$. Multiply each term in the parentheses by mnp: $mnp(3m^2n)-mnp(4n^3p^2)-mnp(mp)=3m^3n^2p-4mn^4p^3-m^2np^2$.

5. $\mathbf{8x+12y-20z}$.

$$4(2x+3y-5z)=4(2x)+4(3y)-4(5z)=8x+12y-20z$$

Another way to deal with multiplying the third, negative term is to change $-5z$ to adding a negative term $+(-5z)$:

$$4(2x+3y-5z)=4(2x)+4(3y)+4(-5z)=8x+12y+(-20z)=8x+12y-20z$$

6. $\mathbf{-27+12a-6b+3c}$.

$$-3(9-4a+2b-c)=-3(9)-(-3)(4a)+(-3)(2b)-(-3)(c)=-27+12a-6b+3c$$

If you would rather change the expression in the parentheses into all addition before distributing, it would look like this:

$$-3(9-4a+2b-c)=-3[9+(-4a)+2b+(-c)]=-3(9)+(-3)(-4a)+(-3)(2b)+(-3)(-c)=$$
$$-27+12a-6b+3c$$

7. $\mathbf{3x-2y+11z-\frac{3}{2}}$.

$$\frac{1}{2}(6x-4y+22z-3)=\frac{1}{\cancel{2}}(\cancel{6}^3x)-\frac{1}{\cancel{2}}(\cancel{4}^2y)+\frac{1}{\cancel{2}}(\cancel{22}^{11}z)-\frac{1}{2}(3)$$
$$=3x-2y+11z-\frac{3}{2}$$

8. $\mathbf{-4a-10b+\frac{28c}{3}+d}$.

$$-12\left(\frac{a}{3}+\frac{5b}{6}-\frac{7c}{9}-\frac{d}{12}\right)$$
$$=-\cancel{12}^4\left(\frac{a}{\cancel{3}}\right)+\left(-\cancel{12}^2\right)\left(\frac{5b}{\cancel{6}}\right)-\left(-\cancel{12}^4\right)\left(\frac{7c}{\cancel{9}^3}\right)-\left(-\cancel{12}\right)\left(\frac{d}{\cancel{12}}\right)$$
$$=-4a-10b+\frac{28c}{3}+d$$

9. $\mathbf{8x^4-3x^3+2x^2-5x}$.

$$x(8x^3-3x^2+2x-5)=x(8x^3)-x(3x^2)+x(2x)-x(5)=8x^4-3x^3+2x^2-5x$$

10 $\mathbf{2x^3y^3 + 3x^3y^2z + x^2y^3z^3}$.

$$x^2y\left(2xy^2 + 3xyz + y^2z^3\right) = x^2y\left(2xy^2\right) + x^2y\left(3xyz\right) + x^2y\left(y^2z^3\right)$$
$$= 2x^3y^3 + 3x^3y^2z + x^2y^3z^3$$

11 $\mathbf{-12y^5 + 8y^3 - 20y^2 + 20y}$.

$$-4y\left(3y^4 - 2y^2 + 5y - 5\right) = \left(-4y\right)\left(3y^4 - 2y^2 + 5y - 5\right)$$
$$= \left(-4y\right)\left(3y^4\right) - \left(-4y\right)\left(2y^2\right) + \left(-4y\right)\left(5y\right) - \left(-4y\right)\left(5\right)$$
$$= -12y^5 + 8y^3 - 20y^2 + 20y$$

12 $\mathbf{x^2 + xy}$.

$$x^{1/2}y^{-1/2}\left(x^{3/2}y^{1/2} + x^{1/2}y^{3/2}\right) = x^{1/2}y^{-1/2}\left(x^{3/2}y^{1/2}\right) + x^{1/2}y^{-1/2}\left(x^{1/2}y^{3/2}\right)$$
$$= x^{1/2+3/2}y^{-1/2+1/2} + x^{1/2+1/2}y^{-1/2+3/2} = x^2y^0 + x^1y^1 = x^2 + xy$$

13 $\mathbf{3x^2 + 6x - 2}$.

$$x^{-3}\left(3x^5 + 6x^4 - 2x^3\right) = 3x^{-3+5} + 6x^{-3+4} - 2x^{-3+3}$$
$$= 3x^2 + 6x^1 - 2x^0 = 3x^2 + 6x - 2$$

After multiplying, the last term has the exponent $x^{-3+3} = x^0 = 1$.

14 $\mathbf{4a^2 - \frac{a^3}{b} + 3ab}$.

$$ab^{-2}\left(4ab^2 - a^2b + 3b^3\right) = 4a^{1+1}b^{-2+2} - a^{1+2}b^{-2+1} + 3ab^{-2+3}$$
$$= 4a^2b^0 - a^3b^{-1} + 3ab^1$$
$$= 4a^2 - \frac{a^3}{b} + 3ab$$

After multiplying, the first term has the exponent on b of $b^{-2+2} = b^0 = 1$.

15 $\mathbf{12y^2z^2 + 8yz - 16}$.

$$4y^{-1}z^{-4}\left(3y^3z^6 + 2y^2z^5 - 4yz^4\right) = 12y^{-1+3}z^{-4+6} + 8y^{-1+2}z^{-4+5} - 16y^{-1+1}z^{-4+4}$$
$$= 12y^2z^2 + 8y^1z^1 - 16y^0z^0$$
$$= 12y^2z^2 + 8yz - 16$$

After multiplying, the last term has the exponents $y^{-1+1}z^{-4+4} = y^0z^0 = 1$.

16 $\mathbf{\frac{1}{m^6n^4} + \frac{1}{m^5n^5}}$.

$$m^{-3}n^{-2}\left(m^{-3}n^{-2} + m^{-2}n^{-3}\right) = m^{-3+(-3)}n^{-2+(-2)} + m^{-3+(-2)}n^{-2+(-3)}$$
$$= m^{-6}n^{-4} + m^{-5}n^{-5}$$
$$= \frac{1}{m^6n^4} + \frac{1}{m^5n^5}$$

17. $\mathbf{x^{7/2} + x^2 + 3x + 2x^{3/2}}$. Adding the exponents when multiplying gives you:

$$x^{3/2+4/2} + x^{3/2+1/2} + 3x^{3/2-1/2} + 2x^{3/2} = x^{7/2} + x^{4/2} + 3x^{2/2} + 2x^{3/2}.$$

18. $\mathbf{3y^2 - 2y}$. Adding the exponents when multiplying gives you:

$$3y^{4/3+2/3} - 2y^{4/3-1/3} = 3y^{6/3} - 2y^{3/3}.$$

19. $\mathbf{1 + 11z - 6z^2}$. Adding the exponents when multiplying gives you:

$$z^{-1/4+1/4} + 11z^{-1/4+5/4} - 6z^{-1/4+9/4} = z^0 + 11z^{4/4} - 6z^{8/4}.$$

20. $\mathbf{a^{1/2} + a^{1/3}}$. Adding the exponents when multiplying gives you:

$$a^{5/6-1/3} + a^{5/6-1/2} = a^{5/6-2/6} + a^{5/6-3/6} = a^{3/6} + a^{2/6}.$$

21. $\mathbf{x^3 + x^2 - 5x + 3}$. Distribute the x over the three terms in the second parentheses and then the 3 over the same three terms. Simplify by combining like terms.

$$x(x^2 - 2x + 1) + 3(x^2 - 2x + 1) = x^3 - 2x^2 + x + 3x^2 - 6x + 3 = x^3 - 2x^2 + 3x^2 + x - 6x + 3 =$$
$$x^3 + x^2 - 5x + 3$$

22. $\mathbf{-2 - 4y + 9y^2 - 12y^3 - 4y^4 + 7y^5}$. Distribute the y^2 over the four terms in the second parentheses and then the -2 over the same four terms. Simplify by combining like terms.

$$y^2(1 + 2y - 4y^2 + 7y^3) - 2(1 + 2y - 4y^2 + 7y^3) = y^2 + 2y^3 - 4y^4 + 7y^5 - 2 - 4y + 8y^2 - 14y^3$$
$$= -2 - 4y + y^2 + 8y^2 + 2y^3 - 14y^3 - 4y^4 + 7y^5 = -2 - 4y + 9y^2 - 12y^3 - 4y^4 + 7y^5$$

23. $\mathbf{z^{12} + 1}$. Distribute the z^4 over the three terms in the second parentheses and then the 1 over the same three terms. Simplify by combining like terms.

$$z^4(z^8 - z^4 + 1) + 1(z^8 - z^4 + 1) = z^{12} - z^8 + z^4 + z^8 - z^4 + 1 = z^{12} - z^8 + z^8 + z^4 - z^4 + 1 = z^{12} + 1$$

24. $\mathbf{x^4 - 2x^3 - 6x^2 + 22x - 15}$. Distribute the x^2 over the three terms in the second parentheses, then the $2x$ over the same three terms, and, finally, the -3 over those same three terms. Simplify by combining like terms.

$$x^2(x^2 - 4x + 5) + 2x(x^2 - 4x + 5) - 3(x^2 - 4x + 5) = x^4 - 4x^3 + 5x^2 + 2x^3 - 8x^2 + 10x - 3x^2 + 12x - 15$$
$$= x^4 - 4x^3 + 2x^3 + 5x^2 - 8x^2 - 3x^2 + 10x + 12x - 15 = x^4 - 2x^3 - 6x^2 + 22x - 15$$

25. $\mathbf{6x^2 - x - 2}$.

$$(2x + 1)(3x - 2) = 6x^2 - 4x + 3x - 2 = 6x^2 - x - 2$$

26. $\mathbf{3x^2 - 16x - 35}$.

$$(x - 7)(3x + 5) = 3x^2 + 5x - 21x - 35 = 3x^2 - 16x - 35$$

27. $\mathbf{x^4 - 6x^2 + 8}$.

$$(x^2 - 2)(x^2 - 4) = x^4 - 4x^2 - 2x^2 + 8 = x^4 - 6x^2 + 8$$

Specializing in Multiplication Matters

(28) $\mathbf{12x^2 + 7xy - 12y^2}$.

$$(3x+4y)(4x-3y) = 12x^2 - 9xy + 16xy - 12y^2 = 12x^2 + 7xy - 12y^2$$

(29) $\mathbf{x^2 + 6x + 9}$

$$(x+3)^2 = x^2 + 2(x)(3) + 3^2 = x^2 + 6x + 9$$

(30) $\mathbf{4y^2 - 4y + 1}$

$$(2y-1)^2 = (2y)^2 - 2(2y)(1) + 1^2 = 4y^2 - 4y + 1$$

(31) $\mathbf{9a^2 - 12ab + 4b^2}$

$$(3a-2b)^2 = (3a)^2 - 2(3a)(2b) + (2b)^2 = 9a^2 - 12ab + 4b^2$$

(32) $\mathbf{25x^2y^2 + 10xyz + z^2}$

$$(5xy+z)^2 = (5xy)^2 + 2(5xy)(z) + z^2 = 25x^2y^2 + 10xyz + z^2$$

(33) $\mathbf{x^2 - 9}$. The product of the two last terms is a negative number.

(34) $\mathbf{4x^2 - 49}$. Don't forget to square both the 2 and the x.

(35) $\mathbf{a^6 - 9}$. Raise a power to a power.

$$(a^3-3)(a^3+3) = (a^3)^2 - 3^2 = a^6 - 9$$

(36) $\mathbf{4x^4h^2 - 81}$. All three factors in the first term are squared.

$$(2x^2h+9)(2x^2h-9) = (2x^2h)^2 - 9^2 = 4x^4h^2 - 81$$

(37) $\mathbf{x^3 + 3x^2 + 3x + 1}$.

$$(x+1)^3 = x^3 + 3(x^2)1 + 3(x)1^2 + 1^3 = x^3 + 3x^2 + 3x + 1$$

(38) $\mathbf{y^3 - 6y^2 + 12y - 8}$.

$$\begin{aligned}(y-2)^3 &= y^3 + 3(y^2)(-2) + 3(y)(-2)^2 + (-2)^3 \\ &= y^3 - 6y^2 + 12y - 8\end{aligned}$$

(39) $\mathbf{27z^3 + 27z^2 + 9z + 1}$.

$$(3z+1)^3 = (3z)^3 + 3(3z)^2(1) + 3(3z)(1)^2 + 1^3 = 27z^3 + 27z^2 + 9z + 1$$

(40) $\mathbf{125 - 150y + 60y^2 - 8y^3}$.

$$\begin{aligned}(5-2y)^3 &= 5^3 + 3(5)^2(-2y) + 3(5)(-2y)^2 + (-2y)^3 \\ &= 125 - 150y + 60y^2 - 8y^3\end{aligned}$$

41. $\mathbf{x^4 + 4x^3 + 6x^2 + 4x + 1.}$

$$(x+1)^4 = x^4 + 4x^3(1) + 6x^2(1)^2 + 4x(1)^3 + 1^4 = x^4 + 4x^3 + 6x^2 + 4x + 1$$

42. $\mathbf{16y^4 - 32y^3 + 24y^2 - 8y + 1.}$

$$\begin{aligned}(2y-1)^4 &= [2y+(-1)]^4\\ &= (2y)^4 + 4(2y)^3(-1) + 6(2y)^2(-1)^2 + 4(2y)(-1)^3 + (-1)^4\\ &= 16y^4 - 32y^3 + 24y^2 - 8y + 1\end{aligned}$$

43. $\mathbf{z^5 - 5z^4 + 10z^3 - 10z^2 + 5z - 1.}$

$$(z-1)^5 = [z+(-1)]^5$$

$$= z^5 + 5z^4(-1) + 10z^3(-1)^2 + 10z^2(-1)^3 + 5z(-1)^4 + (-1)^5$$

(Use Pascal line 1-5-10-10-5-1.)

$$= z^5 - 5z^4 + 10z^3 - 10z^2 + 5z - 1$$

44. $\mathbf{243z^5 + 810z^4 + 1{,}080z^3 + 720z^2 + 240z + 32.}$

$$\begin{aligned}(3z+2)^5 &= (3z)^5 + 5(3z)^4(2) + 10(3z)^3(2)^2 + 10(3z)^2(2)^3 + 5(3z)(2)^4 + 2^5\\ &= 243z^5 + 10(81z^4) + 40(27z^3) + 80(9z^2) + 80(3z) + 32\\ &= 243z^5 + 810z^4 + 1{,}080z^3 + 720z^2 + 240z + 32\end{aligned}$$

45. $\mathbf{x^3 - 8.}$ It's the cube of x and the cube of -2.

46. $\mathbf{y^3 + 1.}$ It's the cube of y and the cube of 1.

47. $\mathbf{8z^3 + 125.}$

$$(2z+5)(4z^2 - 10z + 25) = (2z)^3 + 5^3 \text{ by } (a+b)(a^2 - ab + b^2) = a^3 + b^3$$

with $a = 2z$ and $b = 5$.

$$= 8z^3 + 125$$

48. $\mathbf{27x^3 - 8.}$

$$(3x-2)(9x^2 + 6x + 4) = (3x)^3 - 2^3 \text{ by} (a-b)(a^2 + ab + b^2) = a^3 - b^3$$

with $a = 3x$ and $b = 2$.

$$= 27x^3 - 8$$

49. $\mathbf{5 - x^2.}$ Multiply by $\sqrt{5} + x$ to get $(\sqrt{5} - x)(\sqrt{5} + x) = 5 - x^2$.

50. $\mathbf{y - 2z.}$ Multiply by $\sqrt{y} - \sqrt{2z}$ to get $(\sqrt{y} + \sqrt{2z})(\sqrt{y} - \sqrt{2z}) = y - 2z$.

51. $\mathbf{\dfrac{x^2 - x\sqrt{10}}{x^2 - 10}}$. Multiply by $\dfrac{x - \sqrt{10}}{x - \sqrt{10}}$.

$$\frac{x}{x+\sqrt{10}} \cdot \frac{x-\sqrt{10}}{x-\sqrt{10}} = \frac{x(x-\sqrt{10})}{(x+\sqrt{10})(x-\sqrt{10})} = \frac{x^2 - x\sqrt{10}}{x^2 - 10}$$

If you're ready to test your skills a bit more, take the following chapter quiz that incorporates all the chapter topics.

Whaddya Know? Chapter 9 Quiz

Quiz time! Complete each problem to test your knowledge on the various topics covered in this chapter. You can then find the solutions and explanations in the next section.

1. Distribute and simplify. Leave no negative exponents. $3x^2\left(x^2-2x+7\right)=$
2. Use FOIL to find the product. $(y-3)(y+7)=$
3. Raise the binomial to the indicated power. $(x+4)^3=$
4. Distribute and simplify. Leave no negative exponents. $\left(x^2+4x+16\right)(x-4)=$
5. Distribute and simplify. Leave no negative exponents. $6x(x-1)-(x-1)=$
6. Use FOIL to find the product. $(3pq+7)(3pq-7)=$
7. Distribute and simplify. Leave no negative exponents. $5a^{-1}b^{-2}c\left(-4a^3b^3+3ab^4c^2\right)=$
8. Distribute and simplify. Leave no negative exponents. $\left(2a^2b+1\right)\left(3ab^2-1\right)=$
9. Create an equivalent fraction with no radical in the denominator: $\frac{7}{\sqrt{x}-2}$.
10. Use FOIL to find the product. $(3x+4)(x+1)=$
11. Distribute and simplify. Leave no negative exponents. $x^{-3}y^{-2}\left(4x^2+3x^3y^3-2xy\right)=$
12. Raise the binomial to the indicated power. $(2a-3c)^2=$
13. Distribute and simplify. Leave no negative exponents. $-5ab\left(a^2+ab+b^3+1\right)=$
14. Distribute and simplify. Leave no negative exponents. $\sqrt{xy}\left(\sqrt{x^3y}-\sqrt{xy^5}\right)=$
15. Raise the binomial to the indicated power. $(2y-3)^4=$
16. Distribute and simplify. Leave no negative exponents. $x^{1/3}\left(3x^{2/3}-x^{4/3}+2x^{-1/3}\right)=$

Answers to Chapter 9 Quiz

1. $\mathbf{3x^4 - 6x^3 + 21x^2}$.

2. $\mathbf{y^2 + 4y - 21}$. $(y-3)(y+7) = y^2 + 7y - 3y - 21 = y^2 + 4y - 21$

3. $\mathbf{x^3 + 12x^2 + 48x + 64}$.

   $$(x+4)^3 = 1x^3 + 3x^2 \cdot 4^1 + 3x^1 \cdot 4^2 + 1 \cdot 4^3 = x^3 + 12x^2 + 48x + 64$$

4. $\mathbf{x^3 - 64}$.

   $$\begin{aligned}(x^2+4x+16)(x-4) &= x^2(x-4) + 4x(x-4) + 16(x-4) \\ &= x^3 - 4x^2 + 4x^2 - 16x + 16x - 64 \\ &= x^3 - 64\end{aligned}$$

5. $\mathbf{6x^2 - 7x + 1}$.

   $$6x(x-1) - (x-1) = 6x^2 - 6x - x + 1 = 6x^2 - 7x + 1$$

6. $\mathbf{9p^2q^2 - 49}$. $(3pq+7)(3pq-7) = 9p^2q^2 - 21pq + 21pq - 49 = 9p^2q^2 - 49$

7. $\mathbf{-20a^2bc + 15b^2c^3}$.

   $$\begin{aligned}5a^{-1}b^{-2}c\left(-4a^3b^3 + 3ab^4c^2\right) &= -20a^{-1+3}b^{-2+3}c + 15a^{-1+1}b^{-2+4}c^{1+2} \\ &= -20a^2b^1c + 15a^0b^2c^3 = -20a^2bc + 15b^2c^3\end{aligned}$$

8. $\mathbf{6a^3b^3 - 2a^2b + 3ab^2 - 1}$.

   $$(2a^2b+1)(3ab^2-1) = 2a^2b(3ab^2-1) + 1(3ab^2-1) = 6a^3b^3 - 2a^2b + 3ab^2 - 1$$

9. $\mathbf{\dfrac{7\left(\sqrt{x}+2\right)}{x-4}}$.

   $$\frac{7}{\sqrt{x}-2} \cdot \frac{\sqrt{x}+2}{\sqrt{x}+2} = \frac{7\left(\sqrt{x}+2\right)}{x-4} = \frac{7\sqrt{x}+14}{x-4}$$

10. $\mathbf{3x^2 + 7x + 4}$. $(3x+4)(x+1) = 3x^2 + 3x + 4x + 4 = 3x^2 + 7x + 4$

11. $\mathbf{\dfrac{4}{xy^2} + 3y - \dfrac{2}{x^2y}}$.

   $$\begin{aligned}x^{-3}y^{-2}\left(4x^2 + 3x^3y^3 - 2xy\right) &= 4x^{-3+2}y^{-2} + 3x^{-3+3}y^{-2+3} - 2x^{-3+1}y^{-2+1} \\ &= 4x^{-1}y^{-2} + 3x^0y^1 - 2x^{-2}y^{-1} \\ &= \frac{4}{xy^2} + 3y - \frac{2}{x^2y}\end{aligned}$$

12. $\mathbf{4a^2 - 12ac + 9c^2}$.

13. $\mathbf{-5a^3b - 5a^2b^2 - 5ab^4 - 5ab}$.

14) $x^2y^1 - x^1y^3 = x^2y - xy^3$.

$$\begin{aligned}\sqrt{xy}\left(\sqrt{x^3y} - \sqrt{xy^5}\right) &= (xy)^{1/2}\left((x^3y)^{1/2} - (xy^5)^{1/2}\right)\\ &= x^{1/2}y^{1/2}\left(x^{3/2}y^{1/2} - x^{1/2}y^{5/2}\right)\\ &= x^{1/2}y^{1/2}x^{3/2}y^{1/2} - x^{1/2}y^{1/2}x^{1/2}y^{5/2}\\ &= x^{1/2+3/2}y^{1/2+1/2} - x^{1/2+1/2}y^{1/2+5/2}\\ &= x^2y^1 - x^1y^3 = x^2y - xy^3\end{aligned}$$

15) $16y^4 - 96y^3 + 216y^2 - 216y + 81$.

$$\begin{aligned}(2y-3)^4 &= 1(2y)^4 + 4(2y)^3(-3)^1 + 6(2y)^2(-3)^2 + 4(2y)^1(-3)^3 + 1(-3)^4\\ &= 1\left(16y^4\right) + 4\left(8y^3\right)(-3) + 6\left(4y^2\right)(9) + 4(2y)(-27) + 1(81)\\ &= 16y^4 - 96y^3 + 216y^2 - 216y + 81\end{aligned}$$

16) $3x - x^{5/3} + 2$.

$$x^{1/3}\left(3x^{2/3} - x^{4/3} + 2x^{-1/3}\right) = 3x^{1/3+2/3} - x^{1/3+4/3} + 2x^{1/3-1/3} = 3x^1 - x^{5/3} + 2x^0 = 3x - x^{5/3} + 2$$

» Working with binomial and trinomial divisors

» Simplifying the process with synthetic division

Chapter 10
Dividing the Long Way to Simplify Algebraic Expressions

Using long division to simplify algebraic expressions with variables and constants has many similarities to performing long division with just numbers. The variables do add an interesting twist (besides making everything look like alphabet soup) — with the exponents and different letters to consider. But the division problem is still made up of a divisor, dividend, and quotient (what divides in, what's divided into, and the answer). And one difference between traditional long division and algebraic division is that, in algebra, you usually write the remainders as algebraic fractions.

Dividing by a Monomial

Dividing an expression by a *monomial* (one term) can go one of two ways.

» Every term in the expression is evenly divisible by the divisor.

» One or more terms in the expression don't divide evenly.

If a fraction divides evenly — if every term can be divided by the divisor — then the denominator and numerator have a common factor. For instance, in the first example in this section, the denominator, $6y$, divides every term in the numerator. To emphasize the common factor business, I first factor the numerator by dividing out the $6y$, and then I reduce the fraction.

As nice as it would be if algebraic expressions divided evenly every time, that isn't always the case. Often, you have one or more terms in the expression — in the fraction's numerator — that don't contain all the factors in the divisor (denominator). When this happens, the best strategy is to break up the problem into as many fractions as there are terms in the numerator. In the end, though, the method you use is pretty much dictated by what you want to do with the expression when you're done.

EXAMPLE

Q. Perform the division: $\frac{24y^2-18y^3+30y^4}{6y}=$

A. Each term in the numerator contains a factor matching the denominator.

$$\begin{aligned}\frac{24y^2-18y^3+30y^4}{6y}&=\frac{6y^2\left(4-3y+5y^2\right)}{6y}\\&=\frac{\cancel{6y}\cdot y\left(4-3y+5y^2\right)}{\cancel{6y}}\\&=y\left(4-3y+5y^2\right)=4y-3y^2+5y^3\end{aligned}$$

Q. Perform the division: $\frac{40x^4-32x^3+20x^2-12x+3}{4x}=$

A. The last term doesn't have a factor of $4x$, so you break up the numerator into separate fractions for the division.

$$\begin{aligned}&=\frac{40x^4-32x^3+20x^2-12x+3}{4x}\\&=\frac{40x^4}{4x}-\frac{32x^3}{4x}+\frac{20x^2}{4x}-\frac{12x}{4x}+\frac{3}{4x}\\&=\frac{{}^{10}\cancel{40}\cancel{x^4}^{3}}{\cancel{4}\cancel{x}}-\frac{{}^{8}\cancel{32}\cancel{x^3}^{2}}{\cancel{4}\cancel{x}}+\frac{{}^{5}\cancel{20}\cancel{x^2}^{1}}{\cancel{4}\cancel{x}}-\frac{{}^{3}\cancel{12}\cancel{x}}{\cancel{4}\cancel{x}}+\frac{3}{4x}\\&=10x^3-8x^2+5x-3+\frac{3}{4x}\end{aligned}$$

YOUR TURN

1 Perform the division: $\frac{4x^3-3x^2+2x}{x}=$

2 Perform the division: $\frac{8y^4+12y^5-16y^6+40y^8}{4y^4}=$

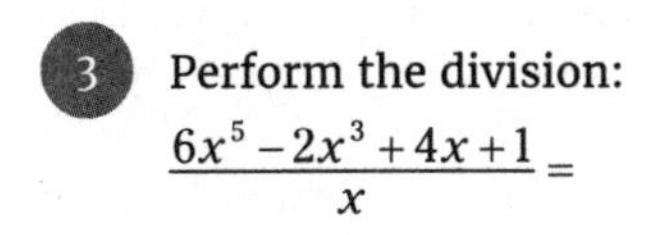

3 Perform the division:

$$\frac{6x^5 - 2x^3 + 4x + 1}{x} =$$

4 Perform the division:

$$\frac{15x^3y^4 + 9x^2y^2 - 12xy}{3xy^2} =$$

Dividing by a Binomial

Dividing by a *binomial* (two terms) in algebra means that those two terms, as a unit or grouping, have to divide into another expression. After dividing, if you find that the division doesn't have a remainder, then you know that the divisor was actually a factor of the original expression. When dividing a binomial into another expression, you always work toward getting rid of the *lead term* (the first term — the one with the highest power) in the original polynomial and then the new terms in the division process.

The example later in this section offers a clearer picture of this concept; it shows a dividend that starts with a third-degree term and is followed by terms in decreasing powers (second degree, first degree, and zero degree, which is a *constant* — just a number with no variable). If your dividend is missing any powers that are lower than the lead term, then you need to fill in the spaces with zeros to keep your division lined up.

Also, if you have a remainder, remember to write that remainder as the numerator of a fraction with the divisor in the denominator.

To divide by a binomial or trinomial (or higher degree), follow these steps:

1. **Put the terms of your divisor in order of decreasing powers.**
2. **Put the terms of your dividend in order of decreasing powers.** If a power is missing in the arrangement, then put in a term of 0 as a placeholder.
3. **Determine what must multiply the lead term in the divisor in order to obtain the current lead term in the dividend.** This becomes the first term of your quotient.
4. **Multiply the quotient term times the divisor and subtract from the dividend.**
5. **Continue until there are no remaining terms in the quotient.**

Q. $x-4\overline{)x^3-9x^2+27x-28} =$

EXAMPLE **A.** Follow these steps.

1. **The first term in the quotient is x^2. Multiply the divisor by x^2 and subtract from the dividend.**

$$
\begin{array}{rl}
 & x^2 \\
x-4 & \overline{)x^3-9x^2+27x-26} \\
 & \underline{x^3-4x^2} \\
 & \quad -5x^2
\end{array}
$$

2. **Bring down the rest of the dividend; then multiply the divisor by $-5x$, subtract, and bring down the last term in the quotient.**

$$
\begin{array}{rl}
 & x^2-5x \\
x-4 & \overline{)x^3-9x^2+27x-26} \\
 & \underline{x^3-4x^2} \\
 & \quad -5x^2+27x-26 \\
 & \quad \underline{-5x^2+20x} \\
 & \qquad\qquad 7x-26
\end{array}
$$

3. **Multiply the divisor by 7 and subtract.**

$$
\begin{array}{rl}
 & x^2-5x+\ 7 \\
x-4 & \overline{)x^3-9x^2+27x-26} \\
 & \underline{x^3-4x^2} \\
 & \quad -5x^2+27x-26 \\
 & \quad \underline{-5x^2+20x} \\
 & \qquad\qquad 7x-26 \\
 & \qquad\qquad \underline{7x-28} \\
 & \qquad\qquad\qquad 2
\end{array}
$$

4. **The remainder is 2, so you write your answer as $x^2-5x+7+\frac{2}{x-4}$.**

5. $x+2\overline{)x^3+7x^2+3x-14}=$

6. $x-3\overline{)x^4-2x^3-5x^2+7x-3}=$

7. $3x-4\overline{)12x^3-10x^2-17x+12}=$

8. $x^2+1\overline{)x^6-3x^5+x^4-2x^3-3x^2+x-3}=$

Dividing the Long Way to Simplify Algebraic Expressions

Dividing by Polynomials with More Terms

Even though dividing by monomials or binomials is the most commonly found division task in algebra, you may run across the occasional opportunity to divide by a *polynomial* with three or more terms.

The process isn't really much different from that used to divide binomials. You just have to keep everything lined up correctly and fill in the blanks if you find missing powers. Any remainder is written as a fraction.

Q. $(9x^6 - 4x^5 + 3x^2 - 1) \div (x^2 - 2x + 1) =$

EXAMPLE **A.** The fourth, third, and first powers are missing, so you put in zeros. Don't forget to distribute the negative sign when subtracting each product.

$$9x^4 + 14x^3 + 9x^2 + 24x + 32 + \frac{40x - 33}{x^2 - 2x + 1}$$

You write the final answer with the remainder as a fraction. Here's what the division looks like:

$$\begin{array}{r}
9x^4 + 14x^3 + 19x^2 + 24x + 32 \\
x^2 - 2x + 1 \overline{)\, 9x^6 - 4x^5 + 0 + \quad 0 + \quad 3x^2 + 0 - 1} \\
-\left(9x^6 - 18x^5 + 9x^4\right) \qquad\qquad\qquad\quad \\
\hline
14x^5 - 9x^4 + 0 + \; 3x^2 + 0 - 1 \\
-\left(14x^5 - 28x^4 + 14x^3\right) \qquad\qquad\quad \\
\hline
+19x^4 - 14x^3 + 3x^2 + 0 - 1 \\
-\left(19x^4 - 38x^3 + 19x^2\right) \qquad\quad \\
\hline
+24x^3 - 16x^2 + \; 0 \; - 1 \\
-\left(24x^3 - 48x^2 + 24x\right) \quad \\
\hline
32x^2 - 24x \; - 1 \\
-\left(32x^2 - 64x + 32\right) \\
\hline
+40x - 33
\end{array}$$

YOUR TURN

 $(x^4 - 2x^3 + x^2 - 7x - 2) \div (x^2 + 3x - 1) =$

10 $(x^6 + 6x^4 - 4x^2 + 21) \div (x^4 - x^2 + 3) =$

Simplifying Division Synthetically

Dividing polynomials by binomials is a very common procedure in algebra. The division process allows you to determine factors of an expression and roots of an equation. A quick, easy way of dividing a polynomial by a binomial of the form $x+a$ or $x-a$ is called *synthetic division.* Notice that in these two binomials, each variable has a coefficient of 1, the variable to the first degree, and a number being added or subtracted. For example, you can divide by $x+4$ or $x-7$.

If the coefficient of the variable isn't 1, then divide both terms by that coefficient to create the format you want.

These are the steps to use when performing synthetic division:

1. **Place the opposite of the number in the divisor (−1 in this case) in front of the problem, in a little offset that looks like ⌋.**

   Then write the coefficients of the dividend in order, using zeros to hold the place(s) of any powers that are missing.

2. **Bring down the first coefficient (put it below the horizontal line), and then multiply it times the number in front.**

   Write the result under the second coefficient and add the two numbers; put the result on the bottom.

3. **Take this new result (the −4 in this problem) and multiply it times the number in front; then add the answer to the next coefficient.**
4. **Repeat this multiply-add process all the way down the line.**
5. **The result on the bottom of your work is the list of coefficients in the answer — plus the remainder, if you have one.**

REMEMBER

To perform *synthetic division,* you use just the *opposite* of the number in the binomial.

EXAMPLE

Q. $(x^4 - 3x^3 + x - 4) \div (x + 1) =$

A. Follow the steps to solve:

1. The opposite of +1 goes in front. The coefficients are written in order.

   $\underline{-1}|\ 1\ \ -3\ \ \ 0\ \ \ 1\ \ -4$

2. Bring down the first coefficient; multiply it times the number in front. Write the result under the second coefficient, and add.

$$\begin{array}{r|rrrrr} -1 & 1 & -3 & 0 & 1 & -4 \\ & & -1 & & & \\ \hline & 1 & -4 & & & \end{array}$$

3. Take the -4 and multiply it by the number in front; add the answer to the next coefficient.

$$\begin{array}{r|rrrrr} -1 & 1 & -3 & 0 & 1 & -4 \\ & & -1 & 4 & & \\ \hline & 1 & -4 & 4 & & \end{array}$$

4. Repeat until finished.

$$\begin{array}{r|rrrrr} -1 & 1 & -3 & 0 & 1 & -4 \\ & & -1 & 4 & -4 & 3 \\ \hline & 1 & -4 & 4 & -3 & -1 \end{array}$$

The answer uses the coefficients, in order, starting with one degree less than the polynomial that was divided. The last number is the remainder, and it goes over the divisor in a fraction. So, to write the answer, the first 1 below the line corresponds to a 1 in front of x^3, then a -4 in front of x^2, and so on. The last -1 is the remainder, which is written in the numerator over the divisor, $x+1$:

$$x^3-4x^2+4x-3-\frac{1}{x+1}$$

See Chapter 13 for more on synthetic division and how it's used in factoring.

YOUR TURN

$(x^4-2x^3-4x^2+x+6)\div(x-3)=$

$(2x^4+x^3-7x^2+5)\div(x+2)=$

Practice Questions Answers and Explanations

1. $\mathbf{4x^2 - 3x + 2.}$

$$\frac{4x^3 - 3x^2 + 2x}{x} = \frac{\cancel{x}(4x^2 - 3x + 2)}{\cancel{x}} = 4x^2 - 3x + 2$$

2. $\mathbf{2 + 3y - 4y^2 + 10y^4.}$

$$\frac{8y^4 + 12y^5 - 16y^6 + 40y^8}{4y^4} = \frac{\cancel{4y^4}(2 + 3y - 4y^2 + 10y^4)}{\cancel{4y^4}}$$
$$= 2 + 3y - 4y^2 + 10y^4$$

3. $\mathbf{6x^4 - 2x^2 + 4 + \frac{1}{x}.}$

$$\frac{6x^5 - 2x^3 + 4x + 1}{x} = \frac{6x^5}{x} - \frac{2x^3}{x} + \frac{4x}{x} + \frac{1}{x}$$
$$= 6x^4 - 2x^2 + 4 + \frac{1}{x}$$

The last term does not have the common factor. Writing it as a fraction indicates a remainder.

4. $\mathbf{5x^2y^2 + 3x - \frac{4}{y}.}$

$$\frac{15x^3y^4 + 9x^2y^2 - 12xy}{3xy^2} = \frac{15x^3y^4}{3xy^2} + \frac{9x^2y^2}{3xy^2} - \frac{12xy}{3xy^2}$$
$$= \frac{{}^{5}\cancel{15}x^{\cancel{3}^{2}}y^{\cancel{4}^{2}}}{\cancel{3}\cancel{x}\cancel{y^2}} + \frac{{}^{3}\cancel{9}x^{\cancel{2}^{1}}\cancel{y^2}}{\cancel{3}\cancel{x}\cancel{y^2}} - \frac{{}^{4}\cancel{12}\cancel{x}\cancel{y}}{\cancel{3}\cancel{x}y^{\cancel{2}^{1}}} = 5x^2y^2 + 3x - \frac{4}{y}$$

The last term does not have the common factor. Writing it as a fraction indicates a remainder.

5. $\mathbf{x^2 + 5x - 7.}$

$$\begin{array}{r l}
 & \quad x^2 + 5x - 7 \\
x + 2 & \overline{\big)\, x^3 + 7x^2 + 3x - 14} \\
 & \underline{-(x^3 + 2x^2)} \\
 & \qquad 5x^2 + 3x \\
 & \qquad \underline{-(5x^2 + 10x)} \\
 & \qquad\qquad -7x - 14 \\
 & \qquad\qquad \underline{-(-7x - 14)} \\
 & \qquad\qquad\qquad 0
\end{array}$$

6) $\mathbf{x^3 + x^2 - 2x + 1}$.

$$
\begin{array}{r}
x^3 + x^2 - 2x + 1 \\
x-3 \,\overline{\big)\, x^4 - 2x^3 - 5x^2 + 7x - 3} \\
\underline{-(x^4 - 3x^3)} \qquad\qquad\qquad\; \\
x^3 - 5x^2 \qquad\qquad\; \\
\underline{-(x^3 - 3x^2)} \qquad\qquad\; \\
-2x^2 + 7x \qquad \\
\underline{-(-2x^2 + 6x)} \qquad \\
x - 3 \\
\underline{-(-x - 3)} \\
0
\end{array}
$$

7) $\mathbf{4x^2 + 2x - 3}$.

$$
\begin{array}{r}
4x^2 + 2x - 3 \\
3x-4 \,\overline{\big)\, 12x^3 - 10x^2 - 17x + 12} \\
\underline{-(12x^3 - 16x^2)} \qquad\qquad\quad \\
6x^2 - 17x \qquad \\
\underline{-(6x^2 - 8x)} \qquad \\
-9x + 12 \\
\underline{-(-9x + 12)} \\
0
\end{array}
$$

8) $\mathbf{x^4 - 3x^3 + x - 3}$.

$$
\begin{array}{r}
x^4 - 3x^3 + x - 3 \\
x^2+1 \,\overline{\big)\, x^6 - 3x^5 + x^4 - 2x^3 - 3x^2 + x - 3} \\
\underline{-(x^6 \qquad\quad + x^4)} \qquad\qquad\qquad\qquad\qquad\qquad \\
-3x^5 \qquad\quad - 2x^3 \qquad\qquad\qquad\qquad \\
\underline{-(-3x^5 \qquad\quad - 3x^3)} \qquad\qquad\qquad\qquad \\
x^3 - 3x^2 + x \qquad\quad \\
\underline{-(x^3 \qquad\quad + x)} \qquad\quad \\
-3x^2 \qquad\quad - 3 \\
\underline{-(-3x^2 \qquad\quad - 3)} \\
0
\end{array}
$$

9. $\mathbf{x^2 - 5x + 17 + \dfrac{-63x + 15}{x^2 + 3x - 1}}$.

$$
\begin{array}{r|l}
 & x^2 - 5x + 17 \\
\hline
x^2 + 3x - 1 & x^4 - 2x^3 + x^2 - 7x - 2 \\
 & -\left(x^4 + 3x^3 - x^2\right) \\
\hline
 & -5x^3 + 2x^2 - 7x \\
 & -\left(-5x^3 - 15x^2 + 5x\right) \\
\hline
 & 17x^2 - 12x - 2 \\
 & -\left(17x^2 + 51x - 17\right) \\
\hline
 & -63x + 15
\end{array}
$$

10. $\mathbf{x^2 + 7}$.

$$
\begin{array}{r|l}
 & x^2 + 7 \\
\hline
x^4 - x^2 + 3 & x^6 + 0 + 6x^4 + 0 - 4x^2 + 0 + 21 \\
 & -\left(x^6 - x^4 + 3x^2\right) \\
\hline
 & 7x^4 - 7x^2 + 21 \\
 & -\left(7x^4 - 7x^2 + 21\right) \\
\hline
 & 0
\end{array}
$$

11. $\mathbf{x^3 + x^2 - x - 2}$.

$$
\begin{array}{r|rrrrr}
3 & 1 & -2 & -4 & 1 & 6 \\
 & & 3 & 3 & -3 & -6 \\
\hline
 & 1 & 1 & -1 & -2 & 0
\end{array}
$$

$$
\text{So, } x^3 + x^2 - x - 2 + \frac{0}{x - 3}
$$
$$
= x^3 + x^2 - x - 2
$$

12. $\mathbf{2x^3 - 3x^2 - x + 2 + \dfrac{1}{x + 2}}$.

Use the following breakdown to solve the problem:

$$
\begin{array}{r|rrrrr}
-2 & 2 & 1 & -7 & 0 & 5 \\
 & & -4 & 6 & 2 & -4 \\
\hline
 & 2 & -3 & -1 & 2 & 1
\end{array}
$$

Write the remainder as a fraction with the divisor in the denominator of the fraction.

If you're ready to test your skills a bit more, take the following chapter quiz that incorporates all the chapter topics.

Whaddya Know? Chapter 10 Quiz

Quiz time! Complete each problem to test your knowledge on the various topics covered in this chapter. You can then find the solutions and explanations in the next section.

1. Use synthetic division: $\frac{x^4+4x^3-2x^2-4x+33}{x+3}=$
2. $\frac{6x^3-4x^2+10x-1}{2x}=$
3. $\frac{2x^4-x^3-12x^2+13x-2}{2x^2+5x-1}=$
4. Use synthetic division: $\frac{x^3-8}{x-2}=$
5. $\frac{x^3+8x^2+12x-9}{x+3}=$

Answers to Chapter 10 Quiz

1. $x^3 + x^2 - 5x + 11.$

   Change the +3 to −3. Drop the first 1 and then multiply, add, multiply, add. The remainder is 0. Write the quotient, starting with one degree smaller than that in the dividend.

$$\begin{array}{r|rrrrr} -3 & 1 & 4 & -2 & -4 & 33 \\ & & -3 & -3 & 15 & -33 \\ \hline & 1 & 1 & -5 & 11 & 0 \end{array}$$

2. $3x^2 - 2x + 5 - \frac{1}{2x}.$

$$\frac{6x^3 - 4x^2 + 10x - 1}{2x} = \frac{6x^3}{2x} - \frac{4x^2}{2x} + \frac{10x}{2x} - \frac{1}{2x} = 3x^2 - 2x + 5 - \frac{1}{2x}$$

3. $x^2 - 3x + 2.$

$$\begin{array}{r} x^2 - 3x + 2 \\ 2x^2 + 5x - 1 \overline{\smash{\big)} 2x^4 - x^3 - 12x^2 + 13x - 2} \\ \underline{2x^4 + 5x^3 - x^2} \\ -6x^3 - 11x^2 + 13x - 2 \\ \underline{-6x^3 - 15x^2 + 3x} \\ 4x^2 + 10x - 2 \\ \underline{4x^2 + 10x - 2} \\ 0 \end{array}$$

4. $x^2 + 2x + 4.$

   Change the −2 to +2. The second power and first power of x are missing, so put in 0's as place holders.

$$\begin{array}{r|rrrr} 2 & 1 & 0 & 0 & -8 \\ & & 2 & 4 & 8 \\ \hline & 1 & 2 & 4 & 0 \end{array}$$

5. $x^2 + 5x - 3.$

$$\begin{array}{r} x^2 + 5x - 3 \\ x + 3 \overline{\smash{\big)} x^3 + 8x^2 + 12x - 9} \\ \underline{x^3 + 3x^2} \\ 5x^2 + 12x - 9 \\ \underline{5x^2 + 15x} \\ -3x - 9 \\ \underline{-3x - 9} \\ 0 \end{array}$$

Dividing the Long Way to Simplify Algebraic Expressions

4
Factoring

Contents at a Glance

» Factoring all at once or in stages

» Using factors to reduce algebraic fractions

Chapter 11

Figuring on Factoring

You may believe in the bigger-is-better philosophy, which can apply to salaries, cookies, or houses, but it doesn't really work for algebra. For the most part, the opposite is true in algebra: Smaller numbers are easier and more comfortable to deal with than larger numbers.

In this chapter, you discover how to get to those smaller-is-better terms. You find out the basics of factoring and how factoring is related to division. The factoring patterns you see here carry over somewhat in more complicated expressions.

Factoring out the Greatest Common Factor

Factoring is another way of saying, "Rewrite this so everything is all multiplied together." You usually start out with two or more terms and have to determine how to rewrite them so they're all multiplied together in some way or another. And, oh yes, the two expressions have to be equal! Why all this fuss? You rewrite expressions as products — keeping the new results equivalent to the old — so that you can perform operations on the results. Fractions reduce more easily, equations are solved more easily, and answers are observed more easily when you can factor.

REVIEWING THE TERMS AND RULES

You'll understand factoring better if you have a firm handle on what things are called when I talk about factoring.

- **Term:** A group of number(s) and/or variable(s) connected to one another by multiplication and/or division and separated from other terms by addition or subtraction.
- **Factor:** Any of the values involved in a multiplication expression (term) that, when multiplied together, produce a result.
- **Coefficient:** A number that multiplies a variable and tells how many of the variable there are.
- **Constant:** A number or variable that never changes in value.
- **Relatively prime:** Terms that have no factors in common. If the only factor that numbers share in common is 1, then they're considered *relatively prime.*

EXAMPLE

Q. In the expression $5xy+4z-6$, identify the parts by the words that describe them.

A. In the expression $5xy+4z-6$, you see three *terms.* In the first term, $5xy$, three *factors* are all multiplied together. The 5 is usually referred to as the *coefficient.* The second term has two factors, 4 and z, and the third term contains just a *constant.* The first and second terms are *relatively prime* because they have no factors in common. (The number 4 is not prime, but it's *relatively prime* to 5.)

Factoring out numbers

Factoring is the opposite of distributing — think of it as "undistributing" (see Chapter 9 for more on distribution). When performing distribution, you multiply a series of terms by a common multiplier. Now, by factoring, you seek to find what a series of terms has in common and then take it away, dividing the common factor or multiplier out from each term. Think of each term as a numerator of a fraction, where you're finding the same denominator for each one. By factoring out, the common factor is put outside parentheses or brackets and all the results of the divisions are left inside.

An expression can be written as the product of the largest value that divides all the terms evenly times the results of the divisions: $ab+ac+ad=a(b+c+d)$.

Writing factoring as division

In the trinomial $16a-8b+40c^2$, 2 is a common factor. But 4 is also a common factor, as is 8. Here are the divisions of the terms by 2, 4, and 8:

$$\frac{16a}{2}-\frac{8b}{2}+\frac{40c^2}{2}=\frac{^{8}\cancel{16}a}{\cancel{2}}-\frac{^{4}\cancel{8}b}{\cancel{2}}+\frac{^{20}\cancel{40}c^2}{\cancel{2}}=8a-4b+20c^2$$

$$\frac{16a}{4}-\frac{8b}{4}+\frac{40c^2}{4}=\frac{^{4}\cancel{16}a}{\cancel{4}}-\frac{^{2}\cancel{8}b}{\cancel{4}}+\frac{^{10}\cancel{40}c^2}{\cancel{4}}=4a-2b+10c^2$$

$$\frac{16a}{8}-\frac{8b}{8}+\frac{40c^2}{8}=\frac{{}^{2}\cancel{16}a}{\cancel{8}}-\frac{{}^{1}\cancel{8}b}{\cancel{8}}+\frac{{}^{5}\cancel{40}c^2}{\cancel{8}}=2a-b+5c^2$$

You see that the final result, in each case, does not contain a fraction. For a number to be a *factor*, it should divide all the terms evenly. To show the results of factoring, you write the factor outside parentheses and the results of the division inside:

$$16a-8b+40c^2=2\left(8a-4b+20c^2\right)$$
$$16a-8b+40c^2=4\left(4a-2b+10c^2\right)$$
$$16a-8b+40c^2=8\left(2a-b+5c^2\right)$$

Outlining the factoring method

The absolutely *proper* way to factor an expression is to write the prime factorization of each of the numbers and look for the GCF. What's really more practical and quicker in the end is to look for the biggest factor that *you can easily recognize.* Factor it out and then see if the numbers in the parentheses need to be factored again. Repeat the division until the terms in the parentheses are relatively prime.

Here's how to use the repeated-division method to factor the expression $450x+540y-486z+216$. You see that the coefficient of each term is even, so divide each term by 2:

$$450x+540y-486z+216=2\left(225x+270y-243z+108\right)$$

The numbers in the parentheses are a mixture of odd and even, so you can't divide by 2 again. The numbers in the parentheses are all divisible by 3, but there's an even better choice: You may have noticed that the digits in the numbers in all the terms add up to 9. That's the rule for divisibility by 9, so 9 can divide each term evenly. (You find rules of divisibility in Chapter 8.) Thus,

$$2\left(225x+270y-243z+108\right)=2\left[9\left(25x+30y-27z+12\right)\right]$$

Now, multiply the 2 and 9 together to get

$$450x+540y-486z+216=18\left(25x+30y-27z+12\right)$$

You could have divided 18 into each term in the first place, but not many people know the multiplication table of 18. (It's a stretch even for me.) What about the coefficients of the numbers in the parentheses? None is a prime number. And several have factors in common. But there's no single factor that divides *all* the coefficients equally. The four coefficients are *relatively prime,* so you're finished with the factoring. You find more on repeated-division factoring later in this chapter, under "Unlocking combinations of numbers and variables."

FACTORING IN THE REAL WORLD

You usually use factoring when you need to reduce fractions or solve a quadratic equation. But one type of factoring comes to the rescue in several real-life situations.

For example, Stephanie just bought a mega-box of packaged healthy snacks and wants to put some in her room, some in her office, and some in her car. She has 18 packs of cherry bars, 24 packs of cranberry bars, 30 packs of peanut bars, and 42 packs of oatmeal-raisin bars. How can she divide the bars so there's a nice balance in each location?

Writing the numbers of bars as a sum, Stephanie has $18+24+30+42$ total. Each of the numbers is divisible by 2, 3, and 6, so the factorizations could be any of the following:

- $2(9+12+15+21)$
- $3(6+8+10+14)$
- $6(3+4+5+7)$

The last factorization shows you four relatively prime numbers within the parentheses. But Stephanie has only three locations to put her snacks in, so she'll go with 6 cherry, 8 cranberry, 10 peanut, and 14 oatmeal-raisin bars in each place.

EXAMPLE

Q. Determine the GCF of the coefficients of the terms $8a+12b+32c$. Then rewrite the expression as a product of that GCF and the factored terms.

A. The GCF of 8, 12, and 32 is 4. Writing the expression in factored form, you get $\mathbf{4(2a+3b+8c)}$. Note that the coefficients of the terms in the parentheses are now relatively prime.

Q. Determine the GCF of the coefficients of the terms $24xy-60xz+108yz$. Then rewrite the expression as a product of that GCF and the factored terms.

A. The GCF of 24, 60, and 108 is 12. Writing the expression in factored form, you get $12(2xy-5xz+9yz)$. Note that the coefficients of the terms in the parentheses are now relatively prime.

YOUR TURN

Determine the GCF of the coefficients of the terms. Then rewrite the expression as a product of that GCF and the factored terms.

1. $15m + 18n - 24p$

2. $50x - 75y + 125z + 250w$

3. $18abc + 27abd + 45cde$

4. $24x^2 - 32x + 40$

Figuring on Factoring

Factoring out variables

Variables represent numerical values; variables with exponents represent the powers of those same values. For that reason, variables as well as numbers can be factored out of the terms in an expression, and in this section you can find out how.

When factoring out powers of a variable, the smallest power that appears in any one term is the most that can be factored out. For example, in an expression such as $a^4b + a^3c + a^2d + a^3e^4$, the smallest power of a that appears in any term is the second power, a^2. So you can factor out a^2

from all the terms because a^2 is the greatest common factor. You can't factor anything else out of each term: $a^4b + a^3c + a^2d + a^3e^4 = a^2\left(a^2b + ac + d + ae^4\right)$.

REMEMBER

When performing algebraic operations or solving equations, always take the time to check your work. Sometimes the check involves no more than just seeing if the answer makes sense. In the case of factoring expressions, a good visual check is to multiply the factor through all the terms in the parentheses to see if you get what you started with before factoring.

To perform checks on your factoring:

- Multiply through (distribute) your answer in your head to be sure that the factored form is equivalent to the original form.
- Visually scan the terms in parentheses to make sure that they don't all share the same variable.

EXAMPLE

Q. Perform the quick checks on the following factored expression: $x^2y^3 + x^3y^2z^4 + x^4yz = x^2y\left(y^2 + xyz^4 + x^2z\right)$.

A. Does your answer multiply out to become what you started with? Multiply in your head:

$$x^2y \cdot y^2 = x^2y^3 \qquad \text{Check!}$$
$$x^2y \cdot xyz^4 = x^3y^2z^4 \qquad \text{Check!}$$
$$x^2y \cdot x^2z = x^4yz \qquad \text{Check!}$$

Those are the three terms in the original problem.

Now, for the second part of the quick check: Look at what's in the parentheses in your answer. The first two terms have y and the second two have x and z, but no variable occurs in all three terms. The terms in the parentheses are relatively prime. Check!

Q. Factor out the greatest common variable factor; then rewrite the expression in factored form: $8x^2 - 15x^3 + 9x^5$.

A. The GCF is x^2, because it's the smallest power of x found in the three terms. Rewrite in factored form: $8x^2 - 15x^3 + 9x^5 = x^2(8 - 15x + 9x^3)$.

Remember that when factoring variables from terms, you're dividing. So, subtract the exponents to determine the resulting power.

Q. Factor out the greatest common variable factor, then rewrite the expression in factored form: $24a^4b^6 + 45a^6b^5 + 5a^8b^4$.

A. The GCF is a^4b^4, because a^4 is the smallest power of a found in the three terms, and b^4 is the smallest power of b found in the three terms. Now, rewrite in factored form:

$$24a^4b^6 + 45a^6b^5 + 5a^8b^4 = a^4b^4\left(24b^2 + 45a^2b + 5a^4\right)$$

YOUR TURN

Factor out the greatest common variable factor; then rewrite the expression in factored form.

5. $11y^5 - 10y^4 + 9y^3$

6. $27zw^4 + 35z^2w^6 - 40z^3w^8$

7. $16a^5 + 13a^{-1} + 12a^{-4}$

8. $100x^{1/2} + 79x^{3/2} + 42x^{5/2} + 11x^{7/2}$

9. $6a^4b^2 - 9a^3b^3 - 12a^2b^4$

10. $36xy^3z^4 + 48x^2y^2z^5 + 60x^3yz^6$

Figuring on Factoring

11 $39mnp - 26m^2np + 39mn$

12 $480wx^6 + 440wx^8 - 520wx^{10}$

Unlocking combinations of numbers and variables

The real test of the factoring process is combining numbers and variables, finding the GCF, and factoring successfully. Sometimes you may miss a factor or two, but a second sweep-through can be done and is nothing to be ashamed of when doing algebra problems. If you do your factoring in more than one step, it really doesn't matter in what order you pull out the factors. You can do numbers first or variables first. It'll come out the same.

DIOPHANTUS

The mathematician Diophantus, the first to use symbols to abbreviate his thoughts systematically, lived some time between A.D. 100 and 400. Some consider him the "father of algebra." Using symbols allowed him to categorize numbers of particular types and then symbolically study their properties. One of Diophantus's followers summarized his life in terms of an algebra riddle:

> Diophantus's youth lasted one-sixth of his life. He grew a beard after one-twelfth more. After one-seventh more of his life, Diophantus married. Five years later, he had a son. The son lived exactly one-half as long as his father, and Diophantus died just four years after his son. All this adds up to the years Diophantus lived.

Just in case you're dying to know the answer: Diophantus lived 84 years.

EXAMPLE

Q. Factor $12x^2y^3z+18x^3y^2z^2-24xy^4z^3$.

A. Each term has a coefficient that's divisible by 2, 3, and 6. You select 6 as the largest of those common factors.

Each term has a factor of x. The powers on x are 2, 3, and 1. You have to select the smallest exponent when looking for the greatest common factor, so the common factor is just x.

Each term has a factor of y. The exponents are 3, 2, and 4. The smallest exponent is 2, so the common factor is y^2.

Each term has a factor of z, and the exponents are 1, 2, and 3. The number 1 is smallest, so you can pull out a z from each term.

Put all the factors together, and you get that the GCF is $6xy^2z$. So,

$$12x^2y^3z+18x^3y^2z^2-24xy^4z^3=6xy^2z\left(2xy+3x^2z-4y^2z^2\right)$$

Doing a quick check, you multiply through by the GCF in your head to be sure that the products match the original expression. You then do a sweep to be sure that there isn't a common factor among the terms within the parentheses.

Q. Factor $100a^4b-200a^3b^2+300a^2b^2-400$.

A. The greatest common factor of the coefficients is 100. Even though the powers of a and b are present in the first three terms, none of them occurs in the last term. So you're out of luck finding any more factors. Doing the factorization:

$$100a^4b-200a^3b^2+300a^2b^2-400=100\left(a^4b-2a^3b^2+3a^2b^2-4\right)$$

Q. Factor $26mn^3-25x^2y+21a^4b^4mnxy$.

A. Even though each of the numbers is a composite (each can be divided by values other than itself), the three have no factors in common. The expression cannot be factored. It's considered prime.

Q. Factor $484x^3y^2+132x^2y^3-88x^4y^5$.

A. Even if you don't divide through by the GCF the first time, all is not lost. A second run takes care of the problem. Often, doing the factorizations in two steps is easier because the numbers you're dividing through each time are smaller, and you can do the work in your head.

Assume that you determined that the GCF of the expression in this example is $4x^2y$. Then $484x^3y^2+132x^2y^3-88x^4y^5=4x^2y\left(121xy+33y^2-22x^2y^4\right)$.

Looking at the expression in the parentheses, you can see that each of the numbers is divisible by 11 and that there's a y in every term. The terms in the parentheses have a GCF of $11y$.

$$\begin{aligned}
&4x^2y\left[121xy+33y^2-22x^2y^4\right]=\\
&4x^2y\left[11y\left(11x+3y-2x^2y^3\right)\right]=\\
&\left(4x^2y\right)\left(11y\right)\left(11x+3y-2x^2y^3\right)=\\
&44x^2y^2\left(11x+3y-2x^2y^3\right)
\end{aligned}$$

You can do this factorization all at the same time, using the GCF $44x^2y^2$, but not everyone recognizes the multiples of 44. Also, the factorization could have been done in two or more steps in a different order with different factors each time. The result always comes out the same in the end.

Q. Factor $-4ab-8a^2b-12ab^2$.

A. Each term in the expression is negative; dividing out the negative from all the terms in the parentheses makes them positive.

$$-4ab-8a^2b-12ab^2=-4ab\left(1+2a+3b\right)$$

When factoring out a negative factor, be sure to change the signs of each of the terms.

Q. Find the GCF and factor the expression: $30x^4y^2-20x^5y^3+50x^6y$.

A. If you divide each term by the GCF, which is $10x^4y$, and put the results of the divisions in parentheses, the factored form is $30x^4y^2-20x^5y^3+50x^6y=10x^4y(3y-2xy^2+5x^2)$. It's like doing this division, with each fraction reducing to become a term in the parentheses:

$$\frac{30x^4y^2-20x^5y^3+50x^6y}{10x^4y}=\frac{30x^4y^2}{10x^4y}-\frac{20x^5y^3}{10x^4y}+\frac{50x^6y}{10x^4y}$$
$$30x^4y^2-20x^5y^3+50x^6y=\mathbf{10x^4y(3y-2xy^2+5x^2)}$$

Q. Factor out the GCF: $8a^{3/2}-12a^{1/2}$.

A. Dealing with fractional exponents can be tricky. Just remember that the same rules apply to fractional exponents as with whole numbers. You subtract the exponents.

$$\frac{8a^{3/2}}{4a^{1/2}}-\frac{12a^{1/2}}{4a^{1/2}}=\frac{{}^{2}\not{8}a^{3/2}}{{}_{1}\not{4}a^{1/2}}-\frac{{}^{3}\not{12}a^{1/2}}{{}_{1}\not{4}a^{1/2}}$$
$$=2a^{3/2-1/2}-3a^{1/2-1/2}=2a^1-3a^0=2a-3$$

Now, write the common factor, $4a^{1/2}$, outside the parentheses and the results of the division inside: $4a^{1/2}\left(2a-3\right)$.

YOUR TURN

 Factor out the GCF: $24x^2y^3 - 42x^3y^2$.

14 Factor out the GCF: $9z^{-4} + 15z^{-2} - 24z^{-1}$.

15 Factor out the GCF: $16a^2b^3c^4 - 48ab^4c^2$.

 Factor out the GCF: $16x^{-3}y^4 + 20x^{-4}y^3$.

Using the Box Method

You've been introduced to several methods of factoring. And everyone has their favorite. Just to give you another option — something to use when the expressions are pretty large and complicated — I now show you the Box Method.

When using the Box Method, you write the terms you're factoring in an upside-down division box. Then you write a common factor of the terms outside on the left. Divide each term by the common factor and put the division results below. Repeat the process on the results until there's no longer a common factor. The common factors along the left are multiplied together to give you the GCF. And, as a bonus, you have the terms that belong in the parentheses!

EXAMPLE

Q. Factor and rewrite: $336x^4y^3 + 432x^3y^4 + 528x^2y^5$.

A. Start with the number coefficients.

- You recognize that they're all divisible by 4. Divide each and write the result underneath.
- The results are all divisible by 4, also.
- Now you have three numbers, all divisible by 3.
- The coefficients 7, 9, and 11 are relatively prime.
- Each term has a factor of x^2.
- And each term has a factor of y^3.

$$\begin{array}{l|lll} 4 & 336x^4y^3 & 432x^3y^4 & 528x^2y^5 \\ \hline 4 & 84x^4y^3 & 108x^3y^4 & 132x^2y^5 \\ \hline 3 & 21x^4y^3 & 27x^3y^4 & 33x^2y^5 \\ \hline x^2 & 7x^4y^3 & 9x^3y^4 & 11x^2y^5 \\ \hline y^3 & 7x^2y^3 & 9xy^4 & 11y^5 \\ \hline & 7x^2 & 9xy^1 & 11y^2 \end{array}$$

- What you have here are all the factors of the GCF going down the left side and the factored terms on the bottom. Multiply the factors to get $48x^2y^3$ and write the expression in factored form: $48x^2y^3\left(7x^2 + 9xy + 11y^2\right)$. It doesn't matter what order you do the dividing — the product will always come out the same.

YOUR TURN

Factor each expression and rewrite it in factored form.

17 $252x^4y^4 - 273x^4y^3 - 504x^4y^2$

18 $576ab^5 + 456a^2b^6 + 384a^3b^7$

19 **$44n^5(14m^2-12mn+13n^2)$.** Each coefficient is divisible by 4 and 11. The GCF of the original expression is $44n^5$.

20 **$72z^{-5}w^{-5}(9z^2+16zw^2-17w^4)$.** Each coefficient is divisible by 8 and 9 and 12 — and so on. There are lots of ways to factor this. And the GCF of the variable factors is $z^{-5}w^{-5}$.

21 $\mathbf{\dfrac{2x(x+4)}{(x+2)^2}}$. All three terms have the common factor $(x+2)$.

$$\frac{4x(x+2)^2-2x^2(x+2)}{(x+2)^3}=\frac{(x+2)\left[4x(x+2)-2x^2\right]}{(x+2)^3}$$

$$=\frac{\cancel{(x+2)}\left[4x^2+8x-2x^2\right]}{(x+2)^{\cancel{3}2}}=\frac{2x^2+8x}{(x+2)^2}=\frac{2x(x+4)}{(x+2)^2}$$

22 **$30a^7b$.**

$$\frac{6!a^4b^{-1}}{4!a^{-3}b^{-2}}=\frac{6\cdot5\cdot4\cdot3\cdot2\cdot1a^4b^{-1}}{4\cdot3\cdot2\cdot1a^{-3}b^{-2}}=\frac{6\cdot5\cdot\cancel{4}\cdot\cancel{3}\cdot\cancel{2}\cdot\cancel{1}\cdot a^{4-(-3)}b^{-1-(-2)}}{\cancel{4}\cdot\cancel{3}\cdot\cancel{2}\cdot\cancel{1}\cdot\cancel{a^{-3}}\,\cancel{b^{-2}}}=\frac{30a^7b^1}{1}$$

For a refresher on factorials, refer to Chapter 7.

23 $\mathbf{\dfrac{2ab-3}{4b^2}}$.

$$\frac{14a^2b-21a}{28ab^2}=\frac{7a(2ab-3)}{28ab^2}=\frac{\cancel{7}\cancel{a}(2ab-3)}{{}_4\cancel{28}\cancel{a}b^2}=\frac{2ab-3}{4b^2}$$

24 $\mathbf{\dfrac{3-4w(w+1)^2}{5w^2(w+1)^2}}$.

$$\frac{6w^3(w+1)-8w^4(w+1)^3}{10w^5(w+1)^3}=\frac{\cancel{2}\cancel{w^3}\cancel{(w+1)}\left[3-4w(w+1)^2\right]}{{}_5\cancel{10}w^{\cancel{5}2}(w+1)^{\cancel{3}2}}=\frac{3-4w(w+1)^2}{5w^2(w+1)^2}$$

WARNING

Even though the answer appears to have a common factor in the numerator and denominator, you can't reduce it. The numerator has two terms, and the first term, the 3, doesn't have that common factor in it.

25 $\mathbf{\dfrac{x(9y-7x)}{4}}$.

$$\frac{9{,}009x^{4/3}y^2-7{,}007x^{7/3}y}{4{,}004x^{1/3}y}=\frac{1{,}001x^{4/3}y(9y-7x^1)}{4(1{,}001)x^{1/3}y}$$

WARNING

When factoring the terms in the numerator, be careful with the subtraction of the fractions. These fractional exponents are found frequently in higher mathematics and behave just as you see here. I put the exponent of 1 on the x in the numerator just to emphasize the result of the subtraction of exponents. Continuing:

$$=\frac{\cancel{1{,}001}x^{\cancel{4/3}\,1}\cancel{y}(9y-7x^1)}{4\cancel{(1{,}001)}\cancel{x^{1/3}}\cancel{y}}=\frac{x(9y-7x)}{4}$$

26 $\dfrac{4b(c^2+1)^2-3ab(c^2+1)+7a^2}{ab\left[2-5ab(c^2+1)\right]}$.

Factor the numerator and denominator separately and then reduce by dividing by the common factors in each:

$$\frac{8a^2b^3(c^2+1)^4-6a^3b^2(c^2+1)^3+14a^4b(c^2+1)^2}{4a^3b^2(c^2+1)^2-10a^4b^3(c^2+1)^3}$$

$$=\frac{2a^2b(c^2+1)^2\left[4b(c^2+1)^2-3ab(c^2+1)+7a^2\right]}{2a^3b^2(c^2+1)^2\left[2-5ab(c^2+1)\right]}$$

When reducing, be sure to only consider the factors in front of the bracketed expressions.

$$=\frac{\cancel{2}\cancel{a^2}\cancel{b}\cancel{(c^2+1)^2}\left[4b(c^2+1)^2-3ab(c^2+1)+7a^2\right]}{\cancel{2}a^{\cancel{3}1}b^{\cancel{2}1}\cancel{(c^2+1)^2}\left[2-5ab(c^2+1)\right]}$$

$$=\frac{4b(c^2+1)^2-3ab(c^2+1)+7a^2}{ab\left[2-5ab(c^2+1)\right]}$$

If you're ready to test your skills a bit more, take the following chapter quiz that incorporates all the chapter topics.

Whaddya Know? Chapter 11 Quiz

Quiz time! Complete each problem to test your knowledge on the various topics covered in this chapter. You can then find the solutions and explanations in the next section.

1. Find the GCF and determine the factored expression $36a^{-1}b^{-2}c+48ab^{-3}=$
2. Reduce the fractions to lowest terms. $\dfrac{18x^5y^4}{24x^6y^3}=$
3. Find the GCF and determine the factored expression $y^6-2y^5+4y^3+y^2=$
4. Reduce the fractions to lowest terms. $\dfrac{3m^{-4}-6m^{-3}}{12m^{-2}}=$
5. Find the GCF and determine the factored expression $324m^2n^3p^4+432mn^2p^3+594np^2=$
6. Find the GCF and determine the factored expression $25x^2y^3z^4-50x^3y^2z^3-100x^4yz^2=$

7. Reduce the fractions to lowest terms. $\frac{8!a^5b^2}{6!2!a^6b^3} =$

8. Find the GCF and determine the factored expression $28x^2 + 35y^2 - 56z^2 =$

9. Reduce the fractions to lowest terms. $\frac{14x^2y(x+y)^3 - 21xy^2(x+y)^2}{35xy(x+y)^4} =$

10. Find the GCF and determine the factored expression $18a^{1/2} + 16a^{3/2} =$

11. Reduce the fractions to lowest terms. $\frac{111x^{1/2}y^{1/3} + 222x^{3/2}y^{4/3}}{333x^{1/2}y^{1/3}} =$

12. Find the GCF and determine the factored expression $14z^{-4} + 5z^{-3} + 2z^{-2} =$

13. Find the GCF and determine the factored expression $15abc(a^2+11)^4 + 45a^2c^3(a^2+11)^2 - 60a^3c^5(a^2+11) =$

Figuring on Factoring

Answers to Chapter 11 Quiz

1. $12a^{-1}b^{-3}\left(3bc+4a^2\right)$**.** The smaller exponents are: –1 on a and –3 on b. The GCF of the numbers is 12.

$$36a^{-1}b^{-2}c+48ab^{-3}=12a^{-1}b^{-3}\left(3a^0b^1c+4a^2b^0\right)=12a^{-1}b^{-3}\left(3bc+4a^2\right)$$

Writing this with no negative exponents, you have $\dfrac{12\left(3bc+4a^2\right)}{ab^3}$.

2. $\dfrac{3y}{4x}$**.** The GCF is $6x^5y^3$. $\dfrac{18x^5y^4}{24x^6y^3}=\dfrac{\cancel{18}^3\cancel{x^5}\,y^{\cancel{4}^1}}{\cancel{24}^4x^{\cancel{6}^1}\cancel{y^3}}=\dfrac{3y}{4x}$

3. y^2**.** $y^6-2y^5+4y^3+y^2=y^2\left(y^4-2y^3+4y+1\right)$

4. $\dfrac{1-2m}{4m^2}$**.** The GCF is $3m^{-4}$. You choose the smallest exponent.

$$\frac{3m^{-4}-6m^{-3}}{12m^{-2}}=\frac{\cancel{3}\cancel{m^{-4}}-\cancel{6}^2m^{\cancel{-3}^1}}{\cancel{12}^4m^{\cancel{-2}^2}}=\frac{1-2m}{4m^2}$$

5. $54np^2\left(6m^2n^2p^2+8mnp+11\right)$**.** The coefficients are fairly large, so you may want to use the Box Method.

$$\begin{array}{l|lll}
6 & 324m^2n^3p^4 & 432mn^2p^3 & 594np^2 \\
3 & 54m^2n^3p^4 & 72mn^2p^3 & 99np^2 \\
3 & 18m^2n^3p^4 & 24mn^2p^3 & 33np^2 \\
n & 6m^2n^3p^4 & 8mn^2p^3 & 11np^2 \\
p^2 & 6m^2n^2p^4 & 8mn^1p^3 & 11p^2 \\
 & 6m^2n^2p^2 & 8mn^1p^1 & 11
\end{array}$$

The GCF is $6\cdot3\cdot3\cdot n\cdot p^2=54np^2$.

$$324m^2n^3p^4+432mn^2p^3+594np^2=54np^2\left(6m^2n^2p^2+8mnp+11\right)$$

6. $25x^2yz^2$**.** $25x^2y^3z^4-50x^3y^2z^3-100x^4yz^2=25x^2yz^2\left(y^2z^2-2xyz-4x^2\right)$

7. $\dfrac{28}{ab}$**.** $\dfrac{8!a^5b^2}{6!2!a^6b^3}=\dfrac{\cancel{8}^4\cdot7\cdot\cancel{6\cdot5\cdot4\cdot3\cdot2\cdot1}\cdot\cancel{a^5}\cancel{b^2}}{\cancel{6\cdot5\cdot4\cdot3\cdot2\cdot1}\cdot\cancel{2}\cdot1\cdot a^{\cancel{6}^1}b^{\cancel{3}^1}}=\dfrac{28}{ab}$

8. **7.** $28x^2+35y^2-56z^2=7\left(4x^2+5y^2-8z^2\right)$

9. $\dfrac{2x(x+y)-3y}{5(x+y)^2}$**.** The GCF is $7xy(x+y)^2$.

$$\frac{14x^2y(x+y)^3-21xy^2(x+y)^2}{35xy(x+y)^4}=\frac{\cancel{14}^2x^{\cancel{2}^1}\cancel{y}(x+y)^{\cancel{3}^1}-\cancel{21}^3\cancel{x}y^{\cancel{2}^1}\cancel{(x+y)^2}}{\cancel{35}^5\cancel{x}\cancel{y}(x+y)^{\cancel{4}^2}}=\frac{2x(x+y)-3y}{5(x+y)^2}$$

(10) $2a^{1/2}$. The smaller exponent is $1/2$. $18a^{1/2} + 16a^{3/2} = 2a^{1/2}(9 + 8a)$

(11) $\frac{1+2xy}{3}$. The GCF is $111x^{1/2}y^{1/3}$.

$$\frac{111x^{1/2}y^{1/3} + 222x^{3/2}y^{4/3}}{333x^{1/2}y^{1/3}} = \frac{\cancel{111}\cancel{x^{1/2}}\cancel{y^{1/3}} + \cancel{222}^{2}x^{\cancel{3/2}\,1}y^{\cancel{4/3}\,1}}{\cancel{333}^{3}\cancel{x^{1/2}}\cancel{y^{1/3}}} = \frac{1+2xy}{3}$$

(12) z^{-4}. The smallest exponent is -4. $14z^{-4} + 5z^{-3} + 2z^{-2} = z^{-4}\left(14 + 5z + 2z^{2}\right)$

(13) $15ac(a^2 + 11)$.

$$\begin{aligned} &15abc(a^2+11)^4 + 45a^2c^3(a^2+11)^2 - 60a^3c^5(a^2+11) \\ &= 15ac(a^2+11)\left[b(a^2+11)^3 + 3ac^2(a^2+11) - 4a^2c^4\right] \end{aligned}$$

Figuring on Factoring

5 $9x^2y^4z^{16} - 2{,}500$.

6 $a^2b^2c^{-4} - d^{-8}e^4$.

Factoring Differences and Sums of Cubes

A *perfect cube* is the number you get when you multiply a number times itself and then multiply the answer times the first number again. A cube is the third power of a number or variable. The difference of two cubes is a binomial expression $a^3 - b^3$ and the sum of two cubes is written $a^3 + b^3$.

The most well-known perfect cubes are those whose roots are integers, not decimals. Here's a short list of some positive integers cubed.

Integer	Cube
1	1
2	8
3	27
4	64
5	125
6	216

Integer	Cube
7	343
8	512
9	729
10	1,000
11	1,331
12	1,728

Becoming familiar with and recognizing these cubes in an algebra problem can save you time and improve your accuracy.

REMEMBER

When cubing variables and numbers that already have an exponent, you multiply the exponent by 3. When cubing the product of numbers and variables in parentheses, you raise each factor to the third power. (Refer to Chapter 6 if you need more information on this.) For example, $(a^2)^3 = a^6$ and $(2yz)^3 = 8y^3z^3$.

Variable cubes are relatively easy to spot because their exponents are always divisible by 3. When a number is cubed and multiplied out, you can't always tell it's a cube.

Look at the following binomials. These expressions are the sum or difference of cubes and can be factored. Each variable power is a cube — and so are the coefficients and constants. So all the variables, coefficients, and constants have cube roots. The variables all have powers that are multiples of 3:

$$m^3 - 8 \qquad 1{,}000 + 27z^3 \qquad 64x^6 - 125y^{15}$$

To factor the *difference* of two perfect cubes, use the following pattern:

$$a^3 - b^3 = (a-b)(a^2 + ab + b^2).$$

Here are the results of factoring the difference of the perfect cubes $a^3 - b^3$:

- The binomial factor $(a-b)$ is made up of the two cube roots of the perfect cubes separated by a minus sign.
- The trinomial factor $(a^2 + ab + b^2)$ is made up of the squares of the two cube roots from the first factor added to the product of the cube roots in the middle.

REMEMBER

A trinomial has three terms, and this one contains all plus signs.

To factor the *sum* of two perfect cubes, use the following pattern: $a^3 + b^3 = (a+b)(a^2 - ab + b^2)$.

Here are the results of factoring the sum of the perfect cubes $a^3 + b^3$:

- The binomial factor $(a+b)$ is made up of the two cube roots of the perfect cubes separated by a plus sign.
- The trinomial factor $(a^2 - ab + b^2)$ is made up of the squares of the two cube roots from the first factor with the product of the cube roots subtracted in the middle.

REMEMBER

When you have two perfect squares, you can use the special factoring rule if the operation is subtraction. With cubes, though, both sums and differences factor into the product of a binomial and a trinomial.

$$a^3 - b^3 = (a-b)(a^2 + ab + b^2) \text{ and } a^3 + b^3 = (a+b)(a^2 - ab + b^2)$$

Here's the pattern: First, you write the sum or difference of the two cube roots corresponding to the sum or difference of cubes; second, you multiply the binomial containing the roots by a trinomial composed of the squares of those two cube roots and the *opposite* of the product of them. If the binomial has a + sign, then the middle term of the trinomial is –. If the binomial has a – sign, then the middle term in the trinomial is +. The two squares in the trinomial are always positive.

GREAT LEADERS MAKE GREAT MATHEMATICIANS

Two famous leaders, Napoleon Bonaparte and U.S. President James Garfield, were drawn to the mysteries of mathematics. Bonaparte fancied himself an amateur geometer and liked to hang out with mathematicians — they're such party animals!

Napoleon's theorem, which he named for himself, says that if you take any triangle and construct equilateral triangles on each of the three sides and find the center of each of these three triangles and connect them, the connecting segments always form another equilateral triangle. Not bad for someone who met his Waterloo!

Garfield, the twentieth U.S. president, also dabbled in mathematics and discovered a new proof for the Pythagorean Theorem, which is done with a trapezoid consisting of three right triangles and some work with the areas of the triangles.

EXAMPLE

Q. Factor: $x^3 - 27$.

A. $x^3 - 27 = (x-3)(x^2 + x \cdot 3 + 3^2) = (x-3)(x^2 + 3x + 9)$

Q. Factor: $125 + 8y^3$.

A. $125 + 8y^3 = (5+2y)(5^2 - 5 \cdot 2y + [2y]^2) = (5+2y)(25 - 10y + 4y^2)$

Q. Factor $64x^3 - 27y^6$.

A. The cube root of $64x^3$ is $4x$, and the cube root of $27y^6$ is $3y^2$. The square of $4x$ is $16x^2$, the square of $3y^2$ is $(3y^2)^2 = 9y^4$, and the product of $(4x)(3y^2)$ is $12xy^2$.

$$64x^3 - 27y^6 = \left(4x - 3y^2\right)\left(16x^2 + 12xy^2 + 9y^4\right)$$

Q. Factor $a^3b^6c^9 - 1{,}331d^{300}$.

A. The cube root of $a^3b^6c^9$ is ab^2c^3, and the cube root of $1{,}331d^{300}$ is $11d^{100}$. The square of ab^2c^3 is $a^2b^4c^6$, and the square of $11d^{100}$ is $121d^{200}$. The product of $(ab^2c^3)(11d^{100})$ is $11ab^2c^3d^{100}$.

$$a^3b^6c^9 - 1{,}331d^{300} = \left(ab^2c^3 - 11d^{100}\right)\left(a^2b^4c^6 + 11ab^2c^3d^{100} + 121d^{200}\right)$$

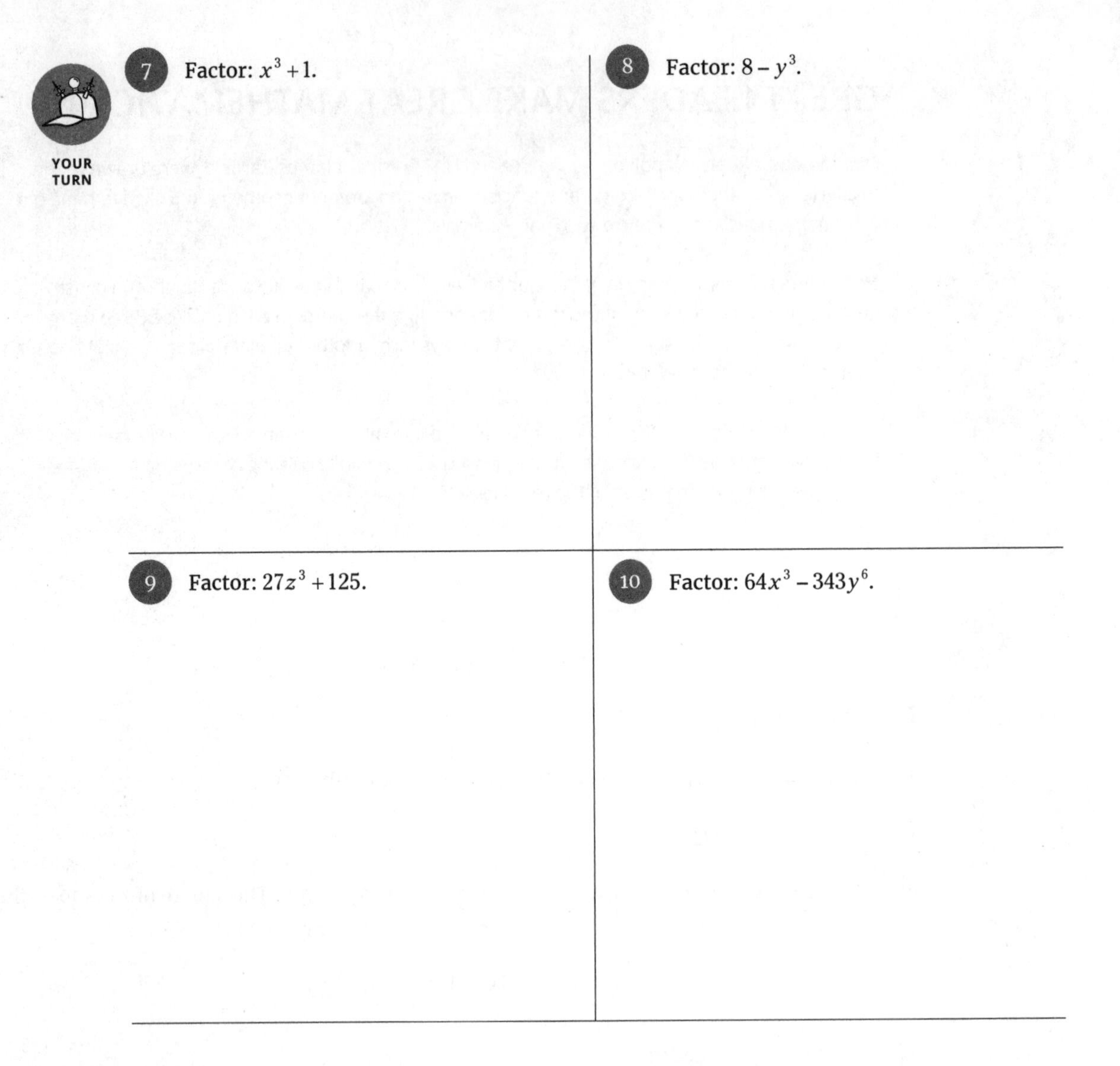

YOUR TURN

7 Factor: $x^3 + 1$.

8 Factor: $8 - y^3$.

9 Factor: $27z^3 + 125$.

10 Factor: $64x^3 - 343y^6$.

Making Factoring a Multiple Mission

Many factorization problems in mathematics involve more than one type of factoring process. You may find a GCF in the terms, and then you may recognize that what's left is the difference of two cubes. You sometimes factor the difference of two squares just to find that one of those binomials is the difference of two new squares.

TIP

Solving these problems is really like figuring out a gigantic puzzle. You discover how to conquer it by applying the factorization rules. In general, first look for a GCF. Life is much easier when the numbers and powers are smaller because they're easier to deal with and work out in your head.

EXAMPLE

Q. Factor: $4x^6 + 108x^3$.

A. First, deal with the GCF. You see that both terms are divisible by $4x^3$, so factor that out.

$$4x^6 + 108x^3 = 4x^3(x^3 + 27)$$

The binomial in the parentheses is the sum of two cubes. Factoring that sum, you have

$$4x^3(x^3 + 27) = 4x^3(x + 3)(x^2 - 3x + 9)$$

One of my favorite scenes from the movie *The Agony and the Ecstasy,* which chronicles Michelangelo's painting of the Sistine Chapel, comes when the pope enters the Sistine Chapel, looks up at the scaffolding, dripping paint, and Michelangelo perched up near the ceiling, and yells, "When will it be done?" Michelangelo's reply: "When I'm finished!"

The pope's lament can be applied to factoring problems: "When is it done?"

Factoring is done when no more parts can be factored. If you refer to the listing of ways to factor two, three, four, or more terms, then you can check off the options, discard those that don't fit, and stop when none works. After doing one type of factoring, you should then look at the values in parentheses to see if any of them can be factored.

EXAMPLE

Q. Factor $x^4 - 104x^2 + 400$.

A. There's no GCF, so the only other option when there are three terms is to unFOIL or find the two binomials whose product is the trinomial (see Chapter 13 for more on unFOILing):

$$x^4 - 104x^2 + 400 = (x^2 - 4)(x^2 - 100)$$

There are now two factors, but each of them is the difference of perfect squares:

$$(x^2 - 4)(x^2 - 100) = (x + 2)(x - 2)(x + 10)(x - 10)$$

You're finished!

Q. Factor $3x^5 - 18x^3 - 81x$.

A. First pull out the GCF: $3x^5 - 18x^3 - 81x = 3x(x^4 - 6x^2 - 27)$

The trinomial can be factored into two binomials: $3x(x^4 - 6x^2 - 27) = 3x(x^2 - 9)(x^2 + 3)$

And now you have a binomial that's the difference of squares. Factor that first binomial:

$$3x(x^2 - 9)(x^2 + 3) = 3x(x + 3)(x - 3)(x^2 + 3)$$

You're finished!

YOUR TURN

Completely factor the following.

11 $3x^3y^3 - 27xy^3$.

12 $36x^2 - 100y^2$.

13 $80y^4 - 10y$.

14 $10{,}000x^4 - 1$.

15 $x^{-4} + x^{-7}$.

16 $125a^3b^3 - 125c^6$.

17 $(x^2 - 49)(a^2b^2c^4) - (x^2 - 49)(d^6e^8)$.

Practice Questions Answers and Explanations

1. $\mathbf{(x+5)(x-5)}$.

2. $\mathbf{(8a+y)(8a-y)}$.

3. $\mathbf{(7xy+3zw^2)(7xy-3zw^2)}$.

4. $\mathbf{(10x^{1/4}+9y^{1/8})(10x^{1/4}-9y^{1/8})}$.

   When looking at the exponents, you see that $\frac{1}{2}$ is twice the fraction $\frac{1}{4}$, and $\frac{1}{4}$ is twice the fraction $\frac{1}{8}$.

5. $\mathbf{(3xy^2z^8+50)(3xy^2z^8-50)}$. The first term in the problem is the square of $3xy^2z^8$. Be careful, here. Don't write the exponent of z as a 4 because of the perfect square. The second term is the square of 50. So the answer is $(3xy^2z^8+50)(3xy^2z^8-50)$.

6. $\mathbf{(abc^{-2}+d^{-4}e^2)(abc^{-2}-d^{-4}e^2)}$. The first term is the square of abc^{-2}, and the second term is the square of $d^{-4}e^2$. So the answer is $(abc^{-2}+d^{-4}e^2)(abc^{-2}-d^{-4}e^2)$.

7. $\mathbf{(x+1)(x^2-x+1)}$. The cube of x is x^3 and the cube of 1 is 1.

8. $\mathbf{(2-y)(4+2y+y^2)}$. $8-y^3=(2-y)(2^2+2y+y^2)=(2-y)(4+2y+y^2)$

TIP

It's nice to have a list of the first ten cubes handy when factoring the sum or difference of cubes: 1, 8, 27, 64, 125, 216, 343, 512, 729, 1,000.

9. $\mathbf{(3z+5)(9z^2-15z+25)}$.

$$\begin{aligned}27z^3+125&=(3z+5)\left[(3z)^2-3z(5)+5^2\right]\\&=(3z+5)(9z^2-15z+25)\end{aligned}$$

10. $\mathbf{(4x-7y^2)(16x^2+28xy^2+49y^4)}$.

    Did you remember that $7^3=343$?

$$\begin{aligned}64x^3-343y^6&=(4x-7y^2)\left[(4x)^2+(4x)(7y^2)+(7y^2)^2\right]\\&=(4x-7y^2)(16x^2+28xy^2+49y^4)\end{aligned}$$

11. $\mathbf{3xy^3(x+3)(x-3)}$. First factor out the GCF.

$$3x^3y^3-27xy^3=3xy^3(x^2-9)=3xy^3(x+3)(x-3)$$

12. $\mathbf{4(3x+5y)(3x-5y)}$. Go for the GCF first.

$$36x^2-100y^2=4(9x^2-25y^2)=4(3x+5y)(3x-5y)$$

Taking the Bite out of Binomial Factoring

(13) $\mathbf{10y(2y-1)(4y^2+2y+1)}$.

$$\begin{aligned}80y^4-10y&=10y(8y^3-1)\\&=10y(2y-1)\left[(2y)^2+2y(1)+1^2\right]\\&=10y(2y-1)(4y^2+2y+1)\end{aligned}$$

(14) $\mathbf{(100x^2+1)(10x+1)(10x-1)}$. **You're starting with the difference of squares.**

$$10{,}000x^4-1=(100x^2+1)(100x^2-1)=(100x^2+1)(10x+1)(10x-1)$$

(15) $\mathbf{x^{-7}(x+1)(x^2-x+1)}$. **The GCF involves the smaller exponent.**

$$x^{-4}+x^{-7}=x^{-7}(x^3+1)=x^{-7}(x+1)(x^2-x+1)$$

REMEMBER

The GCF, which involves the smaller exponent, is the most negative exponent. The resulting binomial in the parentheses is the sum of two perfect cubes.

(16) $\mathbf{125(ab-c^2)(a^2b^2+abc^2+c^4)}$.

$$\begin{aligned}125a^3b^3-125c^6&=125(a^3b^3-c^6)\\&=125(ab-c^2)\left[(ab)^2+(ab)c^2+(c^2)^2\right]\\&=125(ab-c^2)(a^2b^2+abc^2+c^4)\end{aligned}$$

You may have been tempted to go right into the difference of cubes because 125 is a perfect cube. It's always more desirable, though, to factor out large numbers when possible.

(17) $\mathbf{(x+7)(x-7)(abc^2+d^3e^4)(abc^2-d^3e^4)}$. **Factor out the GCF, even though you may be tempted to factor the difference of the squares first.**

$$(x^2-49)(a^2b^2c^4)-(x^2-49)(d^6e^8)=(x^2-49)(a^2b^2c^4-d^6e^8)$$

You now have two factors that are each the difference of squares.

$$(x^2-49)(a^2b^2c^4-d^6e^8)=(x+7)(x-7)(abc^2+d^3e^4)(abc^2-d^3e^4)$$

If you're ready to test your skills a bit more, take the following chapter quiz that incorporates all the chapter topics.

Whaddya Know? Chapter 12 Quiz

Quiz time! Complete each problem to test your knowledge on the various topics covered in this chapter. You can then find the solutions and explanations in the next section.

Factor each completely.

1. $27xya^2 - 27xyb^2$
2. $125 - 64y^3$
3. $y^{-8} - y^{-11}$
4. $40a^2c - 360a^2cz^2$
5. $16x^2 - 25z^2$
6. $36x^2y^4z^8 - 49$
7. $64x^6 - y^9$
8. $a^{13/4}b^{1/4} + a^{1/4}b^{13/4}$
9. $x^4y^3 - 16y^3$
10. $30a^2b^3 + 35a^3b^4$
11. $80x^3y^2 + 128x^2yz^3$
12. $8a^3 + 27b^3$
13. $x^{-2} - 4x^{-4}$
14. $1{,}000a^3 + 343b^6c^{12}$

Answers to Chapter 12 Quiz

1. $\mathbf{27xy(a+b)(a-b)}$. First factor out the GCF. Then factor the difference of the squares.

2. $\mathbf{(5-4y)(25+20y+16y^2)}$. Factor as the difference of two cubes.

3. $\mathbf{y^{-11}(y-1)(y^2+y+1)}$. First, factor out the GCF. Then factor the difference of cubes.

$$y^{-8}-y^{-11}=y^{-11}(y^3-1)=y^{-11}(y-1)(y^2+y+1)$$

4. $\mathbf{40a^2c(1+3z)(1-3z)}$.

5. $\mathbf{(4x+5z)(4x-5z)}$.

6. $\mathbf{(6xy^2z^4+7)(6xy^2z^4-7)}$.

7. $\mathbf{(4x^2-y^3)(16x^4+4x^2y^3+y^6)}$. Remember that $\left(x^2\right)^3=x^6$ and $\left(y^3\right)^3=y^9$.

8. $\mathbf{a^{1/4}b^{1/4}(a+b)(a^2-ab+b^2)}$. First, factor out the GCF and then factor the sum of cubes.

$$a^{13/4}b^{1/4}+a^{1/4}b^{13/4}=a^{1/4}b^{1/4}\left(a^{12/4}+b^{12/4}\right)$$

9. $\mathbf{y^3(x+2)(x-2)(x^2+4)}$. First, factor out the GCF, y^3. This gives you the difference of squares. One of the factors in that difference can also be factored as the difference of squares.

10. $\mathbf{5a^2b^3(6+7ab)}$. The GCF is $5ab^2$.

11. $\mathbf{16x^2y(5xy+8z^3)}$. Factor out the GCF.

12. $\mathbf{(2a+3b)(4a^2-6ab+9b^2)}$. This is the sum of two cubes.

13. $\mathbf{x^{-4}(x+2)(x-2)}$. First factor out the GCF. Then factor the difference of squares.

$$x^{-2}-4x^{-4}=x^{-4}(x^2-4)=x^{-4}(x+2)(x-2)$$

14. $\mathbf{(10a+7b^2c^4)(100a^2-70ab^2c^4+49b^4c^8)}$.

» Reversing FOIL for a factorization

» Assigning terms to groups to factor them

» Making use of multiple methods to factor

Chapter 13
Factoring Trinomials and Special Polynomials

In Chapter 12, you find the basic ways to factor a *binomial* (an expression with two terms). Factoring means to change the expression from several terms to one expression connected by multiplication and division. When dealing with a polynomial with four terms, such as $x^4 - 4x^3 - 11x^2 - 6x$, the four terms become one when you write the factored form using multiplication: $x(x+1)^2(x-6)$. The factored form has many advantages, especially when you want to simplify fractions, solve equations, or graph functions.

When working with an algebraic expression with three terms (a *trinomial*) or more, you have a number of different methods available for factoring it. You generally start with the greatest common factor (GCF) and then apply one or more of the other techniques, if necessary. This chapter covers these different methods and even shows you how to use synthetic division.

Recognizing the Standard Quadratic Expression

The quadratic, or second-degree, expression in x has an x variable that is squared, and no x terms with powers higher than 2. The coefficient on the squared variable is not equal to 0. The standard quadratic form is $ax^2 + bx + c$.

You may notice that the following examples of quadratic expressions both have a variable raised to the second degree:

$$4x^2+3x-2 \qquad a^2+116y^2-5y$$

Quadratics are usually written in terms of a variable represented by an x, y, z, or w. The letters at the end of the alphabet are used more frequently for the variable, while those at the beginning of the alphabet are usually used for a number or constant. This isn't always the case, but it's the standard convention.

In a quadratic expression, the a — the coefficient of the variable raised to the second power — can't be 0. If a were allowed to be 0, then the x^2 would be multiplied by 0, and it wouldn't be a quadratic expression anymore. The variables b or c can be 0, but a can't.

Quadratics don't necessarily have all positive terms either. The standard form, ax^2+bx+c, is written with all positives for convenience. But if a, b, or c represents a negative number, then that term is negative. A mathematical convention is that the terms are written with the second-degree term first, the first-degree term next, and the number last. Another convention also has to do with the order of the terms in a quadratic expression. If you find more than one variable, decide which variable makes it a quadratic expression (look for the variable that's squared) and write the expression in terms of that variable. This means, after you find the variable that's squared, write the rest of the expression in decreasing powers of that variable.

EXAMPLE

Q. Rewrite $aby+cdy^2+ef$ using the standard convention involving order. This can be a second-degree expression in y.

A. Written in the standard form for quadratics, ax^2+bx+c, where the second-degree term comes first, it looks like $(cd)y^2+(ab)y+ef$. The parentheses aren't necessary around the cd or the ab, and they don't change anything, but they're used sometimes for emphasis. The parentheses just make seeing the different parts easier.

Q. Rewrite $a^2bx+cdx^2+aef$ using the standard convention involving order. This can be a second-degree expression in terms of either a or x.

A. Writing as a second degree in a:

$$(bx)a^2+(ef)a+cdx^2$$

Even though there's a second-degree factor of x in the last term, that term is thought of as a constant, a value that doesn't change, rather than a variable if the expression is a to the second degree. Now, changing roles, with the second-degree expression in x, you have:

$$(cd)x^2+\left(a^2b\right)x+aef$$

2. **Determine all the ways you can multiply two numbers to get 4.**

   $4 = 4 \times 1 = 2 \times 2.$

3. **The 4 is positive, so you want the sum of the outer and inner products.**

   To get a sum of 9, use the 2×1 and the 4×1 factors, multiplying (2)(4) to get 8, and multiplying the two ones together to get 1. The sum of the 8 and the 1 is 9.

4. **Arrange your choices as binomials so the results are those you want.**

   $(2x \quad 1)(1x \quad 4)$

5. **Placing the signs, both binomials have to have subtraction so that the sum is −9 and the product is +4.**

   $(2x-1)(1x-4) = 2x^2 - 8x - 1x + 4 = 2x^2 - 9x + 4$

In the next example, all the terms are positive. The sum of the outer and inner products will be used. And there are several choices for the multipliers.

Factor: $10x^2 + 31x + 15$.

1. **Determine all the ways you can multiply two numbers to get 10.**

   The 10 can be written as 10×1 or 5×2.

2. **Determine all the ways you can multiply two numbers to get 15.**

   The 15 can be written as 15×1 or 5×3.

3. **The last term is +15, so you want the sum of the products to be 31.**

   Using the 5×2 and the 5×3, multiply (2)(3) to get 6, and multiply (5)(5) to get 25. The sum of 6 and 25 is 31.

4. **Arrange your choices in the binomials so the factors line up the way you want to give you the products.**

   $(2x \quad 5)(5x \quad 3)$

5. **Placing the signs is easy because everything is positive.**

   $(2x+5)(5x+3) = 10x^2 + 6x + 25x + 15 = 10x^2 + 31x + 15$

Coming to the end of the FOIL roll

This last example looks, at first, like a great candidate for factoring by this method. You'll see, though, that not everything can factor. Also, I get to make the point that using this method assures you that you've "left no stone unturned" and can be confident when claiming that a trinomial is *prime* (can't be factored).

Factor: $18x^2 - 27x - 4$.

1. **Determine all the ways you can multiply two numbers to get 18.**

   The 18 can be written as 18×1, 9×2, or 6×3.

2. **Determine all the ways you can multiply two numbers to get 4.**

   The 4 can be written as 4×1 or 2×2.

3. **Look at the sign of −4, and you see that you want a difference.**

   And the difference of the products is to be 27.

   You can't seem to find any combination that gives you a difference of 27. Run through all of them to be sure that you haven't missed anything.

   Using the 18×1, cross it with the following:

   - 4×1, which gives you a difference of either 14, using the (1)(4) and (18)(1), or 71, using the (1)(1) and the (18)(4).
   - 2×2, which gives you a difference of 34, using (1)(2) and (18)(2); there's only one choice because both of the second factors are 2.

   Using the 9×2, cross it with the following:

   - 4×1, which gives you a difference of either 34, using (2)(1) and (9)(4), or 1, using (2)(4) and (9)(1).
   - 2×2, which gives you a difference of 14, only.

   Using the 6×3, cross it with the following:

   - 4×1, which gives you a difference of either 21, using (3)(1) and (6)(4), or 6, using (3)(4) and (6)(1).
   - 2×2, which gives you a difference of 6, only.

   Because you've exhausted all the possibilities and you haven't been able to create a difference of 27, you can assume that this quadratic can't be factored. It's prime.

EXAMPLE

Q. Factor $2x^2 - 5x - 3$.

A. Use these steps.

1. The trinomial is already written with descending powers.
2. The only possible factors for $2x^2$ are $2x$ and x.
3. The only possible factors for the last term are 3 and 1.
4. Work on creating the middle term.

Because the last term is *negative*, you want to find a way to arrange the factors so that the outer and inner products have a *difference* of $5x$. You do this by placing the $2x$ and 3 so they multiply by one another: $(2x\ 1)(x\ 3)$. When deciding on the placement of the signs, use *a* + and *a* −, and situate them so that the middle term is negative. Putting the − sign in front of the 3 results in a $-6x$ and a $+1x$. Combining them gives you the $-5x$.

The factorization is $(2x+1)(x-3)$.

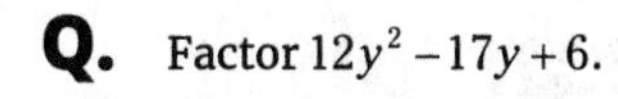

Q. Factor $12y^2 - 17y + 6$.

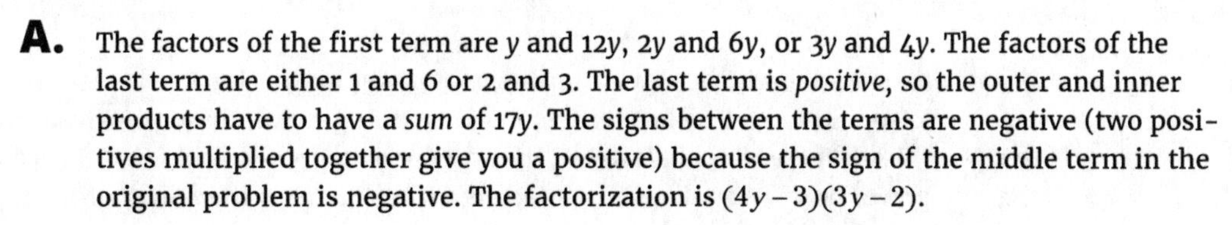

A. The factors of the first term are *y* and 12*y*, 2*y* and 6*y*, or 3*y* and 4*y*. The factors of the last term are either 1 and 6 or 2 and 3. The last term is *positive*, so the outer and inner products have to have a *sum* of 17*y*. The signs between the terms are negative (two positives multiplied together give you a positive) because the sign of the middle term in the original problem is negative. The factorization is $(4y-3)(3y-2)$.

YOUR TURN

Factor.

5 $x^2 - 8x + 15$	**6** $y^2 - 6y - 40$
7 $2x^2 + 3x - 2$	**8** $4z^2 + 12z + 9$
9 $w^2 - 16$	**10** $12x^2 - 8x - 15$

Factoring Quadratic-Like Trinomials

A *quadratic-like* trinomial has a first term whose power on the variable is twice that of the variable in the second term. The last term is a constant. In general, these trinomials are of the form $ax^{2n}+bx^{n}+c$. If these trinomials factor, then the factorizations look like $\left(dx^{n}+e\right)\left(fx^{n}+g\right)$. Notice that the power on the variables in the factored form matches the power of the middle term in the original trinomial.

To factor a quadratic-like trinomial, you treat it as if it were ay^2+by+c, with the same rules applying to unFOILing and just using the higher powers on the variables.

EXAMPLE

Q. Factor $6x^4+13x^2-28$.

A. Treat this problem as if it were the trinomial $6y^2+13y-28$. You have to find factors for the first term, involving the 6, and factors for the last term, involving the 28. The middle term, with the 13, has to be the difference between the outer and inner products. Using $3y^2$ from the first term and 7 from the last term, you get a product of $21y^2$. Then, using $2y^2$ from the first and 4 from the last, you get $8y^2$. The difference between 21 and 8 is 13. Get the factors aligned correctly and the signs inserted in the right places, and the factorization of $6y^2+13y-28$ is $(3y-4)(2y+7)$. Replace the y's with x^2 and the factored form is $\left(3x^2-4\right)\left(2x^2+7\right)$.

Q. Factor $5x^{-6}-36x^{-3}+36$.

A. Don't let the negative exponents throw you. The middle term has an exponent that's half the first term's exponent. Think of the trinomial as being like $5y^2-36y+36$. It works! The factored form is $\left(5x^{-3}-6\right)\left(x^{-3}-6\right)$.

YOUR TURN

Factor.

11 $x^{10}+4x^5+3$

12 $4y^{16}-9$

Factor $2x^2y^2 + 6x^2y + 2x^2 - 3y^2 - 9y - 3$. Here are your choices:

- Group the first three terms together, factoring out $2x^2$; group the second three terms together, factoring out -3.
- Group the first and fourth terms together, factoring out y^2; group the second and fifth terms, factoring out $3y$; and group the third and sixth terms, just showing a multiplication of 1.

Using the first choice:

$$2x^2y^2 + 6x^2y + 2x^2 - 3y^2 - 9y - 3 =$$
$$2x^2\left(y^2 + 3y + 1\right) - 3\left(y^2 + 3y + 1\right)$$

The common factor of the two terms is then factored out.

$$\left(y^2 + 3y + 1\right)\left(2x^2 - 3\right)$$

Now, using the second method, I first have to rearrange the terms:

$$2x^2y^2 + 6x^2y + 2x^2 - 3y^2 - 9y - 3 =$$
$$\left(2x^2y^2 - 3y^2\right) + \left(6x^2y - 9y\right) + \left(2x^2 - 3\right) =$$
$$y^2\left(2x^2 - 3\right) + 3y\left(2x^2 - 3\right) + 1\left(2x^2 - 3\right)$$

You see that the three terms now all have a common factor of $(2x^2 - 3)$, which can be factored out.

$$\left(2x^2 - 3\right)\left(y^2 + 3y + 1\right)$$

The order of the two factors is different from what you get using the other method, but multiplication is commutative, so they're equivalent.

EXAMPLE

Q. Factor $4ab^2 - 8ac^2 + 5x^2b - 10x^2c$.

A. Grouping and factoring,

$$\left(4ab^2 - 8ac^2\right) + \left(5x^2b - 10x^2c\right) =$$
$$4a\left(b^2 - 2c^2\right) + 5x^2\left(b - 2c\right)$$

The expressions in the parentheses look similar, but they aren't the same. Changing the order won't help in this case. There are now two terms, but they don't have a common factor. This expression is as simple as it can be. In other words, it's prime (in the algebraic sense).

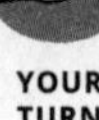

YOUR TURN

Factor by grouping.

23 $ab^2 + 2ab + b + 2$

24 $xz^2 - 5z^2 + 3x - 15$

25 $m^2n - 3m^2 - 4n + 12$

26 $n^{4/3} - 2n^{1/3} + n - 2$

27 $ax - 3x + ay - 3y + az - 3z$

28 $x^2y^2 + 3y^2 + x^2y + 3y - 6x^2 - 18$

33 $8a^3b^2 - 32a^3 - b^2 + 4$

34 $z^8 - 97z^4 + 1{,}296$ (***Hint:*** $1{,}296 = 81 \times 16$)

35 $4m^5 - 4m^4 - 36m^3 + 36m^2$

36 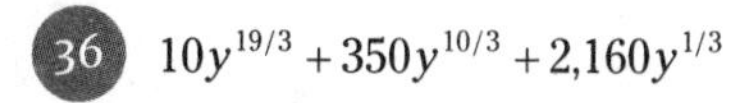
$10y^{19/3} + 350y^{10/3} + 2{,}160y^{1/3}$

Incorporating the Remainder Theorem

The *Remainder Theorem* is used heavily when you're dealing with polynomials of high degrees, and you want to graph them or find solutions for equations involving the polynomials. I go into these processes in great detail in *Algebra II For Dummies* (John Wiley & Sons, Inc.). For now, I pick out just the best part (lucky you) and show you how to make use of the Remainder Theorem and synthetic division to help you with your factoring chores.

The Remainder Theorem of algebra says that when you divide a polynomial by some linear binomial, the remainder resulting from the division is the same number as you'd get if you evaluated the polynomial using the opposite of the constant in the binomial.

"What?" you ask. Okay, in other words, if you were to divide the polynomial x^3+x^2-3x+4 by the binomial $x+1$ (which you find in Chapter 10), you could use synthetic division or do long division, as shown here:

$$
\begin{array}{r}
x^2 \quad -3 \\
x+1\overline{)x^3+x^2-3x+4} \\
\underline{-(x^3+x^2)} \\
0-3x+4 \\
\underline{-(-3x-3)} \\
7
\end{array}
$$

The remainder 7 is the same result you would get if you evaluated the polynomial $P(x)=x^3+x^2-3x+4$ for $x=-1$.

$$P(-1)=(-1)^3+(-1)^2-3(-1)+4=-1+1+3+4=7$$

(When using the Remainder Theorem, you really do want to use synthetic division instead of long division.)

So, formally, the Remainder Theorem says: The remainder, R, resulting from dividing $P(x)=a_nx^n+a_{n-1}x^{n-1}+a_{n-2}x^{n-2}+\cdots+a_1x^1+a_0$ by $x+b$ is equal to $P(-b)$.

What you prefer, in factoring polynomials, is that the remainder is a 0 — no remainder means that the factor divided evenly. Long division can be tedious, and even the evaluation of polynomials can be a bit messy. So, *synthetic division* comes to the rescue.

Synthesizing with synthetic division

Synthetic division is a way of dividing a polynomial by a first-degree binomial without all the folderol (refer to Chapter 10 to see how it's used in division). In this case, the folderol is all the variables — you just use coefficients and constants. To divide $P(x)=a_nx^n+a_{n-1}x^{n-1}+a_{n-2}x^{n-2}+\cdots+a_1x^1+a_0$ by $x+a$, you list all the coefficients, a_i, putting in zeros for missing terms in the decreasing powers, and then put an upside-down division sign in front of your work. You change the a in the binomial to its opposite and place it in the division sign. Then you multiply, add, multiply, add, and so on until all the coefficients have been added. The last number is your remainder.

Suppose you want to divide $x^4+5x^3-2x^2-28x-12$ by $x+3$ using synthetic division.

To do this, you would write the coefficients in a row and a -3 in front.

$$-3 \quad 1\ 5-2-28-12$$

Now bring the 1 down, multiply it times −3, put the result under the 5, and add. Multiply the sum by the −3, put it under the −2, and add. Multiply the sum times the −3, put the product under the 20, and so on.

$$\begin{array}{r|rrrrr} -3 & 1 & 5 & -2 & -28 & -12 \\ & & -3 & -6 & 24 & 12 \\ \hline & 1 & 2 & -8 & -4 & 0 \end{array}$$

The first four numbers along the bottom are the coefficients of the quotient, and the 0 is the remainder. When using synthetic division to help you with factoring, the 0 remainder is what you're looking for. It means that the binomial divides evenly and is a factor. The polynomial can now be written as follows:

$$=(x+3)(x^3+2x^2-8x-4)$$

WARNING

When rewriting a polynomial in factored form after applying synthetic division, be sure to change the sign of the number you used in the division to its opposite in the binomial.

Choosing numbers for synthetic division

Synthetic division is quick, neat, and relatively painless. But even quick, neat, and painless becomes tedious when you apply it without good results. When determining what may factor a particular polynomial, you need some clues. For example, you may be wondering if $(x-1)$, $(x+4)$, $(x-3)$, or some other binomials are factors of $x^4-x^3-7x^2+x+6$. I can tell just by looking that the binomial $(x+4)$ won't work and that the other two factors are possibilities. How can I do that?

The *rational root theorem* says that if a *rational number* (a number that can be written as a fraction) is a solution, *r*, of the equation

$$a_nx^n+a_{n-1}x^{n-1}+a_{n-2}x^{n-2}+\cdots+a_1x^1+a_0=0, \text{ then } r=\frac{\text{some factor of } a_0}{\text{some factor of } a_n}.$$

Using the rational root theorem for my factoring, I just find these possible solutions of the equation and do the synthetic division using only these possibilities.

Suppose I want to factor $x^4-x^3-7x^2+x+6$ using synthetic division, the rational root theorem, and the factor theorem. First, I make a list of the possible solutions if this were an equation. All the factors of the constant, a_0, are ±1, ±2, ±3, and ±6.

Next, I divide each of the factors by the factors of the lead coefficient, a_n. I caught a break here. The lead coefficient is a 1, so the divisions are just the original numbers.

Now I use synthetic division to see if I get a remainder of 0 using any of these numbers:

$$\begin{array}{r|rrrrr} 1 & 1 & -1 & -7 & 1 & 6 \\ & & 1 & 0 & -7 & -6 \\ \hline & 1 & 0 & -7 & -6 & 0 \end{array}$$

The number 1 is a solution, so $(x-1)$ is a factor. Dividing again, into the result:

$$\begin{array}{r|rrrr} -1 & 1 & 0 & -7 & -6 \\ & & -1 & 1 & 6 \\ \hline & 1 & -1 & -6 & 0 \end{array}$$

The number -1 is a solution, so $(x+1)$ is a factor. The numbers across the bottom are the coefficients of the trinomial factor multiplying the two binomial factors, so you can write

$$x^4 - x^3 - 7x^2 + x + 6 = (x-1)(x+1)(x^2 - x - 6)$$

What's even nicer is that the trinomial is easily factored, giving you an end result of

$$= (x-1)(x+1)(x-3)(x+2)$$

EXAMPLE

Q. Factor $4x^4 - 5x^3 - 99x^2 + 125x - 25$ using synthetic division, the rational root theorem, and the factor theorem.

A. Listing all the possible solutions: $\pm 1, \pm 5, \pm 25, \pm\frac{1}{2}, \pm\frac{1}{4}, \pm\frac{5}{2}, \pm\frac{5}{4}, \pm\frac{25}{2}, \pm\frac{25}{4}$.

Use synthetic division to look for remainders of 0. (Realistically, I won't guess all the correct answers without guessing some wrong ones, but I'm only showing you the ones that work.)

First, trying $x = 1$:

1	4	–5	–99	125	–25
		4	–1	–100	25
	4	–1	–100	25	0

The remainder is 0, so $(x-1)$ is a factor.

Next, trying 5 in the coefficients representing the quotient:

5	4	–5	–99	125	–25
		20	75	–120	25
	4	15	–24	5	0

The remainder is 0, so $(x-5)$ is a factor.

At this point, I would factor the trinomial represented by the quotient, $4x^2 + 19x - 5$, and show you that the remaining factors are $(4x-1)(x+5)$, but I want to take the opportunity to show you how the division looks when you try a fraction. I will try the number ¼ in the synthetic division. As a rule, when you're doing synthetic division using a fractional divisor, if you start getting fractions in the bottom row, it will not get better. Just stop — it's not an answer.

Trying in the coefficients representing the quotient,

¼	4	–5	–99	125	–25
		1	–1	–25	25
	4	–4	–100	100	0

Going back to the problem, and using the two factors from the division and the factors from the trinomial, the factorization of $4x^4 - 5x^3 - 99x^2 + 125x - 25 = (x-1)(x-5)(4x-1)(x+5)$

YOUR TURN

Factor the polynomials.

37 $x^3 + 9x^2 + 23x + 15$

 $2x^3 - 9x^2 - 200x + 900$

 $x^4 - 7x^3 + 36x$

Practice Questions Answers and Explanations

1. $\mathbf{2xy^2\left(4x^2 - 2xy + 7y^2\right)}$. The GCF is $2xy^2$.

2. $\mathbf{12w^2(3w^2 - 2w - 4)}$. The GCF is $12w^2$.

3. $\mathbf{5(x-3)(12x^5 - 36x^4 + 3x^2 - 18x + 28)}$.

$$15(x-3)^3 + 60x^4(x-3)^2 + 5(x-3) = 5(x-3)\left[3(x-3)^2 + 12x^4(x-3) + 1\right]$$

The trinomial in the brackets doesn't factor as a quadratic-like expression, so you need to expand each term by multiplying and simplifying.

$$5(x-3)\left[\left(3(x-3)^2 + 12x^4(x-3) + 1\right)\right] = 5(x-3)\left[3\left(x^2 - 6x + 9\right) + 12x^4(x-3) + 1\right]$$
$$= 5(x-3)\left[3x^2 - 18x + 27 + 12x^5 - 36x^4 + 1\right] = 5(x-3)\left[12x^5 - 36x^4 + 3x^2 - 18x + 28\right]$$

4. $\mathbf{5bcd(a+2a^2+6e+4b^2c)}$. The GCF is $5bcd$.

5. $\mathbf{(x-5)(x-3)}$, considering 15 and 1, or 5 and 3 for the factors of 15.

6. $\mathbf{(y-10)(y+4)}$, which needs opposite signs with factors of 40 to be 40 and 1, 20 and 2, 10 and 4, or 8 and 5.

7. $\mathbf{(2x-1)(x+2)}$. Both the first and the last numbers are prime, so the choices for the factors are limited.

8. $\mathbf{(2z+3)^2}$. $4z^2+12z+9=(2z+3)(2z+3)=(2z+3)^2$

9. $\mathbf{(w+4)(w-4)}$, which is a difference of squares.

10. $\mathbf{(6x+5)(2x-3)}$, considering $6=1\times6=2\times3$ and $15=1\times15=3\times5$.

11. $\mathbf{(x^5+3)(x^5+1)}$. Work from the corresponding quadratic, y^2+4y+3.

12. $\mathbf{(2y^8+3)(2y^8-3)}$, which is a difference of squares.

13. $\mathbf{(x^{-4}-8)(x^{-4}+1)}$. Use the corresponding quadratic, y^2-7y-8.

14. $\mathbf{(2z^{1/6}-1)(z^{1/6}-3)}$. Twice $\frac{1}{6}$ is $\frac{1}{3}$. Work from the quadratic, $2y^2-7y+3$. Remember that you add exponents when multiplying.

15. $\mathbf{3(z-2)^2}$. First, take out the GCF, 3. Then factor the trinomial in the parentheses.

$$3z^2-12z+12=3(z^2-4z+4)=3(z-2)(z-2)=3(z-2)^2$$

16. $\mathbf{5y(y-2)(y+1)}$. The GCF is $5y$. Factor that out first.

$$5y^3-5y^2-10y=5y(y^2-y-2)=5y(y-2)(y+1)$$

17. $\mathbf{x^4(x-9)^2}$.

$$x^6-18x^5+81x^4=x^4(x^2-18x+81)=x^4(x-9)(x-9)=x^4(x-9)^2$$

18. $\mathbf{(w+3)(w-3)(w+1)(w-1)}$. This is a quadratic-like expression. After factoring the trinomial, you find that both binomials are the difference of perfect squares.

$$w^4-10w^2+9=(w^2-9)(w^2-1)=(w+3)(w-3)(w+1)(w-1)$$

19. $\mathbf{3(x-2)^2(x+4)(x-1)}$. First, divide out the GCF, $3(x-2)^2$. Then you have a factorable trinomial in the parentheses.

$$3x^2(x-2)^2+9x(x-2)^2-12(x-2)^2=3(x-2)^2(x^2+3x-4)=3(x-2)^2(x+4)(x-1)$$

20. $\mathbf{(x+5)(x-5)(a-14)(a-1)}$. The GCF is (x^2-25), which is the difference of squares.

$$a^2(x^2-25)-15a(x^2-25)+14(x^2-25)=(x^2-25)(a^2-15a+14)$$
$$=(x+5)(x-5)(a-14)(a-1)$$

21. $\left(8y^{1/8}+5\right)\left(5y^{1/8}-3\right)$**.** This is a quadratic-like expression. Treat it like $40x^2+x-15$, which factors into $(8x+5)(5x-3)$. Using the variables and exponents in this problem, $40y^{1/4}+y^{1/8}-15=\left(8y^{1/8}+5\right)\left(5y^{1/8}-3\right)$.

22. $x^{-3}(6x-1)(4x-1)$**.** First, divide by the GCF to get $x^{-3}\left(1-10x^1+24x^2\right)$. You can factor the trinomial as it's written or rearrange the expression to read $x^{-3}\left(24x^2-10x+1\right)$. Now the trinomial factors, and you get $x^{-3}(6x-1)(4x-1)$.

23. $(b+2)(ab+1)$**.**

$$ab^2+2ab+b+2=ab(b+2)+1(b+2)=(b+2)(ab+1)$$

24. $(x-5)\left(z^2+3\right)$**.**

$$xz^2-5z^2+3x-15=z^2(x-5)+3(x-5)=(x-5)\left(z^2+3\right)$$

25. $(n-3)(m-2)(m+2)$**.** Be sure to factor -4 out of the second two terms to make the binomials match.

$$m^2n-3m^2-4n+12=m^2(n-3)-4(n-3)=(n-3)\left(m^2-4\right)$$

Now you factor the difference of squares. $(n-3)\left(m^2-4\right)=(n-3)(m-2)(m+2)$

26. $(n-2)\left(n^{1/3}+1\right)$**.** The GCF of the first two terms is $n^{1/3}$. The GCF of the last two terms is just 1. So $n^{4/3}-2n^{1/3}+n-2=n^{1/3}\left(n^1-2\right)+1(n-2)=(n-2)\left(n^{1/3}+1\right)$.

27. $(a-3)(x+y+z)$**.**

$$\begin{aligned}ax-3x+ay-3y+az-3z&=x(a-3)+y(a-3)+z(a-3)\\&=(a-3)(x+y+z)\end{aligned}$$

28. $\left(x^2+3\right)(y+3)(y-2)$**.**

$$\begin{aligned}x^2y^2+3y^2+x^2y+3y-6x^2-18&=y^2\left(x^2+3\right)+y\left(x^2+3\right)-6\left(x^2+3\right)\\&=\left(x^2+3\right)\left(y^2+y-6\right)\\&=\left(x^2+3\right)(y+3)(y-2)\end{aligned}$$

29. $5x(x+4)(x-4)$**.** First, factor 5x out of each term: $5x\left(x^2-16\right)$. Then you can factor the binomial as the sum and difference of the same two terms.

30. $(y+3)(y-3)(y+1)\left(y^2-y+1\right)$**.** Start by grouping the terms.

$$\begin{aligned}y^5-9y^3+y^2-9&=y^3\left(y^2-9\right)+1\left(y^2-9\right)\\&=\left(y^2-9\right)\left(y^3+1\right)\\&=(y+3)(y-3)(y+1)\left(y^2-y+1\right)\end{aligned}$$

31) $\mathbf{3x(x+5)(x-5)(x^2+3)}$. **First, take out the GCF.**

$$\begin{aligned}3x^5-66x^3-225x &= 3x\left(x^4-22x^2-75\right)\\ &= 3x\left(x^2-25\right)\left(x^2+3\right)\\ &= 3x(x+5)(x-5)\left(x^2+3\right)\end{aligned}$$

32) $\mathbf{(z+2)(z^2-2z+4)(z-2)(z^2+2z+4)}$. **First, factor the difference of squares.**

$$z^6-64=\left(z^3+8\right)\left(z^3-8\right)=(z+2)\left(z^2-2z+4\right)(z-2)\left(z^2+2z+4\right)$$

When you do this problem as the difference of squares rather than the difference of cubes, the second part of the factoring is easier.

33) $\mathbf{(b+2)(b-2)(2a-1)(4a^2+2a+1)}$. **There are four terms, so start with grouping.**

$$\begin{aligned}8a^3b^2-32a^3-b^2+4 &= 8a^3\left(b^2-4\right)-1\left(b^2-4\right)\\ &= \left(b^2-4\right)\left(8a^3-1\right)\\ &= (b+2)(b-2)(2a-1)\left(4a^2+2a+1\right)\end{aligned}$$

34) $\mathbf{(z^2+9)(z+3)(z-3)(z^2+4)(z+2)(z-2)}$. **Start with a quadratic-like trinomial.**

$$\begin{aligned}z^8-97z^4+1{,}296 &= \left(z^4-81\right)\left(z^4-16\right)\\ &= \left(z^2+9\right)\left(z^2-9\right)\left(z^2+4\right)\left(z^2-4\right)\\ &= \left(z^2+9\right)(z+3)(z-3)\left(z^2+4\right)(z+2)(z-2)\end{aligned}$$

35) $\mathbf{4m^2(m-1)(m+3)(m-3)}$.

$$\begin{aligned}4m^5-4m^4-36m^3+36m^2 &= 4m^2\left(m^3-m^2-9m+9\right)\\ &= 4m^2\left[m^2(m-1)-9(m-1)\right]\\ &= 4m^2\left[(m-1)\left(m^2-9\right)\right]\\ &= 4m^2(m-1)(m+3)(m-3)\end{aligned}$$

36) $\mathbf{10y^{1/3}(y+2)(y^2-2y+4)(y+3)(y^2-3y+9)}$. **Start with the GCF.**

$$\begin{aligned}10y^{19/3}+350y^{10/3}+2{,}160y^{1/3} &= 10y^{1/3}\left(y^6+35y^3+216\right)\\ &= 10y^{1/3}\left(y^6+35y^3+8\times 27\right)\\ &= 10y^{1/3}\left(y^3+8\right)\left(y^3+27\right)\\ &= 10y^{1/3}(y+2)\left(y^2-2y+4\right)(y+3)\left(y^2-3y+9\right)\end{aligned}$$

37) $(x+3)(x+1)(x+5)$**.** Listing the possible solutions, ±1, ±3, ±15. Using synthetic division and trying $x=-3$,

$$\begin{array}{r|rrrr} -3 & 1 & 9 & 23 & 15 \\ & & -3 & -18 & -15 \\ \hline & 1 & 6 & 5 & 0 \end{array}$$

The remainder is 0, so $(x+3)$ is a factor. Writing the trinomial from the bottom row, $x^2+6x+5=(x+1)(x+5)$. So, the factorization of $x^3+9x^2+23x+15=(x+3)(x+1)(x+5)$.

38) $(x-10)(2x-9)(x+10)$**.** Listing the possible solutions, ±1, ±2, ±3, ±4, ±5, ±6, ±9, ±10, ±12, ±15, ±18, ±20, ±25, ±30, ±36, ±45, ±50, ±60, ±75, ±90, ±100, ±150, ±180, ±225, ±300, ±450, ±900. Whew! The expression is to the third degree, so there will only be three factors. So many to choose from. And I haven't even divided each of these by 2, which will introduce 18 fractional choices. In a case like this, just try some numbers and hope you can avoid having to use fractions.

Trying $x=10$,

$$\begin{array}{r|rrrr} 10 & 2 & -9 & -200 & 900 \\ & & 20 & 110 & -900 \\ \hline & 2 & 11 & -90 & 0 \end{array}$$

The remainder is 0, so $(x-10)$ is a factor. The trinomial formed from the bottom row is $2x^2+11x-90$, which factors into $(2x-9)(x+10)$. So the factorization of $2x^3-9x^2-200x+900=(x-10)(2x-9)(x+10)$.

39) $x(x+2)(x-3)(x-6)$**.** First, factor out x to create a constant last term: $x^4-7x^3+36x=x(x^3-7x^2+36)$. When using synthetic division, be sure to put a 0 in the position of the x^1 term. The factors of 36 are ±1, ±2, ±3, ±4, ±6, ±9, ±12, ±18, ±36.

Trying $x=-2$,

$$\begin{array}{r|rrrr} -2 & 1 & -7 & 0 & 36 \\ & & -2 & 18 & -36 \\ \hline & 1 & -9 & 18 & 0 \end{array}$$

The remainder is 0, so $(x+2)$ is a factor. The trinomial formed from the bottom row is $x^2-9x+18$, which factors into $(x-3)(x-6)$. So the factorization of $x^4-7x^3+36x=x(x+2)(x-3)(x-6)$.

If you're ready to test your skills a bit more, take the following chapter quiz that incorporates all the chapter topics.

Whaddya Know? Chapter 13 Quiz

Quiz time! Complete each problem to test your knowledge on the various topics covered in this chapter. You can then find the solutions and explanations in the next section.

Factor each completely.

1. $10y^6 + 260y^3 - 270$
2. $12x^5 + 32x^4 + 20x^3$
3. $12x^2 + 7x - 12$
4. $y^{-4} + 13y^{-2} + 22$
5. $x^3 - 2x^2 - 11x + 12$
6. $5a^6 - 5a^5 - 8a^3 + 8a^2$
7. $x^4 + 3x^3 + 7x^2 + 15x + 10$
8. $4z^5 - 136z^3 + 900z$
9. $18x^4 - 12x^3 - 18x^2 + 12x$
10. $2x^3 - 5x^2 - 53x - 70$
11. $6y^3 - 24y^2 + 18y - 72$
12. $a^6 - 1$
13. $z^2 + 2z - 35$
14. $6b^{-2} - 7b^{-3} - 3b^{-4}$
15. $2x - 5x^{1/2} - 3$

Answers to Chapter 13 Quiz

1. **$10(y-1)(y^2+y+1)(y+3)(y^2-3y+9)$.** First, factor out the GCF.

   $10y^6+260y^3-270=10\left(y^6+26y^3-27\right)$. Now factor the quadratic-like trinomial.

   $10\left(y^6+26y^3-27\right)=10\left(y^3-1\right)\left(y^3+27\right)$. The two binomials are both factorable as the difference and sum of cubes.

2. **$4x^3\left(3x^2+8x+5\right)$.** The GCF is $4x^3$. First, factor out $4x^3$. The trinomial is not factorable.

3. **$(4x-3(3x+4)$.** Use unFOIL. The factors of 12 can be 1 and 12, 2 and 6, or 3 and 4. Using 3 and 4 for both 12s in the problem, you can create the products of 16 and 9, which have a difference of 7.

   $$12x^2+7x-12=(4x-3)(3x+4)$$

4. **$\left(y^{-2}+2\right)\left(y^{-2}+11\right)$.** Factor the quadratic-like trinomial by comparing it to $x^2+13x+22$. After performing the factoring, replace the x's with the y terms.

5. **$(x-1)(x+3)(x-4)$.** There is no GCF, and the four terms don't group. Use synthetic division to test the possible divisors (that will become part of the factorization). The choices are $\pm1,\ \pm2,\ \pm3,\ \pm4,\ \pm6,\ \pm12$.

   $$\begin{array}{r|rrrr} 1 & 1 & -2 & -11 & 12 \\ & & 1 & -1 & -12 \\ \hline & 1 & -1 & -12 & \end{array}$$

   The resulting quotient, x^2-x-12, factors into $(x+3)(x-4)$. Include the factor $x-1$, using the opposite sign of the divisor.

6. **$a^2(a-1)(5a^3-8)$.** Group the terms and find the GCF of each.

   $5a^6-5a^5-8a^3+8a^2=5a^5(a-1)-8a^2(a-1)$. Now factor out the GCF of the two terms.

   $5a^5(a-1)-8a^2(a-1)=a^2(a-1)(5a^3-8)$. If you had factored $+8a^2$ out of the last two terms, you could change the signs of the GCF and terms in the binomial.

7. **$(x^2+5)(x+1)(x+2)$.** There is no GCF, and the five terms don't group. Use synthetic division to test the possible divisors (that will become part of the factorization). The choices are $\pm1,\ \pm2,\ \pm5,\ \pm10$.

   $$\begin{array}{r|rrrrr} -1 & 1 & 3 & 7 & 15 & 10 \\ & & -1 & -2 & -5 & -10 \\ \hline & 1 & 2 & 5 & 10 & \end{array} \text{ and then } \begin{array}{r|rrrr} -2 & 1 & 2 & 5 & 10 \\ & & -2 & 0 & -10 \\ \hline & 1 & 0 & 5 & \end{array}$$

   The resulting quotient doesn't factor. So you have $(x+1)(x+2)(x^2+5)$.

8. **$4z\left(z+3\right)(z-3)(z+5)(z-5)$.** First, factor out the GCF.

   $4z^5-136z^3+900z=4z\left(z^4-34z^2+225\right)$. Next, factor the quadratic-like trinomial using the form $x^2-34x+225$: $4z\left(z^4-34z^2+225\right)=4z\left(z^2-9\right)\left(z^2-25\right)$. Both of the binomials factor.

9 $\mathbf{6x(3x-2)(x+1)(x-1)}$**.** First, factor out the GCF.

$18x^4-12x^3-18x^2+12x=6x\left(3x^3-2x^2-3x+2\right)$. The polynomial factors by grouping.

$$6x\left(3x^3-2x^2-3x+2\right)=6x\left[x^2(3x-2)-1(3x-2)\right]=6x\left[(3x-2)(x^2-1)\right]$$

Finally, you can factor the difference of squares.

10 $\mathbf{(2x+5)(x-7)(x+2)}$**.** There is no GCF, and the four terms don't group. Use synthetic division to test the possible divisors (that will become part of the factorization). The choices are $\pm1,\ \pm2,\ \pm5,\ \pm7,\ \pm10,\ \pm14,\ \pm35,\ \pm70,\ \pm\frac{1}{2},\ \pm\frac{5}{2},\ \pm\frac{7}{2},\ \pm\frac{35}{2}$.

$$\begin{array}{r|rrrr} 7 & 2 & -5 & -53 & -70 \\ & & 14 & 63 & 70 \\ \hline & 2 & 9 & 10 & \end{array}$$

The resulting quotient, $2x^2+9x+10=(2x+5)(x+2)$. Include the factor $x-7$, using the opposite sign of the divisor.

11 $\mathbf{6\left(y-4\right)\left(y^2+3\right)}$**.** First, factor out the GCF.

$6y^3-24y^2+18y-72=6\left(y^3-4y^2+3y-12\right)$. The polynomial can be factored by grouping:

$$6\left(y^3-4y^2+3y-12\right)=6\left[y^2(y-4)+3(y-4)\right]=6\left[(y-4)(y^2+3)\right]$$

12 $\mathbf{(a+1)(a^2-a+1)(a-1)(a^2+a+1)}$**.** First, factor the binomial as the difference of squares. $a^6-1=\left(a^3+1\right)\left(a^3-1\right)$

Then factor the two resulting binomials — sum and difference of cubes.

13 $\mathbf{(z+7)(x-5)}$**.** Use unFOIL. The factors of 35 are 5 and 7, and their difference is 2. The 7 needs to be positive.

$$z^2+2z-35=(z+7)(x-5)$$

14 $\mathbf{b^{-4}\left(3b+1\right)\left(2b-3\right)}$**.** First, factor out the GCF.

$6b^{-2}-7b^{-3}-3b^{-4}=b^{-4}\left(6b^2-7b-3\right)$. Now factor the trinomial:
$b^{-4}\left(6b^2-7b-3\right)=b^{-4}\left(3b+1\right)\left(2b-3\right)$.

15 $\mathbf{\left(x^{1/2}-3\right)\left(2x^{1/2}+1\right)}$**.** Factor the quadratic-like trinomial by comparing it to $2y^2-5y-3$. Position the factors so that the 2 multiplies the 3 and the 1 multiplies the 1 to have a difference of 5. $2y^2-5y-3=(y-3)(2y+1)$. Then replace the y's with the x terms.

5

Solving Linear and Polynomial Equations

Contents at a Glance

» Keeping your balance while performing equation manipulations

» Doing a reality check and reconciling your work

» Recognizing practical applications and using them

Chapter 14

Establishing Ground Rules for Solving Equations

In this chapter, you find many different considerations involving solving equations in algebra. In earlier chapters, I cover the mechanics of working with algebraic expressions correctly. Now I put those rules to work by introducing the equal sign (=). Just as a verb makes a phrase into a sentence, an equal sign makes an expression into an equation. In this chapter, instead of dealing with expressions, such as $3x+2$, I show you how to prepare for solving equations, such as $3x+2=11$.

Two-term equations, such as $3x=-15$, unlike two-term presidents, are pretty simple. Master the easier equations, and you can apply the techniques you use on these equations to those more complicated ones. Introduce more than two terms or make the exponent on the variable bigger than 1, and you have many possibilities for solutions of simple equations.

Creating the Correct Setup for Solving Equations

Different types of equations take different types of handling in order to solve for the correct solution, all the solutions, and not too many solutions. The setup of the equation depends on which type of equation you're dealing with at that time.

Setting up equations for further action

Here's a list of the most common algebraic equation types and their most general format.

- **Linear equation:** $ax+b=c$
- **Quadratic equation:** $ax^2+bx+c=0$
- **Cubic equation:** $ax^3+bx^2+cx+d=0$
- **General polynomial equation:** $a_nx^n+a_{n-1}x^{n-1}+a_{n-2}x^{n-2}+\cdots+a_1x+a_0=0$
- **Radical equation:** $\sqrt{ax+b}=c$
- **Rational equation:** $\frac{ax+b}{x+c}+\frac{dx+e}{x+f}+\cdots=0$

The convention is to use the letters toward the beginning of the alphabet as numbers or constants and the letters toward the end of the alphabet for variables.

In general, equations are usually set equal to 0 or some constant. I cover the differences and similarities in detail in the next few chapters.

Making plans for solving equations

In the previous section, I show you the usual setup for the types of equations you'll be solving. Now I show you how to change each of the equation types so you can determine the solution.

- **Linear:** $ax+b=c \rightarrow ax=c-b \rightarrow x=\frac{c-b}{a}$. Isolate the term with the *x*, and then multiply or divide to solve for *x*.
- **Quadratic:** $ax^2+bx+c=0 \rightarrow a(x+d)(x+e)=0$ where $c=ade$ and $b=a(d+e) \rightarrow x=-d$ or $x=-e$. Factor the expression that's set equal to 0. Then use the *multiplication property of zero* to solve for the two solutions. If the trinomial doesn't factor, then use the *quadratic formula.*
- **Cubic:** $ax^3+bx^2+cx+d=0 \rightarrow a(x+e)(x+f)(x+g)=0$ where $d=aefg$ and *b* and *c* are the results of the corresponding multiplications $\rightarrow x=-e$ or $x=-f$ or $x=-g$. Factor the expression that's set equal to 0. Then use the *multiplication property of zero* to solve for the three solutions.

» **General polynomial:** $a_nx^n + a_{n-1}x^{n-1} + a_{n-2}x^{n-2} + \ldots + a_1x + a_0 = 0 \rightarrow a_n(x+b_n)(x+b_{n-1})(x+b_{n-2})\ldots(x+b_0) = 0 \rightarrow x = -b_n$ or $x = -b_{n-1}$ or $x = -b_{n-2}\ldots$ or $x = -b_0$. Factor the expression that's set equal to 0. Then use the *multiplication property of zero* to solve for the *n* solutions.

» **Radical:** $\sqrt{ax+b} = c \rightarrow ax+b = c^2 \rightarrow ax = c^2 - b \rightarrow x = \dfrac{c^2-b}{a}$. Square both sides, and then solve the linear equation that results. Be sure to check for extraneous roots.

» **Rational:** $\dfrac{ax+b}{x+c} + \dfrac{dx+e}{x+f} + \cdots = 0 \rightarrow \dfrac{(x+g)(x+h)(x+i)\cdots}{(x+j)(x+k)(x+l)\cdots} = 0 \rightarrow x = -g$ or $x = -h$ or $x = -i$, and so on. You set only the factors in the numerator equal to zero to solve for the solutions.

EXAMPLE

Q. Check the solution of $2x^2 + 5x - 3 = 0$.

A. The quadratic equation $2x^2 + 5x - 3 = 0$ becomes $(2x-1)(x+3) = 0$, which solves as $x = \frac{1}{2}$ or $x = -3$. Checking $x = \frac{1}{2}$ in the original equation, $2\left(\frac{1}{2}\right)^2 + 5\left(\frac{1}{2}\right) - 3 = 2\left(\frac{1}{4}\right) + 5\left(\frac{1}{2}\right) - 3 = \frac{1}{2} + \frac{5}{2} - 3 = 3 - 3 = 0$, which is a true statement. Checking -3 in the original equation, $2(-3)^2 + 5(-3) - 3 = 2(9) + 5(-3) - 3 = 18 - 15 - 3 = 0$, which is a true statement. You'll find the steps necessary to create the solution in this and later chapters.

Q. Check the solution of $\sqrt{4x+8} = 2$.

A. The radical equation $\sqrt{4x+8} = 2$ becomes $4x + 8 = 4$, which becomes $4x = -4$, which solves as $x = -1$. Checking this in the original equation, $\sqrt{4(-1)+8} = 2$ becomes $\sqrt{4} = 2$, which is a true statement. You'll find the steps necessary to create the solution in this and later chapters.

YOUR TURN

Show that the given solution or solutions create a true statement.

1 $x = 2$ in the linear equation $5x - 6 = 4$.

2 $x = 0$ or $x = 3$ or $x = -4$ in the cubic equation $x^3 + x^2 - 12x = 0$.

3 $x = 4$ or $x = -4$ in the radical equation $\sqrt{x^2 - 7} = 3$.

4 $x = 3$ or $x = 6$ in the rational equation $\dfrac{5}{x-1} + \dfrac{x}{2} = 4$.

Keeping Equations Balanced

When presented with an algebraic equation, your usual task is to solve the equation. *Solving* an equation means to find the value(s) that replace the unknown(s) to make the equation a true statement. You may be able to just guess the answer, but you can't rely on that method for all equations. Some answers are just too darned hard to find by guessing.

In this section, I tell you about the tried-and-true, approved methods of changing the original equation so that it's in a format that shows you the solution.

Balancing with binary operations

One of the most efficient and easiest methods of changing the format of an equation is to perform arithmetic operations on each side of the equation. Picture a teeter-totter or balance scale. When the teeter-totter is in balance, the two ends are at the same level, and the board is parallel to the ground. With a balance scale, you weigh items by placing the object in one tray and then adding known weights to the other tray until the two trays are in balance. Algebraic equations start out in balance, and your task is to keep them that way.

Adding to each side or subtracting from each side

When changing the format of an equation, if you add some amount (or subtract some amount) from one side of the equation, then you must do the same thing to the other side.

The following examples show how to add or subtract from each side of the equation.

EXAMPLE

Q. Subtract 10 from each side of $x^2+5x=10$.

A. If $x^2+5x=10$, then $x^2+5x-10=10-10$, or $x^2+5x-10=0$.

Q. Add 10 to each side of $\sqrt{5x-2}-10=x$.

A. If $\sqrt{5x-2}-10=x$, then $\sqrt{5x-2}-10+10=x+10$, or $\sqrt{5x-2}=x+10$.

Multiplying each side by the same number

You can multiply each side of an equation by any number and not change the equality of the statement. You can even multiply each side by 0 and not change the equality (because you'd have $0=0$), but you wouldn't have much to work with if you did that. And you can change a false statement into a true statement by multiplying each side by 0. That's not playing fair.

EXAMPLE

Q. Multiply both sides of $\frac{3x}{10}+2=x$ by 10.

$$10\left(\frac{3x}{10}+2\right)=10(x)$$

A. $$\cancel{10}\left(\frac{3x}{\cancel{10}}\right)+10(2)=10x$$

$$3x+20=10x$$

Remember: You have to multiply each of the three terms by 10.

To show you that a reality check can save you from making a big error, pretend that you didn't really think this through and decided to solve the problem with the following equation:

$$\frac{c}{11} = 330$$

The letter c represents the number of soccer clubs. You divide c by the number of players in each club and set it equal to the total number of players.

You used the variable and the two numbers in the problem. Does it matter what you use where? Will the equation give you a reasonable answer?

Multiply each side by 11 to solve for c:

$$11 \cdot \frac{c}{11} = 330 \cdot 11$$
$$\cancel{11} \cdot \frac{c}{\cancel{11}} = 330 \cdot 11$$
$$c = 3{,}630 \text{ clubs}$$

Humph. This can't be right. The answer doesn't make any sense — only 330 players are involved. The answer may satisfy your equation, but if it doesn't make sense, the equation could be wrong.

A quick look at the equation shows that it should have read: 11 players per club $\times$ the number of clubs $=$ the total number of players: $11c = 330$.

Now, solve this:

$$11c = 330$$

Divide each side by 11:

$$\frac{11}{11}c = \frac{330}{11}$$
$$\frac{\cancel{11}}{\cancel{11}}c = \frac{330}{11}$$
$$c = 30 \text{ clubs}$$

That makes much more sense.

REMEMBER

You can solve an equation correctly, but that doesn't mean you chose the right equation to solve the problem in the first place. Make sure that your answer makes sense.

Thinking like a car mechanic when checking your work

A more complete check of algebraic processes involves checking the computations and algebraic operations. When a car mechanic has a spiffy-doodle computer to run a diagnostic check on your car, they find all the problems very efficiently. If your car is older, though, or they don't have that kind of electronic setup, then it's a step-by-step, point-by-point check of all the essential parts. This is more like an algebraic check.

I want to see how good you are at checking work. Here's a problem that a student did, and the answer is wrong. It's more helpful to someone who's made an error when you can point out where the error is in their computations. Can you find it? (The answer is −2.)

$$\frac{-6\left[3^2+4-5(2)\right]}{\sqrt{16}+5}=\frac{-6\left(3^2\right)-6(4)-6\left[5(2)\right]}{4+5}=$$
$$\frac{-6(9)-6(4)-6(10)}{9}=\frac{-54-24-60}{9}=-\frac{138}{9}=-\frac{46}{9}$$

You probably spotted the error right away because one of the most common mistakes in distributing is in not distributing the negative signs correctly. Yes, you're right, the third term in the top-right fraction should be $-6[-5(2)]$. (I show you the distribution of signed numbers in Chapter 9, if you need a refresher.)

EXAMPLE

Q. A student tried to solve the proportion $\frac{x-3}{4}=\frac{1}{2}$ and used the equation $2x-3=4$ to get the answer $x=\frac{7}{2}$. The answer is $x=5$. What went wrong?

A. When solving proportions, you can cross-multiply. When multiplying the $x-3$ by 2, the student forgot to distribute the 2 over both the x and the 3.

Q. You're putting a new tile floor in your basement rec room. The room measures 30 feet by 40 feet, and the tile you want costs $8 per square foot. You estimate that your cost will be about $960. Is this right?

A. No. The estimate is way off. An area of 30 by 40 feet comes to 1200 square feet. Multiplying by 8, you have $9,600 dollars. The decimal point was in the wrong place.

YOUR TURN

21 You recently went to the store and bought ten packs of gum and ten cans of soda. The gum was 79 cents per pack, and the soda was 81 cents per can. The clerk rang up the order and said that the total, without tax, was $8.71. What's wrong with this result?

22 You're calculating the area of a circular garden area that's being built in your back yard. You know that the area of a circle is found with $A=\pi r^2$, where r is the radius of the circle and π is about 3.14. Your dad measures the radius to be 10 feet and says that the area must be about 990 square feet. You know that the estimate is incorrect; what happened here?

23 There are 70 people in your club, and each brought 70 cookies to the party. The hostess decided to divide the cookies up and display them on 7 tables, putting 100 cookies on each table. With that plan in mind, she didn't have enough platters to display the cookies, because it was more than 100 per table. What happened to the math?

24 You're working at a bank and have a barrel with 1 million nickels in it. Each nickel weighs 5 grams, and you need to determine how many kilograms the nickels weigh. A fellow employee says there are 50 kilograms of nickels. You're pretty sure that's off by quite a bit. What happened here? (Recall: 1 kilogram = 1,000 grams.)

Practice Problems Answers and Explanations

1. Replacing the x with 2 in the equation, $5(2)-6=4$ gives you $10-6=4$, which is a true statement.

2. Replacing the x with 0 in the equation, $0^3+0^2-12(0)=0$ gives you $0+0-0=0$, which is true.

   Replacing the x with 3, $3^3+3^2-12(3)=0$ gives you $27+9-36=0$ or $36-36=0$, which is true.

   Replacing the x with -4, $(-4)^3+(-4)^2-12(-4)=0$ gives you $-64+16+48=0$ or $-64+64=0$, which is true.

3. Replacing the x with 4 in the equation, $\sqrt{4^2-7}=3$ gives you $\sqrt{16-7}=3$ or $\sqrt{9}=3$, which is true.

   Replacing the x with -4 in the equation, $\sqrt{(-4)^2-7}=3$ gives you $\sqrt{16-7}=3$ or $\sqrt{9}=3$, which is true.

4. Replacing the x with 3 in the equation, $\frac{5}{3-1}+\frac{3}{2}=4$ becomes $\frac{5}{2}+\frac{3}{2}=\frac{8}{2}=4$, which is true.

   Replacing x with 6 in the equation, $\frac{5}{6-1}+\frac{6}{2}=4$ becomes $\frac{5}{5}+\frac{6}{2}=1+3=4$, which is true.

5. Adding 5 to both sides of $x^2-5=4$, you have $x^2-5+5=4+5$ or $x^2=9$. Replace the x with -3 to get $3^2=9$ or $9=9$. This is a true statement.

6. Multiplying both sides of the equation $\frac{2x}{3}=\frac{6}{x}$ by 3, you have $\frac{3}{1}\cdot\frac{2x}{3}=\frac{3}{1}\cdot\frac{6}{x}$, which becomes $\frac{\cancel{3}}{1}\cdot\frac{2x}{\cancel{3}}=\frac{3}{1}\cdot\frac{6}{x}$ or $2x=\frac{18}{x}$. Replace the x with -3 to get $2(-3)=\frac{18}{-3}$ or $-6=-6$, a true statement.

7. Dividing both sides of the equation $4\sqrt{x+19}=13-x$ by 4, you have $\frac{\cancel{4}\sqrt{x+19}}{\cancel{4}}=\frac{13-x}{4}$ or $\sqrt{x+19}=\frac{13-x}{4}$. Note that both terms on the right are divided by 4. Replacing the x with -3, you have $\sqrt{-3+19}=\frac{13-(-3)}{4}$ or $\sqrt{16}=\frac{16}{4}$, which says $4=4$, a true statement.

8. Subtracting 15 from both sides of the equation $x^3+x^2-x+15=0$, you have $x^3+x^2-x+15-15=0-15$, which becomes $x^3+x^2-x=-15$. Replacing each x with -3, you have $(-3)^3+(-3)^2-(-3)=-15$ or $-27+9+3=-15$ or $-15=-15$, a true statement.

9. Squaring both sides of the equation $\sqrt{x+1}=\sqrt{9}$ gives you $\left(\sqrt{x+1}\right)^2=\left(\sqrt{9}\right)^2$ or $x+1=9$. Replacing the x with 8, you have $8+1=9$, a true statement.

10. Squaring both sides of the equation $\sqrt{5x-5}=\sqrt{25}$ gives you $\left(\sqrt{5x-5}\right)^2=\left(\sqrt{25}\right)^2$ or $5x-5=25$. Replacing the x with 6, you have $5(6)-5=30-5=25$, which is a true statement.

11. Squaring both sides of the equation $\sqrt{3-2x}=\sqrt{6-x}$ gives you $\left(\sqrt{3-2x}\right)^2=\left(\sqrt{6-x}\right)^2$ or $3-2x=6-x$. Replacing the x's with -3, you have $3-2(-3)=6-(-3)$ or $3+6=6+3$, which is a true statement.

12. Squaring both sides of the equation $\sqrt{x^2-16}=\sqrt{x+4}$ gives you $\left(\sqrt{x^2-16}\right)^2=\left(\sqrt{x+4}\right)^2$ or $x^2-16=x+4$. Replacing the x's with 5, you have $5^2-16=5+4$ or $25-16=9$, which is a true statement.

13. $\boldsymbol{x=\pm 9}$**.** $\sqrt{x^2}=\pm\sqrt{81}$ so $x=\pm 9$.

14. $\boldsymbol{x=11}$ **and** $\boldsymbol{x=-9}$**.** $\sqrt{(x-1)^2}=\pm\sqrt{100}$, so $x-1=\pm 10$. You have two situations: some number minus 1 equals 10, and that would be $x=11$. The other situation is that a number minus 1 is -10. One less than -9 is -10, so $x=-9$.

15. $x = 3$. $\sqrt[3]{x^3} = \sqrt[3]{27}$, so $x = 3$.

16. $x = 0$. $\sqrt[3]{(1-x)^3} = \sqrt[3]{1}$ becomes $1 - x = 1$. The only value x can have is 0.

17. $x = 12$. The reciprocal of $\frac{2}{3}$ is $\frac{3}{2}$. Multiplying, $\frac{\cancel{3}}{\cancel{2}} \cdot \frac{\cancel{2}}{\cancel{3}} x = \frac{3}{\cancel{2}} \cdot \cancel{8}^4$, which gives you $x = 3 \times 4 = 12$.

18. $y = 10$. The reciprocal of $-\frac{4}{5}$ is $-\frac{5}{4}$. Multiplying, $\left(-\frac{\cancel{5}}{\cancel{4}}\right)\left(-\frac{\cancel{4}}{\cancel{5}}\right)y = \left(-\frac{5}{\cancel{4}}\right)\left(-\cancel{8}^2\right)$, which gives you $y = (-5)(-2) = 10$.

19. $z = 700$. The fractional equivalent of 0.01 is $\frac{1}{100}$, and the reciprocal of that fraction is 100. Multiplying, $100(0.01)z = (100)(7)$, which gives you $z = 700$.

20. $w = -1.25$. The fractional equivalent of 6.4 is $6\frac{4}{10} = 6\frac{2}{5} = \frac{32}{5}$. The reciprocal of the fraction is $\frac{5}{32}$. Multiplying, $\frac{\cancel{5}}{\cancel{32}} \cdot \frac{\cancel{32}}{\cancel{5}} w = \left(\frac{5}{{}_4\cancel{32}}\right)(-\cancel{8})$, which gives you $w = -\frac{5}{4} = -1\frac{1}{4} = -1.25$.

21. You were estimating the cost in your head and figured that the bill would be something less than $20. You did that by just rounding up each item to one dollar and multiplying 20 items times $1. The $8.71 seems a bit too low. It appears that the clerk didn't use the distributive property correctly (yes, this can happen with incorrect entries in the cash register). The total amount should have been $10(0.79 + 0.81) = 7.90 + 8.10 = 16.00$. If the distributive property isn't performed correctly, and the second number in the parentheses isn't multiplied, then the incorrect total can come from $7.90 + 0.81 = 8.71$.

22. You figured that the area would be 314 square feet, because 10 squared is 100, and 100 times 3.14 is 314. How did your dad get a number three times as large? It looks like he used the order of operations incorrectly. If he took $A = \pi r^2$ and multiplied π times 10 first, to get 31.4, and then he squared the product, he would get 985.96 for an answer. Not right.

23. It looks like division gone awry. One explanation is that she divided both 70's by 7: $\frac{\cancel{7}0 \cdot \cancel{7}0}{\cancel{7}}$. This would produce an answer of 100. The actual number of cookies per table should be 10(70) or 700.

24. This is a case of decimal-dopiness. Watch the decimal points! One million has six zeros. Multiply by 5, and the result is 5,000,000 grams. Divide by 1,000 to get 5,000 kilograms. The decimal place was misplaced.

If you're ready to test your skills a bit more, take the following chapter quiz that incorporates all the chapter topics.

Whaddya Know? Chapter 14 Quiz

Quiz time! Complete each problem to test your knowledge on the various topics covered in this chapter. You can then find the solutions and explanations in the next section.

1. The distance you ride your bike every day is about 239,000 feet (about 45 miles). Which is the best estimate of that distance in inches: 3,000,000 inches or 30,000,000 inches or 300,000,000 inches?

2. Determine which is the correct solution to $\sqrt{2-x}=8$: $x=-62$ or $x=-10$.

3. Rewrite the equation $\frac{\sqrt{x-4}}{6}=2$ by multiplying each side by 6.

4. What is the reciprocal of $1\frac{3}{4}$?

5. Determine which is the correct solution to $\frac{4x-3}{7}=3$: $x=6$ or $x=-3$.

6. You worked 10 hours per day for 14 days and earned about 10 dollars per hour. Which is a more accurate estimate of your total wages (before taxes, and so on): $x=\$140$ or $x=\$1,400$ or $x=\$14,000$?

7. Rewrite the equation $3x^2-1=47$ by adding 1 to each side.

8. Square both sides of the equation $\sqrt{x^2-1}=\sqrt{15}$.

9. Determine which is the correct solution to $3x-1=-10$: $x=3$ or $x=-3$.

10. What is the reciprocal of -3?

Answers to Chapter 14 Quiz

1. **3,000,000 inches.** Multiplying 200,000 times 12 gives you 2,400,000 inches. Multiplying 300,000 times 12 gives you 3,600,000 inches. You want something between 2.4 million and 3.6 million inches.

2. $\mathbf{x = -62}$**.** Substituting into $\sqrt{2-x}=8$, $\sqrt{2-(-62)}=8$ becomes $\sqrt{2+62}=8$ and $\sqrt{64}=8$.

3. $\mathbf{\sqrt{x-4}=12}$**.** $\frac{6}{1}\cdot\frac{\sqrt{x-4}}{6}=2\cdot 6 \rightarrow \frac{\cancel{6}}{1}\cdot\frac{\sqrt{x-4}}{\cancel{6}}=2\cdot 6 \rightarrow \sqrt{x-4}=12$.

4. $\mathbf{\frac{4}{7}}$**.** $1\frac{3}{4}=\frac{7}{4}$. Then "flip" the improper fraction.

5. $\mathbf{x = 6}$**.** Substituting into $\frac{4x-3}{7}=3$, $\frac{4(6)-3}{7}=3$ becomes $\frac{24-3}{7}=\frac{21}{7}=3$.

6. **$1,400.** Multiply 10 times 14 for the total number of hours. Then multiply by 10 for the amount earned.

7. $\mathbf{3x^2=48}$**.** $3x^2-1+1=47+1$.

8. $\mathbf{x^2-1=15}$**.** $\left(\sqrt{x^2-1}\right)^2=\left(\sqrt{15}\right)^2 \rightarrow x^2-1=15$.

9. $\mathbf{x = -3}$**.** Substituting into the equation, $3x-1=-10$ becomes $3(-3)-1=-10$ and simplifies to $-9-1=-10$ or $-10=-10$.

10. $\mathbf{-\frac{1}{3}}$.

» Making grouping symbols and fractions work

» Putting proportions in their place

Chapter 15

Lining Up Linear Equations

Linear equations consist of some terms that have variables and others that are constants. A standard form of a linear equation is $ax + b = c$. What distinguishes linear equations from the rest of the pack is the fact that the variables are always raised to the first power. If you're looking for squared variables or variables raised to higher or more exciting powers, turn to Chapters 16 and 17 for information on dealing with those types of equations.

In this chapter, I take you through many different types of opportunities for dealing with linear equations. Most of the principles you use with these first-degree equations are applicable to the higher-order equations, so you don't have to start from scratch later on.

When you use algebra in the real world, more often than not you turn to a formula to help you work through a problem. Fortunately, when it comes to algebraic formulas, you don't have to reinvent the wheel: You can make use of standard, tried-and-true formulas to solve some common, everyday problems. I show you how to change the format or adapt many of your favorite formulas to make them more usable for your particular situation.

Playing by the Rules

When you're solving equations with just two terms or three terms or even more than three terms, the big question is: "What do I do first?" Actually, as long as the equation stays balanced, you can perform any operations in any order. But you also don't want to waste your time performing operations that don't get you anywhere or even make matters worse.

The basic process behind solving equations is to use the *reverse* of the order of operations.

REMEMBER

The order of operations (see Chapter 7) is powers or roots first, then multiplication and division, and addition and subtraction last. Grouping symbols override the order. You perform the operations inside the grouping symbols to get rid of them first.

So, reversing the order of operations, the basic process is:

1. **Do all the addition and subtraction.**

   Combine all terms that can be combined both on the same side of the equation and on opposite sides using addition and subtraction.

2. **Do all multiplication and division.**

   This step is usually the one that isolates or solves for the value of the variable or some power of the variable.

3. **Multiply exponents and find the roots.**

   Powers and roots aren't found in these linear equations — they come in quadratic and higher-powered equations. But these would come next in the reverse order of operations.

When solving linear equations, the goal is to isolate the variable you're trying to find the value of. Isolating it, or getting it all alone on one side, can take one step or many steps. And it has to be done according to the rules — you can't just move things willy-nilly, helter-skelter, hocus-pocus (you get the idea).

Using the Addition/Subtraction Property

One of the most basic properties of equations is that you can add or subtract the same amount from each side of the equation and not change the balance or *equality*. The equation is still a true statement (as long as it started out that way) after adding or subtracting the same from each side. You use this property to get all the terms with the variable you want to solve for to one side and all the other letters and numbers to the other side so that you can solve the equation for the value of the variable.

TIP

You can check the solution by putting the answer back in the original equation to see whether it gives you a true statement.

EXAMPLE

Q. Solve for x: $x+7=11$.

A. Subtract 7 from each side (same as adding −7 to each side), like this:

$$\begin{array}{r} x+7=11 \\ \underline{-7\ \ -7} \\ x\quad\ =4 \end{array}$$

You can do a quick check and see that, indeed, $4+7$ does equal 11.

Q. Solve for y: $8y-2=7y-10$.

A. First, add $-7y$ to each side to get rid of the variable on the right (this moves the variable to the left with the other variable term), and then add 2 to each side to get rid of the -2 (this gets the numbers together on the right). This is what the process looks like:

$$\begin{array}{r} 8y-2=7y-10 \\ \underline{-7y\quad -7y\quad\ } \\ y-2=\quad -10 \\ \underline{+2\qquad +2} \\ y\quad =\quad -8 \end{array}$$

Checking your answer in the original equation, $8y-2=7y-10$, use −8 in place of y.

$$\begin{aligned} 8(-8)-2&=7(-8)-10 \\ -64-2&=-56-10 \\ -66&=-66 \end{aligned}$$

YOUR TURN

1. Solve for x: $x+4=15$.

2. Solve for y: $y-2=11$.

3. Solve for x: $5x+3=4x-1$.

4. Solve for y: $2y+9+6y-8=4y+5+3y-11$.

Using the Multiplication/Division Property

The following equations are all examples of linear equations in two terms:

$14x = 84$ $\quad -64 = 8y$ $\quad \frac{9z}{5} = 18$ $\quad \frac{7w}{6} = \frac{35}{9}$

Linear equations that contain just two terms are solved with multiplication, division, reciprocals, or some combinations of the operations.

Devising a method using division

One of the most basic methods for solving equations is to divide each side of the equation by the same number. Many formulas and equations include a *coefficient* (multiplier) with the variable. To get rid of the coefficient and solve the equation, you divide. The following examples take you step by step through solving with division.

EXAMPLE

Q. Solve for x in $20x = 170$.

A. $x = 8.5$.

1. **Determine the coefficient of the variable and divide both sides by it.**

   Because the equation involves multiplying by 20, undo the multiplication in the equation by doing the opposite, which is division. Divide each side by 20:

   $$\frac{20x}{20} = \frac{170}{20}$$

2. **Reduce both sides of the equal sign.**

   $$\frac{\cancel{20}x}{\cancel{20}} = \frac{170}{20}$$
   $$x = 8.5$$

   Remember: Do unto one side of the equation what the other side has had done unto it.

Q. You need to buy 300 donuts for a big meeting. How many dozen donuts is that?

A. Let d represent the number of dozen donuts you need. There are 12 donuts in a dozen, so $12d = 300$. Twelve times the number of donuts you need has to equal 300.

1. **Determine the coefficient of the variable and divide both sides by it.**

   Divide each side by 12.

   $$\frac{12d}{12} = \frac{300}{12}$$

2. **Reduce both sides of the equal sign.**

   $d = 25$ dozen donuts

In Chapter 4, I show you how to change a repeating decimal to a fraction by putting as many 9's as there are repeating digits. Place the 9's under the repeating digits and then reduce the fraction you created. But what if not all the digits repeat? See the following example.

EXAMPLE

Q. Change the decimal $0.1388\overline{8}$ to a fraction.

A. You want to create a subtraction problem with the repeating digits, only, on the right side of the decimal point.

First, name the decimal number N, so you have $N = 0.1388\overline{8}$.

Next, multiply N by 100 to move the 1 and 3 to the left of the decimal point.

$$100N = 13.88\overline{8}$$

Now, multiply N by 1,000 to move the 1, the 3, and one of the 8's to the left of the decimal point.

$$1{,}000N = 138.88\overline{8}$$

Subtract $1{,}000N - 100N$.

$$\begin{aligned} 1{,}000N &= 138.88\overline{8} \\ -100N &= 13.88\overline{8} \\ \hline 900N &= 125 \end{aligned}$$

Now solve for N by multiplying by the reciprocal of 900 and reducing the fraction.

$$\frac{125}{900} = \frac{\cancel{125}^{5}}{\cancel{900}_{36}} = \frac{5}{36}$$

This tells you that $0.1388\overline{8} = \frac{5}{36}$.

YOUR TURN

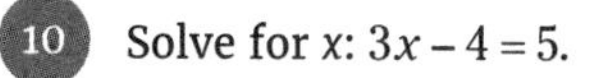

10 Solve for x: $3x - 4 = 5$.

11 Solve for y: $8 - \frac{y}{2} = 7$.

12 Solve for x: $5x-3=8x+9$.

13 Solve for z: $\frac{z}{6}-3=z+7$.

14 Solve for y: $4y+16-3y=7+3y$.

15 Solve for x: $\frac{3x}{4}-2=\frac{9x}{4}+13$.

16 Find the fraction equivalent for the repeating decimal $0.722\overline{2}$.

17 Find the fraction equivalent for the repeating decimal $0.6722\overline{2}$.

Solving Linear Equations with Grouping Symbols

The most general procedure to use when solving linear equations is to add and subtract first and then multiply or divide. This general rule is interrupted when the problem contains grouping symbols such as ( ), [ ], or { }. (See Chapter 3 for more on grouping symbols.) Linear equations don't always start out in the nice, $ax+b=c$ form. Sometimes, because of the complexity of the application, a linear equation can contain multiple variable and constant terms and lots of grouping symbols, such as in this equation:

$$3\left[4x+5(x+2)\right]+6=1-2\left[9-2(x-4)\right]$$

The different types of grouping symbols are used for *nested expressions* (one inside the other), and the rules regarding *order of operations* (see Chapter 7) apply as you work toward figuring out what the variable *x* represents.

REMEMBER

If you perform an operation on the grouping symbol, then every term in the grouping symbol has to have that operation performed on it.

Nesting isn't for the birds

When you have a number or variable that needs to be multiplied by every value inside parentheses, brackets, braces, or a combination of those grouping symbols, you distribute that number or variable. *Distributing* means that the number or variable next to the grouping symbol multiplies every term inside the grouping symbol. If two or more of the grouping symbols are inside one another, they're nested. Nested expressions are written within parentheses, brackets, and braces to make the intent clearer.

REMEMBER

The following conventions are used when nesting:

- When using nested expressions, every opening grouping symbol — such as left parenthesis (, bracket [, or brace { — has to have a closing grouping symbol — a right parenthesis), bracket], or brace }.
- When simplifying nested expressions, work from the inside to the outside. The innermost expression is the one with no grouping symbols inside it. Simplify that expression or distribute over it so the innermost grouping symbols can be dropped. Then go to the next innermost grouping.

Distributing first

Equations containing grouping symbols offer opportunities for making wise decisions. In some cases, you need to distribute, working from the inside out, and in other cases it's wise to multiply or divide first. In general, you'll distribute first if you find more than two terms in the entire equation.

EXAMPLE

Q. Solve for y in $8(3y-5)=9(y-6)-1$.

A. $y=-1$.

The equation has two terms involving grouping symbols. Distribute the 8 and 9 first:

$$24y-40=9y-54-1$$

Combine the two constant terms on the right. Then subtract $9y$ from each side of the equation:

$$\begin{array}{l} 24y-40=9y-55 \\ \underline{-9y \qquad -9y} \\ 15y-40= \quad -55 \end{array}$$

Now add 40 to each side of the equation; then divide each side by 15:

$$\begin{array}{l} 15y-40=-55 \\ \underline{\quad +40 \quad +40} \\ 15y \qquad =-15 \\ \dfrac{\cancel{15}y}{\cancel{15}}=\dfrac{-\cancel{15}}{\cancel{15}} \\ \qquad y=-1 \end{array}$$

Now let me show you the solution of the example I give at the beginning of this section.

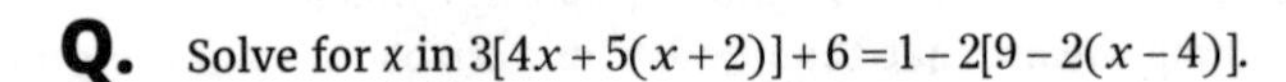

Q. Solve for x in $3[4x+5(x+2)]+6=1-2[9-2(x-4)]$.

A. The best way to sort through all these operations is to simplify from the inside out. You see parentheses within brackets. The binomials in the parentheses have multipliers. I'll step through this carefully to show you an organized plan of attack.

First, distribute the 5 over the binomial inside the left parentheses and the −2 over the binomial inside the right parentheses:

$$3[4x+5x+10]+6=1-2[9-2x+8]$$

Now combine terms within the brackets:

$$3[9x+10]+6=1-2[17-2x]$$

Distribute the 3 over the two terms in the left brackets and the −2 over the terms in the right brackets:

$$27x+30+6=1-34+4x$$

The constant terms on each side can be combined:

$$27x+36=-33+4x$$

Now subtract $4x$ from each side and subtract 36 from each side:

$$\begin{aligned} 27x + 36 &= -33 + 4x \\ -4x \quad &\quad\quad\;\; -4x \\ \hline 23x + 36 &= -33 \\ -36 &\;\; -36 \\ \hline 23x \quad &= -69 \end{aligned}$$

Now, dividing each side of the equation by 23, you get that $x = -3$.

Q. Solve for x: $8(2x+1)+6=5(x-3)+7$.

A. $x = -2$.

1. **Distribute the 8 over the two terms in the left parentheses and the 5 over the two terms in the right parentheses.**

   You get the equation $16x+8+6=5x-15+7$.

2. **Combine the two numbers on each side of the equation.**

   You get $16x+14=5x-8$.

3. **Subtract 5x and 14 from each side; then divide each side by 11.**

$$\begin{aligned} 16x + 14 &= 5x - 8 \\ -5x - 14 &\;\; -5x - 14 \\ \hline 11x \quad &= \quad -22 \\ \frac{\cancel{11}x}{\cancel{11}} &= \frac{-22}{11} \\ x &= -2 \end{aligned}$$

Q. Solve for x: $\frac{x-5}{4}+3=x+4$.

A. First, multiply each term on both sides of the equation by 4:

$$\begin{aligned} \cancel{4}\left(\frac{x-5}{\cancel{4}}\right)+4(3) &= 4(x)+4(4) \\ x-5+12 &= 4x+16 \end{aligned}$$

Combine the like terms on the left. Then subtract $4x$ from each side and subtract 7 from each side; finally, divide each side by -3.

$$\begin{aligned} x + 7 &= 4x + 16 \\ -4x - 7 &\;\; -4x - 7 \\ \hline -3x \quad &= \quad 9 \\ \frac{\cancel{-3}x}{\cancel{-3}} &= \frac{9}{-3} \\ x &= -3 \end{aligned}$$

Lining Up Linear Equations

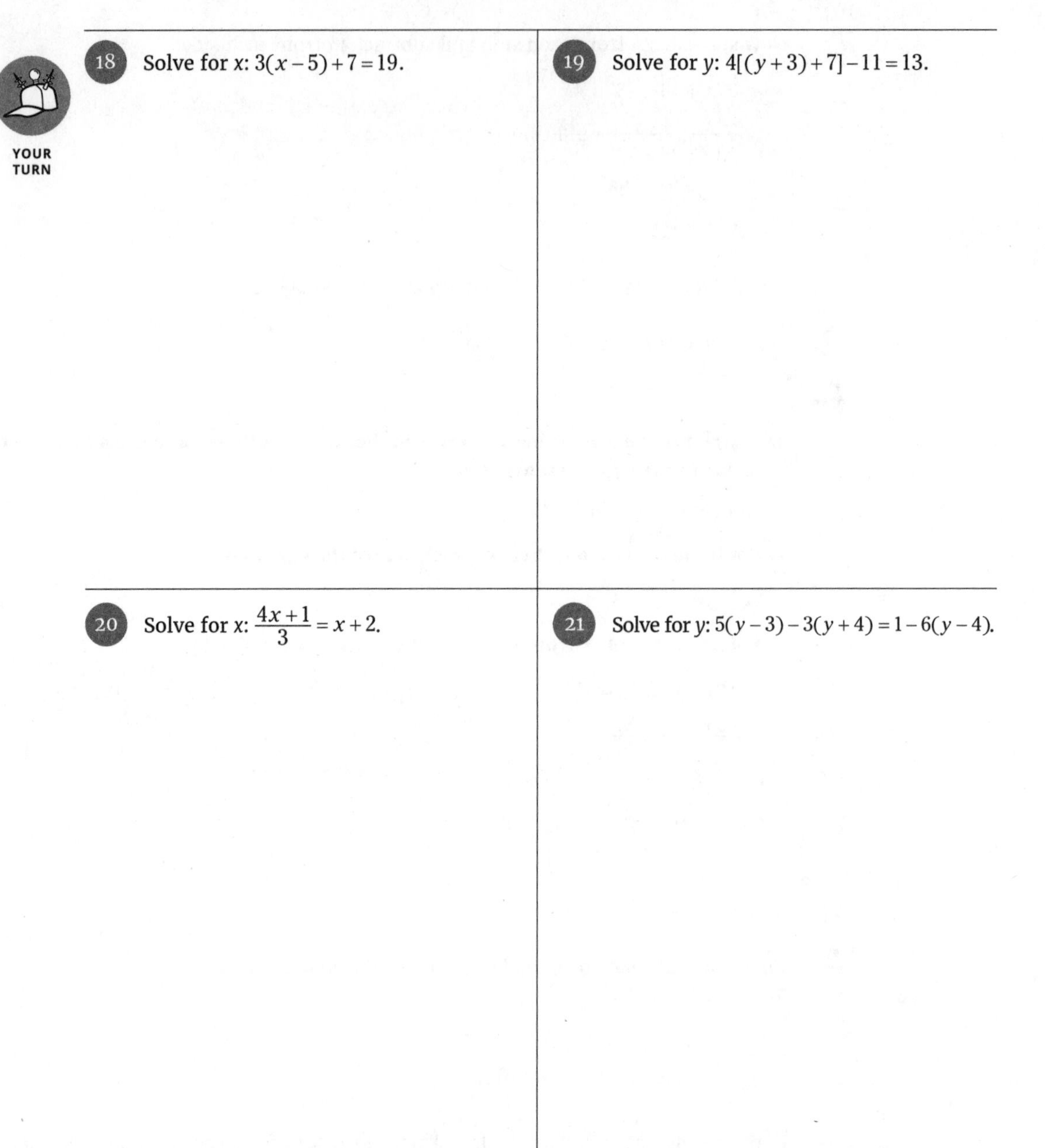

Multiplying or dividing before distributing

In this section, I show you where it may be easier to divide through by a number rather than distribute first. My only caution is that you always divide (or multiply) each term by the same number.

EXAMPLE

Q. Solve for z in $12z-3(z+7)=6(z-1)$.

A. In this equation, you see three terms: two on the left and one on the right. Each term has a multiplier of a multiple of 3. So divide each term by 3:

$$\frac{{}^{4}\cancel{12}z}{\cancel{3}}-\frac{\cancel{3}(z+7)}{\cancel{3}}=\frac{{}^{2}\cancel{6}(z-1)}{\cancel{3}}$$
$$4z-(z+7)=2(z-1)$$

Warning: Notice that the second term has the negative sign in front of the resulting binomial. Be very careful not to lose track of the negative multipliers.

Distribute the negative sign and the 2:

$$4z-z-7=2z-2$$

Combine the two variable terms on the left. Then subtract 2z from each side:

$$\begin{array}{r} 3z-7=2z-2 \\ \underline{-2z \qquad -2z} \\ z-7= \quad -2 \end{array}$$

Finally, add 7 to each side and you get:

$$z=-2+7=5$$

The next example mixes two different situations that are actually the same. The terms in the equation either have a fractional multiplier or are in a fraction themselves. The point of the example is to show when multiplying each term by the same number first is preferable to distributing first.

Q. Solve for x in the following equation: $\frac{3(x-2)}{4}+\frac{1}{2}(5x+2)=\frac{14x+12}{8}+7$.

A. At first glance, the equation looks a bit forbidding. But quick action — in the form of multiplying each term by 8 — takes care of all the fractions. You're left with rather large numbers, but that's still nicer than fractions with different denominators. I choose to multiply by 8 because that's the least common denominator of each term (even the last term). Each of the four terms is multiplied by 8:

$$\frac{{}^{2}\cancel{8}}{1}\left[\frac{3(x-2)}{\cancel{4}}\right]+\frac{{}^{4}\cancel{8}}{1}\left[\frac{1}{\cancel{2}}(5x+2)\right]=\frac{\cancel{8}}{1}\left[\frac{14x+12}{\cancel{8}}\right]+8(7)$$
$$6(x-2)+4(5x+2)=(14x+12)+56$$

Do the multiplication and distribution in steps to avoid errors:

$$6x-12+20x+8=14x+12+56$$

The two variable terms on the left and the two constant terms on the left can be combined. Likewise, combine the two constant terms on the right:

$$26x-4=14x+68$$

Now subtract $14x$ from each side and add 4 to each side:

$$\begin{array}{rcl} 26x-4 & = & 14x+68 \\ -14x & & -14x \\ \hline 12x-4 & = & 68 \\ +4 & & +4 \\ \hline 12x & = & 72 \end{array}$$

Dividing each side of the equation by 12, you see that $x = 6$.

YOUR TURN

22 Solve for x: $5(x-3)-10(2x+1)=10(3x+2)$.

23 Solve for y: 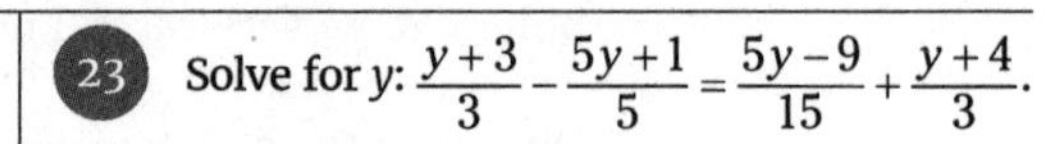

$\frac{y+3}{3}-\frac{5y+1}{5}=\frac{5y-9}{15}+\frac{y+4}{3}$.

24 Solve for z: $\frac{1}{2}(2z+1)+\frac{1}{3}(3z+1)+\frac{1}{6}(6z+1)=-5$.

25 Solve for w: $\frac{2(w+7)}{5}+4=\frac{3(w+6)}{7}-1$.

26 Simplify the rational equation by multiplying each term by y. Then solve the resulting linear equation for y: $\frac{4}{y}-\frac{6}{y}=1$.

27 Simplify the rational equation by multiplying each term by $6z$. Then solve the resulting linear equation for z: $\frac{1}{3z}-\frac{1}{2z}=\frac{1}{6}$.

Working with Proportions

A *proportion* is actually an equation with two fractions set equal to one another. The proportion $\frac{a}{b}=\frac{c}{d}$ has the following properties:

- The cross-products are equal: $ad = bc$.
- If the proportion is true, then the *flip* of the proportion is also true: $\frac{b}{a}=\frac{d}{c}$
- You can reduce vertically or horizontally: $\frac{a\cdot \not{e}}{b\cdot \not{e}}=\frac{c}{d}$ or $\frac{a\cdot \not{e}}{b}=\frac{c\cdot \not{e}}{d}$

These properties make solving equations involving proportions so much nicer and easier.

Using the rules for proportions

You can solve proportions that are algebraic equations by cross-multiplying, flipping, reducing, or using a combination of two or more of the processes. The flipping part of solving proportions usually occurs when you have the variable in the denominator and can do a quick solution by first flipping the proportion and then multiplying by a number.

EXAMPLE

Q. Solve for y: $\frac{27}{6}=\frac{2y+6}{8}$.

A. First, reduce horizontally through the denominators. Then reduce vertically through the left fractions. Then cross-multiply.

$$\frac{27}{\not{6}_3}=\frac{2y+6}{\not{8}_4}$$
$$\frac{\not{27}^9}{\not{3}}=\frac{2y+6}{4}$$
$$\frac{9}{1}=\frac{2y+6}{4}$$
$$9(4)=1(2y+6)$$
$$36=2y+6$$
$$30=2y$$
$$\frac{30}{2}=\frac{2y}{2}$$
$$15=y$$

You could also have reduced the right fraction by dividing by 2, but the numbers weren't really too big to handle.

Q. Solve for y: $\frac{8y-10}{3}-\frac{12y-18}{5}=0$.

A. The first thing to do is change the equation to a proportion. Move the second fraction to the right-hand side by adding that fraction to each side of the equation:

$$\frac{8y-10}{3}=\frac{12y-18}{5}$$

Lining Up Linear Equations

Now factor the terms in the two numerators and "reduce horizontally":

$$\frac{\not{2}(4y-5)}{3} = \frac{{}^{3}\not{6}(2y-3)}{5}$$
$$\frac{4y-5}{3} = \frac{3(2y-3)}{5}$$

Next, cross-multiply and simplify:

$$(4y-5)\cdot 5 = 3\cdot 3(2y-3)$$
$$20y-25 = 18y-27$$

Now, solve the equation by subtracting 18*y* from each side and then adding 25 to each side:

$$\begin{aligned} 20y - 25 &= 18y - 27 \\ -18y \quad &\quad -18y \\ \hline 2y - 25 &= \quad -27 \\ +25 &\quad +25 \\ \hline 2y \quad &= \quad -2 \end{aligned}$$

And, finally, dividing each side by 2, you see that $y = -1$.

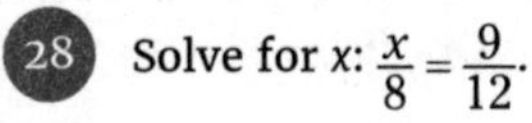

28 Solve for *x*: $\frac{x}{8} = \frac{9}{12}$.

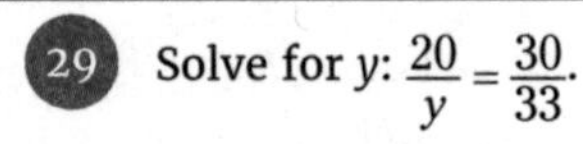

29 Solve for *y*: $\frac{20}{y} = \frac{30}{33}$.

30 Solve for *z*: $\frac{z+4}{32} = \frac{35}{56}$.

31 Solve for *y*: $\frac{6}{27} = \frac{8}{2y+6}$.

Transforming fractional equations into proportions

Proportions are very nice to work with because of their unique properties of reducing and changing into nonfractional equations. Many equations involving fractions must be dealt with in that fractional form, but other equations are easily changed into proportions. When possible, you want to take advantage of the situations where transformations can be done.

EXAMPLE

Q. Solve the following equation for x: $\frac{x+2}{3}-\frac{5x+1}{6}=\frac{3x-1}{2}+\frac{x-9}{8}$.

A. You could solve the problem by multiplying each fraction by the least common factor of all the fractions: 24. Another option is to find a common denominator for the two fractions on the left and subtract them, and then find a common denominator for the two fractions on the right and add them. Your result is a proportion:

$$\frac{2(x+2)}{2\cdot3}-\frac{5x+1}{6}=\frac{4(3x-1)}{4\cdot2}+\frac{x-9}{8}$$
$$\frac{2(x+2)-(5x+1)}{6}=\frac{4(3x-1)+(x-9)}{8}$$
$$\frac{2x+4-5x-1}{6}=\frac{12x-4+x-9}{8}$$
$$\frac{-3x+3}{6}=\frac{13x-13}{8}$$

The proportion can be reduced by dividing by 2 horizontally:

$$\frac{-3x+3}{\not{6}_3}=\frac{13x-13}{\not{8}_4}$$

Now cross-multiply and simplify the products:

$$(-3x+3)\cdot4=3\cdot(13x-13)$$
$$-12x+12=39x-39$$

Add 12*x* to each side, and then add 39 to each side:

$$\begin{array}{r} -12x+12=39x-39 \\ \underline{+12x \qquad +12x} \\ 12=51x-39 \\ \underline{+39 \qquad +39} \\ 51=51x \end{array}$$

The last step consists of just dividing each side by 51 to get $1=x$.

YOUR TURN

Solve the equations by first changing them to proportions and using the appropriate properties.

32 $\frac{x}{5}-\frac{2x+7}{3}=0$

33 $\frac{y+1}{2}+\frac{5y-2}{3}=\frac{3y+3}{2}$

34 $\frac{4(x-5)}{6}-\frac{3(x+1)}{5}=\frac{7x-13}{5}$

35 $\frac{4z-3}{8}+\frac{3z-2}{6}=\frac{7z-1}{8}-\frac{5z-8}{6}$

Solving for Variables in Formulas

A formula is an equation that represents a relationship between some structures or quantities or other entities. It's a rule that uses mathematical computations and can be counted on to be accurate each time you use it when applied correctly. The following are some of the more commonly used formulas that contain only variables raised to the first power.

- $A=\frac{1}{2}bh$: The area of a triangle involves base and height.
- $I=Prt$: The interest earned uses principal, rate, and time.
- $C=2\pi r$: Circumference is twice π times the radius.
- $°F=32°+\frac{9}{5}°C$: Degrees Fahrenheit uses degrees Celsius.
- $P=R-C$: Profit is based on revenue and cost.

ARCHIMEDES: MOVER AND BATHER

Born about 287 B.C., Archimedes, an inspired mathematician and inventor, devised a pump to raise water from a lower level to a higher level. These pumps were used for irrigation, in ships, and in mines, and they're still used today in some parts of the world.

He also made astronomical instruments and designed tools for the defense of his city during a war. Known for being able to move great weights with simple levers, cogwheels, and pulleys, Archimedes determined the smallest-possible cylinder that could contain a sphere and, thus, discovered how to calculate the volume of a sphere with his formula. The sphere/cylinder diagram was engraved on his tombstone.

A favorite legend has it that as Archimedes lowered himself into a bath basin, he had a revelation involving how he could determine the purity of a gold object using a similar water-immersion method. He was so excited at the revelation that he jumped out of the tub and ran naked through the streets of the city shouting, "Eureka! Eureka!" ("I have found it!").

When you use a formula to find the indicated variable (the one on the left of the equal sign), you just put the numbers in, and out pops the answer. Sometimes, though, you're looking for one of the other variables in the equation and end up solving for that variable over and over.

For example, let's say that you're planning a circular rose garden in your backyard. You find edging on sale and can buy a 20-foot roll of edging, a 36-foot roll, a 40-foot roll, or a 48-foot roll. You're going to use every bit of the edging and let the length of the roll dictate how large the garden will be. If you want to know the radius of the garden based on the length of the roll of edging, you use the formula for circumference and solve the following four equations:

$$20 = 2\pi r \quad 36 = 2\pi r \quad 40 = 2\pi r \quad 48 = 2\pi r$$

Another alternative to solving four different equations is to solve for r in the formula and then put the different roll sizes into the new formula. Starting with $C = 2\pi r$, you divide each side of the equation by 2π, giving you:

$$r = \frac{C}{2\pi}$$

The computations are much easier if you just divide the length of the roll by 2π.

Now, I'll show you some examples of solving for one of the variables in an equation. I won't try to come up with any more gardening or other clever scenarios.

EXAMPLE

Q. Solve for b in the formula for the area of a triangle: $A = \frac{1}{2}bh$.

A. First, multiply each side of the equation by 2:

$$2 \cdot A = 2 \cdot \frac{1}{2}bh$$

or $2A = bh$

Now divide each side by h:

$$\frac{2A}{h} = b$$

Q. Solve for x_5 in the following formula for finding the mean average of five test grades:

$$A = \frac{x_1 + x_2 + x_3 + x_4 + x_5}{5}$$

A. Multiply each side of the equation by 5:

$$5(A) = \left(\frac{x_1 + x_2 + x_3 + x_4 + x_5}{\cancel{5}}\right) \cdot \frac{\cancel{5}}{1}$$
$$5A = x_1 + x_2 + x_3 + x_4 + x_5$$

Now, subtract every x_i except the last one:

$$5A - x_1 - x_2 - x_3 - x_4 = x_5$$
$$5A - (x_1 + x_2 + x_3 + x_4) = x_5$$

YOUR TURN

36 Solve for w in $P = 2(l + w)$.

37 Solve for $°F$ in $°C = \frac{5}{9}(°F - 32)$.

38 Solve for h in $A = 2\pi r(r + h)$.

39 Solve for b_1 in $A = \frac{1}{2}h(b_1 + b_2)$.

Practice Questions Answers and Explanations

(1) $\boldsymbol{x = 11.}$

$$\begin{array}{r} x+4=15 \\ \underline{-4\quad -4} \\ x\qquad =11 \end{array}$$

(2) $\boldsymbol{y = 13.}$

$$\begin{array}{r} y-2=11 \\ \underline{+2\quad +2} \\ y\qquad =13 \end{array}$$

(3) $\boldsymbol{x = -4.}$ Subtract $4x$ from each side and subtract 3 from each side.

$$\begin{array}{r} 5x+3=4x-1 \\ \underline{-4x\qquad -4x\quad} \\ x+3=\quad -1 \\ \underline{-3\qquad -3} \\ x=-4 \end{array}$$

(4) $\boldsymbol{y = -7.}$ First, combine all the y terms on each side. You have $8y$ on the left and $7y$ on the right. Then combine the numbers on each side. Isolate the y's on the left by subtracting $7y$ from each side. Finally, subtract 1 from each side.

$$\begin{array}{r} 8y+1=7y-6 \\ \underline{-7y\qquad -7y\quad} \\ y+1=-6 \\ \underline{-1\quad -1} \\ y=-7 \end{array}$$

(5) $\boldsymbol{x = 4.}$

$$\begin{aligned} 6x &= 24 \\ \frac{6x}{6} &= \frac{24}{6} \\ x &= 4 \end{aligned}$$

(6) $\boldsymbol{y = -5.}$

$$\begin{aligned} -4y &= 20 \\ \frac{-4y}{-4} &= \frac{20}{-4} \\ y &= -5 \end{aligned}$$

(7) $z = 33$.

$$\frac{z}{3} = 11$$
$$3\left(\frac{z}{3}\right) = (3)(11)$$
$$z = 33$$

(8) $w = 8$.

$$\frac{w}{-4} = -2$$
$$(-4)\left(\frac{w}{-4}\right) = (-4)(-2)$$
$$w = 8$$

(9) $a = 24$. The reciprocal of $\frac{3}{8}$ is $\frac{8}{3}$.

$$\frac{8}{3} \cdot \frac{3}{8} a = \cancel{9}^{3} \cdot \frac{8}{\cancel{3}}$$
$$a = 24$$

(10) $x = 3$.

$$\begin{aligned} 3x - 4 &= 5 \\ +4 &\ \ +4 \\ \hline 3x &= 9 \\ \frac{3x}{3} &= \frac{9}{3} \\ x &= 3 \end{aligned}$$

(11) $y = 2$. First, get the term with y by itself on the left.

$$\begin{aligned} 8 - \frac{y}{2} &= 7 \\ -8 \quad &\ \ -8 \\ \hline -\frac{y}{2} &= -1 \\ (-2)\left(-\frac{y}{2}\right) &= (-2)(-1) \\ y &= 2 \end{aligned}$$

(12) $x = -4$.

$$\begin{aligned} 5x - 3 &= 8x + 9 \\ -5x - 9 &\ \ -5x - 9 \\ \hline -12 &= 3x \\ \frac{-12}{3} &= x \\ -4 &= x \end{aligned}$$

13 $z = -12.$

$$\begin{array}{rl} \frac{z}{6} - 3 = & z + 7 \\ -\frac{z}{6} - 7 & -\frac{z}{6} - 7 \\ \hline -10 = & \frac{5z}{6} \end{array}$$

Note that $z - \frac{z}{6} = \frac{6z}{6} - \frac{z}{6} = \frac{5z}{6}$.

$$(6)(-10) = \left(\frac{5z}{6}\right)(6)$$
$$-60 = 5z$$
$$\frac{5z}{5} = \frac{-60}{5}$$
$$z = -12$$

14 $y = \frac{9}{2}$**. First, combine the two y terms on the left.**

$$\begin{array}{rl} y + 16 = & 7 + 3y \\ -3y - 16 & -16 - 3y \\ \hline -2y & = -9 \end{array}$$
$$\frac{-2y}{-2} = \frac{-9}{-2}$$
$$y = \frac{9}{2}$$

15 $x = -10.$

$$\begin{array}{rl} \frac{3x}{4} - 2 = & \frac{9x}{4} + 13 \\ -\frac{9x}{4} + 2 & -\frac{9x}{4} + 2 \\ \hline -\frac{6x}{4} = & 15 \end{array}$$
$$-\frac{3x}{2} = 15$$
$$(2)\left(-\frac{3x}{2}\right) = (2)(15)$$
$$-3x = 30$$
$$\frac{-3x}{-3} = \frac{30}{-3}$$
$$x = -10$$

(16) $\mathbf{\frac{13}{18}}$.

Multiply N by both 100 and 10 and subtract $100N - 10N$.

$$\begin{aligned} 100N &= 72.22\overline{2} \\ -10N &= 7.22\overline{2} \\ \hline 90N &= 65 \end{aligned}$$

$$N = \frac{65}{90} = \frac{\cancel{65}^{13}}{\cancel{90}^{18}} = \frac{13}{18}$$

(17) $\mathbf{\frac{121}{180}}$.

Let $N = 0.6722\overline{2}$. Multiply N by both 100 and 1,000 and subtract $1{,}000N - 100N$.

$$\begin{aligned} 1{,}000N &= 672.22\overline{2} \\ -100N &= 67.22\overline{2} \\ \hline 900N &= 605 \end{aligned}$$

$$N = \frac{605}{900} = \frac{\cancel{605}^{121}}{\cancel{900}^{180}} = \frac{121}{180}$$

(18) $\mathbf{x = 9}$**.** Distribute the 3 over the parentheses, and when you add the two constants on the left, $3x - 15 + 7 = 19$ becomes $3x - 8 = 19$. Add 8 to each side to get $3x = 27$. Divide each side of the equation, and $x = 9$.

(19) $\mathbf{y = -4}$**.** Working inside the brackets, first add and simplify to get $4[y + 10] - 11 = 13$. Distribute the 4, and you have $4y + 40 - 11 = 13$. Subtract 11 from 40: $4y + 29 = 13$. Now subtract 29 from each side of the equation: $4y + 29 - 29 = 13 - 29$ becomes $4y = -16$. Dividing each side of the equation by 4, $y = -4$.

(20) $\mathbf{x = 5}$**.** First, multiply each side by 3 to get rid of the fraction.

$$\begin{aligned} \frac{4x+1}{3} &= x + 2 \\ (3)\left(\frac{4x+1}{3}\right) &= (3)(x+2) \\ 4x + 1 &= 3x + 6 \\ -3x - 1 &\quad -3x - 1 \\ \hline x &= 5 \end{aligned}$$

(21) $\mathbf{y = \frac{13}{2}}$**.** First, distribute the 5, –3 and –6 over the respective parenthesees.

$$\begin{aligned} 5(y-3) - 3(y+4) &= 1 - 6(y-4) \\ 5y - 15 - 3y - 12 &= 1 - 6y + 24 \\ 2y - 27 &= -6y + 25 \\ +6y + 27 &\quad +6y + 27 \\ \hline 8y &= 52 \\ \frac{8y}{8} &= \frac{52}{8} \\ y &= \frac{13}{2} \end{aligned}$$

(22) $x = -1$. Each of the three terms in the equation has a multiplier that's divisible by 5. Divide each term by 5 before distributing.

$\frac{\cancel{5}(x-3)}{\cancel{5}} - \frac{\cancel{10}^2(2x+1)}{\cancel{5}} = \frac{\cancel{10}^2(3x+2)}{\cancel{5}}$ becomes $(x-3)-2(2x+1)=2(3x+2)$. Distributing, you have $x-3-4x-2=6x+4$. Simplifying terms on the left, $-3x-5=6x+4$. Adding $3x$ to each side and subtracting 4 from each side, $-3x+3x-5-4=6x+3x+4-4$ or $-9=9x$. Now, dividing each side by 9, you have $-1=x$.

(23) $y = \frac{1}{20}$

A nice way of getting rid of the fractions is to multiply each term by the least common denominator of the fractions, 15.

$$\cancel{15}^5 \cdot \frac{y+3}{\cancel{3}} - \cancel{15}^3 \cdot \frac{5y+1}{\cancel{5}} = \cancel{15} \cdot \frac{5y-9}{\cancel{15}} + \cancel{15}^5 \cdot \frac{y+4}{\cancel{3}}$$

Now the equation reads $5(y+3)-3(5y+1)=5y-9+5(y+4)$. Distributing the terms, you have $5y+15-15y-3=5y-9+5y+20$. Combining like terms on each side of the equation, $12-10y=10y+11$. Now add $10y$ to each side and subtract 11 from each side. $12-11-10y+10y=10y+10y+11-11$, and you have $1=20y$. Divide each side of the equation by 20 to get $\frac{1}{20}=y$.

(24) $z = -2$. Distributing first wouldn't be too bad, because only three of the terms would be fractions, but it still works very nicely to multiply through by the least common denominator of the fractions. Just be sure to multiply the right side by that number, also.

$\cancel{6}^3 \cdot \frac{1}{\cancel{2}}(2z+1) + \cancel{6}^2 \cdot \frac{1}{\cancel{3}}(3z+1) + \cancel{6} \cdot \frac{1}{\cancel{6}}(6z+1) = 6(-5)$, which simplifies to $3(2z+1)+2(3z+1)+6z+1=-30$. Now distribute: $6z+3+6z+2+6z+1=-30$. Combining like terms on the left: $18z+6=-30$. Adding -6 to each side of the equation, $18z+6-6=-30-6$, which simplifies to $18z=-36$. When you divide each side of the equation by 18, you get $z=-2$.

(25) $w = 183$. Before multiplying through by the least common denominator of the fractions, add 1 to each side of the equation, combining the two constants in the equation.

$\frac{2(w+7)}{5}+4+1=\frac{3(w+6)}{7}-1+1$ gives you $\frac{2(w+7)}{5}+5=\frac{3(w+6)}{7}$. Now multiply each term by the least common denominator. Don't forget to multiply the 5, also.

$\cancel{35}^7 \cdot \frac{2(w+7)}{\cancel{5}} + 35 \cdot 5 = \cancel{35}^5 \cdot \frac{3(w+6)}{\cancel{7}}$. Simplifying, you have $14(w+7)+175=15(w+6)$. Distributing, $14w+98+175=15w+90$. Combine the two constants on the left to get $14w+273=15w+90$. Now subtract $14w$ from each side and subtract 90 from each side. $14w-14w+273-90=15w-14w+90-90$, which simplifies to $183=w$.

(26) $y = -2$. The common denominator is y, so

$$\begin{aligned} \cancel{y} \cdot \left(\frac{4}{\cancel{y}}\right) - \cancel{y} \cdot \left(\frac{6}{\cancel{y}}\right) &= (y)(1) \\ 4-6 &= y \\ -2 &= y \end{aligned}$$

Lining Up Linear Equations

(27) $\boldsymbol{z = -1}$**.** A common denominator is 6z, so

$$\cancel{6z}^2 \cdot \left(\frac{1}{\cancel{3z}}\right) - \cancel{6z}^3 \cdot \left(\frac{1}{\cancel{2z}}\right) = \cancel{6}z \cdot \left(\frac{1}{\cancel{6}}\right)$$
$$2 - 3 = z$$
$$-1 = z$$

(28) $\boldsymbol{x = 6}$**. Reduce by dividing the denominators by 4 and the right fractions by 3. Then cross-multiply.**

$$\frac{x}{{}_2\cancel{8}} = \frac{9}{\cancel{12}_3} \rightarrow \frac{x}{2} = \frac{\cancel{9}^3}{\cancel{3}_1} \rightarrow \frac{x}{2} = \frac{3}{1} \rightarrow x = 6$$

(29) $\boldsymbol{y = 22}$**. Reduce through the numerators. Then flip.**

$$\frac{{}^2\cancel{20}}{y} = \frac{\cancel{30}^3}{33}$$
$$\frac{y}{2} = \frac{33}{3}$$
$$\frac{y}{2} = \frac{\cancel{33}^{11}}{\cancel{3}}$$
$$y = 22$$

(30) $\boldsymbol{z = 16}$**. Reduce the fraction on the right by dividing the numerator and denominator by 7. Then reduce through the denominators before you cross-multiply:**

$$\frac{z+4}{32} = \frac{35}{56} = \frac{\cancel{35}^5}{\cancel{56}_8} = \frac{5}{8}$$
$$\frac{z+4}{{}_4\cancel{32}} = \frac{5}{{}_1\cancel{8}}$$
$$(z+4) \cdot 1 = 4 \cdot 5$$
$$z + 4 = 20$$
$$z = 16$$

(31) $\boldsymbol{y = 15}$**. Flip to get** $\frac{27}{6} = \frac{2y+6}{8}$ **and solve by reducing the fraction on the left and then cross-multiplying:**

$$\frac{\cancel{27}^9}{\cancel{6}_2} = \frac{2y+6}{8}$$
$$9(8) = 2(2y+6)$$
$$72 = 4y + 12$$
$$60 = 4y$$
$$\frac{60}{4} = \frac{4y}{4}$$
$$15 = y$$

Yes, you could also have reduced through the denominators, but when the numbers are small enough, it's just as quick to skip that step.

(32) $x = -5$. Move the second term to the right by adding it to each side of the equation.

$$\frac{x}{5} = \frac{2x+7}{3}$$

Now cross-multiply to get $3x = 5(2x+7)$. Distributing on the right, the equation becomes $3x = 10x + 35$. Subtract 10x from each side, so $3x - 10x = 10x - 10x + 35$ becomes $-7x = 35$. Dividing each side of the equation by -7 results in $x = -5$.

(33) $y = \frac{5}{2}$. Two of the terms have denominators of 2. Subtract the first term from each side, and then subtract the fractions on the right.
$\frac{y+1}{2} - \frac{y+1}{2} + \frac{5y-2}{3} = \frac{3y+3}{2} - \frac{y+1}{2}$ becomes $\frac{5y-2}{3} = \frac{3y+3-(y+1)}{2}$, which simplifies to $\frac{5y-2}{3} = \frac{2y+2}{2}$. You can reduce the fraction on the right by factoring 2 out of the terms in the numerator: $\frac{5y-2}{3} = \frac{\cancel{2}(y+1)}{\cancel{2}}$ or $\frac{5y-2}{3} = \frac{y+1}{1}$. Now cross-multiply to get $5y - 2 = 3y + 3$. Subtracting 3y from each side and adding 2 to each side, $5y - 3y - 2 + 2 = 3y - 3y + 3 + 2$ simplifies to $2y = 5$. Divide each side by 2 and $y = \frac{5}{2}$.

(34) $x = -1$. Two of the terms have denominators of 5. Add the second term to each side, and then add the fractions on the right: $\frac{4(x-5)}{6} - \frac{3(x+1)}{5} + \frac{3(x+1)}{5} = \frac{7x-13}{5} + \frac{3(x+1)}{5}$ becomes $\frac{4(x-5)}{6} = \frac{7x-13+3(x+1)}{5} = \frac{7x-13+3x+3}{5}$, which simplifies to $\frac{4(x-5)}{6} = \frac{10x-10}{5}$. Factor the 10 out of the two terms in the denominator, and reduce the fraction: $\frac{4(x-5)}{6} = \frac{\cancel{10}^2(x-1)}{\cancel{5}}$ or $\frac{4(x-5)}{6} = \frac{2(x-1)}{1}$. The two numerators have factors of 2, so you can reduce horizontally. $\frac{\cancel{4}^2(x-5)}{6} = \frac{\cancel{2}(x-1)}{1}$ is now $\frac{2(x-5)}{6} = \frac{x-1}{1}$. The left-side fraction reduces: $\frac{\cancel{2}(x-5)}{\cancel{6}^3} = \frac{x-1}{1}$ or $\frac{x-5}{3} = \frac{x-1}{1}$. Finally, cross-multiply to get $x - 5 = 3(x-1)$. Distributing, $x - 5 = 3x - 3$. Now subtract x from each side and add 3 to each side, giving you $x - x - 5 + 3 = 3x - x - 3 + 3$, which simplifies to $-2 = 2x$. Dividing each side of the equation by 2, you have $-1 = x$.

(35) $z = 2$. You find two pairs of fractions with the same denominator. Subtract the second fraction from each side, and subtract the first fraction on the right from each side.

$\frac{4z-3}{8} - \frac{7z-1}{8} + \frac{3z-2}{6} - \frac{3z-2}{6} = \frac{7z-1}{8} - \frac{7z-1}{8} - \frac{5z-8}{6} - \frac{3z-2}{6}$, which simplifies to $\frac{4z-3-(7z-1)}{8} = \frac{-(5z-8)-(3z-2)}{6}$. Distribute in the numerators and simplify the fractions. $\frac{4z-3-7z+1}{8} = \frac{-5z+8-3z+2}{6}$ simplifies to $\frac{-3z-2}{8} = \frac{-8z+10}{6}$. The two denominators have a common factor of 2, so, reducing horizontally, $\frac{-3z-2}{\cancel{8}^4} = \frac{-8z+10}{\cancel{6}^3}$ or $\frac{-3z-2}{4} = \frac{-8z+10}{3}$.
Now cross-multiply to get $3(-3z-2) = 4(-8z+10)$. Distributing, the equation becomes $-9z - 6 = -32z + 40$. Add 32z to each side and add 6 to each side to get $-9z + 32z - 6 + 6 = -32z + 32z + 40 + 6$ or $23z = 46$. Dividing each side of the equation by 23 gives you $z = 2$.

(36) $\boldsymbol{w = \frac{P-2l}{2}}$. One method is to divide each side of the equation by 2, first: $\frac{P}{2} = \frac{\cancel{2}(l+w)}{\cancel{2}}$, giving you $\frac{P}{2} = l + w$. Now subtract l from each side, giving you $\frac{P}{2} - l = l - l + w$ or $\frac{P}{2} - l = w$. This can be written $w = \frac{P}{2} - l$, or you can combine the two terms on the right with the common denominator 2: $w = \frac{P}{2} - \frac{2l}{2} = \frac{P-2l}{2}$.

(37) $\boldsymbol{\frac{9}{5}°C + 32 = °F}$. First, multiply each side of the equation by $\frac{9}{5}$: $\frac{9}{5}°C = \frac{\cancel{9}}{\cancel{5}} \cdot \frac{\cancel{5}}{\cancel{9}}(°F - 32)$, which now reads $\frac{9}{5}°C = °F - 32$. Add 32 to each side of the equation, and $\frac{9}{5}°C + 32 = °F$.

(38) $\boldsymbol{h = \frac{A - 2\pi r^2}{2\pi r}}$. First, divide each side of the equation by $2\pi r$: $\frac{A}{2\pi r} = \frac{\cancel{2\pi r}(r+h)}{\cancel{2\pi r}}$, simplifying to $\frac{A}{2\pi r} = r + h$. Then, subtract r from each side to get $\frac{A}{2\pi r} - r = h$. If you want h to be expressed in just one term, then rewrite the expression using the common denominator $2\pi r$: $h = \frac{A}{2\pi r} - r \cdot \frac{2\pi r}{2\pi r} = \frac{A - 2\pi r^2}{2\pi r}$. You would get this single fraction directly if you started out by distributing the $2\pi r$ over the two terms in the parentheses, subtracting $2\pi r^2$ from each side, and then dividing by $2\pi r$. Same result; your choice.

(39) $\boldsymbol{\frac{2A - hb_2}{h} = b_1}$. First, multiply each side by 2: $2 \cdot A = \cancel{2} \cdot \frac{1}{\cancel{2}} h(b_1 + b_2)$, giving you $2A = h(b_1 + b_2)$. Now distribute the h over the two terms in the parentheses: $2A = hb_1 + hb_2$. Subtract hb_2 from each side, giving you $2A - hb_2 = hb_1$. Finally, divide each side of the equation by h, which results in $\frac{2A - hb_2}{h} = b_1$.

If you're ready to test your skills a bit more, take the following chapter quiz that incorporates all the chapter topics.

Whaddya Know? Chapter 15 Quiz

Quiz time! Complete each problem to test your knowledge on the various topics covered in this chapter. You can then find the solutions and explanations in the next section.

1. Solve for y: $\frac{y+8}{3}+1=\frac{3(y+1)}{4}+5$
2. Solve for y: $5y+7=3y+5$
3. Solve for x: $6-\frac{x}{4}=8$
4. Solve for w: $\frac{2w-3}{7}=w-4$
5. Solve for z: $\frac{z}{6}=-2$
6. Solve for z: $\frac{6}{z}-\frac{13}{z}=\frac{9}{2}$
7. Solve for w: $-\frac{4}{5}w=12$
8. Solve for y_1: $M=\frac{y_2-y_1}{x_2-x_1}$
9. Solve for w: $\frac{25}{w}=\frac{75}{9}$
10. Solve for z: $6[(z-2)+3]=5z+8$
11. Find the fraction equivalent to the repeating decimal: $0.277\bar{7}$
12. Solve for x: $\frac{3(x+1)}{5}-\frac{x}{3}=\frac{2}{3}$
13. Solve for x: $\frac{3+x}{4}=2(x-11)$
14. Solve for x: $2x-3=11$
15. Solve for y: $\frac{3y}{2}+4=\frac{5y}{3}+3$

Lining Up Linear Equations

Answers to Chapter 15 Quiz

1. $y = -5$. Subtract 1 from each side. Then multiply each term by 12.

$$\begin{array}{rcl} \frac{y+8}{3} + 1 & = & \frac{3(y+1)}{4} + 5 \\ -1 & & -1 \\ \hline \frac{y+8}{3} & = & \frac{3(y+1)}{4} + 4 \end{array} \rightarrow \begin{array}{l} \cancel{12}^4 \cdot \frac{y+8}{\cancel{3}} = \cancel{12}^3 \cdot \frac{3(y+1)}{\cancel{4}} + 12(4) \\ 4(y+8) = 9(y+1) + 48 \end{array}$$

Distribute the 4 and 9 and simplify on the right. Then subtract $4y$ from each side and 57 from each side.

$$\begin{array}{l} 4y + 32 = 9y + 9 + 48 \\ 4y + 32 = 9y + 57 \end{array} \rightarrow \begin{array}{rcl} 4y + 32 & = & 9y + 57 \\ -4y \quad -57 & & -4y \quad -57 \\ \hline -25 & = & 5y \end{array}$$

Divide each side by 5.

$$\frac{-\cancel{25}^5}{\cancel{5}} = \frac{\cancel{5}y}{\cancel{5}}$$
$$-5 = y$$

2. $y = -1$. Subtract 7 and $3y$ from each side. Then divide by 2.

$$\begin{array}{rcl} 5y + 7 & = & 3y + 5 \\ -3y \quad -7 & & -3y \quad -7 \\ \hline 2y & = & -2 \end{array} \rightarrow \begin{array}{l} \frac{2y}{2} = \frac{-2}{2} \\ y = -1 \end{array}$$

3. $x = -8$. Subtract 6 from each side. Then multiply each side by -4.

$$\begin{array}{rcl} 6 - \frac{x}{4} & = & 8 \\ -6 & & -6 \\ \hline -\frac{x}{4} & = & 2 \end{array} \rightarrow \begin{array}{l} (-\cancel{4})\left(-\frac{x}{\cancel{4}}\right) = 2(-4) \\ x = -8 \end{array}$$

4. $w = 5$. Multiply each side by 7. Then subtract $2w$ from each side and add 28 to each side.

$$\begin{array}{l} \cancel{7} \cdot \frac{2w-3}{\cancel{7}} = 7 \cdot (w-4) \\ 2w - 3 = 7w - 28 \end{array} \rightarrow \begin{array}{rcl} 2w - 3 & = & 7w - 28 \\ -2w \quad +28 & & -2w \quad +28 \\ \hline 25 & = & 5w \end{array}$$

Now divide each side by 5.

$$\frac{\cancel{25}^5}{\cancel{5}} = \frac{\cancel{5}w}{\cancel{5}}$$
$$5 = w$$

5. $z = -12$. Multiply each side by 6.

$$\cancel{6} \cdot \frac{z}{\cancel{6}} = -2 \cdot 6$$
$$z = -12$$

(6) $z = -\frac{14}{9}$. Simplify the left by subtracting. Then cross-multiply in the proportion.

$$\frac{6}{z} - \frac{13}{z} = \frac{9}{2} \quad \rightarrow \quad -\frac{7}{z} = \frac{9}{2} \quad \rightarrow \quad \begin{aligned} -7 \cdot 2 &= 9 \cdot z \\ -14 &= 9z \end{aligned}$$

Divide each side by 9.

$$\frac{-14}{9} = \frac{\cancel{9}z}{\cancel{9}} \quad \rightarrow \quad \frac{-14}{9} = z$$

(7) $w = -15$. Multiply each side by $-\frac{5}{4}$.

$$\left(-\frac{5}{4}\right)\left(-\frac{4}{5}\right)w = \cancel{12}^{3}\left(-\frac{5}{\cancel{4}}\right)$$
$$w = -15$$

(8) $y_1 = y_2 - (x_2 - x_1) \cdot M$. Multiply each side by $x_2 - x_1$. Then subtract y_2 from each side.

$$(x_2 - x_1) \cdot M = \frac{y_2 - y_1}{\cancel{x_2 - x_1}} \cdot \cancel{(x_2 - x_1)}$$
$$(x_2 - x_1) \cdot M = y_2 - y_1$$

$$\rightarrow \quad \begin{array}{rcrr} (x_2 - x_1) \cdot M & = & y_2 & - \; y_1 \\ -y_2 & & -y_2 & \\ \hline (x_2 - x_1) \cdot M - y_2 & = & & - \; y_1 \end{array}$$

Multiply each side by -1 to change the sign of y_1.

$$-1[(x_2 - x_1) \cdot M - y_2] = -1(-y_1)$$
$$-(x_2 - x_1) \cdot M + y_2 = y_1$$

You can reverse the order of the two terms so that the expression starts out with a positive value.

(9) $w = 3$. Reduce across the top. Then reduce the right fraction. Finally, cross-multiply.

$$\frac{\cancel{25}}{w} = \frac{\cancel{75}^{3}}{9} \rightarrow \frac{1}{w} = \frac{\cancel{3}}{\cancel{9}^{3}} \quad \rightarrow \quad w = 3$$

(10) $z = 2$. Simplify in the brackets and then distribute the 6. Subtract $5z$ and 6 from each side.

$$\begin{aligned} 6[(z-2)+3] &= 5z + 8 \\ 6[z+1] &= 5z + 8 \\ 6z + 6 &= 5z + 8 \end{aligned} \quad \rightarrow \quad \begin{array}{rrcrr} 6z & + \; 6 & = & 5z & + \; 8 \\ -5z & -6 & & -5z & -6 \\ \hline z & & = & & 2 \end{array}$$

(11) $N = \frac{5}{18}$. Write the equation $N = 0.277\bar{7}$. Multiply each side by 100 and by 10 and then subtract.

$$\begin{array}{r} 100N = 27.777\bar{7} \\ -10N = 2.777\bar{7} \\ \hline 90N = 25 \end{array}$$

Divide each side by 90.

$$\frac{\cancel{90}N}{\cancel{90}} = \frac{\cancel{25}^{5}}{\cancel{90}^{18}}$$
$$N = \frac{5}{18}$$

Lining Up Linear Equations

12 $x = \frac{1}{4}$. Add the second fraction on the left to each side. Solve for x: $\frac{3(x+1)}{5} - \frac{x}{3} = \frac{2}{3}$

$$\begin{array}{rcrcr} \frac{3(x+1)}{5} & - & \frac{x}{3} & = & \frac{2}{3} \\ & & +\frac{x}{3} & & +\frac{x}{3} \\ \hline \frac{3(x+1)}{5} & & & = & \frac{2+x}{3} \end{array}$$

Cross-multiply and simplify. Then subtract 5x from each side and 9 from each side.

$$\begin{array}{c} 3 \cdot 3(x+1) = 5(2+x) \\ 9x + 9 = 10 + 5x \end{array} \rightarrow \begin{array}{rcrcrcr} 9x & + & 9 & = & 10 & + & 5x \\ -5x & & -9 & & -9 & & -5x \\ \hline 4x & & & = & 1 & & \end{array}$$

Divide each side by 4.

$$\frac{\cancel{4}x}{\cancel{4}} = \frac{1}{4} \rightarrow x = \frac{1}{4}$$

13 $x = 13$. Multiply each side by 4. Then subtract x from each side and add 88 to each side.

$$\begin{array}{c} \cancel{4} \cdot \frac{3+x}{\cancel{4}} = 2(x-11) \cdot 4 \\ 3 + x = 8x - 88 \end{array} \rightarrow \begin{array}{rcrcrcr} 3 & + & x & = & 8x & - & 88 \\ +88 & & -x & & -x & & +88 \\ \hline 91 & & & = & 7x & & \end{array}$$

Now divide each side by 7.

$$\frac{\cancel{91}^{13}}{\cancel{7}} = \frac{\cancel{7}x}{\cancel{7}}$$
$$13 = x$$

14 $x = 7$. Add 3 to each side and then divide each side by 2.

$$\begin{array}{rcrcr} 2x & - & 3 & = & 11 \\ & + & 3 & & +3 \\ \hline 2x & & & = & 14 \end{array} \rightarrow \begin{array}{c} \frac{2x}{2} = \frac{14}{2} \\ x = 7 \end{array}$$

15 $y = 6$. Multiply each term by 6 to eliminate the fractions.

$$\cancel{6}^3\left(\frac{3y}{\cancel{2}}\right) + 6(4) = \cancel{6}^2\left(\frac{5y}{\cancel{3}}\right) + 6(3)$$
$$9y + 24 = 10y + 18$$

Subtract 9y and 18 from each side.

$$\begin{array}{rcrcrcr} 9y & + & 24 & = & 10y & + & 18 \\ -9y & & -18 & & -9y & & -18 \\ \hline & & 6 & = & y & & \end{array}$$

» Solving quadratic equations by factoring

» Enlisting the quadratic formula

» Completing the square to solve quadratics

» Dealing with the impossible

Chapter 16
Muscling Up to Quadratic Equations

Quadratic (second-degree) equations are nice to work with because they're manageable. Finding the solution or deciding whether a solution exists is relatively easy — easy, at least, in the world of mathematics.

A quadratic equation is an equation that is usually written as $ax^2 + bx + c = 0$, where b, c, or both b and c may be equal to 0, but a is never equal to 0. The solutions of quadratic equations can be two real numbers, one real number, or no real number at all. (Real numbers are all the whole numbers, fractions, negatives and positives, radicals, and irrational decimals. Imaginary numbers are something else again!)

When solving quadratic equations, the most useful form for the equation is one in which the equation is set equal to 0, and the terms are written in decreasing powers of the variable. When set is equal to zero, you can factor for a solution or use the quadratic formula. An exception to this rule, though, is when you have just a squared term and a number, and you want to use the *square-root rule.* (Say *that* three times quickly.) I go into further detail about all these procedures in this chapter.

Quadratic equations are important to algebra and many sciences. Some quadratic equations say that what goes up must come down. Other equations describe the paths that planets and comets take. In all, quadratic equations are fascinating — and just dandy to work with.

Using the Square-Root Rule

The general quadratic equation has the form $ax^2 + bx + c = 0$, and b or c or both of them can be equal to 0. This section shows you how nice it is — and how easy it is to solve equations — when b is equal to 0.

The following is the rule for some special quadratic equations — the ones where $b = 0$. They start out looking like $ax^2 + c = 0$, but the c is usually negative, giving you $ax^2 - c = 0$, and the equation is rewritten as $ax^2 = c$.

If $x^2 = k$, then $x = \pm\sqrt{k}$ or if $ax^2 = c$, then $x = \pm\sqrt{\frac{c}{a}}$. If the square of a variable is equal to the number k, then the variable is equal to the principal square root of k or its opposite.

The following examples show you how to use this square-root rule on quadratic equations where $b = 0$.

EXAMPLE

Q. Solve for x in $x^2 = 49$.

A. Using the square-root rule, $x = \pm\sqrt{49} = \pm 7$. Checking, $(7)^2 = 49$ and $(-7)^2 = 49$.

Q. Solve for m in $3m^2 + 4 = 52$.

A. This equation isn't quite ready for the square-root rule. Add −4 to each side: $3m^2 = 48$.

Now divide each side by 3:

$$m^2 = 16$$

So $m = \pm\sqrt{16} = \pm 4$.

Q. Solve for x in $x^2 = 29$.

A. Using the square-root rule, $x = \pm\sqrt{29}$. And that's as far as you can go. The number 29 isn't a perfect square, so all you can do is estimate. $\pm\sqrt{29} \approx \pm 5.4$ or ± 5.39 or ± 5.385 or ± 5.3852 and so on, depending on the number of decimal places you need.

Q. Solve for p in $p^2 + 11 = 7$.

A. Add −11 to each side to get $p^2 = -4$. Oops! What number times itself is equal to −4? The answer is: "No real number that you can imagine!"

Mathematicians have created numbers that don't actually exist so that these problems can be finished. The numbers are called *imaginary numbers*, but this section is concerned with the less-heady numbers. So, this problem doesn't have an answer, if you're looking for a real number.

YOUR TURN

Use the square-root rule to solve.

1. $x^2 = 9$

2. $5y^2 = 80$

3. $z^2 - 100 = 0$

4. $20w^2 - 125 = 0$

5. $(x+6)^2 = 64$

6. $4(3-z)^2 = 196$

Muscling Up to Quadratic Equations

Factoring for a Solution

This section is where running through all the factoring methods can really pay off. (Refer to Chapters 11, 12, and 13 for all the details.) In most quadratic equations, factoring is used rather than the square-root rule method covered in the preceding section. The square-root rule is used only when $b = 0$ in the quadratic equation $ax^2 + bx + c = 0$. Factoring is used when $c = 0$ or when neither b nor c is 0.

A very important property used along with the factoring to solve these equations is the multiplication property of zero. This is a very straightforward rule — and it even makes sense. Use the greatest common factor and the multiplication property of zero when solving quadratic equations that aren't in the form for the square-root rule.

Zeroing in on the multiplication property of zero

Before you get into factoring quadratics for solutions, you need to know about the multiplication property of zero. You may say, "What's there to know? Zero multiplies anything and leaves nothing. It wipes out everything!" True enough, but there's this other nice property of 0 that is the basis of much equation-solving in algebra. By itself, 0 is nothing. Put it as the result of a multiplication problem, and you really have something: the *multiplication property of zero.*

The *multiplication property of zero* (MPZ) states that if $p \times q = 0$, then either $p = 0$ or $q = 0$. Thus, at least one of them must be equal to 0.

This may seem obvious, but think about it. No other number has such a power over all other numbers. If you say that $p \times q = 12$, you can't predict a thing about p or q alone. These variables could be any number at all — positive, negative, fractional, radical, or a mixture of these. A product of 0, however, leads to one conclusion: One of the multipliers must be 0. No other means of arriving at a 0 product exists. Why is this such a big deal? Let me show you a few equations and how the MPZ works.

EXAMPLE

Q. Find the value of x and y if $xy = 0$.

A. You have two possibilities in this equation. If $x = 0$, then y can be any number, even 0. If $x \neq 0$, then y must be 0, according to the MPZ.

Q. Solve for x in $x(x-5) = 0$.

A. Again, you have two possibilities. If $x = 0$, then the product of $0(-5) = 0$. The other choice is when $x = 5$. Then you have $5(0) = 0$.

YOUR TURN

7 Find the value of x if $-2x = 0$.

8 Find the value of y and z if $yz = 0$.

9 Find the value of w if $(w+4)(w-7) = 0$.

10 Find the value of x if $x^2(x+9)^2 = 0$.

11 Find the value of x if $x\left(x^2+9\right) = 0$.

GETTING THE QUADRATIC SECOND-DEGREE

The word *quadratic* is used to describe equations that have a second-degree term. Why, then, is the prefix *quad-*, which means "four," used in a second-degree equation? It appears that this came about because a square is a regular four-sided figure whose sides are the same. The area of a square with sides x long is x^2. So "squaring" in this case is raising to the second power.

Assigning the greatest common factor and multiplication property of zero to solving quadratics

Factoring is relatively simple when there are only two terms and they have a common factor. This is true in quadratic equations of the form $ax^2 + bx = 0$ (where $c = 0$). The two remaining terms have the common factor of x, at least. You find the greatest common factor (GCF) and factor that out, and then use the MPZ to solve the equation. The following examples make use of the fact that the constant term is 0, and there's a common factor of at least an x in the two terms.

Q. Use factoring to solve for x in $x^2 - 7x = 0$.

A. The GCF of the two terms is x, so write the left side in factored form: $x(x-7) = 0$.

Use the MPZ to say that either $x = 0$ or $x - 7 = 0$. The first equation gives you $x = 0$, and the second solves to give you $x = 7$.

Q. Solve for x in $6x^2 + 18x = 0$.

A. The GCF of the two terms is 6x, so write the left side in factored form: $6x(x+3) = 0$.

Use the MPZ to say that $6x = 0$ or $x + 3 = 0$, which gives you the two solutions, $x = 0$ or $x = -3$.

Technically, I could have written three different equations from the factored form:

$$6 = 0 \quad x = 0 \quad x + 3 = 0$$

The first equation, $6 = 0$, makes no sense — it's an impossible statement. So you either ignore setting the constants equal to 0 or combine them with the factored-out variable, where they'll do no harm.

Missing the $x = 0$, a full half of the solution, is an amazingly frequent occurrence. You don't notice the lonely little x in the front of the parentheses when you have $x(ax + b) = 0$. Don't forget that the x gives you one of the two answers. Be careful.

12 Use factoring to solve for y in $y^2 + 9y = 0$.

13 Use factoring to solve for x in $-x^2 - 2x = 0$.

14 Use the rule that the two solutions of $ax^2 + bx = 0$ are $x = 0$ and $x = -\frac{b}{a}$ to solve for x in $4x^2 - 12x = 0$.

15 Use the rule that the two solutions of $ax^2 + bx = 0$ are $x = 0$ and $x = -\frac{b}{a}$ to solve for x in $-x^2 + 5x = 0$.

Solving Quadratics with Three Terms

Quadratic equations are basic not only to algebra, but also to physics, business, astronomy, and many other applications. By solving a quadratic equation, you get answers to questions such as, "When will the rock hit the ground?" or "When will the profit be greater than 100 percent?" or "When, during the year, will the Earth be closest to the Sun?"

In the two previous sections, either b or c has been equal to 0 in the quadratic equation $ax^2 + bx + c = 0$. This time I won't let anyone skip out. In this section, each of the letters, a, b, and c, is a number that is not 0.

To solve a quadratic equation, moving everything to one side with 0 on the other side of the equal sign is the most efficient method. Factor the equation if possible, and use the MPZ after you factor. If there aren't three terms in the equation, then refer to the previous sections.

The following example lists the steps you use for solving a quadratic trinomial by factoring.

Solve for x in $x^2 - 3x = 28$. Follow these steps:

1. **Move all the terms to one side. Get 0 alone on the right side.**

   In this case, you can subtract 28 from each side:

   $x^2 - 3x - 28 = 0$

   Remember: The standard form for a quadratic equation is $ax^2 + bx + c = 0$.

2. **Determine all the ways you can multiply two numbers to get *a*.**

   In $x^2 - 3x - 28 = 0$, $a = 1$, which can only be 1 times itself.

3. **Determine all the ways you can multiply two numbers to get *c* (ignore the sign for now).**

   28 can be 1×28, 2×14, or 4×7.

Muscling Up to Quadratic Equations

4. **Factor.**

   If c is positive, find an operation from your Step 2 list and an operation from your Step 3 list that match so that the sum of their cross-products is the same as b.

   If c is negative, find an operation from your Step 2 list and an operation from your Step 3 list that match so that the difference of their cross-products is the same as b.

   In this problem, c is negative, and the difference of 4 and 7 is 3. Factoring, you get $(x-7)(x+4)=0$.

5. **Use the MPZ.**

   Either $x-7=0$ or $x+4=0$; now try solving for x by getting x alone to one side of the equal sign.

   - $x-7+7=0+7$ gives you that $x=7$.
   - $x+4-4=0-4$ gives you that $x=-4$.

   So the two solutions are $x=7$ or $x=-4$.

6. **Check your answer.**

   If $x=7$, then $(7)^2-3(7)=49-21=28$.

   If $x=-4$, then $(-4)^2-3(-4)=16+12=28$.

   They both check.

Factoring to solve quadratics sounds pretty simple on the surface. But factoring *trinomial equations* — those with three terms — can be a bit less simple. You can refer to Chapter 13 if you need a review of factoring trinomials. If a quadratic with three terms can be factored, then the product of two binomials is that trinomial. If the quadratic equation with three terms can't be factored, then use the quadratic formula (see "Using the Quadratic Formula" later in this chapter).

REMEMBER

The product of the two binomials $(ax+b)(cx+d)$ is equal to the trinomial $acx^2+(ad+bc)x+bd$. This is a fancy way of showing what you get from using FOIL when multiplying the two binomials together.

Now, on to using unFOIL. If you need more of a review of FOIL and unFOIL, check out Chapter 13.

The following examples all show how factoring and the MPZ allow you to find the solutions of a quadratic equation with all three terms showing.

Solve for x in $x^2-5x-6=0$. Follow these steps:

1. **The equation is in standard form, so you can proceed.**
2. **Determine all the ways you can multiply to get a.**

   $a=1$, which can only be 1 times itself. If there are two binomials that the left side factors into, then they must each start with an x because the coefficient of the first term is 1.

   $(x\quad)(x\quad)=0$

3. **Determine all the ways you can multiply to get c.**

   $c = -6$, so, looking at just the positive factors, you have 1×6 or 2×3.

4. **Factor.**

   To decide which combination should be used, look at the sign of the last term in the trinomial, the 6, which is negative. This tells you that you have to use the *difference* of the absolute value of the two numbers in the list (think of the numbers without their signs) to get the middle term in the trinomial, the –5. In this case, one of the 1 and 6 combinations work, because their difference is 5. If you use the +1 and –6, then you get the –5 immediately from the cross-product in the FOIL process. So $(x-6)(x+1)=0$.

5. **Use the MPZ.**

   Using the MPZ, $x-6=0$ or $x+1=0$. This tells you that $x=6$ or $x=-1$.

6. **Check.**

   If $x=6$, then $(6)^2-5(6)-6=36-30-6=0$.

   If $x=-1$, then $(-1)^2-5(-1)-6=1+5-6=0$.

   They both work!

Solve for x in $6x^2+x=12$. Follow these steps:

1. **Put the equation in standard form.**

   The first thing to do is to add –12 to each side to get the equation into the standard form for factoring and solving:

   $6x^2+x-12=0$

   This one will be a bit more complicated to factor because the 6 in the front has a couple of choices of factors, and the 12 at the end also has several choices. The trick is to pick the correct combination of choices.

2. **Find all the combinations that can be multiplied to get a.**

   You can get 6 with 1×6 or 2×3.

3. **Find all the combinations that can be multiplied to get c.**

   You can get 12 with 1×12, 2×6, or 3×4.

4. **Factor.**

   You have to choose the factors to use so that the difference of their cross-products (outer and inner) is 1, the coefficient of the middle term. How do you know this? Because the 12 is negative, in this standard form, and the value multiplying the middle term is assumed to be 1 when there's nothing showing.

   Looking this over, you can see that using the 2 and 3 from the 6 and the 3 and 4 from the 12 will work: $2 \times 4 = 8$ and $3 \times 3 = 9$. The difference between the 8 and the 9 is, of course, 1. You can worry about the sign later.

Fill in the binomials and line up the factors so that the 2 multiplies the 4 and the 3 multiplies the 3, and you get a 6 in the front and 12 at the end. Whew!

$(2x\ \ 3)(3x\ \ 4)=0$

The quadratic has a + on the term in the middle, so I need the bigger product of the outer and inner to be positive. I get this by making the 9x positive, which happens when the 3 is positive and the 4 is negative.

$(2x+3)(3x-4)=0$

5. **Use the MPZ to solve the equation.**

   The trinomial has been factored. The MPZ tells you that either $2x+3=0$ or $3x-4=0$. If $2x+3=0$, then $2x=-3$ or $x=-\frac{3}{2}$. If $3x-4=0$, then $3x=4$ or $x=\frac{4}{3}$.

6. **Check your work.**

   If $x=\frac{3}{2}$, then $6\left(-\frac{3}{2}\right)^2+\left(-\frac{3}{2}\right)=12$ and $6\left(\frac{9}{4}\right)-\frac{3}{2}=\frac{27}{2}-\frac{3}{2}=\frac{24}{2}=12$.

   If $x=\frac{4}{3}$, then $6\left(\frac{4}{3}\right)^2+\left(\frac{4}{3}\right)=12$ and $6\left(\frac{16}{9}\right)+\frac{4}{3}=\frac{32}{3}+\frac{4}{3}=\frac{36}{3}=12$.

This checking wasn't nearly as fun as some, but it sure does show how well this factoring business can work.

Solve for y in $9y^2-12y+4=0$. Follow these steps:

1. **This is already in standard form.**
2. **Find all the numbers that multiply to get a.**

   The factors for the 9 are 1×9 or 3×3.

3. **Find all the numbers that multiply to get c.**

   The factors for c are 1×4 or 2×2.

4. **Factor.**

   Using the 3's and the 2's is what works because both cross-products are 6, and you need a sum of 12 in the middle. So,

   $9y^2-12y+4=(3y-2)(3y-2)=0$

   Notice that I put the negative signs in because the 12 needs to be a negative sum.

5. **Use the MPZ to solve the equation.**

   The two factors are the same here. That means that using the MPZ gives you the same answer twice. When $3y-2=0$, solve this for y. First, add the 2 to each side, and then divide by 3. The solution is $y=\frac{2}{3}$. This is a *double root*, which, technically, has only one solution, but it occurs twice.

A double root occurs in quadratic trinomial equations that come from perfect-square binomials. Perfect-square binomials are discussed in Chapter 9, if you need a refresher. These perfect-square binomials are no more than the result of multiplying a binomial times itself. That's why, when they're factored, there's only one answer — it's the same one for each binomial.

Solve for z in $12z^2 - 4z - 8 = 0$. Follow these steps:

1. **This quadratic is already in standard form.**

   You can start out by looking for combinations of factors for the 12 and the 8, but you may notice that all three terms are divisible by 4. To make things easier, take out that GCF first, and then work with the smaller numbers in the parentheses.

   $$12x^2 - 4z - 8 = 4\left(3z^2 - z - 2\right) = 0$$

   You're now concentrating on factoring the quadratic $3z^2 - z - 2 = 0$.

2. **Find the numbers that multiply to get 3.**

   $3 = 1 \times 3$

3. **Find the numbers that multiply to get 2.**

   $2 = 1 \times 2$

4. **Factor.**

   This is really wonderful, especially because the 3 and 2 are both prime and can be factored only one way. Your only chore is to line up the factors so there will be a difference of 1 between the cross-products.

   $$4\left(3z^2 - z - 2\right) = 4(3z \quad 2)(z \quad 1) = 0$$

   Because the middle term is negative, you need to make the larger product negative, so put the negative sign on the 1.

   $$4(3z + 2)(z - 1) = 0$$

5. **Use the MPZ to solve for the value of z.**

   This time, when you use the MPZ, there are three factors to consider: $4 = 0$, $3z + 2 = 0$, or $z - 1 = 0$. The first equation is impossible; 4 doesn't ever equal 0. But the other two equations give you answers. If $3z + 2 = 0$, then $z = -\frac{2}{3}$. If $z - 1 = 0$, then $z = 1$.

6. **Check.**

   If $z = -\frac{2}{3}$, then $12\left(-\frac{2}{3}\right)^2 - 4\left(-\frac{2}{3}\right) - 8 = 0$ and $12\left(\frac{4}{9}\right) + \frac{8}{3} - 8 = \frac{16}{3} + \frac{8}{3} - 8 = \frac{24}{3} - 8 = 8 - 8 = 0$.

   If $z = 1$, then $12(1)^2 - 4(1) - 8 = 12 - 4 - 8 = 0$.

REMEMBER

When checking your solution(s) — always use the original equation (the version before you did anything to it).

Muscling Up to Quadratic Equations

EXAMPLE

Q. Solve for x by factoring: $18x^2 + 21x - 60 = 0$.

A. The factored form of the original equation is $3(6x^2 + 7x - 20) = 3(3x - 4)(2x + 5) = 0$. When the product of three factors is 0, one or more of the factors must be equal to 0. The first factor here, the 3, is certainly not equal to 0, so you move on to the next factor, $3x - 4$, and set it equal to 0 to solve for the solution. You use the same process with the last factor, $2x + 5$. Only the variables can take on a value to make the factor equal to 0. That's why, when you set the last two factors equal to 0, you solve those equations and get the answers.

$$3x - 4 = 0 \quad \text{or} \quad 2x + 5 = 0$$
$$3x = 4 \qquad\qquad 2x = -5$$
$$x = \frac{4}{3} \qquad\qquad x = \frac{-5}{2}$$

YOUR TURN

16 Solve for x by factoring: $x^2 - 2x - 15 = 0$.

17 Solve for x by factoring: $3x^2 - 25x + 28 = 0$.

18 Solve for y by factoring: $4y^2 - 9 = 0$.

19 Solve for z by factoring: $z^2 + 64 = 16z$.

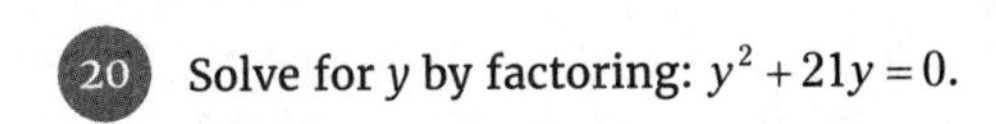

20 Solve for y by factoring: $y^2 + 21y = 0$.

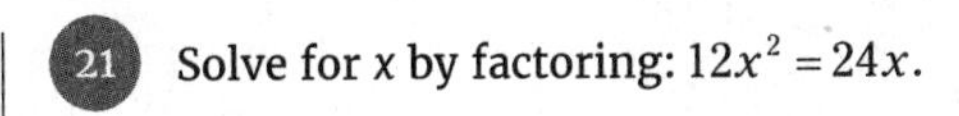

21 Solve for x by factoring: $12x^2 = 24x$.

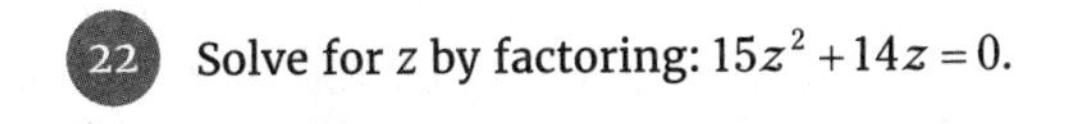

22 Solve for z by factoring: $15z^2 + 14z = 0$.

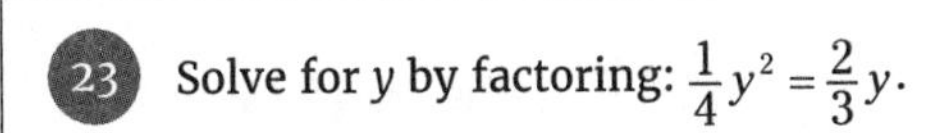

23 Solve for y by factoring: $\frac{1}{4}y^2 = \frac{2}{3}y$.

Using the Quadratic Formula

The quadratic formula is special to quadratic equations. A quadratic equation, $ax^2 + bx + c = 0$, can have as many as two solutions, but there may be only one solution or even no solution at all.

The variables a, b, and c are any real numbers. The a can't equal 0, but the b or c can equal 0.

The quadratic formula allows you to find solutions when the equations aren't very nice. Numbers aren't *nice* when they're funky fractions, indecent decimals with no end, or raucous radicals.

The quadratic formula says that if an equation is in the form $ax^2 + bx + c = 0$, then its solutions, the values of x, can be found with the following:

$$x = \frac{-b \pm \sqrt{b^2 - 4ac}}{2a}$$

You see an operation symbol, ±, in the formula. The symbol is shorthand for saying that the equation can be broken into two separate equations, one using the plus sign and the other using the minus sign. They look like the following:

$$x = \frac{-b + \sqrt{b^2 - 4ac}}{2a} \qquad x = \frac{-b - \sqrt{b^2 - 4ac}}{2a}$$

Can you see the difference between the two equations? The only difference is the change from the plus sign to the minus sign before the radical.

You can apply this formula to *any* quadratic equation to find the solutions — whether it factors or not. Let me show you some examples of how the formula works.

Use the quadratic formula to solve $2x^2 + 7x - 4 = 0$.

Refer to the standard form of a quadratic equation where the coefficient of x^2 is a, the coefficient of x is b, and the constant is c. In this case, $a = 2$, $b = 7$, and $c = -4$. Inserting those numbers into the formula, you get

$$x = \frac{-7 \pm \sqrt{7^2 - 4(2)(-4)}}{2(2)}$$

Now, simplifying, and paying close attention to the order of operations, you get

$$x = \frac{-7 \pm \sqrt{49 - (-32)}}{4} = \frac{-7 \pm \sqrt{81}}{4} = \frac{-7 \pm 9}{4}$$

The two solutions are found by applying the + in front of the 9 and then the − in front of the 9.

$$x = \frac{-7 + 9}{4} = \frac{2}{4} = \frac{1}{2}$$
$$x = \frac{-7 - 9}{4} = \frac{-16}{4} = -4$$

Whenever the answers you get from using the quadratic formula come out as integers or fractions, it means that the trinomial could have been factored. It doesn't mean, though, that you shouldn't use the quadratic formula on factorable problems. Sometimes it's easier to use the formula if the equation has really large or nasty numbers. In general, though, it's quicker to factor using unFOIL and then the MPZ when you can. Just to illustrate this, look at the previous example when it's solved using factoring and the MPZ: $2x^2 + 7x - 4 = (2x - 1)(x + 4) = 0$.

Then, using the MPZ, you get $2x - 1 = 0$ or $x + 4 = 0$, so $x = \frac{1}{2}$ or $x = -4$.

So, what do the results look like when the equation can't be factored? The next example shows you.

WARNING

Here are two things to watch out for when using the quadratic formula:

- **Don't forget that *–b* means to use the *opposite* of *b*.** If the coefficient *b* in the standard form of the equation is a positive number, change it to a negative number before inserting it into the formula. If *b* is negative, then change it to positive in the formula.
- **Be careful when simplifying under the radical.** The order of operations dictates that you square the value of *b* first, and then find the product of the 4 and the other two factors before subtracting them from the square of *b*. Some sign errors can occur if you're not careful.

Solve for *x* using the quadratic formula in $2x^2+8x+7=0$.

In this problem, you let $a=2$, $b=8$, and $c=7$ when using the formula:

$$x=\frac{-8\pm\sqrt{8^2-4(2)(7)}}{2(2)}=\frac{-8\pm\sqrt{64-56}}{4}=\frac{-8\pm\sqrt{8}}{4}$$

The radical can be simplified because $\sqrt{8}=\sqrt{4}\cdot\sqrt{2}=2\sqrt{2}$, so

$$x=\frac{-8\pm2\sqrt{2}}{4}=\frac{-{}^{4}\cancel{8}\pm\cancel{2}\sqrt{2}}{{}^{2}\cancel{4}}=\frac{-4\pm\sqrt{2}}{2}$$

REMEMBER

Be careful when simplifying this expression: $\frac{\left(-4+\sqrt{2}\right)}{2}\neq-2+\sqrt{2}$. Both terms in the numerator of the fraction have to be divided by the 2.

Here are the decimal equivalents of the answers rounded to three decimal places:

$$\frac{-4+\sqrt{2}}{2}\approx\frac{-4+1.414}{2}=\frac{-2.586}{2}=-1.293$$
$$\frac{-4-\sqrt{2}}{2}\approx\frac{-4-1.414}{2}=\frac{-5.414}{2}=-2.707$$

When you check these answers, what do the estimates do? If $x=-1.293$, then $2(-1.293)^2+8(-1.293)+7=3.343698-10.344+7=-0.000302$.

That isn't 0! What happened? Is the answer wrong? No, it's okay. The rounding caused the error — it didn't come out exactly right. This happens when you use a rounded value for the answer, rather than the exact radical form. An estimate was used for the answer because the square root of a number that is not a perfect square is an irrational number, and the decimal never ends. Rounding the decimal value to three decimal places seemed like enough decimal places.

REMEMBER

When you round decimal values, you shouldn't expect the check to come out to be *exactly* 0. In general, if you round the number you get from your check to the same number of places that you rounded your estimate of the radical, then you should get the 0 you're aiming for.

You can use the quadratic formula to solve a quadratic equation, regardless of whether the terms can be factored. Factoring and using the MPZ is almost always easier, but you'll find the quadratic formula most useful when the equation can't be factored or the equation has numbers too large to factor in your head.

EXAMPLE

Q. Use the quadratic formula to solve: $2x^2+11x-21=0$.

A. With $a=2$, $b=11$, and $c=-21$, fill in the formula and simplify to get:

$$\begin{aligned} x &= \frac{-11\pm\sqrt{11^2-4(2)(-21)}}{2(2)} \\ &= \frac{-11\pm\sqrt{121-(-168)}}{4} \\ &= \frac{-11\pm\sqrt{289}}{4} = \frac{-11\pm 17}{4} \\ &= \frac{-11+17}{4} \text{ or } \frac{-11-17}{4} \\ &= \frac{6}{4} \text{ or } \frac{-28}{4} \\ &= \frac{3}{2} \text{ or } -7 \end{aligned}$$

Because the 289 under the radical is a perfect square, you could have solved this problem by factoring.

Q. Use the quadratic formula to solve: $x^2-8x+2=0$.

A. The quadratic equation is already in standard form, with $a=1$, $b=-8$, and $c=2$. Fill in the formula with those values and simplify:

$$\begin{aligned} x &= \frac{-(-8)\pm\sqrt{(-8)^2-4(1)(2)}}{2(1)} \\ &= \frac{8\pm\sqrt{64-(8)}}{2} \\ &= \frac{8\pm\sqrt{56}}{2} = \frac{8\pm\sqrt{4}\sqrt{14}}{2} \\ &= \frac{8\pm 2\sqrt{14}}{2} = \frac{{}^{4}\cancel{8}\pm\cancel{2}\sqrt{14}}{\cancel{2}} \\ &= 4\pm\sqrt{14} \end{aligned}$$

YOUR TURN

Use the quadratic formula to solve.

$x^2-5x-6=0$

$6x^2+13x=-6$

$x^2-4x-6=0$

27 $2x^2+9x=2$

28 $3x^2 - 5x = 0$

29 $4x^2 - 25 = 0$

Completing the Square

An alternative to using factoring or the quadratic formula to solve a quadratic equation is a method called *completing the square*. Admittedly, this is seldom the method of choice, but the technique is used when dealing with equations of conic sections in higher mathematics courses. I include it here to give you some practice using it on some nice quadratic equations.

Here are the basic steps you follow to complete the square:

1. **Rewrite the quadratic equation in the form $ax^2 + bx = -c$.**
2. **Divide every term by a (if a isn't equal to 1).**
3. **Add $\left(\frac{b}{2a}\right)^2$ to each side of the equation.**

   This is essentially just half the x term's newest coefficient squared.
4. **Factor on the left (it's now a perfect square trinomial).**
5. **Take the square root of each side of the equation.**
6. **Solve for x.**

EXAMPLE

Q. Use completing the square to solve $x^2 - 4x - 5 = 0$.

A. First, rewrite the equation as $x^2 - 4x = 5$. Then add $\left(\frac{-4}{2}\right)^2 = 4$ to each side to get $x^2 - 4x + 4 = 5 + 4 = 9$. Factor on the left: $(x-2)^2 = 9$. Next, find the square root of each side:

$$\sqrt{(x-2)^2} = \pm\sqrt{9}$$
$$x - 2 = \pm 3$$

You only need to use the ± in front of the radical on the right. Technically, it belongs on both sides, but because you end up with two pairs of the same answer, just one ± is enough. Solve for x by adding 2 to each side of the equation: $x = 2 \pm 3$, giving you $x = 5$ or $x = -1$.

Q. Use completing the square to solve $2x^2 + 10x - 3 = 0$.

A. Rewrite the equation as $2x^2 + 10x = 3$. Then divide each term by 2: $x^2 + 5x = \frac{3}{2}$. Add $\left(\frac{5}{2}\right)^2 = \frac{25}{4}$ to each side to get $x^2 + 5x + \frac{25}{4} = \frac{3}{2} + \frac{25}{4} = \frac{31}{4}$. Now factor on the left to get: $\left(x + \frac{5}{2}\right)^2 = \frac{31}{4}$. Notice that the constant in the binomial is the number that got squared earlier. Find the square root of each side: $\sqrt{\left(x + \frac{5}{2}\right)^2} = \pm\sqrt{\frac{31}{4}}$ becomes $x + \frac{5}{2} = \pm\frac{\sqrt{31}}{2}$ Now solve for x:

$$x = -\frac{5}{2} \pm \frac{\sqrt{31}}{2} = \frac{-5 \pm \sqrt{31}}{2}.$$

YOUR TURN

Use completing the square to solve for x.

30 $3x^2 + 11x - 4 = 0$

31 $x^2 + 6x + 2 = 0$

Imagining the Worst with Imaginary Numbers

At one time, an imaginary number was something that didn't exist — well, at least until some enterprising mathematicians had their way. Not being happy with having to halt progress in solving some equations because of negative numbers under the radical, mathematicians came up with the imaginary number *i*.

The square root of −1 is designated as i — $\sqrt{-1} = i$ — and $i^2 = -1$. Because of the declaration of the value of *i*, all sorts of neat mathematics and applications have cropped up. Sorry, I can't cover all that good stuff in this book, but I at least give you a little preview of what *complex numbers* are all about.

You're apt to run into these imaginary numbers when using the quadratic formula. In the next example, the quadratic equation doesn't factor and doesn't have any *real* solutions (the only possible answers are *imaginary*).

Use the quadratic formula to solve $5x^2 - 6x + 5 = 0$.

In this quadratic, $a = 5$, $b = -6$, and $c = 5$. Putting the numbers into the formula:

$$x = \frac{-(-6) \pm \sqrt{(-6)^2 - 4(5)(5)}}{2(5)} = \frac{6 \pm \sqrt{36 - 100}}{10} = \frac{6 \pm \sqrt{-64}}{10}$$

You see a −64 under the radical. Only positive numbers and 0 have square roots. So you use the definition of the imaginary number where $i = \sqrt{-1}$ and apply it after simplifying the radical:

$$\frac{6 \pm \sqrt{-64}}{10} = \frac{6 \pm \sqrt{-1}\sqrt{64}}{10} = \frac{6 \pm i \cdot 8}{10} = \frac{3 \pm 4i}{5}$$

Applying this new *imaginary* number allows mathematicians to finish their problems. You have two answers — although both are imaginary. (It's sort of like having an imaginary friend as a child.)

REMEMBER

The square root of a negative number doesn't exist — as far as real numbers are concerned. When you encounter the square root of a negative number as you're solving a quadratic equation, it means that the equation doesn't have a real solution.

You can report your answers when you end up with negatives under the radical by using *imaginary numbers* (or numbers that are indicated with an *i* to show that they aren't real). With imaginary numbers, you let $\sqrt{-1} = i$. Then you can simplify the radical using the imaginary number as a factor. Don't worry if you're slightly confused about imaginary numbers. This concept is more advanced than most of the algebra problems you'll encounter in this book and during Algebra I. For now, just remember that using *i* allows you to make answers more complete; imaginary numbers let you finish the problem.

EXAMPLE

Q. Rewrite $\sqrt{-9}$, using imaginary numbers.

A. Split up the value under the radical into two factors:

$$\sqrt{-9} = \sqrt{(-1)9} = \sqrt{-1}\sqrt{9}$$
$$= i \cdot 3 \text{ or } 3i$$

Q. Rewrite $-6 \pm \sqrt{-48}$, using imaginary numbers.

A. You can write the value under the radical as the product of three factors:

$$-6 \pm \sqrt{-48} = -6 \pm \sqrt{-1}\sqrt{16}\sqrt{3}$$
$$= -6 \pm i \cdot 4\sqrt{3}$$
$$= -6 \pm 4i\sqrt{3}$$

The last two lines of the equation look almost identical. Putting the *i* after the 4 is just a mathematical convention, a preferred format. It doesn't change the value at all.

32 Rewrite $\sqrt{-4}$ as a product with a factor of i.

33 Rewrite $6 \pm \sqrt{-96}$ as an expression with a factor of i.

Answers to Chapter 16 Quiz

1. **7 or −5.** $x^2 - 2x - 35 = (x-7)(x+5) = 0$
2. **1 or −9.** $x = \frac{-8 \pm \sqrt{8^2 - 4(1)(-9)}}{2(1)} = \frac{-8 \pm \sqrt{64+36}}{2} = \frac{-8 \pm \sqrt{100}}{2} = \frac{-8 \pm 10}{2} = -4 \pm 5$
3. **−4 or −6.** Subtract 24 from each side. Then add the square of 5 (half of 10) to each side.

   $x^2 + 10x + 24 = 0 \rightarrow x^2 + 10x = -24 \rightarrow x^2 + 10x + 25 = -24 + 25$

   Factor on the left and simplify on the right. Then take the square root of each side.

   $(x+5)^2 = 1 \rightarrow x + 5 = \pm 1$

   Subtract 5 from each side.

   $x + 5 = \pm 1 \rightarrow x = -5 \pm 1$
4. **0.** The binomial is never equal to 0; there is no real solution to the binomial equation created.
5. **0 or −6.** Multiply both sides of the equation by −1, which gives you $y^2 + 6y = y(y+6) = 0$.
6. **±6.** Add 36 to each side. Then take the square root of both sides.

   $x^2 - 36 = 0 \rightarrow x^2 = 36 \rightarrow \sqrt{x^2} = \pm\sqrt{36} \rightarrow x = \pm 6.$
7. **3 or −7.** Using the MPZ, you set each factor equal to zero. When $x - 3 = 0$, $x = 3$, and when $x + 7 = 0$, $x = -7$.
8. **0 or 8.** $y^2 - 8y = y(y-8) = 0$
9. $\mathbf{\frac{3 \pm \sqrt{41}}{4}}$**.** $x = \frac{-(-3) \pm \sqrt{(-3)^2 - 4(2)(-4)}}{2(2)} = \frac{3 \pm \sqrt{9+32}}{4} = \frac{3 \pm \sqrt{41}}{4}$
10. $\mathbf{-3 \pm 2i\sqrt{5}}$**.** $-3 \pm \sqrt{-20} = -3 \pm \sqrt{4(5)(-1)} = -3 \pm \sqrt{4}\sqrt{5}\sqrt{-1} = -3 \pm 2\sqrt{5}i = -3 \pm 2i\sqrt{5}$
11. **4 (a double root).** Subtract 8z from each side and then factor.

    $z^2 + 16 = 8z \rightarrow z^2 - 8z + 16 = 0 \rightarrow (z-4)^2 = 0$
12. **9 or −5.** Find the square root of each side. Then add 2 to each side.

    $\sqrt{(y-2)^2} = \pm\sqrt{49} \rightarrow y - 2 = \pm 7$ then $y = 2 + 7 = 9$ or $y = 2 - 7 = -5$.
13. $\mathbf{-\frac{1}{4}}$ **or** $\mathbf{\frac{2}{3}}$**.** First, subtract 2 from each side: $12y^2 - 5y = 2 \rightarrow 12y^2 - 5y - 2 = 0$.

    Then, factor: $12y^2 - 5y - 2 = (4y+1)(3y-2) = 0$.

Muscling Up to Quadratic Equations

IN THIS CHAPTER

- Solving basic cubic equations
- Tallying up the possible number of roots
- Making educated guesses about solutions
- Finding solutions by factoring
- Recognizing patterns to make factoring and solutions easier

Chapter 17

Yielding to Higher Powers

Polynomial equations are equations involving the sum (or difference) of terms where the variables have exponents that are whole numbers. Cubic polynomials are special types of polynomials with the standard form $ax^3 + bx^2 + cx + d = 0$. Solving polynomial equations involves setting polynomials equal to zero and then figuring out which values create true statements. You can use the solutions to polynomial equations to solve problems in calculus, algebra, and other mathematical areas. When you're graphing polynomials, the solutions show you where the curve intersects with the x-axis — either crossing it or just touching it at that point. Instead of just taking some wild guesses as to what the solutions may be, you can utilize some of the available techniques that help you make more reasonable guesses as to what the solutions are and then confirm your guess with good algebra.

This chapter provides several examples of these techniques and gives you ample opportunities to try them out.

Queuing Up to Cubic Equations

Cubic equations contain a variable term with a power of 3 but no power higher than 3. In these equations, you can expect to find up to three different solutions, but there may not be as many as three. Also, a cubic equation must have at least one solution, even though it may not be a nice one. A *quadratic equation* (a second-degree equation with a term that has an exponent of 2) doesn't offer this guarantee: Quadratic equations don't have to have real solutions.

If second-degree equations can have as many as two different solutions and third-degree equations can have as many as three different solutions, do you suppose that a pattern exists? Can you assume that fourth-degree equations could have as many as four solutions and fifth-degree equations could have as many as five? Yes, indeed you can — this is the general rule. The degree can tell you what the *maximum* number of solutions is. Although the number of solutions *may* be less than the number of the degree, there won't be any more solutions than that number.

Solving perfectly cubed equations

If a cubic equation has just two terms and they're both perfect cubes, then your task is easy. The sum or difference of perfect cubes can be factored into two factors with only one real solution. The first factor, or the *binomial*, gives you a real solution. The second factor, the *trinomial*, does not give you a real solution. (If you can't remember how to factor these cubics, turn to Chapter 12.)

If $x^3 - a^3 = 0$, then $x^3 - a^3 = (x-a)(x^2 + ax + a^2) = 0$ and $x = a$ is the only real solution. Likewise, if $x^3 + a^3 = 0$, then $(x+a)(x^2 - ax + a^2) = 0$ and $x = -a$ is the only real solution. The reason you have only one solution for each of these cubics is because $x^2 + ax + a^2 = 0$ and $x^2 - ax + a^2 = 0$ have no real solutions. The trinomials can't be factored, and the quadratic formula gives you imaginary solutions. (See Chapter 16 for information on imaginary results.)

The key to solving cubic equations that have two terms that are both cubes is in recognizing that that's what you have.

Solve for x in $x^3 - 8 = 0$. Use these steps:.

1. **Factor first.**

   The factorization is $x^3 - 8 = (x-2)(x^2 + 2x + 4)$.

2. **Apply the multiplication property of zero (MPZ).**

   If $(x-2)(x^2 + 2x + 4) = 0$, then $x - 2 = 0$ or $x^2 + 2x + 4 = 0$.

Only the first equation, $x - 2 = 0$, has an answer: $x = 2$. The other equation doesn't have any real numbers that satisfy it. There's only the one real solution.

The next two problems show you some different twists to these special cubic equations.

Solve for y in $27y^3 + 64 = 0$ using factoring. The factorization here is $27y^3 + 64 = (3y+4)\left(9y^2 - 12y + 16\right)$. The first factor offers a solution, so set $3y + 4$ equal to zero to get $3y = -4$ or $y = -\frac{4}{3}$.

Solve for a in $8a^3 - (a-2)^3 = 0$ using factoring. The factorization here works the same as factorizations of the difference between perfect cubes. It's just more complicated because the second term is a binomial:

$$8a^3 - (a-2)^3 = \left[2a - (a-2)\right]\left[4a^2 + 2a(a-2) + (a-2)^2\right] = 0$$

1. **Simplify inside the first bracket by distributing the negative and you get**

   $$\left[2a - (a-2)\right] = \left[2a - a + 2\right] = \left[a + 2\right]$$

2. **Setting the first factor equal to 0, you get**

   $$a + 2 = 0$$
   $$a = -2$$

As usual, the second factor doesn't give you a real solution, even if you distribute, square the binomial, and combine all the like terms.

Another way to solve the equations in the examples I just showed you is to perform an operation that looks something like the square-root rule. (See Chapter 16 for information on that rule.) Move the constant term to one side of the equation and find the cube root of each side. Unlike the square-root rule, there's only one root, not two.

Redoing the three problems from the examples:

$x^3 - 8 = 0$. You add 8 to each side and find the cube roots: $\sqrt[3]{x^3} = \sqrt[3]{8}$, giving you $x = 2$.

$27y^3 + 64 = 0$. You subtract 64 from each side, divide by 27, and then find the cube roots: $\sqrt[3]{y^3} = \sqrt[3]{-\frac{64}{27}}$, giving you $y = -\frac{4}{3}$.

$8a^3 - (a-2)^3 = 0$. You add $(a-2)^3$ to each side and find the cube roots: $\sqrt[3]{8a^3} = \sqrt[3]{(a-2)^3}$, giving you $2a = a - 2$. Subtract a from each side, and you have $a = -2$.

EXAMPLE

Q. Solve for x in $x^3 - 1 = 0$.

A. Factoring the cubic equation you have $(x-1)\left(x^2 + x + 1\right) = 0$. Only $x - 1 = 0$ has a solution: $x = 1$.

Q. Solve for t in $(3t-1)^3 - (t+1)^3 = 0$.

A. Factoring the cubic equation you have $\left(3t - 1 - (t+1)\right)\left((3t-1)^2 + (3t-1)(t+1) + (t+1)^2\right) = 0$.

This simplifies to: $(2t-2)\left(13t^2 - 2t + 1\right) = 0$. Setting the binomial equal to 0, you have $t = 1$. The trinomial has no real solution.

1 Solve for x in $x^3 - 27 = 0$.

2 Solve for y in $y^3 + 1 = 0$.

3 Solve for z in $27z^3 - (z+1)^3 = 0$.

4 Solve for w in $(w+4)^3 + (2w-3)^3 = 0$.

Working with the not-so-perfectly cubed

When you have a cubic equation consisting of just two terms, you can factor the terms if they're both perfect cubes. But what if the variable is cubed and the other term is a constant that's *not* a perfect cube? Are you stuck? Absolutely not — as long as you're willing to work with the irrational.

The solution of the cubic equation $ax^3 - b = 0$ is $x = \sqrt[3]{\frac{b}{a}}$, and the solution of the cubic equation $ax^3 + b = 0$ is $x = -\sqrt[3]{\frac{b}{a}}$.

The cube roots are irrational numbers when you don't have perfect cubes under the radical. Irrational numbers have decimals that go on forever without repeating.

EXAMPLE

Q. Solve for x in the equation $5x^3 - 4 = 0$.

A. According to the rule, the answer is $\boldsymbol{x = \sqrt[3]{\frac{4}{5}}}$, which is about 0.9283177667, and so on. If you prefer not dealing with a special rule to solve these equations, just use a rule something like the *square-root rule* (see Chapter 16), except take a cube root instead of a square root. Rewrite the original equation as $5x^3 = 4$. Then divide each side of the equation by 5 and take the cube root of each side. Both positive and negative numbers have cube roots, so it doesn't matter if you have a negative under the radical.

Q. Solve for x in the equation $2x^3 + 108 = 0$.

A. Subtract 108 from each side, and you have $2x^3 = -108$. Dividing each side by 2, you have $x^3 = -54$. Next, take the cube root of each side. You see that 54 has a factor that is a perfect cube: $\sqrt[3]{x^3} = \sqrt[3]{-54} = \sqrt[3]{-27}\sqrt[3]{2}$. Simplifying, you have $x = -3\sqrt[3]{2}$.

YOUR TURN

5 Solve for x: $2x^3 = 11$.

6 Solve for y: $3y^3 - 4 = 0$.

7 Solve for z: $4z^3 = -81$.

8 Solve for w: $27w^3 + 16 = 0$.

Going for the greatest common factor

Another type of cubic equation that's easy to solve is one in which you can factor out a variable greatest common factor (GCF), leaving a second factor that is linear or quadratic (first or second degree). You apply the MPZ and work to find the solutions — usually three of them.

Factoring out a first-degree variable GCF

When the terms of a three-term cubic equation all have the same first-degree variable as a factor, then factor the variable out. The resulting equation will have the variable as one factor and a quadratic expression as the second factor. The first-degree variable will always give you a solution of 0 when you apply the MPZ. If the quadratic has solutions, you can find them using the methods shown in Chapter 16.

Solve for x in $x^3 - 4x^2 - 5x = 0$. Use the following steps:

1. **Determine that each term has a factor of x and factor that out.**

   The GCF is x. Factor to get $x(x^2 - 4x - 5) = 0$.

   You're all ready to apply the MPZ when you notice that the second factor, the quadratic, can be factored. Do that first and then use the MPZ on the whole thing.

2. **Factor the quadratic expression, if possible.**

   $x(x^2 - 4x - 5) = x(x-5)(x+1) = 0$

3. **Apply the MPZ and solve.**

   Setting the individual factors equal to 0, you get $x = 0$, $x - 5 = 0$, or $x + 1 = 0$. This means that $x = 0$ or $x = 5$ or $x = -1$.

4. **Check the solutions in the original equation.**

   If $x = 0$, then $0^3 - 4(0)^2 - 5(0) = 0 - 0 - 0 = 0$.

   If $x = 5$, then $5^3 - 4(5)^2 - 5(5) = 125 - 4(25) - 25 = 125 - 100 - 25 = 0$.

   If $x = -1$, then $(-1)^3 - 4(-1)^2 - 5(-1) = -1 - 4(1) + 5 = -1 - 4 + 5 = 0$.

   All three work!

Factoring out a second-degree greatest common factor

Just as with first-degree variable greatest common factors, you can also factor out second-degree variables (or third-degree, fourth-degree, and so on). Factoring leaves you with another expression that may have additional solutions. Consider the following example.

Solve for w in $w^3 - 3w^2 = 0$. Use these steps:

1. **Determine that each term has a factor of w2 and factor that out.**

   Factoring out w^2, you get $w^3 - 3w = w^2(w - 3) = 0$.

2. **Use the MPZ.**

   $w^2 = 0$ or $w - 3 = 0$.

3. **Solve the resulting equations.**

   Solving the first equation involves taking the square root of each side of the equation. This process usually results in two different answers: the positive answer and the negative answer. However, this isn't the case with $w^2 = 0$ because 0 is neither positive nor negative. So there's only one solution from this factor: $w = 0$. And the other factor gives you a solution of $w = 3$. So, even though this is a cubic equation, there are only two solutions to it.

EXAMPLE

Q. Solve for z in $z^3 + z^2 + z = 0$.

A. Use these steps:

The way a double root or double solution works in these equations is that the solutions appear twice in the factored form. If you go backward from the MPZ and write the factors that give the solutions to the cubic equation, it looks like this:

$$y^3-4y^2+5y-2=(y-1)(y-1)(y-2)=0$$

Or, showing the double root or solution more distinctly,

$$(y-1)(y-1)(y-2)=(y-1)^2(y-2)=0$$

In these two examples, I magically pull out numbers that work in the equations. The following sections show you how to do this for yourself.

Determining How Many Possible Roots

Mathematician René Descartes came up with his *rule of signs*, which allows you to determine the number of real roots that a polynomial equation may have. The *real roots* are the *real* numbers that make the equation a true statement. This rule doesn't tell you for sure how many roots there are; it just tells you the maximum number there *could* be. (If this number is less than the maximum number of roots, then it's less than that by two or four or six, and so on.)

To use Descartes's rule, first write the polynomial in decreasing powers of the variable; then do the following:

- To determine the maximum possible *positive roots,* count how many times the signs of the terms change from positive to negative or vice versa.
- To determine the possible number of *negative roots,* replace all the *x*'s with negative *x*'s. Simplify the terms and count how many times the signs change.

EXAMPLE

Q. How many possible real roots are there in $3x^5+5x^4-x^3+2x^2-x+4=0$?

A. The sign changes from positive to negative to positive to negative to positive. That's four changes in sign, so you have a maximum of four positive real roots. If it doesn't have four, then it could have two. If it doesn't have two, then it has none. You step down by twos. Now count the number of possible negative real roots in that same polynomial by replacing all the *x*'s with negative *x*'s and counting the number of sign changes:

$$\begin{aligned}&3(-x)^5+5(-x)^4-(-x)^3+2(-x)^2-(-x)+4\\&=-3x^5+5x^4+x^3+2x^2+x+4=0\end{aligned}$$

This version only has one sign change: from negative to positive, which means that it has one negative real root. You can't go down by two from that, so one negative real root is the only choice.

Q. How many possible real roots are there in $6x^4 + 5x^3 + 3x^2 + 2x - 1 = 0$?

A. Count the number of sign changes in the original equation. It has only one sign change, so there's exactly one positive real root. Change the function by replacing all the x's with negative x's and count the changes in sign:

$$\begin{aligned}&6(-x)^4 + 5(-x)^3 + 3(-x)^2 + 2(-x) - 1\\&= 6x^4 - 5x^3 + 3x^2 - 2x - 1 = 0\end{aligned}$$

It has three sign changes, which means that it has three or one negative real roots.

YOUR TURN

17 How many possible positive and negative real roots are in $x^5 - x^3 + 8x^2 - 8 = 0$?

18 How many possible positive and negative real roots are in $8x^5 - 25x^4 - x^2 + 25 = 0$?

Applying the Rational Root Theorem

In the preceding section, you discover that Descartes's rule of signs counts the possible number of *real* roots. Now you see a rule that helps you figure out just what those real roots are, when they're rational numbers.

REMEMBER

Real numbers can be either rational or irrational. *Rational* numbers are numbers that have fractional equivalents; that is, they can be written as fractions. *Irrational* numbers can't be written as fractions; they have decimal values that never repeat and never end.

The *rational root theorem* says that if you have a polynomial equation written in the form $a_n x^n + a_{n-1}x^{n-1} + a_{n-2}x^{n-2} + \ldots + a_1 x^1 + a_0 = 0$, then you can make a list of all the possible *rational* roots by looking at the first term and the last term. Any rational roots must be able to be written as a fraction with a factor of the *constant* (the last term or a_0) in the numerator of the fraction and a factor of the lead coefficient (a_n) in the denominator.

Here's an example: Use the rational root theorem to find the possible rational roots of the equation $4x^4 - 3x^3 + 5x^2 + 9x - 3 = 0$.

The factors of the constant are +3, −3, +1, −1, and the factors of the coefficient of the first term are +4, −4, +2, −2, +1, −1. The following list includes all the ways that you can create a fraction with a factor of the constant in the numerator and a factor of the lead coefficient in the denominator:

$$\pm\frac{3}{4}, \pm\frac{3}{2}, \pm\frac{3}{1}, \pm\frac{1}{4}, \pm\frac{1}{2}, \pm\frac{1}{1}$$

So the total list of possible rational roots is $\pm1, \pm3, \pm\frac{3}{4}, \pm\frac{3}{2}, \pm\frac{1}{4}, \pm\frac{1}{2}$.

Of course, the two fractions with 1 in the denominator are actually whole numbers, when you simplify. This abbreviated listing represents all the possible ways to combine +3 and −3 and +4 and −4, and so on, to create all the possible fractions. It's just quicker to use the ± notation than to write out every single possibility.

Although this new list has 12 candidates for solutions to the equation, it's really relatively short when you're trying to run through all the possibilities. Many of the polynomials start out with a 1 as the coefficient of the first term, which is great news when you're writing your list because that means the only rational numbers you're considering are whole numbers — the denominators are 1.

EXAMPLE

Q. Determine all the possible rational solutions of this equation: $2x^6 - 4x^3 + 5x^2 + x - 30 = 0$.

A. The factors of the constant are $\pm30, \pm15, \pm10, \pm6, \pm5, \pm3, \pm2, \pm1$, and the factors of the lead coefficient are $\pm2, \pm1$. You create the list of all the numbers that could be considered for roots of the equation by dividing each of the factors of the constant by the factors of the lead coefficient. The numbers shown in the answer don't include repeats or unfactored fractions:

$$\pm30, \pm15, \pm10, \pm6, \pm5, \pm3, \pm2, \pm1,$$
$$\pm\frac{15}{2}, \pm\frac{5}{2}, \pm\frac{3}{2}, \pm\frac{1}{2}$$

Q. Determine all the possible rational solutions of the equation $x^6 - x^3 + x^2 + x - 1 = 0$.

A. The factors of the constant are ±1 and the factors of the lead coefficient are ±1. Yes, even though Descartes tells you that there could be as many as three positive real roots and one negative real root, the only possible rational roots are +1 or −1.

YOUR TURN

19 List all the possible rational roots of $2x^4 - 3x^3 - 54x + 81 = 0$.

20 List all the possible rational roots of $8x^5 - 25x^3 - x^2 + 25 = 0$.

Using the Factor/Root Theorem

Algebra has a theorem that says if the binomial $x - c$ is a factor of a polynomial (it divides the polynomial evenly, with no remainder), then c is a root or solution of the polynomial. You may say, "Okay, so what?" Well, this property means that you can use the very efficient method of synthetic division to solve for solutions of polynomial equations.

Use synthetic division to try out all those rational numbers that you listed as possibilities for roots of a polynomial. (See Chapters 10 and 13 for more on synthetic division.) If $x - c$ is a factor (and c is a root), then you won't have a remainder (the remainder is 0) when you perform synthetic division.

Q. Check to see whether the number 2 is a root of the following polynomial:

$$x^6 - 6x^5 + 8x^4 + 2x^3 - x^2 - 7x + 2 = 0$$

A. Yes, it's a root. Use the 2 and the coefficients of the polynomial in a synthetic division problem:

$$\begin{array}{r|rrrrrrr} 2 & 1 & -6 & 8 & 2 & -1 & -7 & 2 \\ & & 2 & -8 & 0 & 4 & 6 & -2 \\ \hline & 1 & -4 & 0 & 2 & 3 & -1 & 0 \end{array}$$

The remainder is 0, so $x - 2$ is a factor, and 2 is a root or solution. The quotient of this division is $x^5 - 4x^4 + 2x^2 + 3x - 1$, which you write using the coefficients along the bottom. When writing the factorization, make sure you start with a variable that's one degree lower than the one that was divided into. This new polynomial ends in a -1 and has a lead coefficient of 1, so the only possible solutions when setting the quotient equal to 0 are 1 or -1.

Q. Check to see whether 1 or -1 is a solution of the new equation $x^5 - 4x^4 + 2x^2 + 3x - 1 = 0$.

A. Neither is a solution. First, try 1:

$$\begin{array}{r|rrrrrr} 1 & 1 & -4 & 0 & 2 & 3 & -1 \\ & & 1 & -3 & -3 & -1 & 2 \\ \hline & 1 & -3 & -3 & -1 & 2 & 1 \end{array}$$

That '1' didn't work; the remainder isn't 0. Now, try -1:

$$\begin{array}{r|rrrrrr} 1 & 1 & -4 & 0 & 2 & 3 & -1 \\ & & -1 & 5 & -5 & 3 & -6 \\ \hline & -1 & -5 & 5 & -3 & 6 & -7 \end{array}$$

It doesn't work, either. The only rational solution of the original equation is 2.

YOUR TURN

21 Check to see whether −3 is a root of $x^4 - 10x^2 + 9 = 0$.

22 Check to see whether $\frac{3}{2}$ is a root of $2x^4 - 3x^3 - 54x + 81 = 0$.

Yielding to Higher Powers

Solving by Factoring

When determining the solutions for polynomials, many techniques are available to help you determine what those solutions are — if any solutions exist. One method that is usually the quickest, though, is factoring and using the *multiplication property of zero* (MPZ). (Check out Chapter 16 for more ways to use the MPZ.) Not all polynomials lend themselves to factoring, but when they do, using this method is to your advantage. And don't forget to try synthetic division and the factor theorem if you think you have a potential solution.

EXAMPLE

Q. Find the real solutions of $x^4 + 2x^3 - 125x - 250 = 0$.

A. The only two solutions are $x = -2$ or $x = 5$. You can first factor by grouping and then factor the difference of cubes to get $x^3(x+2) - 125(x+2) = (x+2)(x^3 - 125) = (x+2)(x-5)(x^2 + 5x + 25) = 0$. The trinomial factor in this factorization never factors any more, so only the first two factors yield solutions: $x + 2 = 0$, $x = -2$ and $x - 5 = 0$, $x = 5$.

Q. Find the real solutions of $x^4 - 81 = 0$.

A. The only two real solutions are $x = 3$ or $x = -3$. Factoring the binomial into the sum and difference of the roots, you get $(x^2 - 9)(x^2 + 9) = 0$. The first factor of this factored form is also the difference of perfect squares. Factoring again, you get $(x-3)(x+3)(x^2+9) = 0$. Now, to use the MPZ, set the first factor equal to 0 to get $x - 3 = 0$, $x = 3$. Set the second factor equal to 0 to get $x + 3 = 0$, $x = -3$.

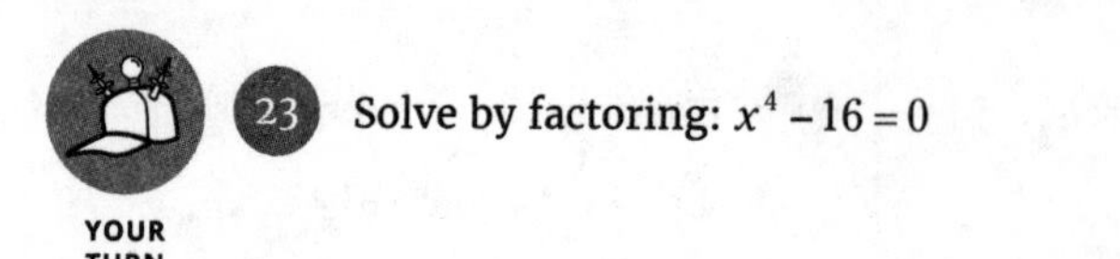

YOUR TURN

23 Solve by factoring: $x^4 - 16 = 0$

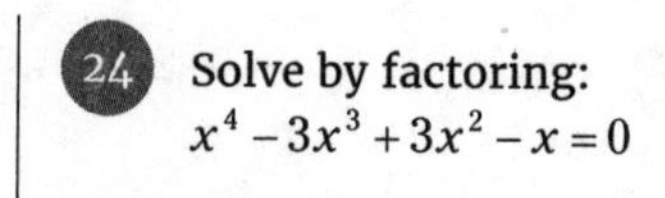

24 Solve by factoring: $x^4 - 3x^3 + 3x^2 - x = 0$

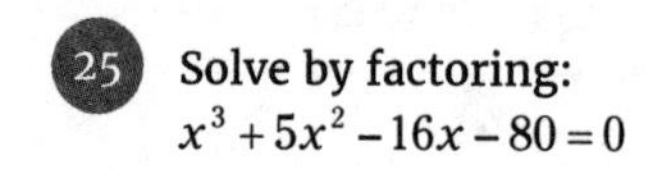

25 Solve by factoring: $x^3 + 5x^2 - 16x - 80 = 0$

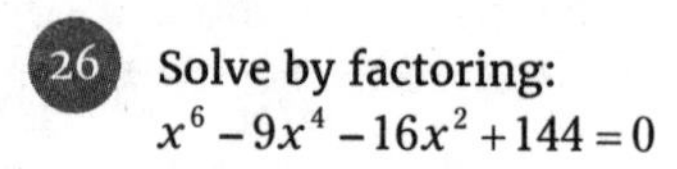

26 Solve by factoring: $x^6 - 9x^4 - 16x^2 + 144 = 0$

Solving Powers That Are Quadratic-Like

Some equations with higher powers or fractional powers are *quadratic-like,* meaning that they have three terms and

- The variable in the first term has an even power (4, 6, 8,) or $\left(\frac{1}{2}, \frac{1}{4}, \frac{1}{6}, \ldots\right)$.
- The variable in the second term has a power that is half that of the first.
- The third term is a constant number.

In general, the format for a quadratic-like equation is: $ax^{2n} + bx^n + c = 0$. Just as in the general quadratic equation, the x is the variable and the a, b, and c are constant numbers. The a can't be 0, but the other two letters have no restrictions. The n is also a constant and can be anything except 0. For example, if $n = 3$, then the equation would read $ax^6 + bx^3 + c = 0$.

To solve a quadratic-like equation, you must first pretend that it's quadratic and use the same methods as you do for those types of equations, and then do a step or two more. The extra steps usually involve taking an extra root or raising to an extra power.

Notice that each of the following quadratic-like equations meets all the requirements:

- $x^4 - 5x^2 + 4 = 0$
- $y^6 + 7y^3 - 8 = 0$
- $z^8 + 7x^4 + 6 = 0$
- $w^{\frac{1}{2}} - 7w^{\frac{1}{4}} + 12 = 0$

Yielding to Higher Powers

When you recognize that you have a quadratic-like equation, solve it by following these steps:

1. **Rewrite the quadratic-like equation as an actual quadratic equation, replacing the actual powers with 2 and 1 by doing a substitution.**

TIP

   Change the letters of the variables so that you don't confuse the rewritten equation with the original.

2. **Factor the new quadratic equation. If the equation doesn't factor, then use the quadratic formula.**
3. **Reverse the substitution and replace the original variables.**
4. **Use the MPZ to find the solutions.**

The highest power of an equation, when it's a whole number, tells you the number of possible solutions; there won't be more than that number.

Here's an example that works through these steps: Solve for x in $x^4 - 5x^2 + 4 = 0$.

Follow the steps:

1. **Rewrite the equation, replacing the actual powers with the numbers 2 and 1.**

   Rewrite this as a quadratic equation using the same *coefficients* (number multipliers) and constant.

TIP

   Change the letter used for the variable, so you won't confuse this new equation with the original. Substitute q for x^2 and q^2 for x^4:

   $q^2 - 5q + 4 = 0$

2. **Factor the quadratic equation.**

   $q^2 - 5q + 4 = 0$ factors nicely into $(q - 4)(q - 1) = 0$.

3. **Reverse the substitution and use the factorization pattern to factor the original equation.**

   Use that same pattern to write the factorization of the original problem. When you replace the variable q in the factored form, use x^2:

   $x^4 - 5x^2 + 4 = (x^2 - 4)(x^2 - 1) = 0$

4. **Solve the equation using the MPZ.**

   Either $x^2 - 4 = 0$ or $x^2 - 1 = 0$. If $x^2 - 4 = 0$, then $x^2 = 4$ and $x = \pm 2$. If $x^2 - 1 = 0$, then $x^2 = 1$ and $x = \pm 1$.

This fourth-degree equation did live up to its reputation and have four different solutions.

This next example presents an interesting problem because the exponents are fractions. But the trinomial fits into the category of quadratic-like, so I'll show you how you can take advantage of this format to solve the equation. And, no, the rule of the number of solutions doesn't work the same way here. There aren't any possible situations where there's half a solution.

Solve $w^{\frac{1}{2}} - 7w^{\frac{1}{4}} + 12 = 0$.

Follow these steps:

1. **Rewrite the equation with powers of 2 and 1. Substitute q for $w^{\frac{1}{4}}$ and q2 for $w^{\frac{1}{2}}$. (Remember: Squaring $w^{\frac{1}{4}}$ gives you $\left(w^{\frac{1}{4}}\right)^2 = w^{\frac{2}{4}} = w^{\frac{1}{2}}$.)**

   Rewrite the equation as $q^2 - 7q + 12 = 0$.

2. **Factor.**

   This factors nicely into $(q-3)(q-4) = 0$.

3. **Replace the variables from the original equation, using the pattern.**

   Replace with the original variables to get $\left(w^{\frac{1}{4}} - 3\right)\left(w^{\frac{1}{4}} - 4\right) = 0$.

4. **Solve the equation for the original variable, w.**

   $$\left(w^{\frac{1}{4}} - 3\right)\left(w^{\frac{1}{4}} - 4\right) = 0$$

   Now, when you use the MPZ, you get that either $w^{\frac{1}{4}} - 3 = 0$ or $w^{\frac{1}{4}} - 4 = 0$. How do you solve these two equations?

   Look at $w^{\frac{1}{4}} - 3 = 0$. Adding 3 to each side, you get $w^{\frac{1}{4}} = 3$. You can solve for w if you raise each side to the fourth power: $\left(w^{\frac{1}{4}}\right)^4 = (3)^4$. This says that $w = 81$.

   Doing the same with the other factor, if $w^{\frac{1}{4}} - 4 = 0$, then $w^{\frac{1}{4}} = 4$ and $\left(w^{\frac{1}{4}}\right)^4 = (4)^4$. This says that $w = 256$.

5. **Check the answers.**

   If $w = 81$, $(81)^{\frac{1}{2}} - 7(81)^{\frac{1}{4}} + 12 = 9 - 7(3) + 12 = 21 - 21 = 0$.

   If $w = 256$, $(256)^{\frac{1}{2}} - 7(256)^{\frac{1}{4}} + 12 = 16 - 7(4) + 12 = 28 - 28 = 0$.

   They both work.

Negative exponents are another interesting twist to these equations, as you see in the next example.

PHYSICAL CHALLENGES

One of the most famous musical composers was Beethoven. His accomplishments were even more incredible when you realize that he was deaf for a good deal of his life and still continued to produce musical masterpieces.

A similar situation occurred with the mathematician Leonhard Euler. Euler was one of the most prolific mathematicians of his generation and produced more than half of his work after he had gone blind. He dictated his findings from memory. Euler showed that a proposed formula for creating prime numbers didn't really work. Fermat conjectured that $2^{(2^x)}+1$ always produced a prime number. Euler showed that the formula failed when $x=5$. (When $x=1$, the formula produces the prime number 5; when $x=2$, the result is 17; when $x=3$, the result is 257; and when $x=4$, the number result is 65,537. *Remember:* This is before computers or even calculators.)

Solve for the value of x in $2x^{-6}-x^{-3}-3=0$.

Follow these steps:

1. **Rewrite the equation using powers of 2 and 1. Substitute q for x^{-3} and q2 for x^{-6}.**

   Rewrite the equation as $2q^2-q-3=0$.

2. **Factor.**

   This factors into $(2q-3)(q+1)=0$.

3. **Go back to the original variables and powers.**

   Use this pattern. Factor the original equation to get:

   $$\left(2x^{-3}-3\right)\left(x^{-3}+1\right)=0$$

4. **Solve.**

   Use the MPZ. The two equations to solve are $2x^{-3}-3=0$ and $x^{-3}+1=0$. These become $2x^{-3}=3$ and $x^{-3}=-1$. Rewrite these equations using the definition of negative exponents:

   $$x^{-n}=\frac{1}{x^n}$$

   So the two equations can be written as $\frac{2}{x^3}=3$ and $\frac{1}{x^3}=-1$. Cross-multiply in each case to get $3x^3=2$ and $x^3=-1$. Divide the first equation through by 3 to get the x^3 alone, and then take the cube root of each side to solve for x:

   $$x=\sqrt[3]{\frac{2}{3}} \text{ or } x=\sqrt[3]{-1}=-1$$

EXAMPLE

Q. Solve for x: $x^8 - 17x^4 + 16 = 0$

A. Factoring, you have: $(x^4 - 1)(x^4 - 16) = 0$. Then factor each binomial to get $\left(x^2 - 1\right)\left(x^2 + 1\right)\left(x^2 - 4\right)\left(x^2 + 4\right) = 0$. There are still two factorable binomials. The final factorization reads: $(x - 1)(x + 1)(x^2 + 1)(x - 2)(x + 2)(x^2 + 4) = 0$. Only the linear factors yield the answers, which are $x = 1$, −1, 2, and −2. The two factors that are the sums of perfect squares don't provide any real roots.

Q. Solve for y: $y^{2/3} + 5y^{1/3} + 6 = 0$

A. First, rewrite the equation using powers of 2. You can say that $q^2 + 5q + 6 = 0$. This factors into $(q + 3)(q + 2) = 0$. Going back to the original variables and powers, you have $\left(y^{1/3} + 3\right)\left(y^{1/3} + 2\right) = 0$. Using the MPZ, $y^{1/3} + 3 = 0$ and $y^{1/3} + 2 = 0$. These become $y^{1/3} = -3$ and $y^{1/3} = -2$. Cube each side of the equations to get $y = -27$ and $y = -8$.

YOUR TURN

Solve the equation.

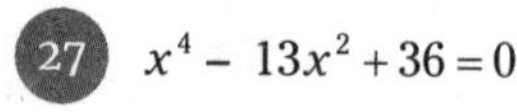

27 $x^4 - 13x^2 + 36 = 0$

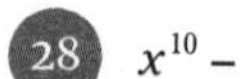

28 $x^{10} - 31x^5 - 32 = 0$

29 $y^{4/3} - 17y^{2/3} + 16 = 0$

30 $z^{-2} + z^{-1} - 12 = 0$

Solving Synthetically

Cubic and higher-degree equations that have nice integer solutions make life easier. But how realistic is that? Many answers to cubic equations that are considered to be rather nice are actually fractions. And fractions are notoriously difficult to work with when you raise them to powers. A method known as *synthetic division* can help out with all these concerns and lessen the drudgery. However, the division looks a little strange because it's synthesized. To *synthesize* means to bring together separate parts. That's what a synthesizer does with music. So turn on the Beethoven and get going.

Synthetic division is a shortcut division process. It takes the coefficients on all the terms in an equation and provides a method for finding the answer to a division problem by only multiplying and adding. It's really pretty neat. I use synthetic division to help find both integer solutions and fractional solutions for polynomial equations.

Earlier in this chapter, in the section, "Solving cubics with integers," I show you how to choose possible solutions for cubic equations whose lead coefficient is a 1. This section expands your capabilities of finding rational solutions. You see how to solve equations with a degree higher than 3, and how to include equations whose lead coefficient is something other than 1.

Refer to Chapter 13 for the specific steps used in synthetic division. In this section, I concentrate on finding the solutions of the polynomials and just ignore the factoring part.

Here's the general process:

1. **Put the terms of the equation in decreasing powers of the variable.**
2. **List all the possible factors of the constant term.**
3. **List all the possible factors of the coefficient of the highest power of the variable (the *lead coefficient*).**
4. **Divide all the factors in Step 2 by the factors in Step 3.**

   This is your list of possible *rational* solutions of the equation.
5. **Use synthetic division to check the possibilities.**

And here's an example that works through these steps: Find the solutions of the equation: $2x^4 + 13x^3 + 4x^2 = 61x + 30$.

Follow the steps to solve:

1. **Put the terms of the equation in decreasing powers of the variable.**

   $2x^4 + 13x^3 + 4x^2 - 61x - 30 = 0$
2. **List all the possible factors of the constant term.**

   The constant term −30 has the following factors: ±1, ±2, ±3, ±5, ±6, ±10, ±15, and ±30.
3. **List all the possible factors of the coefficient of the highest power of the variable (the *lead coefficient*).**

   The lead coefficient 2 has factors ±1 and ±2.
4. **Divide all the factors in Step 2 by the factors in Step 3. This is your list of possible rational solutions of the equation.**

   Dividing the factors of −30 by +1 or −1 doesn't change the list of factors. Dividing by +2 or −2 adds fractions when the number being divided is odd — the even numbers just provide values that are already on the list. So the complete list of possible solutions is ±1, ±2, ±3, ±5, ±6, ±10, ±15, ±30, $\pm\frac{1}{2}, \pm\frac{3}{2}, \pm\frac{5}{2}$, and $\pm\frac{15}{2}$.

5. **Use synthetic division to check the possibilities.**

   I first try the number 2 as a possible solution. The final number in the synthetic division is the value of the polynomial that you get by substituting in the 2, so you want the number to be 0.

$$\begin{array}{r|rrrrr} 2 & 2 & 13 & 4 & -61 & -30 \\ & & 4 & 34 & 76 & 30 \\ \hline & 2 & 17 & 38 & 15 & 0 \end{array}$$

   The 2 is a solution because the final number (what you get in evaluating the expression for 2) is equal to 0.

   Now look at the third row and use the lead coefficient of 2 and the final entry of 15 (ignore the 0). You can now limit your choices to only factors of +15 divided by factors of 2. The new, revised list is ±1, ±3, ±5, ±15, $\pm\frac{1}{2}, \pm\frac{3}{2}, \pm\frac{5}{2}$, and $\pm\frac{15}{2}$.

   I would probably try only integers before trying any fractions, but I want you to see what using a fraction in synthetic division looks like. I choose to try $-\frac{1}{2}$. Use only the numbers in the last row of the previous division.

$$\begin{array}{r|rrrr} -\frac{1}{2} & 2 & 17 & 38 & 15 \\ & & -1 & -8 & -15 \\ \hline & 2 & 16 & 30 & 0 \end{array}$$

   Such a wise choice! The number worked and is a solution. You could go on with more synthetic division, but at this point, I usually stop. The three numbers in the bottom row represent a quadratic trinomial. Write out the trinomial, factor it, use the MPZ, and find the last two solutions.

   The quadratic equation represented by that last row is $2x^2 + 16x + 30 = 0$.

   First, factor 2 out of each term. Then factor the trinomial: $2\left(x^2 + 8x + 15\right) = 2(x+3)(x+5) = 0$.

   The solutions from the factored trinomial are $x = -3$ and $x = -5$. Add these two solutions to $x = 2$ and $x = -\frac{1}{2}$, and you have the four solutions of the polynomial.

   What? You're miffed! You wanted me to finish the problem using synthetic division — not bail out and factor? Okay. I'll pick up where I left off with the synthetic division and show you how it finishes:

$$\begin{array}{r|rrr} -3 & 2 & 16 & 30 \\ & & -6 & -30 \\ \hline & 2 & 10 & 0 \end{array}$$

   And, finally:

$$\begin{array}{r|rr} -5 & 2 & 10 \\ & & -10 \\ \hline & 2 & 0 \end{array}$$

EXAMPLE

Q. Use synthetic division to solve $x^3 + x^2 - 9x - 9 = 0$.

A. The possible solutions are $\pm 1, \pm 3, \pm 9$. Using synthetic division using the 3, you have

$$\begin{array}{r|rrrr} 3 & 1 & 1 & -9 & -9 \\ & & 3 & 12 & 9 \\ \hline & 1 & 4 & 3 & 0 \end{array}$$

So $x = 3$ is a solution. The quotient gives you the expression $x^2 + 4x + 3$, which factors into $(x+1)(x+3)$. Setting this product equal to 0, you get $x = -1$ and $x = -3$, the other two solutions.

Q. Use synthetic division to solve $2x^3 - 3x^2 - 8x + 12 = 0$.

A. The possible solutions are $\pm 1, \pm 2, \pm 3 \pm 4, \pm 6, \pm 12, \pm \frac{1}{2}, \pm \frac{1}{3}$. Using synthetic division using the 2, you have

$$\begin{array}{r|rrrr} 2 & 2 & -3 & -8 & 12 \\ & & 4 & 2 & -12 \\ \hline & 2 & 1 & -6 & 0 \end{array}$$

So $x = 2$ is a solution. The quotient gives you the expression $2x^2 + x - 6$ which factors into $(2x-3)(x+2)$. Setting this product equal to 0, you get $x = \frac{3}{2}$ and $x = -2$, the other two solutions.

YOUR TURN

Solve for *x*.

 31 $6x^3 - 23x^2 - 6x + 8 = 0$

 32 $24x^3 - 10x^2 - 3x + 1 = 0$

 33 $36x^4 - 12x^3 - 23x^2 + 4x + 4 = 0$

 34 $6x^5 + 5x^4 - 12x^3 - 10x^2 + 6x + 5 = 0$

Practice Questions Answers and Explanations

1. **$x = 3$.** Factoring, $x^3 - 27 = (x-3)(x^2+3x+9) = 0$. Using the MPZ, when $x - 3 = 0$, $x = 3$. The alternative is to rewrite the equation as $x^3 = 27$ and take the cube root of each side: $\sqrt[3]{x^3} = \sqrt[3]{27}$ or $x = 3$.

2. **$y = -1$.** Factoring, $y^3 + 1 = (y+1)(y^2 - y + 1) = 0$. Using the MPZ, when $y + 1 = 0$, $y = -1$. The alternative is to rewrite the equation as $y^3 = -1$ and take the cube root of each side: $\sqrt[3]{y^3} = \sqrt[3]{-1}$ or $y = -1$.

3. **$z = \frac{1}{2}$.** Factoring,

$$27z^3 - (z+1)^3 = \left[3z - (z+1)\right]\left[9z^2 + 3z(z+1) + (z+1)^2\right]$$
$$= \left[2z-1\right]\left[z^2 + z^2 + z + z^2 + 2z + 1\right] = \left[2z-1\right]\left[3z^2 + 3z + 1\right]$$

Setting the binomial equal to 0, you have $2z - 1 = 0$ or $z = \frac{1}{2}$. The trinomial doesn't factor, and applying the quadratic formula results in an imaginary number. As expected, you only have the one solution. Using an alternative method, the equation could be rewritten as $27z^3 = (z+1)^3$ and the cube root taken of each side: $\sqrt[3]{27z^3} = \sqrt[3]{(z+1)^3}$ or $3z = z + 1$. Subtracting z from each side, you have $2z = 1$ or $z = \frac{1}{2}$.

4. **$w = -\frac{1}{3}$.** Factoring,

$$(w+4)^3 + (2w-3)^3 = \left[(w+4) + (2w-3)\right]\left[(w+4)^2 - (w+4)(2w-3) + (2w-3)^2\right]$$
$$= \left[3w+1\right]\left[w^2 + 8w + 16 - 3w^2 - 5w + 12 + 4w^2 - 12w + 9\right]$$
$$= \left[3w+1\right]\left[2w^2 - 9w + 37\right]$$

Setting the binomial equal to 0, you have $3w + 1 = 0$ or $w = -\frac{1}{3}$. The trinomial doesn't factor, and applying the quadratic formula results in an imaginary number. As expected, you only have the one solution. Using an alternative method, the equation could be rewritten as $(w+4)^3 = -(2w-3)^3$ and the cube root taken of each side: $\sqrt[3]{(w+4)^3} = \sqrt[3]{-(2w-3)^3}$ or $w + 4 = -(2w-3)$, which simplifies to $w + 4 = -2w + 3$. Adding $2w$ to each side and subtracting 4 from each side, $3w = -1$ or $w = -\frac{1}{3}$.

5. **$x = \sqrt[3]{\frac{11}{2}}$.** Divide each side of the equation by 2, and then find the cube root of each side: $x^3 = \frac{11}{2}$ so $\sqrt[3]{x^3} = x = \sqrt[3]{\frac{11}{2}}$. The decimal value is approximately 1.76517.

6. **$y = \sqrt[3]{\frac{4}{3}}$.** Using the rule $y = \sqrt[3]{\frac{b}{a}}$ and letting $a = 3$ and $b = 4$, you have $y = \sqrt[3]{\frac{4}{3}}$. The decimal value is approximately 1.10064.

7. **$z = \frac{-3\sqrt[3]{3}}{\sqrt[3]{4}}$.** Divide each side of the equation by 4, and then find the cube root of each side: $z^3 = -\frac{81}{4}$ so $\sqrt[3]{z^3} = z = \sqrt[3]{-\frac{81}{4}}$. The decimal value is approximately −2.72568. Another option is to simplify the fraction by writing it first as $z = \frac{\sqrt[3]{-81}}{\sqrt[3]{4}}$. The number 81 is the product of 3 and 27.

The number 27 is a perfect cube, so you write $z=\frac{\sqrt[3]{-27}\sqrt[3]{3}}{\sqrt[3]{4}}=\frac{-3\sqrt[3]{3}}{\sqrt[3]{4}}$. You make this choice when working with multiple factors with cube roots — hoping that you can factor some of those roots. For more on simplifying radicals, see Chapter 18.

8) $\boldsymbol{w=-\frac{2\sqrt[3]{2}}{3}}$**.** Using the rule $w=-\sqrt[3]{\frac{b}{a}}$ and letting $a=27$ and $b=16$, you have $w=-\sqrt[3]{\frac{16}{27}}$. The decimal value is approximately −0.839947.

Another option is to simplify the fraction by writing it first as $w=-\frac{\sqrt[3]{16}}{\sqrt[3]{27}}$. The number 16 is the product of 2 and 8, and 8 is a perfect cube. The number 27 is a perfect cube, so you write $w=-\frac{\sqrt[3]{8}\sqrt[3]{2}}{\sqrt[3]{27}}=-\frac{2\sqrt[3]{2}}{3}$. You make this choice when working with multiple factors with cube roots — hoping that you can factor some of those roots. For more on simplifying radicals, see Chapter 18.

9) $\boldsymbol{x=0, x=9,}$ **and** $\boldsymbol{x=-1}$**.** First, factor x from each term on the left to get: $x(x^2-8x-9)=0$. The trinomial factors, giving you $x(x-9)(x+1)=0$. Setting each of the three factors equal to 0 using the MPZ, you have $x=0$, $x=9$, and $x=-1$.

10) $\boldsymbol{y=0, y=4,}$ **and** $\boldsymbol{y=5}$**.** First, factor y^2 from each term on the left to get $y^2(y^2-9y+20)=0$. The trinomial factors, giving you $y^2(y-4)(y-5)=0$. Setting each of the three factors equal to 0 using the MPZ, you have $y=0$, $y=4$, and $y=5$.

11) $\boldsymbol{z=0, z=-4,}$ **and** $\boldsymbol{z=4}$**.** First, factor $4z$ from each term on the left to get $4z(z^2-16)=0$. The binomial is the difference of perfect squares, so the equation is written as $4z(z+4)(z-4)=0$. When using the MPZ, the factor 4 can never equal 0, but the other factors give you $z=0$, $z=-4$, and $z=4$.

12) $\boldsymbol{w=0, w=-5,}$ **and** $\boldsymbol{w=5}$**.** First, factor $12w^2$ from each term on the left to get $12w^2(w^2-25)=0$. The binomial is the difference of perfect squares, so you can write the equation as $12w^2(w+5)(w-5)=0$. When using the MPZ, the factor 12 can never equal 0, but the other factors give you $w=0$, $w=-5$, and $w=5$.

13) $\boldsymbol{x=-3, x=-5,}$ **and** $\boldsymbol{x=5}$**.** The first two terms are each divisible by x^2, and the last two terms are each divisible by −25: $x^2(x+3)-25(x+3)=0$. The left side of the equation now has two terms, each with a factor of $(x+3)$. Factoring: $(x+3)(x^2-25)=0$ can be factored again, because of the difference of squares in the second binomial. $(x+3)(x+5)(x-5)=0$ is solved using the MPZ, giving you $x=-3$, $x=-5$, and $x=5$.

14) $\boldsymbol{y=2, y=-10,}$ **and** $\boldsymbol{y=10}$**.** The first two terms are each divisible by y^2, and the last two terms are each divisible by −100: $y^2(y-2)-100(y-2)=0$. The left side of the equation now has two terms, each with a factor of $(y-2)$. Factoring, $(y-2)(y^2-100)=0$ can be factored again: $(y-2)(y+10)(y-10)=0$. Applying the MPZ, $y=2$, $y=-10$, and $y=10$.

15) $\boldsymbol{z=-4}$**.** The first two terms are divisible by z^2, and the last two terms are each divisible by 5: $z^2(z+4)+5(z+4)=0$. The left side of the equation now has two terms, each with a factor of $(z+4)$. Factoring, you have $(z+4)(z^2+5)=0$. The second binomial doesn't factor. When you apply the MPZ, the factor $(z+4)$ gives you the solution $z=-4$. The second binomial doesn't have a real solution; applying the binomial theorem shows you there's no real root.

16 $\boldsymbol{w = \frac{4}{3}}$. The first two terms are divisible by w^2, and the last two terms are each divisible by 3: $w^2(3w-4)+3(3w-4)=0$. The left side of the equation now has two terms, each with a factor $(3w-4)$. Factoring, you have $(3w-4)(w^2+3)=0$. The second binomial doesn't factor. When you apply the MPZ, the factor $(3w-4)$ gives you the solution $w=\frac{4}{3}$. The second binomial doesn't have a real solution; applying the quadratic formula shows you there's no real root.

17 **Three or one positive roots and two or no negative roots.** The original equation has three sign changes, so there are three or one possible positive real roots. Substituting $-x$ for each x, you get $-x^5+x^3+8x^2-8=0$, and you have two sign changes, meaning there are two negative roots or none at all.

18 **Two or no positive roots and one negative root.** The original equation has two sign changes (from positive to negative to positive), so there are two or no positive roots. Substituting $-x$ for each x, you get $-8x^5-25x^4-x^2+25=0$, which has one sign change and one negative root.

19 $\boldsymbol{\pm 81, \pm 27, \pm 9, \pm 3, \pm 1, \pm\frac{81}{2}, \pm\frac{27}{2}, \pm\frac{9}{2}, \pm\frac{3}{2}, \pm\frac{1}{2}}$. The constant term is 81. Its factors are $\pm 81, \pm 27, \pm 9, \pm 3, \pm 1$. The lead coefficient is 2 with factors ± 2.

20 $\boldsymbol{\pm 25, \pm 5, \pm 1, \pm\frac{25}{8}, \pm\frac{25}{4}, \pm\frac{25}{2}, \pm\frac{5}{8}, \pm\frac{5}{4}, \pm\frac{5}{2}, \pm\frac{1}{8}, \pm\frac{1}{4}, \pm\frac{1}{2}}$. The constant term is 25, having factors $\pm 25, \pm 5, \pm 1$. The lead coefficient is 8 with factors $\pm 8, \pm 4, \pm 2, \pm 1$.

21 **Yes.** Rewrite the equation with the coefficients showing in front of the variables: $x^4-10x^2+9=1(x^4)+0(x^3)-10(x^2)+0(x)+9$.

$$\begin{array}{r|rrrrr} -3 & 1 & 0 & -10 & 0 & 9 \\ & & -3 & 9 & 3 & -9 \\ \hline & 1 & -3 & -1 & 3 & 0 \end{array}$$

Because the remainder is 0, the equation has a root of -3 and a factor of $(x+3)$.

22 **Yes.** Writing in the coefficients of the terms, you get $2x^4-3x^3+0(x^2)-54x+81=0$. Note that 3 is a factor of 81 and 2 is a factor of 2, so $\frac{3}{2}$ is a possible rational root:

$$\begin{array}{r|rrrrr} \frac{3}{2} & 2 & -3 & 0 & -54 & 81 \\ & & 3 & 0 & 0 & -81 \\ \hline & 2 & 0 & 0 & -54 & 0 \end{array}$$

The remainder is 0, so $\frac{3}{2}$ is a root (solution), and $\left(x-\frac{3}{2}\right)$ or $(2x-3)$ is a factor.

23 $\boldsymbol{x=\pm 2}$. First, factor the binomial as the difference and sum of the same two values; then factor the first of these factors the same way:

$$x^4-16=(x^2-4)(x^2+4)=(x-2)(x+2)(x^2+4)=0$$

$x=2$ or $x=-2$ are the real solutions, so $x=\pm 2$. ***Note:*** $x^2+4=0$ has no real solutions.

(24) $\mathbf{x = 0,\ 1.}$ First, factor out x from each term to get $x\left(x^3-3x^2+3x-1\right)=0$. You may recognize that the expression in the parentheses is a perfect cube. But just in case that hasn't occurred to you, try using synthetic division with the guess, $x=1$.

$$\begin{array}{r|rrrr} 1 & 1 & -3 & 3 & -1 \\ & & 1 & -2 & 1 \\ \hline & 1 & -2 & 1 & 0 \end{array}$$

The factored form now reads $x(x-1)\left(x^2-2x+1\right)=0$. The trinomial in the parentheses is a perfect square, so the final factored form is $x(x-1)^3=0$. So $x=0,\ 1$, and 1 is a *triple root.*

(25) $\mathbf{x = \pm 4,\ -5.}$ Factor by grouping:

$$x^3+5x^2-16x-80=x^2(x+5)-16(x+5)=\left(x^2-16\right)(x+5)=(x-4)(x+4)(x+5)=0$$

so $x-4=0,\ x=4;\ x+4=0,\ x=-4$; or $x+5=0,\ x=-5$. Therefore, $x=\pm 4,\ -5$.

(26) $\mathbf{x = \pm 2,\ \pm 3.}$ First, you get $x^6-9x^4-16x^2+144=x^4\left(x^2-9\right)-16\left(x^2-9\right)=0$ by grouping. Then,

$$\left(x^4-16\right)\left(x^2-9\right)=\left(x^2-4\right)\left(x^2+4\right)\left(x^2-9\right)=(x-2)(x+2)\left(x^2+4\right)(x-3)(x+3)=0$$

So, $x=2,\ x=-2,\ x=3$, and $x=-3$ give the real solutions. Therefore, $x=\pm 2,\ \pm 3$.

(27) $\mathbf{x = \pm 3,\ \pm 2.}$ $x^4-13x^2+36=\left(x^2-9\right)\left(x^2-4\right)=0$. You can continue factoring or use the square root rule: $x^2-9=0,\ x^2=9,\ x=\pm 3$; or $x^2-4=0,\ x^2=4,\ x=\pm 2$.

Therefore, $x=\pm 3,\ \pm 2$.

(28) $\mathbf{x = 2,\ -1.}$ $x^{10}-31x^5-32=\left(x^5-32\right)\left(x^5+1\right)=0$

So, $x^5-32=0,\ x^5=32,\ x=2$; or $x^5+1=0,\ x^5=-1,\ x=-1$.

Therefore, $x=2,\ -1$.

(29) $\mathbf{y = 1, 64.}$ $y^{4/3}-17y^{2/3}+16=\left(y^{2/3}-1\right)\left(y^{2/3}-16\right)=0$

So, $y^{2/3}-1=0,\ y^{2/3}=1,\ \left(y^{2/3}\right)^{3/2}=1^{3/2}$, and so $y=1$

or $y^{2/3}-16=0,\ y^{2/3}=16, \left(y^{2/3}\right)^{3/2}=16^{3/2}=\left(\sqrt{16}\right)^3=64$.

Therefore, $y=1$ or $y=64$.

(30) $\mathbf{z = \frac{1}{3},\ -\frac{1}{4}.}$ $z^{-2}+z^{-1}-12=\left(z^{-1}-3\right)\left(z^{-1}+4\right)=0$

So, $z^{-1}-3=0,\ z^{-1}=3,\ \frac{1}{z}=3,\ 1=3z,\ z=\frac{1}{3}$

or $z^{-1}+4=0,\ z^{-1}=-4,\ \frac{1}{z}=-4,\ 1=-4z,\ z=-\frac{1}{4}$.

Therefore, $z=\frac{1}{3},\ -\frac{1}{4}$.

31 $\boldsymbol{x = 4, x = -\frac{2}{3}, x = \frac{1}{2}}$. Using the rational root theorem, the possible rational roots are:

$\pm 1, \pm 2, \pm 4, \pm 8, \pm\frac{1}{2}, \pm\frac{1}{3}, \pm\frac{2}{3}, \pm\frac{4}{3}, \pm\frac{8}{3}, \pm\frac{1}{6}$

According to Descartes's rule of signs, there are 2 or 0 positive real roots and 1 negative real root.

Using synthetic division with a guess of $x = 4$,

$$\begin{array}{r|rrrr} 4 & 6 & -23 & -6 & 8 \\ & & 24 & 4 & -8 \\ \hline & 6 & 1 & -2 & 0 \end{array}$$

The remainder of 0 means that $x = 4$ is a root. Using the factor theorem and the last line of the division as the quotient, the factorization is $(x-4)(6x^2+x-2)=0$. The trinomial factors, giving you $(x-4)(3x+2)(2x-1)=0$. The MPZ then says that the solutions are $x = 4, x = -\frac{2}{3}, x = \frac{1}{2}$.

32 $\boldsymbol{x = \frac{1}{2}, x = -\frac{1}{3}, x = \frac{1}{4}}$. Using the rational root theorem, the possible rational roots are:

$\pm 1, \pm\frac{1}{2}, \pm\frac{1}{3}, \pm\frac{1}{4}, \pm\frac{1}{6}, \pm\frac{1}{8}, \pm\frac{1}{12}, \pm\frac{1}{24}$

According to Descartes's rule of signs, there are 2 or 0 positive real roots and 1 negative real root. Using synthetic division with a guess of $x = \frac{1}{2}$,

$$\begin{array}{r|rrrr} \frac{1}{2} & 24 & -10 & -3 & 1 \\ & & 12 & 1 & -1 \\ \hline & 24 & 2 & -2 & 0 \end{array}$$

The remainder of 0 means that $x = \frac{1}{2}$ is a root. Using the factor theorem and the last line of the division as the quotient, the factorization is $\left(x-\frac{1}{2}\right)\left(24x^2+2x-2\right)=0$. First, factor out the GCF of 2 from the terms of the trinomial to get $2\left(x-\frac{1}{2}\right)\left(12x^2+x-1\right)=0$. The trinomial factors, giving you $2\left(x-\frac{1}{2}\right)(3x+1)(4x-1)=0$. The MPZ then says that the solutions are $x = \frac{1}{2}, x = -\frac{1}{3}, x = \frac{1}{4}$. You can also rewrite the factorization by distributing the 2 over the terms in the first binomial factor and have $(2x-1)(3x+1)(4x-1)=0$.

REMEMBER

When trying a fractional value as a possible root in synthetic division, if you get a fraction in the bottom row, you should stop. The fractions will never go away, and you will not get the 0 remainder you're looking for.

33 $\boldsymbol{x = \frac{2}{3}, x = \frac{2}{3}, x = -\frac{1}{2}, x = -\frac{1}{2}}$. Using the rational root theorem, the possible rational roots are $\pm 1, \pm 2, \pm 4, \pm\frac{1}{2}, \pm\frac{1}{3}, \pm\frac{2}{3}, \pm\frac{4}{3}, \pm\frac{1}{4}, \pm\frac{1}{6}, \pm\frac{1}{9}, \pm\frac{2}{9}, \pm\frac{4}{9}, \pm\frac{1}{12}, \pm\frac{1}{18}, \pm\frac{1}{36}$.

According to Descartes's rule of signs, there are 2 or 0 positive real roots and 2 or 0 negative real roots.

Using synthetic division with a guess of $x=\frac{2}{3}$,

$$\begin{array}{r|rrrrr} \frac{2}{3} & 36 & -12 & -23 & 4 & 4 \\ & & 24 & 8 & -10 & -4 \\ \hline & 36 & 12 & -15 & -6 & 0 \end{array}$$

$x=\frac{2}{3}$ is a root. Perform synthetic division again using the numbers in the bottom row and a guess of $x=-\frac{1}{2}$.

$$\begin{array}{r|rrrr} -\frac{1}{2} & 36 & 12 & -15 & -6 \\ & & -18 & 3 & 6 \\ \hline & 36 & -6 & -12 & 0 \end{array}$$

The remainder of 0 means that $x=-\frac{1}{2}$ is a root. Using the factor theorem and the last line of the division as the quotient, the factorization is $\left(x-\frac{2}{3}\right)\left(x+\frac{1}{2}\right)\left(36x^2-6x-12\right)=0$. Factor 6 from the terms of the trinomial: $6\left(x-\frac{2}{3}\right)\left(x+\frac{1}{2}\right)\left(6x^2-x-2\right)=0$. The trinomial factors, giving you $6\left(x-\frac{2}{3}\right)\left(x+\frac{1}{2}\right)(3x-2)(2x+1)=0$. The MPZ then says that the solutions are $x=\frac{2}{3}, x=\frac{2}{3}, x=-\frac{1}{2}, x=-\frac{1}{2}$. Distributing the 6 over the first two binomials (multiplying the first binomial by 3 and the second by 2), you can write the factorization as $(3x-2)(2x+1)(3x-2)(2x+1)=(3x-2)^2(2x+1)^2=0$.

(34) $\boldsymbol{x=-\frac{5}{6}, x=1, x=1, x=-1, x=-1}$. Using the rational root theorem, the possible rational roots are: $\pm1, \pm5, \pm\frac{1}{2}, \pm\frac{5}{2}, \pm\frac{1}{3}, \pm\frac{5}{3}, \pm\frac{1}{6}, \pm\frac{5}{6}$.

According to Descartes's rule of signs, there are 2 or 0 positive real roots and 3 or 1 negative real roots.

Using synthetic division with a guess of $x=-\frac{5}{6}$,

$$\begin{array}{r|rrrrrr} -\frac{5}{6} & 6 & 5 & -12 & -10 & 6 & 5 \\ & & -5 & 0 & 10 & 0 & -5 \\ \hline & 6 & 0 & -12 & 0 & 6 & 0 \end{array}$$

The bottom row has an unusual result. Actually, by choosing $x=-\frac{5}{6}$, I created this result to show you a possible technique; I don't usually choose fractions right away when doing synthetic division to search for roots. Writing the factorization of the problem, you have $\left(x+\frac{5}{6}\right)\left(6x^4-12x^2+6\right)=0$. Factor 6 from the terms of the binomial: $6\left(x+\frac{5}{6}\right)\left(x^4-2x^2+1\right)=0$. The fourth-degree trinomial is a quadratic-like trinomial that can be factored. (Refer to Chapter 13 for more on quadratic factoring.) The new factorization is $6\left(x+\frac{5}{6}\right)\left(x^2-1\right)^2=(6x+5)(x-1)^2(x+1)^2=0$. The solutions from the factorization are $x=-\frac{5}{6}, x=1, x=1, x=-1, x=-1$.

If you're ready to test your skills a bit more, take the following chapter quiz that incorporates all the chapter topics.

Whaddya Know? Chapter 17 Quiz

Quiz time! Complete each problem to test your knowledge on the various topics covered in this chapter. You can then find the solutions and explanations in the next section.

1. Solve for x: $x^3 + 64 = 0$

2. Solve for x: $x^{1/2} - 3x^{1/4} - 4 = 0$

3. In the polynomial $x^4 - 3x^3 + 2x^2 + 9x + 11 = 0$, what is the greatest number of possible real roots?

4. List all the possible rational roots of the polynomial $6x^5 - 7x^3 + 2x^2 - 9x - 2 = 0$.

5. Solve for x: $x^4 - 25x^2 + 144 = 0$

6. Solve for x: $x^3 - (2x + 4)^3 = 0$

7. Solve for x: $12x^3 + 25x^2 + x - 2 = 0$

8. Solve for x: $3x^3 + 6 = 0$

9. Solve for x: $x^{-2} + 8x^{-1} + 7 = 0$

10. Solve for x: $x^4 - 2x^3 = 15x^2$

11. In the polynomial $-2x^5 - x^4 - 3x^3 + 2x^2 - 9x - 11 = 0$, what is the greatest number of possible real roots?

12. Solve for x: $x^5 - 5x^3 - x^2 + 5 = 0$

13. List all the possible rational roots of the polynomial $3x^4 - 3x^3 + 2x^2 + 9x + 18 = 0$.

14. Solve for x: $x^3 - 2x^2 - 11x + 12 = 0$

15. Solve for x: $18x^5 - 2x^3 = 0$

Answers to Chapter 17 Quiz

1. $x = -4$**.** Subtract 64 from each side. Then take the cube root of each side: $x^3 = -64$, $\sqrt[3]{x^3} = \sqrt[3]{-64}$.

2. $x = 256$**.** Replace the $x^{1/2}$ with q^2 with and the $x^{1/4}$ with q. Then factor the equation $q^2 - 3q - 4 = 0$, giving you : $q^2 - 3q - 4 = (q-4)(q+1) = \left(x^{1/4} - 4\right)\left(x^{1/4} + 1\right) = 0$. Setting the binomials equal to 0, you have $x^{1/4} - 4 = 0$, $x^{1/4} = 4$; raising both sides to the fourth power, you get $x = 256$. When $x^{1/4} + 1 = 0$, $x^{1/4} = -1$, there's no real solution.

3. **2 or 0 positive; 2 or 0 negative.** There are 2 changes of sign, so there are 2 or 0 possible positive roots. Replacing the x variables with $-x$, there are 2 changes of sign, so there are 2 or 0 possible negatve roots.

4. $\pm 1, \pm 2, \pm \frac{1}{2}, \pm \frac{1}{3}, \pm \frac{2}{3}, \pm \frac{1}{6}$

5. $x = \pm 3, \pm 4$**.** Factor the equation $q^2 - 25q + 144 = 0$, and then replace the q with x^2:

   $$q^2 - 25q + 144 = (q-9)(q-16) = \left(x^2 - 9\right)\left(x^2 - 16\right) = 0$$

   Both quadratics factor as the difference of squares: $x = \pm 3$, $x = \pm 4$.

6. $x = -4$**.** Add the binomial to each side. Take the cube root of each side. Solve for x: $x^3 = (2x+4)^3$, $\sqrt[3]{x^3} = \sqrt[3]{(2x+4)^3}$, $x = 2x + 4$, $-x = 4$, $x = -4$.

7. $x = -2, \frac{1}{4}, -\frac{1}{3}$**.** Using the rational root theorem, the possible roots are: $\pm 1, \pm 2, \pm \frac{1}{2}, \pm \frac{1}{3}, \pm \frac{2}{3}, \pm \frac{1}{4}, \pm \frac{1}{6}, \pm \frac{1}{12}$. Using synthetic division on $x = -2$,

   $$\begin{array}{r|rrrr} -2 & 12 & 25 & 1 & -2 \\ & & -24 & -2 & 2 \\ \hline & 12 & 1 & -1 & 0 \end{array}$$

   so -2 is a root. You can factor the quadratic quotient, $12x^2 + x - 1 = (4x-1)(3x+1) = 0$, to get the roots $\frac{1}{4}$ and $-\frac{1}{3}$. The original equation could also have been factored by grouping.

8. $x = -\sqrt[3]{2}$**.** Add -6 to each side, divide by 3, and then take the cube root of each side: $3x^3 = -6$, $x^3 = -2$, $\sqrt[3]{x^3} = \sqrt[3]{-2}$.

9. $x = -1, -\frac{1}{7}$**.** Factor the equation $q^2 + 8q + 7 = 0$, and then replace the q with x^{-1}:

   $$q^2 + 8q + 7 = x^{-2} + 8x^{-1} + 7 = (x^{-1} + 1)(x^{-1} + 7) = 0$$

   Using the MPZ, $x^{-1} = -1$ and $x^{-1} = -7$. Solving for x, $x^{-1} = \frac{1}{x} = -1$, so $x = -1$ and $x^{-1} = \frac{1}{x} = -7$, so $x = -\frac{1}{7}$.

10. $x = 0, -3, 5$**.** Set the equation equal to 0. Factor out the GCF. Then factor the quadratic.

    $$x^4 - 2x^3 - 15x^2 = 0, x^2\left(x^2 - 2x - 15\right) = x^2(x+3)(x-5) = 0$$

11. **2 or 0 positive; 3 or 1 negative.** There are 2 changes of sign, so there are 2 or 0 possible positive roots. Replacing the x variables with $-x$, there are 3 changes of sign, so there are 3 or 1 possible negatve roots.

(12) $\boldsymbol{x = 1, \ \pm\sqrt{5}}$. Factor by grouping: $x^3(x^2-5)-1(x^2-5)=(x^3-1)(x^2-5)=0$. Use the cube root rule and square root rule to solve for x: $x=1$, $x=\pm\sqrt{5}$.

(13) $\boldsymbol{\pm 1, \ \pm 2, \ \pm 3, \ \pm 6, \ \pm 9, \ \pm 18, \ \pm\frac{1}{3}, \ \pm\frac{2}{3}}$.

(14) $\boldsymbol{x = 1, \ 4, \ -3}$. Using the rational root theorem, the possible roots are $\pm 1, \ \pm 2, \ \pm 3, \ \pm 4, \ \pm 6, \ \pm 12$. Using synthetic division on $x=1$,

$$\begin{array}{r|rrrr} 1 & 1 & -2 & -11 & 12 \\ & & 1 & -1 & -12 \\ \hline & 1 & -1 & -12 & 0 \end{array}$$

so 1 is a root. You can factor the quadratic quotient, $x^2-x-12=(x-4)(x+3)=0$, to get the roots 4 and -3.

(15) $\boldsymbol{x = 0, \ \pm\frac{1}{3}}$. Factor out the GCF. Then factor the difference of squares:

$$2x^3(9x^2-1)=2x^3(3x+1)(3x-1)=0$$

IN THIS CHAPTER

» Squaring radicals once or twice (and making them nice)

» Being absolutely sure of absolute value

» Checking for extraneous solutions

Chapter 18

Reeling in Radical and Absolute Value Equations

Radical equations and absolute value equations are just what their names suggest. Radical equations contain one or more *radicals* (square root or other root symbols), and absolute value equations have an absolute value operation (two vertical bars that say to give the distance from 0). Although they're two completely different types of equations, radical equations and absolute value equations do have something in common: You change both into linear or quadratic equations (see Chapters 15 and 16) and then solve them. After all, going back to something familiar makes more sense than trying to develop (and then remember) a bunch of new rules and procedures.

What's different is *how* you change these two types of equations. I handle each type separately in this chapter and include practice problems for you.

Raising Both Sides to Solve Radical Equations

Some equations have radicals in them. You change those equations to linear or quadratic equations for greater convenience when solving. Radical equations crop up when you do problems involving distance in graphing points and lines. Included in distance problems are those involving the Pythagorean Theorem — that favorite of Pythagoras that describes the relationship between the sides of a right triangle.

The basic process that leads to a solution of equations involving a radical is just getting rid of that radical. Removing the radical changes the problem into something more manageable, but the change also introduces the possibility of a nonsense answer or an error. Checking your answer is even more important in the case of solving radical equations. As long as you're aware that errors can happen, you know to be especially watchful. Even though this may seem a bit of a hassle — that these nonsense things come up — getting rid of the radical is still the most efficient and easiest way to handle these equations.

Powering up by squaring both sides

The main method to use when dealing with equations that contain radicals is to change the equations to those that do not have radicals in them. You accomplish this by raising the radical to a power that changes the fractional exponent (representing the radical) to a 1. If the radical is a square root, which can be written as a power of $\frac{1}{2}$, then the radical is raised to the second power. If the radical is a cube root, which can be written as a power of $\frac{1}{3}$, then the radical is raised to the third power. (Turn to Chapter 5 if you need to review exponents and raising to powers.)

When the fractional power is raised to the reciprocal of that power, the two exponents are multiplied together, giving you a power of 1:

$$\left(x^{\frac{1}{4}}\right)^4 = x^1 \qquad \left(y^{\frac{1}{3}}\right)^3 = y^1 \qquad \left(z^{\frac{1}{7}}\right)^7 = z^1$$

$$\left(\sqrt{x+1}\right)^2 = \left((x+1)^{\frac{1}{2}}\right)^2 = (x+1)^1 = x+1$$

Raising to powers clears out the radicals, but problems can occur when the variables are raised to even powers. Variables can stand for negative numbers or values that allow negatives under the radical, which isn't always apparent until you get into the problem and check an answer. Instead of going on with all this doom and gloom and the problems that occur when powering up both sides of an equation, let me show you some examples of how the process works, what the pitfalls are, and how to deal with any extraneous solutions.

Solve the equation for the value of y: $\sqrt{4-5y}-7=0$.

Follow these steps.

1. **Get the radical by itself on one side of the equal sign.**

   So, if you're solving for y in $\sqrt{4-5y}-7=0$, add 7 to each side to get the radical by itself on the left. Doing that gives you $\sqrt{4-5y}=7$.

2. **Square both sides of the equation to remove the radical.**

   Squaring both sides of the example problem gives you $\left(\sqrt{4-5y}\right)^2=7^2$ or $4-5y=49$.

3. **Solve the resulting linear equation.**

   Subtract 4 from each side to get $-5y=45$, or $y=-9$.

   It may seem strange that the answer is a negative number, but, in the original problem, the negative number is multiplied by another negative, which makes the result under the radical a positive number.

4. **Check your answer. (Always start with the original equation.)**

   If $y=-9$, then $\sqrt{4-5(-9)}-7=0$ or $\sqrt{4+45}-7=0$. That leads to $\sqrt{49}-7=7-7=0$. It checks!

If your radical equation has just one radical term, then you solve it by isolating that radical term on one side of the equation and the other terms on the opposite side, and then squaring both sides. After solving the resulting equation, be sure to check your answer(s) to be sure you actually have a solution.

WARNING

Watch out for *extraneous roots.* These false answers crop up in several situations where you change from one type of equation to another in the course of solving the original equation. In this case, it's the squaring that can introduce extraneous or false roots. These false roots are created because the square of a positive number or its opposite (negative) gives you the same positive number.

Squaring both sides to get these false answers may sound like more trouble than it's worth, but this procedure is still much easier than anything else. You really can't avoid the extraneous roots; just be aware that they can occur so you don't include them in your answer.

REMEMBER

When squaring both sides in radical equations and creating quadratic equations, you can encounter one of three possible outcomes: (1) both of the solutions work, (2) neither solution works, or (3) just one of the two solutions works because the other solution turns out to be extraneous.

Check out the following example to see how to handle an extraneous root in a radical equation.

Solve for x: $\sqrt{x+10}+x=10$

Use these steps.

1. **Isolate the radical term by subtracting x from each side. Then square both sides of the equation.**

$$\sqrt{x+10}=10-x$$
$$\left(\sqrt{x+10}\right)^2=(10-x)^2$$
$$x+10=100-20x+x^2$$

2. **To solve this quadratic equation, subtract x and 10 from each side so that the equation is set equal to zero. Then simplify and factor it.**

$$0=x^2-21x+90$$
$$=(x-15)(x-6)$$

Two solutions appear, $x=15$ or $x=6$.

3. **Check for an extraneous solution.**

In this case, substituting the solutions, you see that the 6 works, but the 15 doesn't.

$$x=15\rightarrow\sqrt{15+10}+15\overset{?}{=}10$$
$$\sqrt{25}+15\neq 10$$
$$x=6\rightarrow\sqrt{6+10}+6\overset{?}{=}10$$
$$\sqrt{16}+6=10$$

Only the $x=6$ works.

EXAMPLE

Q. Solve for x: $2\sqrt{x+15}-3=9$

A. Add 3 to each side, and then square both sides of the equation. $\left(2\sqrt{x+15}\right)^2=(12)^2$ becomes $4(x+15)=144$. Divide each side by 4 to get $x+15=36$. Then, subtracting 15 from both sides you find that $x=21$. Checking your answer in the original equation, $2\sqrt{21+15}-3=9$ becomes $2\sqrt{36}-3=9$ or $2\cdot 6-3=12-3=9$. Good!

Q. Solve for y: $\sqrt{2y+5}-y=1$

A. Add y to each side and then square both sides. $\left(\sqrt{2y+5}\right)^2=(1+y)^2$ becomes $2y+5=1+2y+y^2$. Subtract both 2y and 1 from each side and you have $4=y^2$. Using the square root rule, you have $y=\pm 2$. Checking with $y=2$, $\sqrt{2\cdot 2+5}-2=1$ becomes $\sqrt{9}-2=1$ or $3-2=1$. This works! And then, checking with $y=-2$, $\sqrt{2(-2)+5}-(-2)=1$ becomes $\sqrt{1}+2=1$ or $3+2=1$. No. This is extraneous. Only $y=2$ is a solution.

YOUR TURN

Solve for x.

1. $\sqrt{x-3}=6$

2. $\sqrt{x^2+9}=5$

3. $\sqrt{x+5}+x=1$

4. $\sqrt{x-3}+9=x$

5. $\sqrt{x+7}-7=x$

6. $\sqrt{x-1}=x-1$

Reeling in Radical and Absolute Value Equations

Raising to higher powers

Not all radical equations will be dealing with square roots. You can have cube roots, fourth roots, and so on. But the process is pretty much the same: raise both sides of the equation to the same power and solve the resulting equation.

EXAMPLE

Q. Solve for x in $\sqrt[3]{x+4}=\sqrt[3]{4x+7}$.

A. Use these steps.

1 Cube each side of the equation, and then solve for x.

$$\left(\sqrt[3]{x+4}\right)^3 = \left(\sqrt[3]{4x+7}\right)^3$$
$$x+4 = 4x+7$$
$$-3 = 3x$$
$$x = -1$$

2 **Check:** $\sqrt[3]{-1+4} = \sqrt[3]{4(-1)+7} \rightarrow \sqrt[3]{3} = \sqrt[3]{3}$

Q. Solve for z in $\sqrt[5]{2z+5} - 1 = 0$.

A. Follow these steps.

1 Add 1 to each side and then raise each side to the fifth power.

2 Solve for z.

$$\sqrt[5]{2z+5} = 1$$
$$\left(\sqrt[5]{2z+5}\right)^5 = 1^5$$
$$2z+5 = 1$$
$$2z = -4$$
$$z = -2$$

3 **Check:** $z = -2 \rightarrow \sqrt[5]{2(-2)+5} - 1 = 0 \rightarrow \sqrt[5]{-4+5} - 1 = 0 \rightarrow \sqrt[5]{1} - 1 = 0$. Yes!

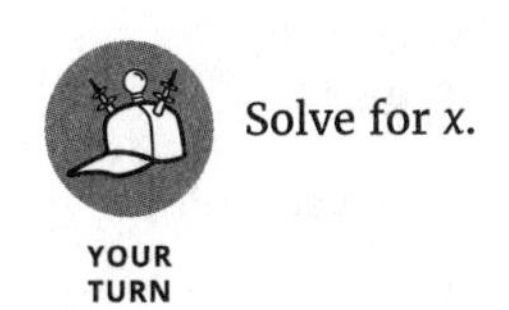

YOUR TURN

Solve for x.

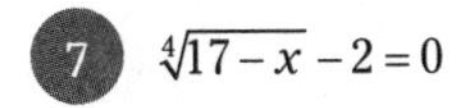

7. $\sqrt[4]{17-x}-2=0$

8. $\sqrt[3]{x^2-11x+4}=4$

Doubling the Fun with Radical Equations

Just when you thought things couldn't get any better, up comes a situation where you have to *get to* square both sides of an equation not once, but twice! This doubling your fun happens when you have more than one radical in an equation and getting them alone on one side of the equation isn't possible.

As you go about solving these particular types of problems, you can't do anything to isolate each radical term by itself on one side of the equation, and you have to square terms twice to get rid of all the radicals. The procedure is a little involved, but nothing too horrible. You see how to go about solving such a problem with the next example.

Solve for the value of x in the equation, $\sqrt{x-3}+4\sqrt{x+6}=12$.

Follow these steps:

1. **Get one radical on each side of the equal sign.**

   Even though you can't get either radical by itself, having them on either side of the equation helps. So subtract the $\sqrt{x-3}$ from each side to put it on the right with the 12: $4\sqrt{x+6}=12-\sqrt{x-3}$.

2. **Square both sides of the equation.**

   On the left side, squaring involves the rule about exponents where you're squaring a product. (This rule is covered in Chapter 5 and in the preceding section.) On the right side, squaring involves squaring a binomial (using FOIL).

   $$\left(4\sqrt{x+6}\right)^2=\left(12-\sqrt{x-3}\right)^2$$
   $$16\left(x+6\right)=144-24\sqrt{x-3}+\left(\sqrt{x-3}\right)^2$$
   $$16x+96=144-24\sqrt{x-3}+x-3$$

3. **Simplify, and get the remaining radical by itself on one side of the equation.**

   Simplifying involves combining the 144 and −3, subtracting x from each side of the equation, and then subtracting 141 from each side.

   $$16x+96=141-24\sqrt{x-3}+x$$
   $$15x-45=-24\sqrt{x-3}$$

4. **Look for a common factor in all the terms of the equation.**

   You can make things a bit easier to deal with by dividing each side by the greatest common factor, 3:

   $$\cancel{3}(5x-15)=\cancel{3}\left(-8\sqrt{x-3}\right)$$
   $$5x-15=-8\sqrt{x-3}$$

   Now you can square both sides more easily (the squares of the numbers are smaller).

5. **Square both sides of the equation.**

   $$(5x-15)^2=\left(-8\sqrt{x-3}\right)^2$$
   $$25x^2-150x+225=64(x-3)$$
   $$25x^2-150x+225=64x-192$$

   These are still some rather large numbers.

6. **Get everything on one side of the equation and factor.**

   You can move everything to the left and see whether you can factor anything out to make the numbers smaller. In this example, you can subtract 64x from each side and add 192 to each side.

   $$25x^2-214x+417=0$$

   This isn't the easiest quadratic to factor, but it does factor, giving you $(25x-139)(x-3)=0$. So, you have two solutions: either $x=\frac{139}{25}$ or $x=3$.

7. **Plug in the solutions to check your answer.**

   If $x=\frac{139}{25}$, then $\sqrt{\frac{139}{25}-3}+4\sqrt{\frac{139}{25}+6}=12$. What are the chances of this being a true statement? You can get out your trusty calculator to see if it works.

   $$\sqrt{\frac{64}{25}}+4\sqrt{\frac{289}{25}}=\frac{8}{5}+4\left(\frac{17}{5}\right)=\frac{76}{5}\neq 12$$

   After all that, the answer doesn't even work! Hope for the 3.

   If $x=3$, then $\sqrt{3-3}+4\sqrt{3+6}=0+4\sqrt{9}=4\cdot 3=12$. Oh, good!

Q. Solve for x: $\sqrt{5x+11}+\sqrt{x+3}=2$.

EXAMPLE

A. Subtract $\sqrt{x+3}$ from each side of the equation and then square both sides.

$$\sqrt{5x+11}=2-\sqrt{x+3}$$
$$\left(\sqrt{5x+11}\right)^2=\left(2-\sqrt{x+3}\right)^2$$
$$5x+11=4-4\sqrt{x+3}+(x+3)$$
$$4x+4=-4\sqrt{x+3}$$

Since each of the terms is divisible by 4. Divide every term by 4 and then square both sides again. $x+1=-\sqrt{x+3}\rightarrow(x+1)^2=\left(-\sqrt{x+3}\right)^2\rightarrow x^2+2x+1=x+3$.

Next, set the quadratic equation equal to 0, factor it, and solve for the solutions to that equation. $x^2+x-2=0\rightarrow(x+2)(x-1)=0\rightarrow x=-2$ or $x=1$.

When you test each solution in the original equation, you find that −2 is a solution but 1 is an extraneous root.

Solve for x.

YOUR TURN

 9 $\sqrt{x}+3=\sqrt{x+27}$

 10 $3\sqrt{x+1}-2\sqrt{x-4}=5$

Solving Absolute Value Equations

Absolute value equations are those involving the absolute value operation. The expression within the absolute value bars must always be equal to a positive number or zero. Linear absolute value equations require one type of process, and quadratic, higher-degree polynomial, and rational equations require another method.

Making linear absolute value equations absolutely wonderful

An equation such as $|x|=7$ is fairly easy to decipher. It's asking for values of x that give you a 7 when you put it in the absolute value symbol. Two answers, 7 and −7, have an absolute value of 7. Those are the only two answers. But what about something a bit more involved, such as $|3x+2|=4$?

To solve an absolute value equation of the form $|ax+b|=c$, change the absolute value equation to two equivalent linear equations and solve them.

So, the equation $|ax+b|=c$ is equivalent to $ax+b=c$ or $ax+b=-c$. Notice that the left side is the same in each equation. The c is positive in the first equation and negative in the second because the expression inside the absolute value symbol can be positive or negative — absolute value makes the expression inside the operation symbols positive when it's performed on them.

Here's an example: Solve for x in $|3x+2|=4$.

Follow the steps to solve:

1. **Rewrite as two linear equations.**

   $3x+2=4$ or $3x+\ 2=-4$

2. **Solve for the value of the variable in each of the equations.**

   Subtract 2 from each side in each equation: $3x=2$ or $3x=-6$.

   Divide each side in each equation by 3: $x=\frac{2}{3}$ or $x=-2$.

3. **Check.**

   If $x=-2$, then $|3(-2)+2|=|-6+2|=|-4|=4$.

   If $x=\frac{2}{3}$, then $\left|3\left(\frac{2}{3}\right)+2\right|=|2+2|=4$.

   They both work.

In the next example, you see the equation set equal to 0. For these problems, though, you don't want the equation set equal to 0. In order to use the rule for changing to linear equations, you have to have the absolute value by itself on one side of the equation.

Solve for x in $|5x-2|+3=0$.

Follow these steps.

1. **Get the absolute value expression by itself on one side of the equation.**

   Adding -3 to each side,

   $|5x-2|=-3$

2. **Rewrite as two linear equations.**

   $5x-2=-3$ or $5x-2=+3$

3. **Solve the two equations for the value of the variable.**

   Add 2 to each side of the equations:

   $5x = -1$ or $5x = 5$

   Divide each side by 5:

   $x = -\frac{1}{5}$ or $x = 1$

4. **Check.**

   If $x = -\frac{1}{5}$ then, $\left|5\left(-\frac{1}{5}\right) - 2\right| + 3 = |-1-2| + 3 = |-3| + 3 = 6$.

   Oops! That's supposed to be a 0. Try the other one.

   If $x = 1$, then $|5(1) - 2| + 3 = |3| + 3 = 6$.

   No, that didn't work either.

Now's the time to realize that the equation was impossible to begin with. (Of course, noticing this before you started would've saved time.) The definition of absolute value tells you that it results in everything being positive. Starting with an absolute value equal to −3 gave you an impossible situation to solve. No wonder you didn't get an answer!

EXAMPLE

Q. Solve for x: $|4x + 5| = 3$.

A. Rewrite the absolute value equation as two different equations: $4x + 5 = 3$ or $4x + 5 = -3$. Solve $4x + 5 = 3$, which gives you $x = -\frac{1}{2}$. Solve $4x + 5 = -3$, which gives you $x = -2$.

Checking the solutions:

$\left|4\left(-\frac{1}{2}\right) + 5\right| = |-2 + 5| = |3| = 3$ and $|4(-2) + 5| = |-8 + 5| = |-3| = 3$

Both $x = -\frac{1}{2}$ and $x = -2$ work.

Q. Solve for x: $|3 + x| - 5 = 1$.

A. Before applying the rule to change the absolute value into linear equations, add 5 to each side of the equation. This gets the absolute value by itself, on the left side: $|3 + x| = 6$.

Now the two equations are $3 + x = 6$ and $3 + x = -6$. The solutions are $x = 3$ and $x = -9$, respectively. Checking these answers in the original equation, $|3 + 3| - 5 = |6| - 5 = 6 - 5 = 1$ and $|3 + (-9)| - 5 = |-6| - 5 = 6 - 5 = 1$.

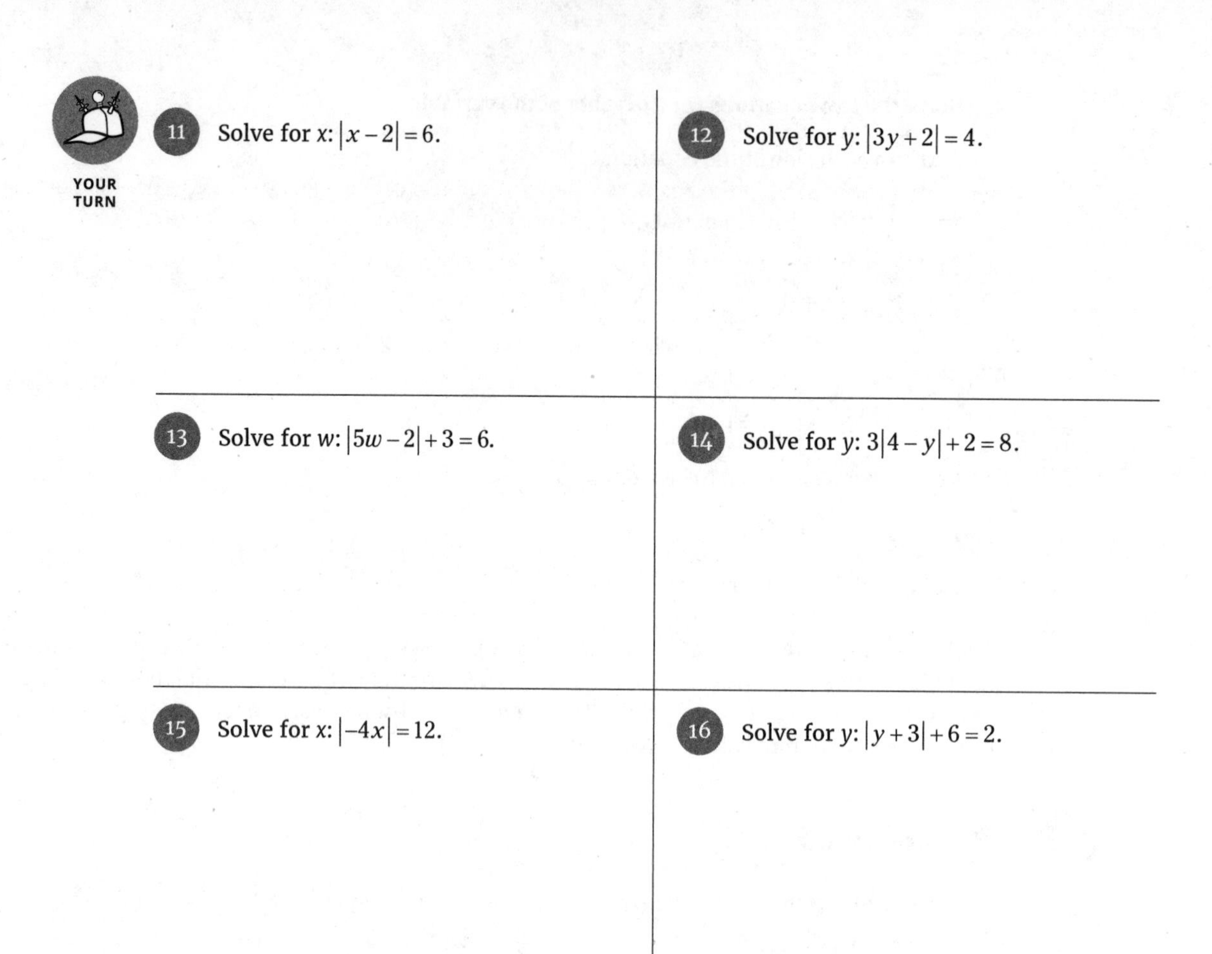

YOUR TURN

11 Solve for x: $|x-2|=6$.

12 Solve for y: $|3y+2|=4$.

13 Solve for w: $|5w-2|+3=6$.

14 Solve for y: $3|4-y|+2=8$.

15 Solve for x: $|-4x|=12$.

16 Solve for y: $|y+3|+6=2$.

Factoring absolute value equations for solutions

The rule for solving linear absolute value equations works very nicely by eliminating the absolute value operation and solving two separate equations. This rule works relatively well with nonlinear absolute value equations, but factoring and careful consideration of the solutions is necessary.

EXAMPLE

Q. Solve for x in $|x^2-x-9|=3$.

A. The expression in the absolute value operation must be equal to either 3 or −3. Starting with $x^2-x-9=3$, you subtract 3 from each side to get $x^2-x-12=0$. The trinomial factors into $(x-4)(x+3)=0$ with solutions $x=4$ and $x=-3$.

Setting the expression equal to −3, $x^2-x-9=-3$. Adding 3 to each side, you get $x^2-x-6=0$. The trinomial factors into $(x-3)(x+2)=0$ with solutions $x=3$ and $x=-2$.

Q. Solve for x in $|x^2+1|=5$.

A. The expression in the absolute value operation must be equal to either 5 or −5. Starting with $x^2+1=5$, you subtract 5 from each side to get $x^2-4=0$. The binomial factors into $(x+2)(x-2)=0$ with solutions $x=-2$ and $x=2$.

Setting the expression equal to −5, you have $x^2+1=-5$. Adding 5 to each side, the equation becomes $x^2+6=0$. The binomial doesn't factor, and using the quadratic formula yields no real number answers. The only two solutions are −2 and 2.

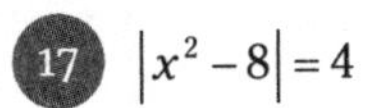

Solve for x.

17 $|x^2-8|=4$

18 $|2x^3-5x^2-12x|=0$

19 $|x^2-1|=8$

20 $\left|\frac{1}{x-2}\right|=3$

Checking for Absolute Value Extraneous Roots

It's always a good idea to check solutions when you've had to change the format of the original problem. In problems involving radicals, the even-numbered roots are the best candidates for finding extraneous roots. And in absolute value equations, these opportunities arise when the absolute value expression is set equal to another expression involving a variable.

EXAMPLE

Q. Solve $|x+6|=2x$ for the value of x.

A. You consider the two equations: $x+6=2x$ and $x+6=-2x$.

Starting with the first equation, $x+6=2x \rightarrow 6=x$, and this checks because $|6+6|=2(6)$.

But in the second equation, $x+6=-2x \rightarrow 3x=-6 \rightarrow x=-2$, and this doesn't work because $|-2+6|=2(-2)$ is not true. Only one of the solutions works.

Q. Solve for y in $|2y-3|=3y-2$.

A. First, solve $2y-3=3y-2$. You get $y=-1$, which doesn't work because $|2(-1)-3|=3(-1)-2 \rightarrow |-5|\neq -5$. And then, when solving $2y-3=-3y+2$, you get $5y=5$ or $y=1$. This one does work, because $|2(1)-3|=3(1)-2 \rightarrow |-1|=1$.

YOUR TURN

Solve for x.

21 $|4x-5|=6x-9$

22 $\frac{1}{3}|2x-1|-3=x$

Practice Questions Answers and Explanations

1 $x = 39$.

1. **Square both sides and then solve for x by adding 3 to each side:**

$$\sqrt{x-3} = 6 \rightarrow \left(\sqrt{x-3}\right)^2 = 6^2 \rightarrow x-3 = 36 \rightarrow x = 39$$

2. **Check:** $\sqrt{39-3} \stackrel{?}{=} 6 \rightarrow \sqrt{36} = 6$

2 $x = \pm 4$.

1. **Square both sides.**

$$\left(\sqrt{x^2+9}\right)^2 = 5^2 \rightarrow x^2 + 9 = 25$$

2. **Subtract 9 from each side.**

$$x^2 + 9 - 9 = 25 - 9$$
$$x^2 = 16$$

3. **Find the square root of each side.**

$$x^2 = 16$$
$$x = 4 \text{ and } x = -4$$

4. **Check to see whether the answers work.**

$$x = 4: \sqrt{(4)^2 + 9} \stackrel{?}{=} 5, \sqrt{16+9} \stackrel{?}{=} 5, \sqrt{25} = 5$$
$$x = -4: \sqrt{(-4)^2 + 9} \stackrel{?}{=} 5, \sqrt{16+9} \stackrel{?}{=} 5, \sqrt{25} = 5$$

3 $x = -1$.

1. **Move x to the right side.**

$$\sqrt{x+5} + x = 1 \rightarrow \sqrt{x+5} = 1 - x$$

2. **Square both sides.**

$$\left(\sqrt{x+5}\right)^2 = (1-x)^2 \rightarrow x+5 = 1-2x+x^2 \rightarrow 0 = x^2 - 3x - 4$$
$$(x-4)(x+1) = 0 \rightarrow x-4 = 0 \rightarrow x = 4 \text{ or } x+1 = 0 \rightarrow x = -1$$

3. **Check.**

$x = 4: \sqrt{4+5} + 4 \stackrel{?}{=} 1, \sqrt{9} + 4 \stackrel{?}{=} 1, 3 + 4 \neq 1$. So 4 isn't a solution; it's extraneous.

$x = -1: \sqrt{(-1)+5} + (-1) \stackrel{?}{=} 1, \sqrt{4} - 1 \stackrel{?}{=} 1, 2 - 1 = 1$. So $x = -1$ is the only solution.

(4) $x = 12.$

1. **Subtract 9 from each side and then square both sides.** Set the quadratic equal to 0 to factor and solve for x:

$$\sqrt{x-3}+9=x \rightarrow \sqrt{x-3}=x-9 \rightarrow \left(\sqrt{x-3}\right)^2=(x-9)^2$$
$$x-3=x^2-18x+81 \rightarrow 0=x^2-19x+84 \rightarrow (x-12)(x-7)=0$$

2. **Using the multiplication property of zero (MPZ; see Chapter 13), you have** $x-12=0$, $x=12$ **or** $x-7=0$, $x=7$.

3. **Check.**

$$x=12: \sqrt{12-3}+9 \stackrel{?}{=} 12 \rightarrow \sqrt{9}+9 \stackrel{?}{=} 12 \rightarrow 3+9=12$$
$$x=7: \sqrt{7-3}+9 \stackrel{?}{=} 7 \rightarrow \sqrt{4}+9 \stackrel{?}{=} 7 \rightarrow 2+9 \neq 7$$

So, the only solution is 12.

(5) **Both** $x=-7$ **and** $x=-6$.

1. **Add 7 to each side. Then square both sides, set the equation equal to 0, and solve for x:**

$$\sqrt{x+7}-7=x \rightarrow \sqrt{x+7}=x+7 \rightarrow \left(\sqrt{x+7}\right)^2=(x+7)^2$$
$$x+7=x^2+14x+49 \rightarrow 0=x^2+13x+42 \rightarrow (x+7)(x+6)=0$$

2. **Using the multiplication property of zero (see Chapter 13), you have** $x+7=0$, $x=-7$ **or** $x+6=0$, $x=-6$.

3. **Check.**

$$x=-7: \sqrt{(-7)+7}-7 \stackrel{?}{=} -7 \rightarrow \sqrt{0}-7 \stackrel{?}{=} -7 \rightarrow 0-7=-7$$
$$x=-6: \sqrt{(-6)+7}-7 \stackrel{?}{=} -6 \rightarrow \sqrt{1}-7 \stackrel{?}{=} -6 \rightarrow 1-7=-6$$

(6) **Both** $x=1$ **and** $x=2$.

1. **Square both sides of the equation. Then set it equal to 0 and factor:**

$$\left(\sqrt{x-1}\right)^2=(x-1)^2 \rightarrow x-1=x^2-2x+1 \rightarrow 0=x^2-3x+2$$
$$0=(x-1)(x-2) \rightarrow x=1 \text{ or } x=2$$

2. **Check.**

$$x=1: \sqrt{1-1} \stackrel{?}{=} 1-1 \rightarrow 0=0$$
$$x=2: \sqrt{2-1} \stackrel{?}{=} 2-1 \rightarrow 1=1$$

(7) $x=1$. Add 2 to each side of the equation and then raise each side to the fourth power. Solve for x in the resulting equation. $\left(\sqrt[4]{17-x}\right)^4=2^4 \rightarrow 17-x=16,\ x=1$

(8) $x = 15$, $x = -4$. Cube each side of the equation, and then solve the quadratic.

$$\left(\sqrt[3]{x^2 - 11x + 4}\right)^3 = 4^3 \rightarrow x^2 - 11x + 4 = 64 \rightarrow x^2 - 11x - 60 = (x - 15)(x + 4) = 0$$

Using the MPZ, $x = 15$ or $x = -4$.

Check: $x = 15$, $\sqrt[3]{15^2 - 11(15) + 4} = \sqrt[3]{225 - 165 + 4} = \sqrt[3]{64} = 4$. Checks.

Check: $x = -4$, $\sqrt[3]{(-4)^2 - 11(-4) + 4} = \sqrt[3]{16 + 44 + 4} = \sqrt[3]{64} = 4$. Checks.

(9) $x = 9$. First, square both sides of the equation. Then keep the radical term on the left and subtract x and 9 from each side. Before squaring both sides again, divide by 6:

$$\sqrt{x} + 3 = \sqrt{x + 27} \rightarrow \left(\sqrt{x} + 3\right)^2 = \left(\sqrt{x + 27}\right)^2$$

$$\left(\sqrt{x}\right)^2 + 6\sqrt{x} + 9 = x + 27 \rightarrow x + 6\sqrt{x} + 9 = x + 27 \rightarrow 6\sqrt{x} = 18 \rightarrow \sqrt{x} = 3$$

Square both sides of the new equation: $\left(\sqrt{x}\right)^2 = 3^2$, $x = 9$

Check: $x = 9$: $\sqrt{9} + 3 \stackrel{?}{=} \sqrt{9 + 27} \rightarrow 3 + 3 = 6 = \sqrt{36}$

(10) $x = 8$. First, move a radical term to the right, square both sides, simplify, and, finally, isolate the radical term on the right. You can then divide each side by 5:

$$3\sqrt{x+1} - 2\sqrt{x-4} = 5 \rightarrow 3\sqrt{x+1} = 2\sqrt{x-4} + 5 \rightarrow \left(3\sqrt{x+1}\right)^2 = \left(2\sqrt{x-4} + 5\right)^2$$

$$9(x+1) = 4(x-4) + 20\sqrt{x-4} + 25 \rightarrow 9x + 9 = 4x - 16 + 20\sqrt{x-4} + 25$$

$$9x + 9 = 4x + 9 + 20\sqrt{x-4} \rightarrow 5x = 20\sqrt{x-4} \rightarrow x = 4\sqrt{x-4}$$

Square both sides again, set the equation equal to 0, and factor:

$$(x)^2 = \left(4\sqrt{x-4}\right)^2 \rightarrow x^2 = 16(x-4) \rightarrow x^2 = 16x - 64 \rightarrow x^2 - 16x + 64 = 0, (x-8)^2 = 0$$

Using the multiplication property of zero (see Chapter 13), you have $x - 8 = 0$, $x = 8$.

Check: $x = 8$: $3\sqrt{8+1} - 2\sqrt{8-4} \stackrel{?}{=} 5 \rightarrow 3\sqrt{9} - 2\sqrt{4} \stackrel{?}{=} 5 \rightarrow 3(3) - 2(2) = 9 - 4 = 5$

(11) $x = 8$ and $x = -4$. First, remove the absolute value symbol by setting what's inside equal to both positive and negative 6. Then solve the two linear equations that can be formed:

$$|x - 2| = 6 \rightarrow x - 2 = \pm 6 \rightarrow x = 2 \pm 6 \rightarrow x = 2 + 6 = 8 \text{ or } x = 2 - 6 = -4$$

Check: $x = 8$: $|8 - 2| = |6| = 6$, and $x = -4$: $|(-4) - 2| = |-6| = 6$

(12) $y = \frac{2}{3}$ and $y = -2$. First, remove the absolute value symbol by setting what's inside equal to both positive and negative 4. Then solve the two linear equations that can be formed:

$$|3y + 2| = 4 \rightarrow 3y + 2 = \pm 4 \rightarrow 3y = -2 \pm 4$$

$$3y = -2 + 4 \rightarrow 3y = 2,\ y = \frac{2}{3} \text{ and } 3y = -2 - 4 \rightarrow 3y = -6,\ y = -2$$

Check: $y = \frac{2}{3}$: $\left|3\left(\frac{2}{3}\right) + 2\right| = |2 + 2| = |4| = 4$, and $y = -2$: $|3(-2) + 2| = |-6 + 2| = |-4| = 4$

(13) $w = 1$ and $w = -\frac{1}{5}$. First, subtract 3 from each side. Then remove the absolute value symbol by setting what's inside equal to both positive and negative 3:

$$|5w - 2| + 3 = 6 \rightarrow |5w - 2| = 6 - 3 = 3 \rightarrow 5w - 2 = \pm 3 \rightarrow 5w = 2 \pm 3$$

The two linear equations that are formed give you two different answers:

$5w=2+3 \rightarrow 5w=5,\ w=1$ and $5w=2-3 \rightarrow 5w=-1,\ w=-\frac{1}{5}$

(14) **$y = 6$ and $y = 2$.** First, subtract 2 from each side. Then divide each side by 3:

$3|4-y|+2=8 \rightarrow 3|4-y|=6 \rightarrow |4-y|=2$

Then rewrite without the absolute value symbol by setting the expression inside the absolute value equal to positive or negative 2: $4-y=\pm 2; 4\pm 2=y$. Then simplify the resulting linear equations:

$y=4+2=6$ or $y=4-2=2$

(15) **$x = 3$ and $x = -3$.** First, rewrite the equation without the absolute value symbol:

$|-4x|=12 \rightarrow -4x=\pm 12$

$-4x=12 \rightarrow x=-\frac{12}{4}=-3$ and $-4x=-12 \rightarrow x=\frac{-12}{-4}=3$

(16) **No answer.** Here's why:

$|y+3|+6=2 \rightarrow |y+3|=-4 \rightarrow y+3=\pm 4 \rightarrow y=-3\pm 4,\ y=1$ or $y=-7$

Check.

$y=1:\ |(1)+3|+6=|4|+6=10\neq 2$

$y=-7:\ |(-7)+3|+6=|-4|+6=4+6=10\neq 2$

(17) **$x = 2\sqrt{3}$, $x = -2\sqrt{3}$, $x = 2$, $x = -2$.** Solve $x^2-8=4$ by adding 8 to each side of the equation. The equation $x^2=12$ can be solved using the square-root rule. (See Chapter 16 for more on that rule.) You get $x=\pm\sqrt{12}$, which simplifies to $x=\pm 2\sqrt{3}$. (Simplifying radical expressions is found in Chapter 5.) Now solve $x^2-8=-4$ by adding 8 to each side and getting $x^2=4$. Taking the square root of each side gives you $x = \pm 2$. You could also have rewritten the equation as $x^2-4=0$ and factored, giving you the same two solutions: $x=2$ and $x=-2$. So the four solutions are $x=2\sqrt{3}$, $x=-2\sqrt{3}$, $x=2$, and $x=-2$.

(18) **$x = 0$, $x = -\frac{3}{2}$, and $x = 4$.** Solve $2x^3-5x^2-12x=0$ by first factoring out x to get $x(2x^2-5x-12)=0$. The trinomial factors, so the completely factored form is $x(2x+3)(x-4)=0$. The multiplication property of zero then gives you the three solutions: $x=0$, $x=-\frac{3}{2}$, and $x=4$. You don't need to solve another equation, because the opposite of 0 is 0. There isn't another choice.

(19) **$x = 3$ and $x = -3$.** Solve $x^2-1=8$ by adding 1 to each side of the equation and applying the square-root rule. You get $x^2=9$, which has solutions $x = \pm 3$. Now solve $x^2-1=-8$. Adding 1 to each side gives you $x^2=-7$. Taking the square root of both sides gives you the square root of a negative number. This has no real answer. So the only solutions are $x=3$ and $x=-3$.

(20) **$x = \frac{7}{3}$ and $x = \frac{5}{3}$.**

Solve $\frac{1}{x-2}=3$ by multiplying both sides of the equation by $x-2$. This gives you $1=3(x-2)$. Distributing the 3, you have $1=3x-6$. Adding 6 to both sides and then dividing by 3, you have $x=\frac{7}{3}$. You check to be sure this value of x doesn't give you a 0 in the denominator of the fraction, and it doesn't. The solution holds. Now, solving $\frac{1}{x-2}=-3$, you multiply each side by $x-2$ and get $1=-3(x-2)$. Distributing the -3, you have $1=-3x+6$. Subtracting 6 from each side and then dividing by -3, the solution is $x=\frac{5}{3}$. This solution also holds when substituted into the fraction. So the two solutions are $x=\frac{7}{3}$ and $x=\frac{5}{3}$.

(21) $\boldsymbol{x=2}$**.** First, solve the equation $4x-5=6x-9$. You get $4=2x$ or $x=2$. When checking, you have $|4(2)-5|=6(2)-9$ or $|3|=3$, so $x=2$ is a solution. Next, solving $4x-5=-6x+9$, you have $10x=14$ or $x=\frac{7}{5}$. Checking this solution, $\left|4\left(\frac{7}{5}\right)-5\right|=6\left(\frac{7}{5}\right)-9 \rightarrow \left|\frac{28}{5}-\frac{25}{5}\right|=\frac{42}{5}-\frac{45}{5}$. This isn't true.

(22) $\boldsymbol{x=-\frac{8}{5}}$**.** First, add 3 to each side and multiply both sides by 3 to get $\frac{1}{3}|2x-1|=x+3$, and then $|2x-1|=3x+9$. The first equation to solve is $2x-1=3x+9$, which gives you $x=-10$. Checking this answer, you have $\frac{1}{3}|2(-10)-1|-3=-10$ or $\frac{1}{3}|-21|-3=-10$, which says that $7-3=-10$. Nope. That doesn't work. And then, when solving $2x-1=-3x-9$, you get $5x=-8$ or $x=-\frac{8}{5}$. Checking this answer, $\frac{1}{3}\left|2\left(-\frac{8}{5}\right)-1\right|-3=-\frac{8}{5}$, which becomes $\frac{1}{3}\left|-\frac{16}{5}-\frac{5}{5}\right|-3=-\frac{8}{5}$ and then $\frac{1}{3}\left|-\frac{21}{5}\right|-3=-\frac{8}{5}$ and then $\frac{7}{5}-\frac{15}{5}=-\frac{8}{5}$, which is true!

If you're ready to test your skills a bit more, take the following chapter quiz that incorporates all the chapter topics.

Whaddya Know? Chapter 18 Quiz

Quiz time! Complete each problem to test your knowledge on the various topics covered in this chapter. You can then find the solutions and explanations in the next section.

Solve for x.

1. $\sqrt{x+2}+\sqrt{x-6}=4$
2. $2|3x+1|=16$
3. $\sqrt[3]{x+3}=-1$
4. $|3x-8|=x+2$
5. $\sqrt[4]{12+x}=2$
6. $|x+3|=7$
7. $\sqrt{x-2}-1=x-5$
8. $|x^2-5x|=6$
9. $\sqrt{x+4}=5$
10. $\left|\frac{6}{x+1}\right|=2$

Answers to Chapter 18 Quiz

1. $\boldsymbol{x=7}$**.** Move a radical term to the right and then square both sides: $\left(\sqrt{x+2}\right)^2=\left(4-\sqrt{x-6}\right)^2 \rightarrow x+2=16-8\sqrt{x-6}+x-6$. Now simplify by isolating the radical term on the right. You get $-8=-8\sqrt{x-6}$. Divide each side by –8 before squaring both sides and simplifying: $1^2=\left(\sqrt{x-6}\right)^2 \quad \rightarrow \quad 1=x-6,\ x=7$.

2. $\boldsymbol{x=\frac{7}{3}}$ **or** $\boldsymbol{x=-3}$**.** First, divide both sides by 2. Then write the two equations and solve for x: in the first equation, $|3x+1|=8$ gives you $3x+1=8 \ \rightarrow \ 3x=7,\ x=\frac{7}{3}$, and in the second equation, $3x+1=-8 \ \rightarrow \ 3x=-9,\ x=-3$.

3. $\boldsymbol{x=-4}$**.** Cube each side and solve for x: $\left(\sqrt[3]{x+3}\right)^3=(-1)^3 \quad \rightarrow \quad x+3=-1,\ x=-4$.

4. $\boldsymbol{x=5}$**.** The first equation, $3x-8=x+2$, becomes $2x=10$, giving you $x=5$. The second equation, $3x-8=-x-2$, becomes $4x=6$ or $x=\frac{3}{2}$. This is extraneous.

5. $\boldsymbol{x=4}$**.** Raise each side to the fourth power and then solve for x: $\left(\sqrt[4]{12+x}\right)^4=2^4 \quad \rightarrow \quad 12+x=16,\ x=4$

6. $\boldsymbol{x=4}$ **or** $\boldsymbol{x=-10}$**.** Write the two equations and solve for x. $x+3=7$ and $x+3=-7$.

7. $\boldsymbol{x=6}$**.** Add 1 to each side: $\sqrt{x-2}=x-4$. Then square both sides and factor the quadratic to solve for x: $\left(\sqrt{x-2}\right)^2=(x-4)^2 \ \rightarrow \ x-2=x^2-8x+16 \ \rightarrow \ 0=x^2-9x+18=(x-6)(x-3)$. Using the MPZ, $x=6$ or $x=3$. But $x=3$ is extraneous.

8. $\boldsymbol{x=6,-1,3,2}$**.** The first equation, $x^2-5x=6$, becomes $x^2-5x-6=0$ and factors into $(x-6)(x+1)=0$. The two solutions are 6 and –1. The second equation, $x^2-5x=-6$, becomes $x^2-5x+6=0$ and factors into $(x-3)(x-2)=0$. The two solutions are 3 and 2.

9. $\boldsymbol{x=21}$**.** Square both sides and solve for x: $\left(\sqrt{x+4}\right)^2=5^5 \ \rightarrow \ x+4=25$.

10. $\boldsymbol{x=2}$ **or** $\boldsymbol{-4}$**.** Solving $\frac{6}{x+1}=2$, you multiply both sides of the equation by the denominator and get $6=2x+2$, which yields $x=2$. Solving $\frac{6}{x+1}=-2$, you multiply both sides by the denominator and get $6=-2x-2$ or $8=-2x$, giving you $x=-4$.

IN THIS CHAPTER

» Playing by the rules when dealing with inequalities

» Solving linear and quadratic inequalities

» Getting the most out of absolute-value inequalities

» Working with compound inequalities

Chapter 19

Getting Even with Inequalities

Algebraic inequalities show relationships between a number and an expression or between two or more expressions. One expression is bigger or smaller than a number or another expression for certain values of a given variable. For example, it could be that Janice has at least four more than twice as many cats as Eloise. There are lots of scenarios that can occur if one value is at least another and not exactly as many.

Many operations involving inequalities work the same as operations on equalities and equations, but you need to pay attention to some important differences that I show you in this chapter.

REMEMBER

The good news about solving inequalities is that nearly all the rules are the same as for solving an equation — with one big difference. The difference in applying rules comes in when you're multiplying or dividing both sides of an inequality by *negative numbers*. If you pay attention to what you're doing, you shouldn't have a problem.

This chapter covers everything from basic inequalities and linear inequalities to the more challenging quadratic, absolute-value, and complex inequalities. Take a deep breath. I offer you plenty of practice problems so you can work out any kinks.

Defining the Inequality Notation

Equations (statements with equal signs) are one type of relation — two things are exactly the same, it says. The *inequality* relation is a bit less precise. One thing can be bigger by a lot or bigger by a little, but there's still that relationship between them — that one is bigger than the other.

Pointing in the right direction

Algebraic operations and manipulations are performed on inequality statements while they're in an inequality format. You see the inequality statements written using the following notations:

- $<$: Less than
- $>$: Greater than
- $\leq$: Less than or equal to
- $\geq$: Greater than or equal to

To keep the direction straight as to which way to point the arrow, just remember that the *itsy-bitsy* part of the arrow is next to the smaller (*itsy-bitsier*) of the two values. You can write $4 < 6$ or $6 > 4$ or $4 \leq 6$ or $6 \geq 4$, and each statement is correct. It's when you start creating inequalities involving numbers and variables that things get even more interesting.

Inequality statements have been around for a long time. The symbols are traditional and accepted by mathematicians around the world. But (weren't you just expecting that qualifying word?), as well as the traditional inequality symbols work, they still have some competition — especially in the publishing and higher-math world. You'll see an alternative notation called *interval notation* in a later section.

Grappling with graphing inequalities

One of the best ways of describing inequalities is with a graph. It's the old "a picture is worth a thousand words" business. A graph or picture isn't always convenient, but it certainly gets the message across. Graphs in the form of number lines are a great help when solving quadratic inequalities (see the sections, "Solving Quadratic Inequalities" and "Dealing with Polynomial and Rational Inequalities," later in this chapter).

A number-line graph of an inequality consists of numbers representing the starting and ending points of any interval described by the inequality, and symbols above each number indicating whether that number is to be included in the answer. The symbols used with inequality notation are hollow circles and filled-in circles.

Q. Graph the inequalities $x > 3$ and $x \le -2$ on a number line.

EXAMPLE

A. Put an empty circle over the number 3 and shade in the number line to the right of 3. Then put a solid circle over the number –2 and shade in the number line to the left of –2. The graph is shown in Figure 19-1.

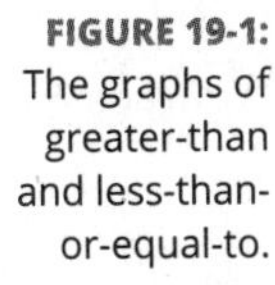

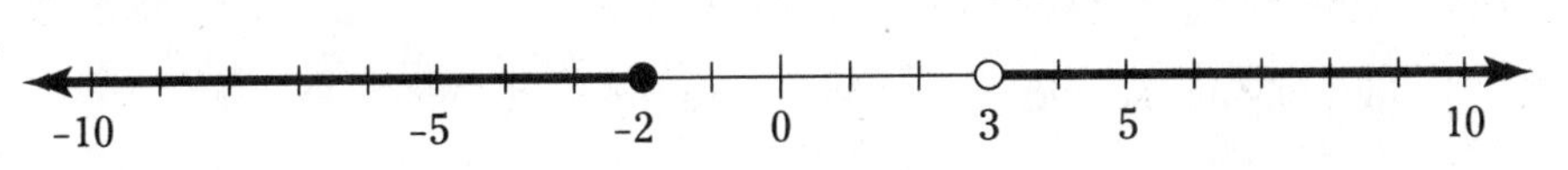

FIGURE 19-1: The graphs of greater-than and less-than-or-equal-to.

Using the Rules to Work on Inequality Statements

REMEMBER

Working with inequalities really isn't that difficult if you just keep a few rules in mind. The following rules deal with inequalities (assume that c is some number):

- If $a > b$, then adding c to each side or subtracting c from each side doesn't change the sense (direction of the inequality), and you get $a \pm c > b \pm c$.
- If $a > b$, then multiplying or dividing each side by a *positive* c doesn't change the sense, and you get $a \cdot c > b \cdot c$ or $\frac{a}{c} > \frac{b}{c}$.
- If $a > b$, then multiplying or dividing each side by a *negative* c does change the sense (reverses the direction), and you get $a \cdot c < b \cdot c$ or $\frac{a}{c} < \frac{b}{c}$.
- If $a > b$, then reversing the terms reverses the sense, and you get $b < a$.

Q. Starting with $-20 < 7$, perform the following operations: Add 5 to each side and multiply each side by 2.

EXAMPLE

A. Here is the solution:

$$-20 < 7$$
$$-20 + 5 < 7 + 5 \rightarrow -15 < 12$$

Adding 5 didn't change the sense.

$$-15(2) > 12(2) \rightarrow -30 > 24$$

Multiplying by 2 didn't change the sense.

$$30 > -24 \rightarrow -24 < 30$$

These processes didn't change the "sense" or the direction of the inequality symbol.

Q. Starting with $7 \geq -3$, first subtract 9 from each side, and then multiply each side by -2.

A. Here is the solution:

$$\begin{aligned} 7 &\geq -3 \\ 7-9 &\geq -3-9 \to -2 \geq -12 \end{aligned}$$

The statement is true.

Multiplying each side by -2 requires reversing the inequality symbol.

$$-2(-2) \ngeq -12(-2) \to 4 \ngeq 24 \to 4 \leq 24$$

The same thing happens when you divide each side by a negative number.

YOUR TURN

1 Starting with $5 > 2$, add 4 to each side and then divide each side by -3; simplify the result.

 Starting with $5 \geq 1$, multiply each side by -4; then divide each side by -2 and simplify the result.

WARNING

In the case of inequalities, you can neither divide nor multiply by 0. Of course, dividing by 0 is always forbidden, but you can usually multiply expressions by 0 (and get a product of 0). However, you can't multiply inequalities by 0.

Look at what happens when each side of an inequality is multiplied by 0:

$$\begin{aligned} 3 &< 7 \\ 0 \times 3 &< 0 \times 7 \\ 0 &< 0 \end{aligned}$$

No! It's just not true: Zero is not less than itself, nor is it greater than itself. So, to keep 0 from getting an inferiority or superiority complex, don't use it to multiply inequalities. If you have $3 \leq 7$ and multiply each side by 0, you get $0 \leq 0$, which is true in the one case.

Rewriting Inequalities Using Interval Notation

Just when you thought it couldn't get any better, here comes another way to write an inequality: in *interval notation.* Interval notation uses parentheses and brackets instead of inequality symbols, and it introduces the infinity symbol. This really is a very handy notation, especially when you're trying to write some mathy explanation. The only drawback or confusing part is when you write numbers using just parentheses; they look like the coordinates of points. You just have to be aware of the situation you're discussing — the context — to catch the difference.

Before defining how interval notation is used, let me first give a couple of examples in terms of writing the same statement in both inequality and interval notation:

- $x > 8$ is written $(8, \infty)$.
- $x < 2$ is written $(-\infty, 2)$.
- $x \geq -7$ is written $[-7, \infty)$.
- $x \leq 5$ is written $(-\infty, 5]$.
- $-4 < x \leq 10$ is written $(-4, 10]$.

So, now that you've seen interval notation in action, let me give you the rules for using it.

Interval notation expresses inequality statements with the following rules:

- Parentheses to show *less than* or *greater than* (but not including)
- Brackets to show *less than or equal to* or *greater than or equal to*
- Parentheses to show *all numbers* with both infinity and negative infinity inside
- Parentheses and/or brackets to show *starting and stopping points* with numbers and symbols written in the same left-to-right order as on a number line

Here are some examples of writing inequality statements using interval notation:

- $-3 \leq x \leq 11$ becomes $[-3, 11]$.
- $-4 \leq x < -3$ becomes $[-4, -3)$.
- $x > -9$ becomes $(-9, \infty)$.
- $5 < x$ becomes $(5, \infty)$. Notice that the variable didn't come first in the inequality statement, and saying 5 must be smaller than some numbers is the same as saying that those numbers are bigger (greater) than 5, or $x > 5$.
- $4 < x < 15$ becomes $(4, 15)$. Here's my biggest problem with interval notation: The notation $(4, 15)$ looks like a point on the coordinate plane, not an interval containing numbers between 4 and 15. You just have to be aware of the context when you come across this notation.

Now here are some examples of writing interval-notation statements using inequalities:

- $[-8, 5]$ becomes $-8 \le x \le 5$.
- $(-\infty, 0)$ becomes $x \le 0$.
- $(44, \infty)$ becomes $x > 44$.

The graph of an inequality consists of numbers representing the starting and ending points and symbols above each number indicating whether that number is to be included in the answer. The symbols used with inequality notation are hollow circles and filled-in circles. The symbols used with interval notation are the same parentheses and brackets used in the statements.

EXAMPLE

Q. Write the statement "all numbers between -3 and 4, including the 4" in inequality notation and interval notation. Then graph the inequality using both types of notation.

A. The inequality notation is $-3 < x \le 4$. The interval notation is $(-3, 4]$. The graphs are shown in Figures 19-2 and 19-3.

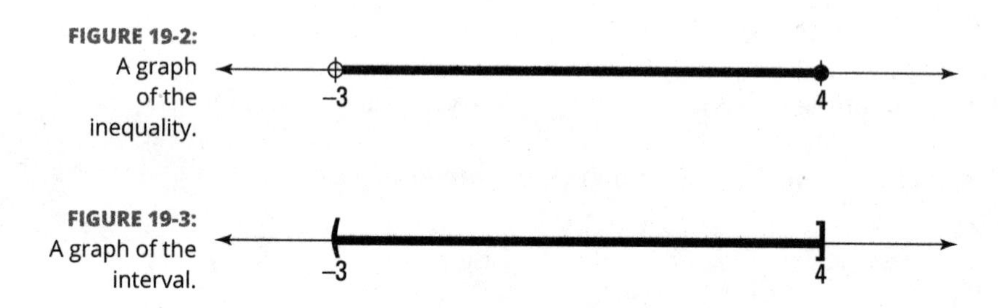

FIGURE 19-2: A graph of the inequality.

FIGURE 19-3: A graph of the interval.

Q. Write the statement "all numbers greater than -5" in inequality notation and interval notation.

A. The inequality notation is $x \ge -5$. The graph is shown in Figure 19-4. The interval notation is $[-5, \infty)$. The graph is shown in Figure 19-5.

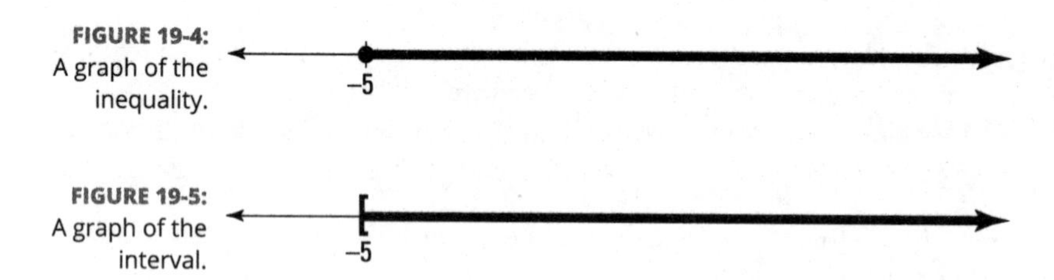

FIGURE 19-4: A graph of the inequality.

FIGURE 19-5: A graph of the interval.

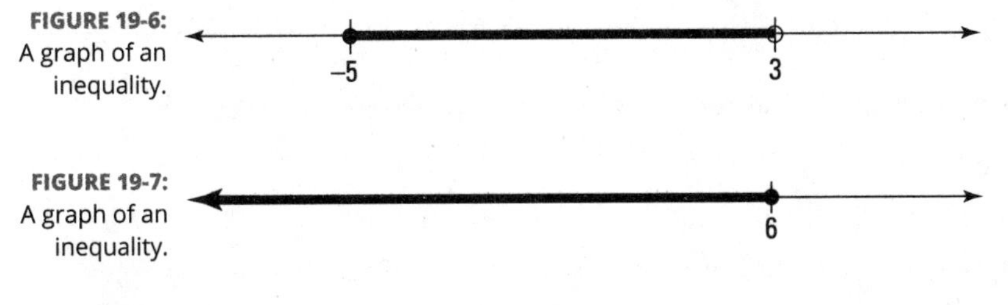

FIGURE 19-6: A graph of an inequality.

FIGURE 19-7: A graph of an inequality.

YOUR TURN

3 Using the graph in Figure 19-6, write statements in both inequality notation and interval notation describing the graph.

4 Using the graph in Figure 19-7, write statements in both inequality notation and interval notation describing the graph.

5 Write $x > 7$, using interval notation.

6 Write $2 \le x \le 21$, using interval notation.

Solving Linear Inequalities

Solving linear inequalities involves pretty much the same process as solving linear equations: Move all the variable terms to one side and the number to the other side. Then multiply or divide to solve for the variable. The tricky part is when you multiply or divide by a negative number. Because this special situation doesn't happen frequently, people tend to forget how to handle it.

REMEMBER

If you multiply both sides of $-x < -3$ by −1, the inequality becomes $x > 3$; you have to reverse the sense.

Another type of linear inequality has the linear expression sandwiched between two numbers, like $-1 < 8 - x \le 4$. The main rule here is that whatever you do to one section of the inequality, you do to all the others. For more on this, go to the section, "Solving Complex Inequalities," later in this chapter.

Q. Solve for x: $5-3x \geq 14+6x$.

EXAMPLE **A.** Subtract 6x from each side and subtract 5 from each side; then divide by −9:

$$\begin{aligned} 5-3x &\geq 14+6x \\ -5-6x \quad & \quad -5-6x \\ -9x &\geq 9 \\ \frac{-9x}{-9} &\leq \frac{9}{-9} \\ x &\leq -1 \end{aligned}$$

$x \leq -1$ or $(-\infty, -1]$.

Note that you can do this problem another way to avoid division by a negative number. See the next example for the alternate method.

Q. Solve for x: $5-3x \geq 14+6x$.

A. Add 3x to each side and subtract 14 from each side. Then divide by 9. This is the same answer, if you reverse the inequality and the numbers.

$$\begin{aligned} 5-3x \quad &\geq 14+6x \\ -14+3x \quad & \quad -14+3x \\ \hline -9 \quad &\geq \quad 9x \\ \frac{-9}{9} &\geq \frac{9x}{9} \\ -1 &\geq x \end{aligned}$$

Just flip the order of the statement, and the answer can be written $x \leq -1$ or $(-\infty, -1]$.

YOUR TURN

7 Solve for y: $4-5y \leq 19$.

8 Solve for x: $3(x+2) > 4x+5$.

9 Solve for x: $-5 < 2x+3 \leq 7$.

10 Solve for x: $3 \leq 7-2x < 11$.

Solving Quadratic Inequalities

A *quadratic inequality* is an inequality that involves a variable term with a second-degree power. When solving quadratic inequalities, the rules of addition, subtraction, multiplication, and division of inequalities still hold, but the final step in the solution is different. These quadratic inequalities are almost like puzzles that fall neatly into place as you work on them. The best way to describe how to solve a quadratic inequality is to use an example and put the rules right in the example.

Suppose you want to solve for *x* in $x^2 + 3x > 4$.

The answers to these inequalities can go in more than one direction — the numbers can be bigger than one number or smaller than another number or both — so I'm going to demonstrate how the solutions work before showing you how to solve them. Start by making some guesses as to what works for *x* in this expression:

- If $x = 2$, then $(2)2 + 3(2)$ is $4 + 6$; $10 > 4$, so 2 works.
- If $x = 5$, then $(5)2 + 3(5)$ is $25 + 15$; $40 > 4$, so 5 works. It looks like the bigger, the better.
- If $x = 0$, then $(0)2 + 3(0)$ is $0 + 0$; 0 is not greater than 4, so, no, 0 doesn't work. But, does anything smaller work? How about negative numbers?
- If $x = -6$, then $(-6)2 + 3(-6)$ is $36 - 18$; $18 > 4$, so, yes, -6 works.

Some negatives work; some positives work. The challenge is to determine where those negative and positive numbers are. There's a method you can use to find which work and which don't work without all this guessing.

To solve quadratic inequalities, follow these steps:

1. **Move all terms to one side of the inequality symbol so that the terms are greater than or less than 0.**
2. **Factor, if possible.**
3. **Find all values of the factored side that make that side equal to 0.**

   These are your *critical numbers.*
4. **Create a number line listing the values (critical numbers), in order, that make the expression equal to 0.**

   Leave spaces between the numbers for signs. Determine the signs (positive and negative) of the factored expression between those values that make it equal 0 and write them on the chart.
5. **Determine which intervals give you solutions to the problem.**

Now, apply this to the problem.

1. **Move all terms to one side.**

   First, move the 4 to the left by subtracting 4 from each side.

   $x^2+3x>4$

   $x^2+3x-4>0$

2. **Factor.**

   Factor the quadratic on the left using unFOIL.

   $(x+4)(x-1)>0$

3. **Find all the values of x that make the factored side equal to 0.**

   In this case, there are two values. Using the multiplication property of zero, you get $x+4=0$ or $x-1=0$, which results in $x=-4$ or $x=1$.

4. **Make a number line listing the values from Step 3, and determine the signs of the expression between the values on the chart.**

   When you choose a number to the left of −4, both factors are negative and the product is positive. Between −4 and 1, the first factor is negative and the second factor is positive, resulting in a negative product. To the right of 1, both factors are positive, giving you a positive product. Just testing one of the numbers in the interval tells you what will happen to all of them. Figure 19-8 shows you a number line with the critical numbers in their places and the signs in the intervals between the points.

5. **Determine which intervals give you solutions to the problem.**

   The values for x that work to make the quadratic $x^2+3x-4>0$ positive are all the negative numbers smaller than −4 down lower to really small numbers, and all the positive numbers bigger than 1 all the way up to really big numbers. The only numbers that don't work are those between −4 and 1. You write your answer as $x<-4$ or $x>1$.

   In interval notation, the answer is $(-\infty, -4) \cup (1, \infty)$. The ∪ symbol is for *union*, meaning everything in either interval (one or the other) works.

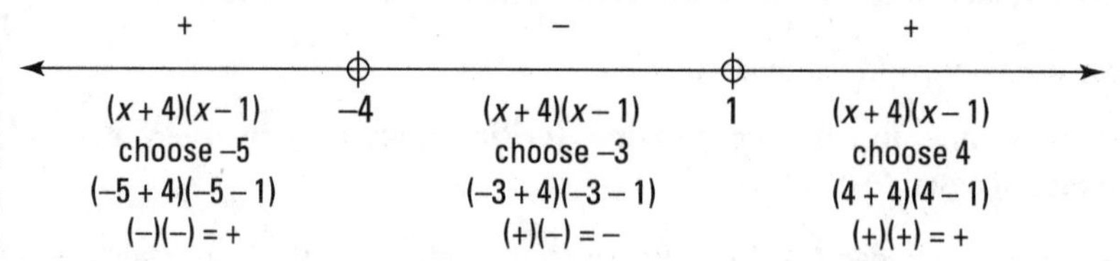

FIGURE 19-8: A number line helps you find the signs of the factors and their products.

In the next example, the end points of the intervals (critical numbers) are included in the answer.

Solve for the values of y in $y^2+15 \le 8y$.

Follow the steps.

1. **Subtract 8y from each side.**

   $y^2 - 8y + 15 \le 0$

2. **Factor.**

   $(y - 3)(y - 5) \le 0$

3. **Find the values of y that make the factored expression equal to 0.**

   The numbers you want are 3 and 5.

4. **Make a number line using the values that make the expression equal to 0.**

   Check for the signs of the factors and their products to determine the signs between the critical numbers. You see how to create the sign line in Figure 19-9.

5. **Determine which intervals give you solutions to the problem.**

   The original statement, $y^2 + 15 \le 8y$, is true when $y^2 - 8y + 15 \le 0$ is equal to 0 or less than 0 (negative). So the numbers 3, 5, and all those between 3 and 5 are solutions of the inequality. The answer is written $3 \le y \le 5$ or, in interval notation, [3, 5].

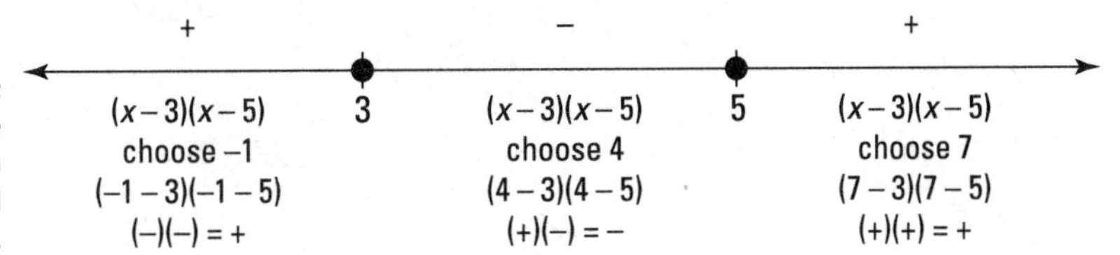

FIGURE 19-9: Filling in the signs between the critical numbers.

REMEMBER

When an inequality involves a quadratic expression, you have to resort to a completely different type of process to solve it than that used for linear inequalities. The quickest and most efficient method to find a solution is to use a number line.

After finding the *critical numbers* (where the expression changes from positive to negative or vice versa), you use a number line and place some + and − signs to indicate what's happening to the factors. (See Chapter 1 for some background info on the number line and where numbers fall on it.)

EXAMPLE

Q. Solve for x: $x^2 - x - 12 \le 0$.

A. First, factor the trinomial: $(x-4)(x+3) \le 0$. Setting the factors equal to 0, you have $x = 4$ or $x = -3$ Place those on the number line using solid points. Testing the intervals created, when $x < -3$, both factors are negative, so the product is positive. When $-3 < x < 4$, the first factor is negative and the second positive, so the product is negative. When $x > 4$, both factors are positive, so the product is positive. The only portion of the line to be shaded is that between −3 and 4. See the following figure.

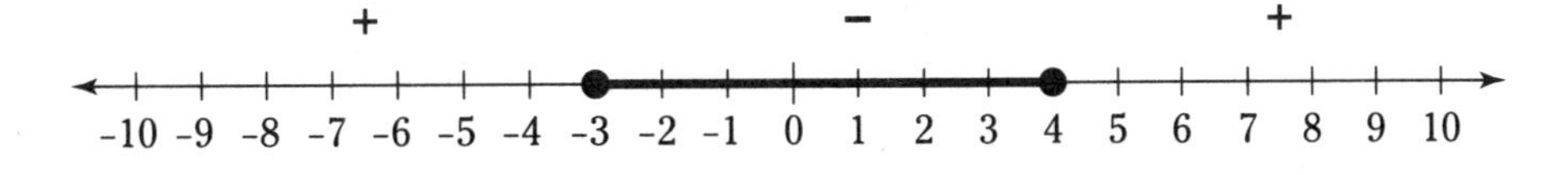

Q. Solve for x: $x^2 > 5x$.

A. First, subtract 5x from each side, and then factor the binomial: $x^2 > 5x \rightarrow x^2 - 5x > 0$ factors into $x(x-5) > 0$. Setting the factors equal to 0, you have $x = 0$ or $x = 5$. Place those on the number line using empty points. Testing the intervals created, when $x < 0$, both factors are negative, so the product is positive. When $0 < x < 5$, the first factor is positive and the second negative, so the product is negative. When $x > 5$, both factors are positive, so the product is positive. The line is shaded to the left of 0 and the right of 5. Refer to the following figure.

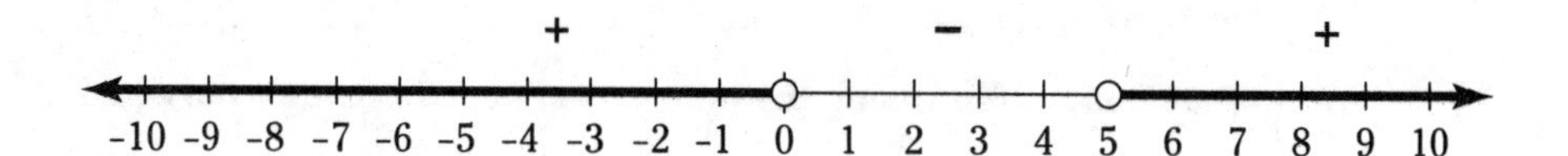

YOUR TURN

11 Solve for x: $(x+7)(x-1) > 0$.

12 Solve for x: $x^2 - x \leq 20$.

13 Solve for x: $3x^2 \geq 9x$.

14 Solve for x: $x^2 - 25 \geq 0$.

Dealing with Polynomial and Rational Inequalities

This section deals with inequalities that are pretty much handled the same way as quadratic inequalities. The difference is that you may have more than two factors in the factorization of the expression. You can really have any number of factors and any arrangement of factors and do the positive-and-negative business to get the answer.

Suppose you want to solve for the values of x that work in $(x-4)(x+3)(x-2)(x+7)>0$.

This problem is already factored, so you can easily determine that the numbers that make the expression equal to 0 (the critical numbers) are $x=4$, $x=-3$, $x=2$, $x=-7$. Put them in order from the smallest to the largest on a number line (see Figure 19-10), and test for the signs of the products in the intervals.

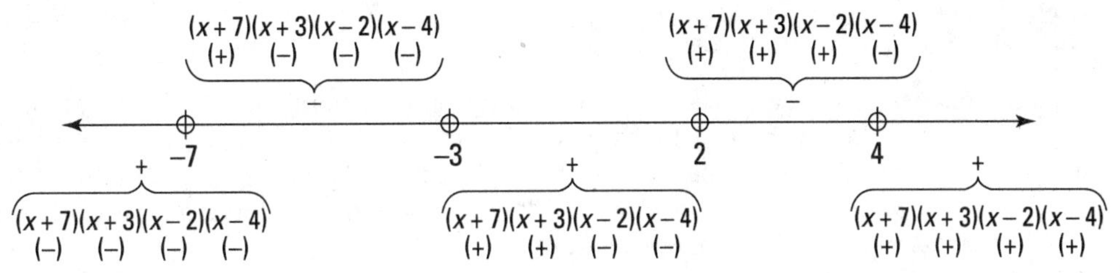

FIGURE 19-10: The sign changes at each critical number in this problem.

REMEMBER

When multiplying or dividing integers, if the number of negative signs in the problem is even, the result is positive. If the number of negative signs in the problem is odd, the result is negative.

Because the original problem is looking for values that make the expression greater than 0, or positive, the solution includes numbers in the intervals that are positive. Those numbers are

- Smaller than -7
- Between -3 and 2
- Bigger than 4

The solution is written $x<-7$ or $-3<x<2$ or $x>4$. In interval notation, the solution is written $(-\infty, -7) \cup (-3, 2) \cup (4, \infty)$.

The same process that gives you solutions to quadratic inequalities is used to solve some other types of inequality problems (see the section, "Solving Quadratic Inequalities," earlier in this chapter for more information). You use the number line process for polynomials (higher-degree expressions) and for rational expressions (fractions where a variable ends up in the denominator).

Inequalities with fractions that have variables in the denominator are another special type of inequality that fits under the general heading of quadratic inequalities; they get to be in this chapter because of the way you solve them.

To solve these rational (fractional) inequalities, you need to do somewhat the same thing as you do with the inequalities dealing with two or more factors: Find where the expression equals 0. Actually, expand that to looking for, separately, what makes the numerator (top) equal to 0 and what makes the denominator (bottom) equal to 0. These are your *critical numbers*. Check the intervals between the zeros, and then write out the answer.

WARNING

The one big caution with rational inequalities is not to include any number in the final answer that makes the denominator of the fraction equal 0. Zero in the denominator makes it an impossible situation, not to mention an impossible fraction. So why look at what makes the denominator 0 at all? The number 0 separates positive numbers from negative numbers. Even though the 0 itself can't be used in the solution, it indicates where the sign changes from positive to negative or negative to positive.

EXAMPLE

Q. Solve for y in $\frac{y+4}{y-3} > 0$.

A. The numbers making the numerator or the denominator equal to 0 are $y = -4$ or $y = 3$. Make a sign line with the two critical numbers in proper order. Determine the sign of the quotient formed by the two binomials. In Figure 19-11, you see the critical numbers and the signs in the intervals. The critical number 3 gets a hollow circle to indicate that it can't be used in the answer.

The problem only asks for values that make the expression greater than 0, or positive, so the solution is: $y < -4$ or $y > 3$. In interval notation, the answer is written as: $(-\infty, -4) \cup (3, \infty)$.

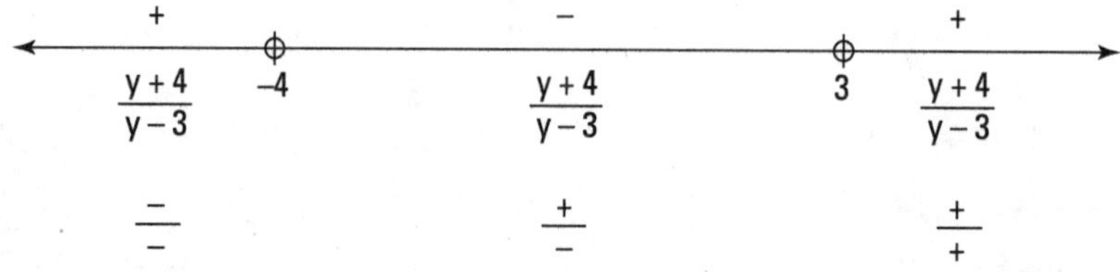

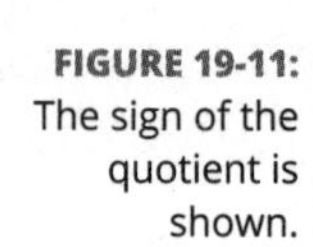

FIGURE 19-11: The sign of the quotient is shown.

Q. Solve for z in $\frac{z^2-1}{z^2-9} \leq 0$.

A. Factor the numerator and denominator to get $\frac{(z+1)(z-1)}{(z+3)(z-3)} \leq 0$. The numbers making the numerator or denominator equal to 0 are $z = +1, -1, +3, -3$. Make a number line that contains the critical numbers and the signs of the intervals (see Figure 19-12).

Because you're looking for values of z that make the expression negative, you want the values between −3 and −1 and those between 1 and 3. Also, you want values that make the expression equal to 0. That can only include the numbers that make the numerator equal to 0, the 1 and −1. The answer is written $-3 < z \leq -1$ or $1 \leq z < 3$.

In interval notation, the solution is written $(-3, -1] \cup [1, 3)$.

Notice that the < symbol is used by the −3 and 3 so those two numbers don't get included in the answer.

+ −3 − −1 + 1 − 3 +

$\frac{(z+1)(z-1)}{(z+3)(z-3)}$: $\frac{(-)(-)}{(-)(-)}$ | $\frac{(-)(-)}{(+)(-)}$ | $\frac{(+)(-)}{(+)(-)}$ | $\frac{(+)(+)}{(+)(-)}$ | $\frac{(+)(+)}{(+)(+)}$

FIGURE 19-12: The 1 and −1 are included in the solution.

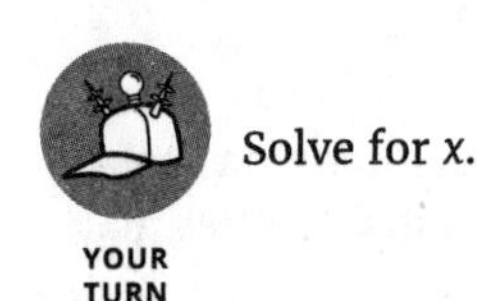

Solve for x.

15 $x(x-1)(x+2) \geq 0$

16 $x^3 - 4x^2 + 4x - 16 \leq 0$

17 $\frac{5+x}{x} > 0$

18 $\frac{x^2-1}{x+3} \leq 0$

Getting Even with Inequalities

Solving Absolute-Value Inequalities

Put the absolute-value function together with an inequality, and you create an *absolute-value inequality.* When solving absolute-value equations (refer to Chapter 18, if you need a refresher), you rewrite the equations without the absolute-value function in them. To solve absolute-value inequalities, you also rewrite the form to create simpler inequality problems — types you already know how to solve. Solve the new problem or problems for the solution to the original.

You need to keep two different situations in mind. (Always assume that the c is a positive number or 0.) In the following list, I show only the rules for $>$ and $<$, but the same holds if you're working with $\leq$ or $\geq$:

- If you have $|ax+b| > c$, change the problem to $ax+b > c$ or $ax+b < -c$ and solve the two inequalities.
- If you have $|ax+b| < c$, change the problem to $-c < ax+b < c$ and solve the one compound inequality.

EXAMPLE

Q. Solve for *x*: $|9x - 5| > 4$.

A. Change the absolute-value inequality to the two separate inequalities $9x - 5 > 4$ or $9x - 5 < -4$. Solving the two inequalities, you get the two solutions $9x > 9$, $x > 1$ or $9x < 1$, $x < \frac{1}{9}$. So numbers bigger than 1 or smaller than $\frac{1}{9}$ satisfy the original statement. The solution is $x < \frac{1}{9}$, $x > 1$ or $\left(-\infty, \frac{1}{9}\right)$, $(1, \infty)$.

Q. Solve for *x*: $|9x - 5| < 4$.

A. This is like the first example, except that the sense has been turned around. Rewrite the inequality as $-4 < 9x - 5 < 4$. Solving it, the solution is $1 < 9x < 9$, $\frac{1}{9} < x < \ 1 < x < 1$. The inequality statement says that all the numbers between $\frac{1}{9}$ and 1 work. Notice that the interval in this answer is what was left out in writing the solution to the problem when the inequality was reversed. The solution is $\frac{1}{9} < x < 1$ or $\left(\frac{1}{9}, 1\right)$.

YOUR TURN

19 Solve for *x*: $|4x - 3| < 5$.

20 Solve for *x*: $|2x + 1| \geq 7$.

21 Solve for *x*: $|x - 3| + 6 \leq 8$.

22 Solve for *x*: $5|7x - 4| + 1 > 6$.

Solving Complex Inequalities

A *complex inequality* — one with more than two *sections* (intervals or expressions sandwiched between inequalities) to it — can just be compound and have variables in the middle, or it can have variables in more than one section where adding and subtracting can't isolate the variable in one place. When this happens, you have to break up the inequality into solvable sections and write the answer in terms of what the different solutions share.

One big advantage that inequalities have over equations is that they can be expanded or strung out into compound statements, and you can do more than one comparison at the same time. Look at this statement: $2<4<7<11<12$. You can create another true statement by pulling out any pair of numbers from the inequality, as long as you write them in the same order. They don't even have to be next to one another. For example: $4<12 \quad 2<11 \quad 2<12$. One thing you can't do, though, is mix up inequalities, going in opposite directions, in the same statement. You can't write $7<12>2$.

The operations on these compound inequality expressions use the same rules as for the linear expressions (refer to the section, "Using the Rules to Work on Inequality Statement," earlier in this chapter). You just extend them to act on each section or part.

- Here's the first statement: $2<4<7<11<12$
- Add 5 to each section: $7<9<12<16<17$
- Multiply each by −1, and reverse the inequality, of course: $-7>-9>-12>-16>-17$

Suppose you want to solve for the values of x in $-3 \le 5x+2<17$. Your steps would be as follows:

1. **The goal is to get the variable alone in the middle. Start by subtracting 2 from each section.**

   $-3-2 \le 5x+2-2<17-2$

   $-5 \le 5x<15$

2. **Now divide each section by 5.**

   The number 5 is positive, so don't turn the inequality signs around.

   $\frac{-5}{5} \le \frac{5x}{5} < \frac{15}{5}$

   $-1 \le x < 3$

   This says that x is greater than or equal to −1, while at the same time, it's less than 3. Some possible solutions are: 0, 1, 2, 2.9.

3. **Check the problem using two of these possibilities.**

   If $x=1$, then $-3<5(1)+2<17$, or $-3<7<17$. That's true.

   If $x=2$, then $-3 \le 5(2)+2<17$, or $-3 \le 12<17$. This also works.

EXAMPLE

Q. Solve for x: $1 \le 4x - 3 < 3x + 7$.

A. Break it up into the two separate problems: $1 \le 4x - 3$ and $4x - 3 < 3x + 7$. Solve the first inequality $1 \le 4x - 3$ to get $1 \le x$. Solve the second inequality $4x - 3 < 3x + 7$ to get $x < 10$. Putting these two answers together, you get that x must be some number both bigger than or equal to 1 and smaller than 10. So, $1 \le x < 10$ or $[1, 10)$.

Q. Solve for x: $2x < 3x + 1 \le 5x - 2$.

A. Break the inequality up into two separate problems. The solution to $2x < 3x + 1$ is $-1 < x$, and the solution to $3x + 1 \le 5x - 2$ is $\frac{3}{2} \le x$. The two solutions overlap, with all the common solutions lying to the right of and including $\frac{3}{2}$. So, $x \ge \frac{3}{2}$ or $\left[\frac{3}{2}, \infty\right)$.

YOUR TURN

23 Solve for x: $6 \le 5x + 1 < 2x + 10$.

24 Solve for x: $-6 \le 4x - 3 < 5x + 1$.

Practice Questions Answers and Explanations

1. **$-3 < -2$.** Starting with $5 > 2$, add 4 to each side and then divide by -3.

$$\begin{aligned} 5 &> 2 \\ 5+4 > 2+4 &\rightarrow 9 > 6 \\ \frac{9}{-3} < \frac{6}{-3} &\rightarrow -3 < -2 \end{aligned}$$

2. **$10 \geq 2$.** Starting with $5 \geq 1$, multiply each side by -4 and then divide each side by -2. The answer is $10 \geq 2$, because you reverse the sense twice.

$$\begin{aligned} 5 &\geq 1 \\ 5(-4) \leq 1(-4) &\rightarrow -20 \leq -4 \\ \frac{-20}{-2} \geq \frac{-4}{-2} &\rightarrow 10 \geq 2 \end{aligned}$$

3. Using inequality notation: **$-5 \leq x < 3$** and in interval notation: **$[-5, 3)$**.

4. Using inequality notation: **$x \leq 6$** and in interval notation: **$(-\infty, 6]$**.

5. **$(7, \infty)$.**

6. **$[2, 21]$.**

7. **$y \geq -3$ or $[-3, \infty)$.** Subtract 4 and then divide by -5:

$$\begin{aligned} 4 - 5y &\leq 19 \\ \underline{-4 \qquad} &\ \underline{-4} \\ -5y &\leq 15 \\ \frac{-5y}{-5} &\geq \frac{15}{-5} \\ y &\geq -3 \end{aligned}$$

8. **$1 > x$ or $(-\infty, 1)$.** Distribute the 3, then subtract $3x$ from each side and 5 from each side.

$$\begin{aligned} 3(x+2) &> 4x+5 \\ 3x+6 &> 4x+5 \\ \underline{-3x-5} &\ \underline{-3x-5} \\ 1 > x &\text{ or } x < 1 \end{aligned}$$

9. **$-4 < x \leq 2$ or $(-4, 2]$.** Subtract 3 from each side and the middle. Then divide by 2.

$$\begin{aligned} -5 < 2x + 3 &\leq 7 \\ \underline{-3 < \quad -3} &\ \underline{-3} \\ -8 < 2x &\leq 4 \\ \frac{-8}{2} < \frac{2x}{2} &\leq \frac{4}{2} \\ -4 < x &\leq 2 \end{aligned}$$

Getting Even with Inequalities

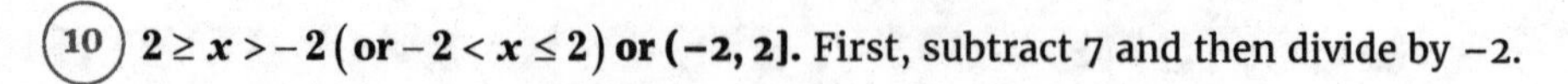

(10) $\mathbf{2 \geq x > -2\left(or -2 < x \leq 2\right) or (-2, 2]}$. First, subtract 7 and then divide by −2.

$$\begin{aligned} -3 \ &\leq 7 - 2x < 11 \\ -7 \ -7 \ & \quad \ \ \ -7 \\ -4 \leq & \quad -2x < 4 \\ \frac{-4}{-2} \geq & \ \frac{-2x}{-2} > \frac{4}{-2} \\ 2 \geq x & > -2 \text{ or } -2 < x \leq 2 \end{aligned}$$

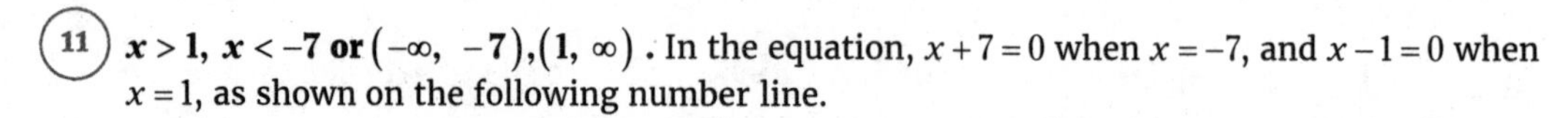

(11) $\mathbf{x > 1,\ x < -7}$ **or** $\mathbf{(-\infty,\ -7), (1,\ \infty)}$. In the equation, $x + 7 = 0$ when $x = -7$, and $x - 1 = 0$ when $x = 1$, as shown on the following number line.

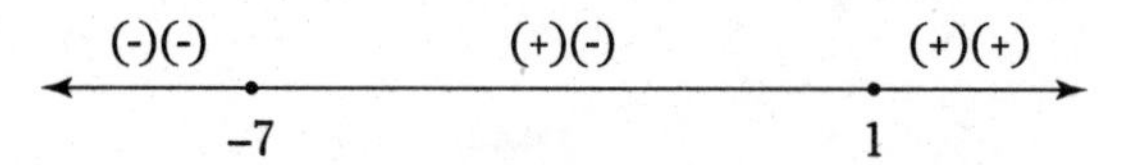

If $x > 1$, then both factors are positive. If $-7 < x < 1$, then $x + 7$ is positive, but $x - 1$ is negative, making the product negative. If $x < -7$, then both factors are negative, and the product is positive. So $x > 1$ or $x < -7$.

(12) $\mathbf{-4 \leq x \leq 5}$ **or** $\mathbf{[-4, 5]}$. First, subtract 20 from both sides; then factor the quadratic and set the factors equal to 0 to find the values where the factors change signs:
$x^2 - x \leq 20 \rightarrow x^2 - x - 20 \leq 0 \rightarrow (x - 5)(x + 4) \leq 0$

$x - 5 = 0$ when $x = 5$, and $x + 4 = 0$ when $x = -4$, as shown on the following number line.

(-)(-) (-)(+) (+)(+)

−4 5

If $x > 5$, then $x - 5 > 0$ and $x + 4 > 0$, making the product positive. If $-4 < x < 5$, then $x - 5 < 0$ and $x + 4 > 0$, resulting in a product that's negative. If $x < -4$, then both factors are negative, and the product is positive. So the product is negative only if $-4 < x < 5$. But the solutions of $(x - 5)(x + 4) = 0$ are $x = 5$, $x = -4$. Including these two values, the solution is then $-4 \leq x \leq 5$.

(13) $\mathbf{x \geq 3,\ x \leq 0}$ **or** $\mathbf{(-\infty,\ 0],\ [3,\ \infty)}$. First, subtract 9x from each side. Then factor the quadratic and set the factors equal to 0: $3x^2 \geq 9x \rightarrow 3x^2 - 9x \geq 0 \rightarrow 3x(x - 3) \geq 0$.

$3x(x - 3) = 0$ when $x = 0$, $x = 3$, as shown on the following number line.

(-)(-) (+)(-) (+)(+)

0 3

If $x > 3$, then both factors are positive, and $3x(x - 3) > 0$. If $0 < x < 3$, then $3x > 0$ and $x - 3 < 0$, making the product negative. If $x < 0$, then both factors are negative, and $3x(x - 3) > 0$. When $x = 0$ or $x - 3$, $3x(x - 3) = 0$. Include those in the solution to get $x \geq 3$ or $x \leq 0$.

(14) $\mathbf{x \geq 5,\ x \leq -5}$ **or** $\mathbf{(-\infty,\ -5],\ [5,\ \infty)}$. First, set the expression equal to 0 to help find the factors: $x^2 - 25 \geq 0 \rightarrow (x - 5)(x + 5) \geq 0$.

The critical numbers are 5 and −5. Place them on the number line.

(-)(-) (-)(+) (+)(+)

−5 5

From the figure, $x^2 - 25 > 0$ when $x > 5$ or $x < -5$. When $x = \pm 5$, $x^2 - 25 = 0$. So $x \geq 5$ or $x \leq -5$.

15 $\boldsymbol{x \geq 1,\ -2 \leq x \leq 0}$ **or** $\boldsymbol{[-2,\ 0],[1,\ \infty)}$**.** First, set the factored expression equal to 0. Put these values on the number line: $x(x-1)(x+2) \geq 0 \rightarrow x(x-1)(x+2)=0$, when $x=0,\ 1,\ -2$.

(-)(-)(-) (-)(-)(+) (+)(-)(+) (+)(+)(+)

−2 0 1

Assign signs to each of the four regions to get the answer $x \geq 1 \text{or} -2 \leq x \leq 0$.

16 $\boldsymbol{x \leq 4}$ **or** $\boldsymbol{(-\infty,\ 4]}$**.** First, factor the expression by grouping. Then set the factors equal to 0.

$$x^3-4x^2+4x-16 \leq 0 \rightarrow x^2(x-4)+4(x-4) \leq 0$$
$$(x^2+4)(x-4) \leq 0$$

The factored form is equal to 0 only if $x=4$. Put this number on the number line.

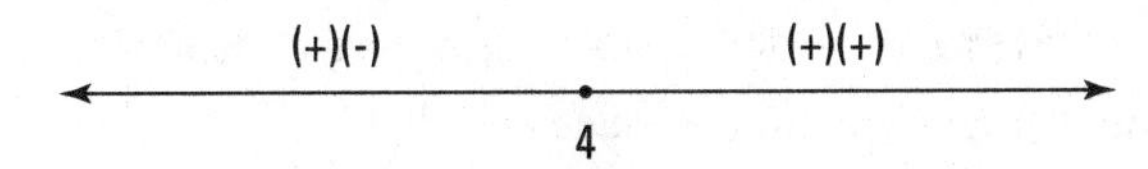

Assign signs to each interval to get the answer $x \leq 4$.

17 $\boldsymbol{x > 0,\ x < -5}$ **or** $\boldsymbol{(-\infty,\ -5),\ (0,\ \infty)}$**.**

In the inequality $\frac{5+x}{x} > 0$, when the numerator $5+x=0$, $x=-5$. When the denominator $x=0$, $\frac{5+x}{x}$ is undefined, so place $x=0$ and $x=-5$ on the number line.

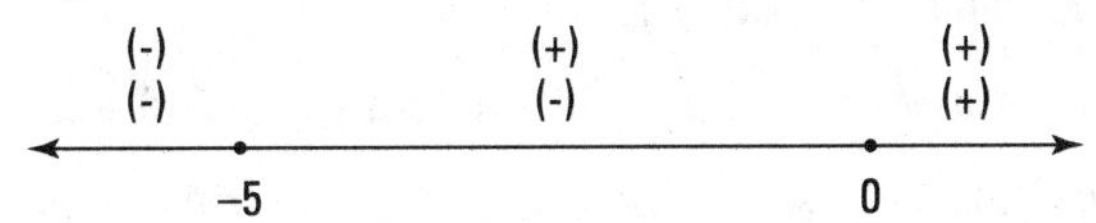

Assign signs to each of the three intervals to get the answer: $x > 0$ or $x < -5$.

18 $\boldsymbol{-1 \leq x \leq 1,\ x < -3}$ **or** $\boldsymbol{(-\infty,\ -3),[-1,\ 1]}$**.**

$$\frac{x^2-1}{x+3} \leq 0 \rightarrow \frac{(x-1)(x+1)}{x+3} \leq 0$$

The fraction equals 0 when $x=1$ or $x=-1$, and it's undefined when $x=-3$.

(-)(-) (-)(-) (-)(+) (+)(+)
(-) (+) (+) (+)

−3 −1 1

Assign signs to each of the four intervals. So $\frac{(x-1)(x+1)}{x+3} < 0$ when $-1 < x < 1$ or $x < -3$. But $\frac{(x-1)(x+1)}{x+3} = 0$ only when $x=1$ or $x=-1$, not at $x=-3$. So the way to write these answers all together with inequality symbols is $-1 \leq x \leq 1$ or $x < -3$.

19 $\boldsymbol{-\frac{1}{2} < x < 2}$ **or** $\boldsymbol{\left(-\frac{1}{2},\ 2\right)}$**.** First, rewrite the absolute value as an inequality. Then use the rules for solving inequalities to isolate *x* in the middle and determine the answer:

$$|4x-3| < 5 \rightarrow -5 < 4x-3 < 5$$

Adding 3 to each section gives you $-2 < 4x < 8$. Then divide each section by 4 to get $-\frac{1}{2} < x < 2$.

Getting Even with Inequalities

20. $x \geq 3, x \leq -4, (-\infty, -4], [3, \infty)$. First, rewrite the absolute value as two inequalities. The x is then isolated to one side of the inequality by adding and dividing.

$|2x+1| \geq 7 \rightarrow 2x+1 \geq 7$ or $2x+1 \leq -7 \rightarrow 2x \geq 6$ or $2x \leq -8$

Solving for x: $x \geq 3$ or $x \leq -4$, by dividing 2 into each side.

21. $1 \leq x \leq 5$ **or** $[1, 5]$. Before rewriting the absolute value, isolate it on one side of the inequality.

Add -6 to each side: $|x-3|+6 \leq 8 \rightarrow |x-3| \leq 2$. Now rewrite the absolute value as an inequality that can be solved. The x gets isolated in the middle, giving you the answer.

$-2 \leq x-3 \leq 2 \rightarrow 1 \leq x \leq 5$ by adding 3 to each section.

22. $x > \frac{5}{7}$ **or** $x < \frac{3}{7}$ **or** $\left(-\infty, \frac{3}{7}\right), \left(\frac{5}{7}, \infty\right)$. Before rewriting the absolute value, it has to be alone, on the left side. First, add -1 to each side: $5|7x-4|+1 > 6 \rightarrow 5|7x-4| > 5$. Then divide each side by 5 to get $|7x-4| > 1$. Now you can rewrite the absolute value as two inequality statements, $7x-4 > 1$ and $7x-4 < -1$. Each statement is solved by performing operations that end up as x greater than or less than some value. Add 4 to each side and divide by 7 to get $x > \frac{5}{7}$ or $x < \frac{3}{7}$.

23. $1 \leq x < 3$ **or** $[1, 3)$. First, separate $6 \leq 5x+1 < 2x+10$ into $6 \leq 5x+1$ and $5x+1 < 2x+10$.

When $6 \leq 5x+1$, subtract 1 from each side to get $5 \leq 5x$ and then divide each side by 5 to get $1 \leq x$.

When $5x+1 < 2x+10$, subtract 1 from each side and subtract $2x$ from each side to get $3x < 9$ and then divide each side by 3 to get $x < 3$. So $1 \leq x$ and $x < 3$, which gives the answer $1 \leq x < 3$. This part of the answer also includes the first part.

24. $-\frac{3}{4} \leq x$ **or** $\left[-\frac{3}{4}, \infty\right)$. First, separate $-6 \leq 4x-3 < 5x+1$ into $-6 \leq 4x-3$ and $4x-3 < 5x+1$.

Solving the first inequality, add 3 to each side to get $-3 \leq 4x$. Then divide each side by 4 to get $-\frac{3}{4} \leq x$. Solving the other inequality, you subtract 1 from each side and subtract $4x$ from each side to get $-4 < x$. This second part of the answer is an overlap of the first — all the answers are already covered in the first inequality.

For x to satisfy both inequalities, the answer is only $-\frac{3}{4} \leq x$ or $x \geq -\frac{3}{4}$.

If you're ready to test your skills a bit more, take the following chapter quiz that incorporates all the chapter topics.

Whaddya Know? Chapter 19 Quiz

Quiz time! Complete each problem to test your knowledge on the various topics covered in this chapter. You can then find the solutions and explanations in the next section.

1. Solve for x. Write your answer in both inequality and interval notation. $\frac{x^2+3x}{x^2-16} \geq 0$
2. Solve for x. Write your answer in both inequality and interval notation. $-3 \leq 4 - x < 6$
3. Write "x is less than or equal to 7" using inequality notation.
4. Write $x > 6$ using interval notation.
5. Write (6, 8) as an inequality involving x.
6. Solve for x. Write your answer in both inequality and interval notation. $3x + 7 > x - 5$
7. Solve for x. Write your answer in both inequality and interval notation. $|3x + 4| < 8$
8. Solve for x. Write your answer in both inequality and interval notation.
   $-10 < 3x - 1 \leq 2x + 7$
9. Write "y is greater than 3 and less than or equal to 51" using inequality notation.
10. Write $x \leq 0$ using interval notation.
11. Solve for x. Write your answer in both inequality and interval notation. $x^2 - 5x \geq 6$
12. Graph $-4 < x \leq 7$.
13. Solve for x. Write your answer in both inequality and interval notation. $|x - 6| > 5$
14. Graph $x \geq -2$.
15. Write $[-5, \infty)$ as an inequality involving x.
16. Solve for x. Write your answer in both inequality and interval notation. $x^3 - 4x^2 < 21x$

Answers to Chapter 19 Quiz

1. $x < -4,\ -3 \le x \le 0,\ x > 4,\ (-\infty,\ -4),\ [-3,\ 0],\ (4,\ \infty)$

$$\frac{x^2+3x}{x^2-16} \ge 0$$
$$\frac{x(x+3)}{(x-4)(x+4)} \ge 0$$

The critical numbers are: $-4,\ -3,\ 0,\ 4$.

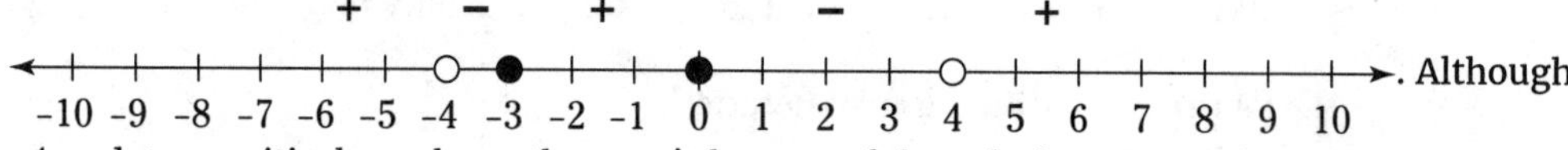

. Although -4 and 4 are critical numbers, they can't be part of the solution. Determine where the quotient is positive or 0.

$x < -4,\ -3 \le x \le 0,\ x > 4$

2. $2 < x \le 7$ or $(2,\ 7]$

$$\begin{array}{l} -3 \le 4 - x < 6 \\ \underline{-4 \quad -4 \quad -4} \\ -7 \le \ -x < 2 \end{array}$$

Now multiply through by -1 and reverse the inequality symbols. Then "flip" the order.

$7 \ge x > -2$ becomes $-2 < x \le 7$.

3. $x \le 7$

4. $(6,\ \infty)$

5. $6 < x < 8$

6. $x > -6$ or $(-6,\ \infty)$

$$\begin{array}{l} 3x + 7 > x - 5 \\ \underline{-x - 7 - x - 7} \\ \dfrac{2x}{2} > \dfrac{-12}{2} \\ x > -6 \end{array}$$

7. $-4 < x < \frac{4}{3}$ or $\left(-4,\ \frac{4}{3}\right)$

Solve the inequality $-8 < 3x + 4 < 8$.

$-8 < 3x + 4 < 8 \ \rightarrow\ -12 < 3x < 4 \ \rightarrow\ -4 < x < \frac{4}{3}$

8 $\mathbf{-3 < x \le 8, (-3,\ 8]}$

Solve the two inequalities $-10 < 3x - 1$ and $3x - 1 \le 2x + 7$.

$-10 < 3x - 1 \;\rightarrow\; -9 < 3x \;\rightarrow\; -3 < x$

$3x - 1 \le 2x + 7 \;\rightarrow\; x \le 8$

The common solution to these two inequalities is: $-3 < x \le 8$.

9 $\mathbf{3 < y \le 51}$

10 $\mathbf{(-\infty,\ 0]}$

11 $\mathbf{x \le -1 \text{ or } x \ge 6, (-\infty,\ -1] \text{ or } [6,\ \infty)}$

$$x^2 - 5x \ge 6$$
$$x^2 - 5x - 6 \ge 0$$
$$(x - 6)(x + 1) \ge 0$$

The critical numbers are 6 and -1.

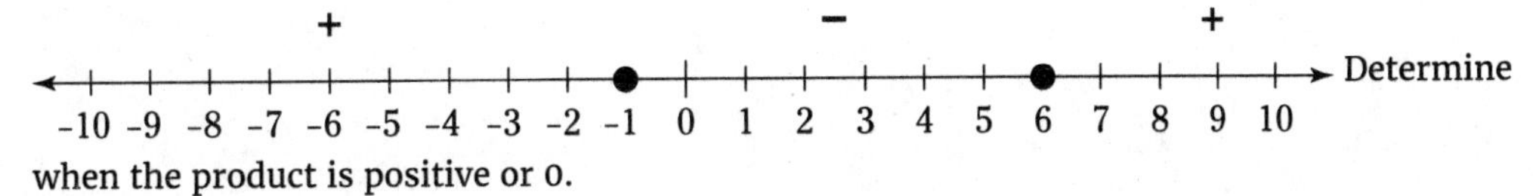

Determine when the product is positive or 0.

$x \le -1$ or $x \ge 6$

12

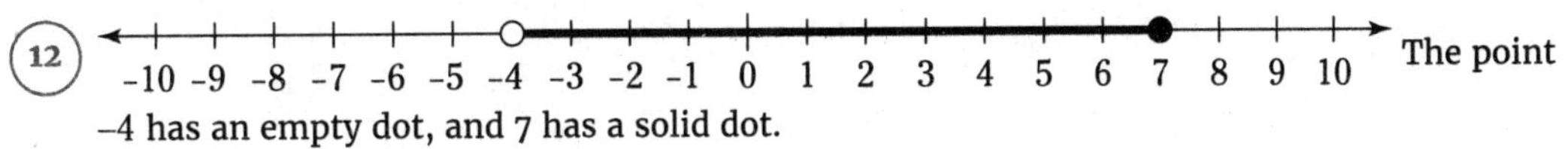

The point -4 has an empty dot, and 7 has a solid dot.

13 $\mathbf{x < 1 \text{ or } x > 11, (-\infty,\ 1) \text{ or } (11,\ \infty)}$

Solve the inequalities $x - 6 > 5$ or $x - 6 < -5$.

$x - 6 > 5 \;\rightarrow\; x > 11$ or $x - 6 < -5 \;\rightarrow\; x < 1$.

14

-10 -9 -8 -7 -6 -5 -4 -3 -2 -1 0 1 2 3 4 5 6 7 8 9 10

The point -2 has a solid dot.

15 $\mathbf{x \ge -5}$

16 $\mathbf{x < -3,\ 0 < x < 7, (-\infty,\ -3) \text{ or } (0,\ 7)}$

$$x^3 - 4x^2 < 21x$$
$$x^3 - 4x^2 - 21x < 0$$
$$x(x - 7)(x + 3) < 0$$

The critical numbers are 0, 7, and -3.

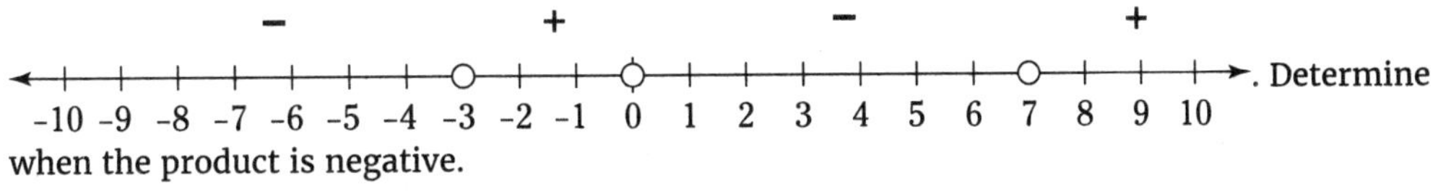

Determine when the product is negative.

$x < -3,\ 0 < x < 7$

7 Evaluating Formulas and Story Problems

Contents at a Glance

IN THIS CHAPTER

» Finding comfort in familiar formulas

» Introducing perimeter, area, and volume formulas

» Computing interest and percentages

» Counting on permutations and combinations

Chapter 20

Facing Up to Formulas

Just as a cook refers to a recipe when preparing delectable concoctions, algebra uses formulas to whip up solutions. In the kitchen, a cook relies on the recipe to turn equal amounts of jalapeños, cream cheese, and beans into a zippy dip. In algebra, a formula is an equation that expresses some relationship you can count on to help you concoct such items as the diagonal distance across a rectangle or the amount of interest paid on a loan.

Formulas such as $I = Prt$, $A = \pi r^2$, and $d = rt$ are much more compact than all the words needed to describe them. And as long as you know what the letters stand for, you can use these formulas to solve problems.

Working with formulas is easy, and you can apply them to so many situations in algebra and in real life. Most formulas become old, familiar friends. Plus, a certain comfort comes from working with formulas, because you know they never change with time, temperature, or relative humidity.

This chapter gives you several chances to work through some of the more common formulas and to tweak areas where you may need a little extra work. The problems are pretty straightforward, and I tell you which formula to use. In Chapter 21, you get to make decisions as to when and whether to use a formula, or whether you get to come up with an equation all on your own.

Working with Formulas

A *formula* is an equation that expresses some known relationship between given quantities. You use formulas to determine how much a dollar is worth when you go to another country. You use a formula to figure out how much paint to buy when redecorating your home.

Sometimes, solving for one of the variables in a formula is advantageous if you have to repeat the same computation over and over again. In each of the examples and practice exercises, you use the same rules from solving equations, so you see familiar processes in familiar formulas. The examples in this section introduce you to various formulas and then show you how to use them to solve problems.

EXAMPLE

Q. Find I if the principal, P, is $10,000, the rate, r is 2%, and the time, t, is 4 years.

A. 800. The simple interest formula is $I = Prt$, where I is the interest earned, P is the principal or what you start with, r is the interest rate written as a decimal, and t is the amount of time in years.

Inserting the numbers into the formula (and changing 2% to the decimal equivalent 0.02), you get $I = 10{,}000(0.02)4 = 800$.

The answer is $800.

Q. What is the temperature in degrees Fahrenheit when it's 25°C?

A. 77. The formula for changing from degrees Celsius, °C, to degrees Fahrenheit, °F, is $F = \frac{9}{5}C + 32$. Replace C with 25, multiply by $\frac{9}{5}$, and then add 32.

$$F = \frac{9}{\cancel{5}}\left(\cancel{25}^{5}\right) + 32 = 45 + 32 = 77$$

The answer is 77°F.

YOUR TURN

1 You are buying a new computer, and the store will allow you to pay for it over the next 12 months for an interest charge of 15%. Use the simple interest formula, $I = Prt$, to determine how much you'll pay in interest (which will be added to the price of the computer). The computer costs $1,200 (this is P); the interest rate, r, is 15%, and the time is one year.

2 What is the temperature in degrees Fahrenheit when your thermometer reads 100°C? Use the formula: $F = \frac{9}{5}C + 32$.

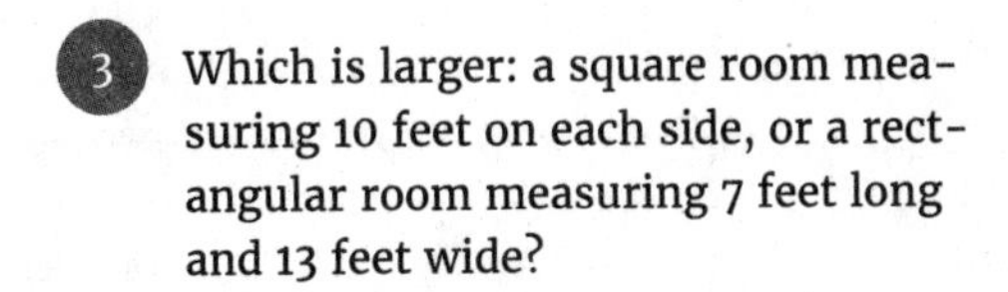

3. Which is larger: a square room measuring 10 feet on each side, or a rectangular room measuring 7 feet long and 13 feet wide?

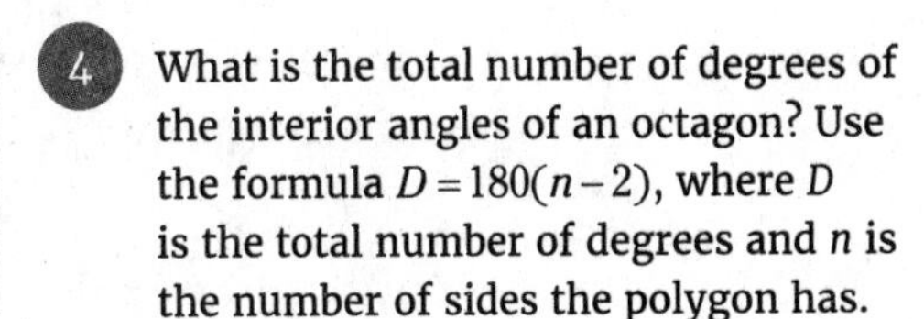

4. What is the total number of degrees of the interior angles of an octagon? Use the formula $D = 180(n-2)$, where D is the total number of degrees and n is the number of sides the polygon has.

Measuring Up

Some universal concerns — some start at an early age — are those dealing with measurements. How far is it? How big are you? How much room do you need, anyway? How much more wrapping paper are you going to need? These questions all have to do with measurements and, usually, formulas.

Finding out how long: Units of length

Before measurements were standardized, they varied according to who was doing the measuring: A yard was the distance from the tip of the nose to the end of an outstretched arm; a foot, well, you can probably guess where that came from; and an inch was often the length of the second bone in the index finger. When measuring fabric to purchase for their shop, a tailor would let their tall brother-in-law with the long arms do the measuring. When selling planks in their lumberyard, a business owner would let Cousin Vinnie, the little guy, be the measurement employee.

The units of measure for length most commonly used in the United States are inches, feet, yards, and miles. Some equivalent measures are $12 \text{ inches} = 1 \text{ foot}$, $3 \text{ feet} = 1 \text{ yard}$, and $5{,}280 \text{ feet} = 1 \text{ mile}$.

You can change the basic length equivalencies into formulas as follows:

- **Feet to inches:** Number of inches = number of feet $\times$ 12
- **Inches to feet:** Number of feet = number of inches $\div$ 12

» **Yards to feet:** Number of feet = number of yards × 3

» **Miles to feet:** Number of feet = number of miles × 5,280

The best way to deal with these and other measures is to write a proportion. (To review the properties of proportions, see Chapter 15.)

When using a proportion to solve a measurement problem, write same units over same units or same units across from same units.

EXAMPLE

Q. How many inches in 8 feet?

A. 96 inches. You know that 12 inches = 1 foot. So, put inches over inches and feet over feet:

$$\frac{12 \text{ inches}}{x \text{ inches}} = \frac{1 \text{ foot}}{8 \text{ feet}}$$

The values in the known relationship are across from one another. The unknown is represented by x. Now cross-multiply:

$$12 \times 8 = x \times 1$$
$$96 \text{ inches} = x$$

Eight feet is the same as 96 inches.

Q. You're in a plane, and the pilot says that you're cruising at 14,000 feet. How high is that in miles?

A. $2\frac{43}{66}$. You know that 5,280 feet = 1 mile, so

$$\frac{5{,}280 \text{ feet}}{14{,}000 \text{ feet}} = \frac{1 \text{ mile}}{x \text{ miles}}$$

Cross-multiply:

$$5{,}280 \times x = 14{,}000 \times 1$$
$$5{,}280x = 14{,}000$$

Divide each side of the equation by 5,280: $x = \frac{14{,}000}{5{,}280} = 2\frac{3{,}400}{5{,}280}$ miles ≈ 2.65 miles up in the air.

A smart way to handle this problem is to do some reducing of numbers before multiplying. In Chapter 15, you're shown how to reduce horizontally and vertically. In this case, you can find common factors for the 5,280 and 14,000. The most obvious is that they are both divisible by 10.

$$\frac{5{,}28\cancel{0} \text{ feet}}{14{,}00\cancel{0} \text{ feet}} = \frac{1 \text{ mile}}{x \text{ miles}} \rightarrow \frac{528 \text{ feet}}{1{,}400 \text{ feet}} = \frac{1 \text{ mile}}{x \text{ miles}}$$

- **Triangle:** $P = a + b + c$, where a, b, and c are the sides
- ***n*-sided Polygon:** $P = a_1 + a_2 + a_3 + a_4 + \quad + a_n$, where each a_i is a side of the polygon

EXAMPLE

Q. If you know that the perimeter of a particular rectangle is 20 yards and that the length is 8 yards, then what is the width?

A. 2 yards. You can find a rectangle's perimeter by using the formula $P = 2(l + w)$, where l and w are the length and width of the rectangle. Substitute what you know into the formula and solve for the unknown. In this case, you know P and l. The formula now reads $20 = 2(8 + w)$. Divide each side of the equation by 2 to get $10 = 8 + w$. Subtract 8 from each side, and you get the width, w, of 2 yards.

Q. An isosceles triangle has a perimeter of 40 yards and two equal sides, each 5 yards longer than the base. How long is the base?

A. The base is 10 yards long. First, you can write the triangle's perimeter as $P = 2s + b$. The two equal sides, s, are 5 yards longer than the base, b, which means you can write the lengths of the sides as $b + 5$. Putting $b + 5$ in for the s in the formula and putting the 40 in for P, the problem now involves solving the equation $40 = 2(b + 5) + b$. Distribute the 2 to get $40 = 2b + 10 + b$. Simplify on the right to get $40 = 3b + 10$. Subtracting 10 from each side gives you $30 = 3b$. Dividing by 3, you get $10 = b$. So the base is 10 yards. The two equal sides are then 15 yards each. If you add the two 15-yard sides to the 10-yard base, you get (drum roll, please) $15 + 15 + 10 = 40$, the perimeter.

YOUR TURN

12 If a rectangle has a length that's 3 inches greater than twice the width, and if the perimeter of the rectangle is 36 inches, then what is its length?

13 You have 400 feet of fencing to fence in a rectangular yard. If the yard is 30 feet wide and you're going to use all 400 feet to fence in the yard, then how long is the yard?

A square and an *equilateral* triangle (all three sides equal in length) have sides that are the same length. If the sum of their perimeters is 84 feet, then what is the perimeter of the square?

A triangle has one side that's twice as long as the shortest side, and a third side that's 8 inches longer than the shortest side. Its perimeter is 60 inches. What are the lengths of the three sides?

Squaring off with area formulas

You measure the area of a figure in square inches, square feet, square yards, and so on. Some of the more commonly found figures, such as rectangles, circles, and triangles, have standard area formulas. Obscure figures even have formulas, but they aren't used very often, especially in an algebra class. Here are the area formulas for some basic figures.

- **Rectangle:** $A = lw$, where *l* and *w* represent the length and width
- **Square:** $A = s^2$, where *s* represents the length of a side
- **Circle:** $A = \pi r^2$, where *r* is the radius
- **Triangle:** $A = \frac{1}{2}bh$, where *b* is the base and *h* is the height
- **Trapezoid:** $A = \frac{1}{2}h(b_1 + b_2)$, where *h* is the height and b_1 and b_2 are the parallel bases

EXAMPLE

Q. Find the area of a circular field with a circumference of 1,256 feet.

A. 125,600. You're told that the distance around the outside *(circumference)* of a circular field is 1,256 feet. The formula for the circumference of a circle is $C = \pi d = 2\pi r$, which says that the circumference is π (about 3.14) times the diameter, or two times π times the radius. To find the area of a circle, you need the formula $A = \pi r^2$. So, to find the area of this circular field, you first find the radius by putting the 1,256 feet into the circumference formula: $1{,}256 = 2\pi r$. Replace the π with 3.14 and solve for *r*:

$$\frac{1{,}256}{2\pi} = \frac{\cancel{2\pi} r}{\cancel{2\pi}}$$

$$r = \frac{1{,}256}{2\pi} \approx \frac{1{,}256}{2(3.14)} = \frac{1{,}256}{6.28} = 200$$

The radius is 200 feet. Putting that into the area formula, you get that the area is 125,600 square feet.

$$A = \pi r^2 \approx 3.14(200)^2 = 3.14(40{,}000) = 125{,}600$$

Q. A builder is designing a house with a square room. If they increase the sides of the room by 8 feet, the area increases by 224 square feet. What are the dimensions of the expanded room?

A. You can find the area of a square with $A = s^2$, where s is the length of the sides. Start by letting the original room have sides measuring s feet. Its area is $A = s^2$. The larger room has sides that measure $s + 8$ feet. Its area is $A = (s+8)^2$. The difference between these two areas is 224 feet, so subtract the smaller area from the larger area and write the equation showing the 224 as a difference: $(s+8)^2 - s^2 = 224$. Simplify the left side of the equation: $s^2 + 16s + 64 - s^2 = 16s + 64 = 224$. Subtract 64 from each side and then divide by 16: $16s = 160 \rightarrow \frac{\cancel{16}s}{\cancel{16}} = \frac{160}{16} = 10$.

The original room has walls measuring 10 feet. Eight feet more than that is 18 feet.

YOUR TURN

16 If a rectangle is 4 inches longer than it is wide and the area is 60 square inches, then what are the dimensions of the rectangle?

17 You can find the area of a trapezoid with $A = \frac{1}{2}h(b_1 + b_2)$. Determine the length of the base b_1 if the trapezoid has an area of 30 square yards, a height of 5 yards, and a base b_2 of 3 yards.

18 The perimeter of a square is 40 feet. What is its area? (***Remember:*** $P = 4s$ and $A = s^2$.)

19 You can find the area of a triangle with $A = \frac{1}{2}bh$, where the base and the height are perpendicular to one another. If a right triangle has legs measuring 10 inches and 24 inches and a hypotenuse of 26 inches, what is its area?

Soaring with Heron's formula

Heron of Alexandria is credited with developing a formula for the area of a triangle using the measures of the three sides, rather than a base and height of the triangle. He proved that the formula worked in his book, written in 60 A.D. It's a great help for those who are working with large triangular areas, such as playgrounds or building sites, and can't construct the perpendicular to a side of the triangle to measure a height.

According to Heron's formula, the area of any triangle is equal to the square root of the product of four values:

- The semi-perimeter (half the perimeter)
- The semi-perimeter minus the length of the first side
- The semi-perimeter minus the second side
- The semi-perimeter minus the third side

Let s represent the semi-perimeter and a, b, and c represent the measures of the sides:

$$A = \sqrt{s(s-a)(s-b)(s-c)}$$

When you're trying to find the area of a huge triangle — say, a big park — or if you can't measure any angles to draw a line perpendicular to one of the sides for the height, then you can find the area simply by measuring the three sides and using Heron's formula.

EXAMPLE

Q. Find the area of a triangle with sides of 10 inches, 17 inches, and 21 inches.

A. 84 square inches. Let $a = 10$, $b = 17$, and $c = 21$. The perimeter is $P = 10 + 17 + 21 = 48$ inches, so the semi-perimeter $s = 24$ inches. Using Heron's formula to find the area:

$$A = \sqrt{s(s-a)(s-b)(s-c)} = \sqrt{24(24-10)(24-17)(24-21)}$$
$$= \sqrt{24(14)(7)(3)} = \sqrt{7{,}056} = 84$$

The area is 84 square inches. I have to admit that I purposely used measurements that would give a nice, whole-number answer. These nice answers are more the exception than the rule. Having a radical in a formula can cause all sorts of complications. The next example shows you what I mean.

Q. Find the area of a triangle with sides 2, 3, and 4 feet.

A. 2.905 square feet. If the sides are 2, 3, and 4, then $a = 2, b = 3, c = 4$, and $s = \frac{1}{2}(2+3+4) = 4.5$. So, using Heron's formula to find the area:

$$A = \sqrt{s(s-a)(s-b)(s-c)} = \sqrt{4.5(4.5-2)(4.5-3)(4.5-4)} =$$
$$= \sqrt{4.5(2.5)(1.5)(.5)} = \sqrt{8.4375} \approx 2.905 \text{ square feet}$$

YOUR TURN

20 Find the area of a triangle whose sides measure 20, 48, and 52 inches.

21 A contractor needs to find the area of a triangular piece of land that measures 1,300 feet, 1,400 feet, and 1,500 feet on its three sides. What is that area?

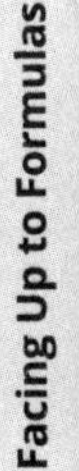

Working with volume formulas

The *volume* of an object is a three-dimensional measurement. In a way, you're asking, "How many little cubes can I fit into this object?" Cubes won't fit into spheres, pyramids, or other structures with slants and curves, so you have to accept that some of these little cubes are getting shaved off or cut into pieces to fit. Having a formula is much easier than actually sitting and trying to fit all those little cubes into an often large or unwieldy object. Here are some important volume formulas.

- **Box (rectangular prism):** $V = lwh$
- **Sphere:** $V = \frac{4}{3}\pi r^3$
- **Cylinder:** $V = \pi r^2 h$
- **Pyramid:** $V = \frac{1}{3}(\text{area of base}) \cdot h$
- **Cone:** $V = \frac{1}{3}\pi r^2 h$

EXAMPLE

Q. Find the volume of an orange traffic cone that's 30 inches tall and has a diameter of 18 inches.

A. Over 2,500 cubic inches. A right circular cone (that's what those traffic cones outlining a construction area look like) has a volume you can find if you know its radius and its height.

The formula is $V = \frac{1}{3}\pi r^2 h$. As you can see, the multiplier π is in this formula because the base is a circle. Use 3.14 as an estimate of π; because the diameter is 18 inches, use 9 inches for the radius. To find this cone's volume, put those dimensions into the formula to get $V = \frac{1}{3}(3.14)(9)^2(30) = 2{,}543.4$. The cone's volume is over 2,500 cubic inches.

Q. Find the original volume of the Great Pyramid, which originally had a square base with each side measuring 756 feet and a height of 480 feet.

A. The base is a square, so the area of the base is s^2: $V = \frac{1}{3}s^2 \cdot h = \frac{1}{3}(756)^2 \cdot 480 = 91{,}445{,}760$ cubic feet.

YOUR TURN

22 You can find the volume of a box (right rectangular prism) with $V = lwh$. Find the height of the box if the volume is 200 cubic feet and the square base is 5 feet on each side (length and width are each 5).

23 The volume of a sphere (ball) is $V = \frac{4}{3}\pi r^3$, where r is the *radius* of the sphere — the measure from the center to the outside. What is the volume of a sphere with a radius of 6 inches?

24 You can find the volume of a right circular cylinder (soda-pop can) with $V = \pi r^2 h$, where r is the radius, and h is the height of the cylinder — the distance between the two circular bases (the top and bottom of the can). Which has the greater volume: a cylinder with a radius of 6 cm and a height of 9 cm, or a cylinder with a radius of 9 cm and a height of 4 cm?

25 The volume of a cube is 216 cubic centimeters. What is the new volume if you double the length of each side?

Getting Interested in Using Percent

Percentages are a part of modern vocabulary. You probably hear or say one of these sentences every day:

- The chance of rain is 40 percent.
- There was a 2 percent rise in the Dow Jones Industrial Average.
- The grade on your test is 99 percent.
- Your height puts you in the 80th percentile.

Percent is one way of expressing fractions as equivalent fractions with a denominator of 100. The percent is what comes from the numerator of the fraction — how many out of 100:

- $80 \text{ percent} = \frac{80}{100} = 0.80$
- $16\frac{1}{2} \text{ percent} = \frac{16.5}{100} = 0.165$
- $2 \text{ percent} = \frac{2}{100} = 0.02$

You use percent and percentages in the formulas that follow. Change the percentages to decimals so that they're easier to multiply and divide. To change from percent to decimal, you move the decimal point in the percent two places to the left. If no decimal point is showing, assume it's to the right of the number.

Percentages are a form of leveling the playing field. They're great for comparing ratios of numbers that have different bases. For example, if you want to compare the fact that 45 men out of 80 bought a Kindle with the fact that 33 women out of 60 bought a Kindle, you can change both of these to percentages to determine who is more likely to buy a Kindle. (In this case, it's $56\frac{1}{4}\%$ men and 55% women.)

To change a ratio or fraction to a percent, divide the part by the whole (numerator by denominator) and multiply by 100. For instance, in the case of the Kindles, you divide 45 by 80 and get 0.5625. Multiplying that by 100, you get 56.25, which you can write as $56\frac{1}{4}\%$.

Percent also shows up in interest formulas because you earn interest on an investment or pay interest on a loan based on a percentage of the initial amount. Both the simple interest formula and the compound interest formula use P for the principal or beginning amount of money and r for the rate of interest, written as a decimal, to do the computation.

Compounding interest formulas

Figuring out how much interest you have to pay, or how much you're earning in interest, is simple with the formulas in this section. You probably want to dig out a calculator, though, to compute compound interest.

Figuring simple interest

Simple interest is used to determine the amount of money earned in interest when you're not using compounding. It's also used to figure the total amount to pay back when buying something on time. Simple interest is basically a percentage of the original amount. It's figured on the beginning amount only — not on any changing total amount that can occur as an investment grows. To take advantage of the growth in an account, use compound interest.

The amount of simple interest earned is equal to the amount of the principal, P (the starting amount), times the rate of interest, r (which is written as a decimal), times the amount of time, t, involved (usually in years). The formula to calculate simple interest is $I = Prt$.

EXAMPLE

Q. What is the amount of simple interest on $10,000 when the interest rate is $2\frac{1}{2}$ percent and the time period is $3\frac{1}{2}$ years?

A. The interest is $875. Insert the numbers in the formula, changing the percent to a decimal.

$$I = Prt$$
$$I = 10{,}000 \times 0.025 \times 3.5 = 875$$

Q. You're going to buy a television "on time." The appliance store will charge you 12 percent simple interest. You add this onto the price of the television and pay back the total amount in "24 easy monthly payments." Twenty-four months is two years, so $t = 2$. The television costs $600. How much is the interest?

A. $31.00. Using the simple interest formula, $I = 600 \times 0.12 \times 2 = 144$

The interest is $144. Now, to compute the "easy payments," add the interest onto the cost of the television and the total is $744. Divide this by 24, and the payments are $\frac{744}{24} = 31$. That's $31 per payment. Such a deal!

Tallying compound interest

Compound interest is used when determining the total amount that you have in your savings account after a certain amount of time. Compound interest gets its name because the interest earned is added to the beginning amount before the next interest is figured on the new total. The amount of times per year the interest is compounded depends on your account, but many savings accounts compound quarterly, or four times per year.

By leaving the earned interest in your account, you're actually earning more money because the interest is figured on the new, bigger sum.

The formula for compound interest is $A = P\left(1 + \frac{r}{n}\right)^{nt}$, where A is the total amount in the account, P is the principal (starting amount), r is the percentage rate (written as a decimal), n is the number of times it's compounded each year, and t is the number of years. Whew!

The following examples show you how the formula works.

EXAMPLE

Q. How much is there in an account that started with \$5,000 and has been earning interest for the last 14 years at the rate of 6 percent, compounded quarterly?

A. \$9,200. Using the compound interest formula, the principal is \$5,000; the rate is 6 percent, or 0.06; the number of times per year it's compounded is 4; the time in years is 14. So, carefully work from the inside out, starting with $A = 5{,}000\left(1+\frac{0.06}{4}\right)^{4 \cdot 14}$, divide the 0.06 by 4 and add it to the 1. At the same time, multiply the 4 and 14 in the exponent to make it simpler: $A = 5{,}000(1.015)^{56}$

By the order of operations, raise to the power first and then multiply the result by 5,000:

$$A = 5{,}000(2.30196) = 11{,}509.82$$

The amount of money more than doubled. Compare this to the same amount of money earning simple interest. Using the simple interest formula:

$$I = Prt = 5{,}000 \times 0.06 \times 14 = \$4{,}200$$

Add this interest onto the original \$5,000, and the total is \$9,200. Using compound interest earns you over \$2,500 in additional revenue.

Q. Here's an even more dramatic example of the power of compounding: Suppose that you get a letter from the Bank of the West Indies, which claims that some ancestor of yours came over with Columbus, deposited a coin equivalent to \$1 with the bank, and then was lost at sea on the way home. Their dollar's worth of deposit has been sitting in the bank, earning interest at the rate of $3\frac{1}{2}$ percent compounded quarterly. The bank claims that your ancestor's account is becoming a nuisance account because fees have to be collected; the bank wants to charge this account the current fee rate of \$25 per year — retroactively. Do you want to claim this account? Pay the fees?

A. At first, you may say, "No way! I'd owe money." Then you get out your trusty calculator and do some figuring. If your ancestor came over with Columbus in 1492, and if you got the letter in the year 2021, what exactly are you looking at?

The principal is \$1; the interest rate is $3\frac{1}{2}$ percent compounded four times per year. This money has been deposited for 529 years, but that means 529 years of \$25 service charges:

$$A = 1\left(1+\frac{0.035}{4}\right)^{4(529)} = 1(1.00875)^{2116} \approx 101{,}388{,}102.90$$

That's over \$101 million for an initial deposit of \$1.

Subtracting the service charges: $25 \times 529 = 13{,}225$

Paying \$13,000 is minor. Take the money!

If 60% of the class has the flu and that 60% is 21 people, then how many are in the class?

How much simple interest will you earn on $4,000 invested at 3% for 10 years? What is the total amount of money at the end of the 10 years?

How much money will be in an account that started with $4,000 and earned 3% compounded quarterly for 10 years?

If you earned $500 in simple interest on an investment that was deposited at 2% interest for 5 years, how much had you invested?

Gauging taxes and discounts

You can figure both the tax charged on an item you're buying and the discount price of sale items with percentages.

- **Total price** = price of item × (1 + tax percent as a decimal)
- **Discounted price** = original price × (1 − discount percent as a decimal)
- **Original price** = discount price ÷ (1 − discount percent as a decimal)

All consumers are faced with taxes on purchases and hope to find situations in which they can buy things on sale. It pays to be a wise consumer.

EXAMPLE

Q. The $24,000 car you want is being discounted by 8 percent. How much will it cost now with the discount? (Be sure to add the 5 percent sales tax.)

A. First, compute the discounted price and then the total price with the sales tax.

$$\begin{aligned} \text{discounted price} &= 24{,}000 \times (1-0.08) = 24{,}000 \times 0.92 = \$22{,}080 \\ \text{total price} &= \text{cost of item} \times (1 + \text{tax percent as a decimal}) \\ \text{total price} &= 22{,}080 \times (1+0.05) = \$23{,}184 \end{aligned}$$

Q. The shoes you're looking at were discounted by 40 percent and then that new price was discounted another 15 percent. What did they cost, originally, if you can buy them for $68 now?

A. If the price now was discounted 15 percent, find the amount they were discounted from first (the first discount price). Solving the discounted price formula for the original price:

$$\text{original price} = \frac{\text{discount price}}{1 - \text{percent discount as decimal}}$$

$$\text{“second discounted price”} = \frac{68}{(1-0.15)} = \frac{68}{.85} = \$80$$

$$\text{“first discounted price”} = \frac{80}{(1-0.40)} = \frac{80}{.60} = \$133.33$$

The discount of 40 percent followed by 15 percent is not the same as a discount of 55 percent. A 55 percent discount would have resulted in $60 shoes.

YOUR TURN

30 You're buying a $40 shirt and bring it to the cashier. There's an 8.25% sales tax charge. What is the total amount you'll be paying for the shirt?

31 You bought a new sofa for $1,200 when it was on sale for 25% off the original price. What was the original price?

32 You have a coupon for 10% off a purchase at a particular store. When you get there to make the purchase, you see that they're having a flash sale of 30% off all items. They'll also accept your coupon. What will you pay for the $80 appliance after the discounts have been applied?

33 Which is the better deal? You've found the same shoes at two different stores. Store A has them listed at $96 with a 16% discount, and Store B has them for $89 with a 10% discount.

Working out the Combinations and Permutations

Combinations and permutations are methods and formulas for counting things. You may think that you have that "counting stuff" mastered already, but do you really want to count the number of ways in the following?

- How many different vacations can you take if you plan to go to three different states on your next trip?
- How many different ways can you rearrange the letters in the word *smart* — and how many of the arrangements actually make words?
- How many different ways can you pick 6 numbers out of 54, and can you bet $1 on each set of 6 numbers to win the lottery?

You could start making lists of the different ways to accomplish the preceding problems, but you'd quickly get overwhelmed and perhaps a little bored. Algebra comes to the rescue with some counting formulas called combinations and permutations.

Counting down to factorials

The main operation in combinations and permutations is the factorial operation. This is really a neat operation that only takes one number to perform. The symbol that tells you to perform the operation is an exclamation point (!). When I write, "6!", I don't mean, "Six, wow!" Well, I suppose I might say that if my dog had six puppies. But, in a math context, the exclamation point has a specific meaning:

$$6! = 6 \times 5 \times 4 \times 3 \times 2 \times 1 = 720$$
$$4! = 4 \times 3 \times 2 \times 1 = 24$$

The factorial of any whole number is the product you get by multiplying that whole number by every counting number smaller than it: $n! = n(n-1)(n-2)(n-3)\ldots3 \cdot 2 \cdot 1$

REMEMBER

The counting numbers are 1, 2, 3, 4, and so on.

Factorial works when n is a whole number; that means that you can use numbers such as 0, 1, 2, 3, 4, and so on.

One surprise, though, is the value of 0!. Try it on a calculator. You get $0! = 1$. The value of 0! doesn't really fit the formula; 0! was "declared" to be a 1 so that the formulas that use it would work.

Counting on combinations

Combinations tell you how many different ways you can choose some of the items from an entire group of items; you can choose anywhere from one item to all the items in the group. For example, you can:

- Figure out how many different ways to choose 3 states out of 50 to visit (if you don't care in what order you visit them).
- Figure out how many ways there are to choose 6 numbers out of a possible 54 numbers.
- Figure out how to choose 8 astronauts for the flight out of a group of 40 candidates.

Combinations don't tell you what is in each of these selections, but they tell you how many ways there are. If you're making a listing, you know when to stop if you know how many should be in the list.

The number of combinations of r items taken from a total possible of n items is

$$_nC_r = \frac{n!}{r!(n-r)!}$$

The subscripts on the C tell you two things:

- To the left, the n indicates how many items are available altogether.
- To the right, the r tells how many are to be chosen from all those available.

12 **13 inches.** Use this figure to help you solve this problem.

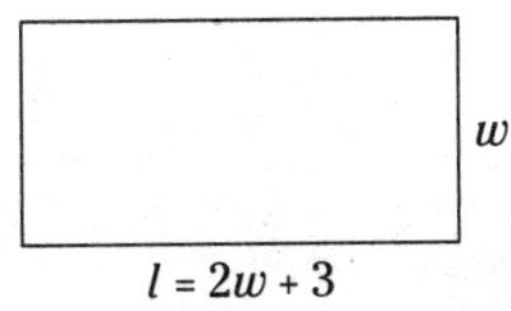

Let w = the width of the rectangle, which makes the length, $l = 3 + 2w$. A rectangle's perimeter is $P = 2(l + w)$. Substituting in $3 + 2w$ for the l in this formula and replacing P with 36, you get $36 = 2(3 + 2w + w)$. Simplifying, you get $36 = 2(3w + 3)$. Now divide each side of the equation by 2 to get $18 = 3w + 3$. Subtract 3 from each side: $15 = 3w$. Divide each side by 3: $5 = w$. The length is $3 + 2w = 3 + 2(5) = 13$. The rectangle is 5 inches by 13 inches.

13 **170 feet.**

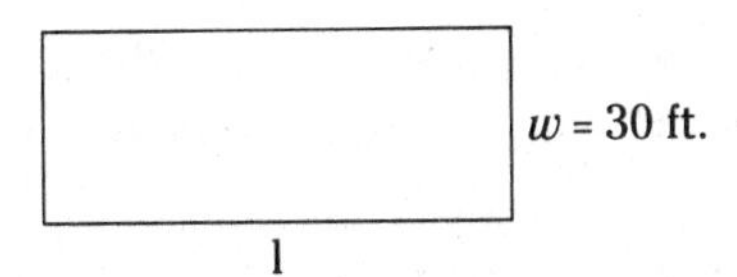

You know that the total distance around the yard (its perimeter) is 400 and the width of the yard is 30, so plug those values into the formula $P = 2(l + w)$ to get $400 = 2(l + 30)$. Divide by 2 to get $200 = l + 30$ and then subtract 30 from each side to get $200 - 30 = l$. So $l = 170$ feet.

14 **48 feet.** The perimeter of the square is $4l$, and the perimeter of the equilateral triangle is $3l$. Adding these together, you get $4l + 3l = 7l = 84$ feet. Dividing each side of $7l = 84$ by 7 gives you $l = 12$ feet. So the perimeter of the square is $4(12) = 48$ feet.

15 **13 inches, 26 inches, 21 inches.** This problem doesn't need a special perimeter formula. The perimeter of a triangle is just the sum of the measures of the three sides. Letting the shortest side be x, twice that is 2x, and 8 more than that is $x + 8$. Add the three measures together to get $x + 2x + x + 8 = 60$, which works out to be $4x + 8 = 60$, or $4x = 52$. Dividing by 4 gives you $x = 13$. Twice that is 26, and eight more than that is 21.

16 **6 inches by 10 inches.** Using the area formula for a rectangle, $A = lw$, and letting $l = w + 4$, you get the equation $60 = (w + 4)w$. Simplifying, setting the equation equal to 0, and then factoring gives you $(w + 10)(w - 6) = 0$. The two solutions of the equation are −10 and 6. Only the +6 makes sense, so the width is 6, and the length is 4 more than that, or 10.

17 **9 yards.** Plug in the values to get $30 = \frac{1}{2}(5)(b_1 + 3)$ and then multiply each side by two: $60 = (5)(b_1 + 3)$. Then divide each side by 5 to get $12 = b_1 + 3$. Do the math to find that $12 - 3 = b_1$, so $b_1 = 9$ yards.

18 **100 square feet.** Dividing each side of $40 = 4s$ by 4, you get $10 = s$. Now find the area with $A = s^2$. Yes, 100 square feet is correct.

19 **120 square inches.** In a right triangle, the hypotenuse is always the longest side, and the two legs are perpendicular to one another. Use the measures of the two legs for your base and height, so $A = \frac{1}{2}(10)(24) = 120$.

20 **480 square inches.** Using Heron's formula, you first find the semi-perimeter. The perimeter of the triangle is $20 + 48 + 52 = 120$ inches.. The semi-perimeter is half of 120, or 60 inches. Putting the semi-perimeter and side measures in the formula, you have

$A = \sqrt{60(60-20)(60-48)(60-52)} = \sqrt{60(40)(12)(8)}$. Multiplying under the radical, $A = \sqrt{230{,}400} = 480$ square inches.

21. **840,000 square feet.** Using Heron's formula, you first find the semi-perimeter. The perimeter of the triangle is $1{,}300 + 1{,}400 + 1{,}500 = 4{,}200$. Half that is 2,100. Putting the semi-perimeter and side measures in the formula, you have:

$$A = \sqrt{2{,}100(2{,}100-1{,}300)(2{,}100-1{,}400)(2{,}100-1{,}500)} = \sqrt{2{,}100(800)(700)(600)}$$

Multiplying under the radical, $A = \sqrt{705{,}600{,}000{,}000} = 840{,}000$ square feet.

22. **8 feet.** Replace the V with 200 and replace the l and w each with 5. Dividing each side of the equation by 25, you get $200 = (5)(5)h = 25h \rightarrow \frac{200}{25} = \frac{25h}{25}$. So the height is 8 feet.

23. **904.32 cubic inches.** Substituting in the 6 for r, you get $V = \frac{4}{3}\pi(6)^3 = 288\pi \approx 904.32$.

24. **Neither; the volumes are the same.** The first volume is $V = \pi(6)^2 \cdot 9 = 324\pi$. The second volume is $V = \pi(9)^2 \cdot 4 = 324\pi$. So they have the same volume, namely 324π or about 1,017.36 cubic centimeters.

25. **1,728 cm³.** First, find the lengths of the edges of the original cube. If s is the length of the edge of the cube, then use the formula for the volume of a cube $\left(V = s^3\right)$.

$$216 = s^3 \rightarrow c = \sqrt[3]{216} = 6$$

Doubling the length of the edge for the new cube results in sides of $2(6) = 12$ centimeters. So the volume of the new cube is $V = 12^3 = 1{,}728 \text{ cm}^3$.

26. **35 people.** Let x = the number of people in the class. Then 60% of x is 21, which is written $0.60x = 21$. Divide each side by 0.60, and you get that $x = 35$ people.

27. **\$1,200** and **\$5,200.** Use the formula $I = Prt$, where the principal (P) is 4,000, the rate (r) is 0.03, and the time (t) is 10: $I = 4{,}000(0.03)(10) = \$1{,}200$. Add this amount to the original \$4,000 to get a total of \$5,200.

28. **\$5,393.39.** Use the formula for compound interest: $A = P\left(1 + \frac{r}{n}\right)^{nt}$. (***Note:*** Compare this total amount with the total using simple interest, in problem 2. In that problem, the total is $\$4{,}000 + \$1{,}200 = \$5{,}200$.) In the compound interest formula, $P = 4{,}000$, $r = 0.03$, $n = 4$, and $t = 10$.

$$A = 4{,}000\left(1 + \frac{0.03}{4}\right)^{4(10)} = 4{,}000(1 + 0.0075)^{40} = 5{,}393.3944 = \$5{,}393.39$$

So you do slightly better by letting the interest compound.

29. **\$5,000.** Use $I = Prt$ where $I = 500$, $r = 0.02$, and $t = 5$. Putting the numbers into the formula, $500 = P(0.02)(5)$. Divide each side of $500 = P(0.10)$ by 0.10, and you get that $P = 5{,}000$, the amount that was invested.

30. **\$43.30.** Total price $= \$40(1 + 0.0825) = \$40(1.0825) = \$43.30$.

31. **\$1,600.** Original price $= \frac{\$1{,}200}{1 - 0.25} = \frac{\$1{,}200}{0.75} = \$1{,}600$.

32. **\$50.40.** Discounted price $= \$80(1 - 0.10)(1 - 0.30) = \$80(0.90)(0.70) = \$50.40$.

IN THIS CHAPTER

» Working with perimeter, area, and volume

» Tackling geometric structures

» Moving along with distance problems

» Making interest interesting

Chapter 21

Making Formulas Work in Basic Story Problems

Algebra allows you to solve problems. Not all problems — it won't help with that noisy neighbor — but problems involving how to divvy up money equitably or make things fit in a room. In this chapter, you find some practical applications for algebra. You may not be faced with the exact situations I use in this chapter, but you should find some skills that will allow you to solve the story problems or practical applications that are special to your situation.

Algebra students often groan and moan when they see story problems. You "feel their pain," you say? It's time to put the myth to bed that story problems are too challenging. You know you're facing a story problem when you see a bunch of words followed by a question. And the trick to doing story problems is quite simple: Change the words into a solvable equation, solve that equation (now a familiar friend), and then answer the question based on the equation's solution. Coming up with the equation is often the biggest challenge. However, some story problems have a formula built into them to help. Look for those first.

Setting Up to Solve Story Problems

When solving story problems, the equation you should use or how all the ingredients interact isn't always immediately apparent. Sometimes you have to come up with a game plan to get you started. Sometimes, just picking up a pencil and drawing a picture can be a big help.

Other times, you can just write down all the numbers involved; I'm very visual, and I like to see what's going on with a problem.

You don't have to use every suggestion in the following list with every problem, but using as many as possible can make the task more manageable:

- **Draw a picture.** It doesn't have to be particularly lovely or artistic. Many folks respond well to visual stimuli, and a picture can act as one. Label your picture with numbers or names or other information that help you make sense of the situation. Fill it in more or change the drawing as you set up an equation for the problem.
- **Assign a variable(s) to represent *how many* or *number of*.** You may use more than one variable at first and refine the problem to just one variable later.

  A variable can represent only a number; it can't stand in for a person, place, or thing. A variable can represent the length of a boat or the number of people, but it can't represent the boat itself or a person. You can choose the letters so they can help make sense of the problem. For example, you can let k represent Ken's height or Ken's age — just don't let it represent Ken.
- **If you use more than one variable, go back and substitute known relationships for the extra variables.** When it comes to solving the equations, you want to solve for just one variable. You can often rewrite all the variables in terms of just one of them. For example, if you let e represent the number of Ernie's cookies and b represent Bert's cookies, but you know that Ernie has four more cookies than Bert, then e can be replaced with $b+4$.
- **Look at the end of the question or problem statement.** This often gives a big clue as to what's being asked for and what the variables should represent. It can also give a clue as to what formula to use, if a formula is appropriate. For example:

  Marilee and Scott ran in a race. Marilee finished 2 minutes before Scott, but she ran one less kilometer than Scott did. If they ran at the same rate, and the total distance they ran (added together) was 9 kilometers, *then how long did it take them?*

  Just look at all those words. Go to the last sentence — and even the last phrase of the last sentence. It tells you that you're looking for the amount of time it took. The formula that the last sentence suggests is $d = rt$ (distance = rate × time).
- **Translate the words into an equation.** Replace
  - *and, more than,* and *exceeded by* with the plus sign (+)
  - *less than, less,* and *subtract from* with the minus sign (–)
  - *of* and *times as much* with the multiplication sign (×)
  - *twice* with two times (2×)
  - *divided by* with the division sign (÷)
  - *half as much* with one-half times $\left(\frac{1}{2}\times\right)$
  - the verb (*is* or *are,* for example) with the equal sign (=)
- **Plug in a standard formula, if the problem lends itself to one.** When possible, use a formula as your equation or as part of your equation. Formulas are a good place to start to set up relationships. Be familiar with what the variables in the formula stand for.

- **Check to see if the answer makes any sense.** When you get an answer, decide whether it makes sense within the context of the problem. If you're solving for the height of a man, and your answer comes out to be 40 feet, you probably made an error somewhere. Having an answer make sense doesn't guarantee that it's a correct answer, but it's the first check to tell if it isn't correct.
- **Check the algebra.** Do that by putting the solution back into the original equation and checking. If that works, then work your answer through the written story problem to see if it works out with all the situations and relationships.

Applying the Pythagorean Theorem

Pythagoras, the Greek mathematician, is credited with discovering the wonderful relationship between the lengths of the sides of a *right triangle* (a triangle with one 90-degree angle). The Pythagorean Theorem is $a^2 + b^2 = c^2$. If you square the length of the two shorter sides of a right triangle and add them, then that sum equals the square of the longest side (called the *hypotenuse*), which is always across from the right angle. Right triangles show up frequently in story problems, so when you recognize them, you have an instant, built-in equation to work with on the problem.

EXAMPLE

Q. I have a helium-filled balloon attached to the end of a 500-foot string. My friend, Keith, is standing directly under the balloon, 300 feet away from me. (And yes, the ground is perfectly level, as it always is in these hypothetical situations — at least until you get into higher math.) These dimensions form a right triangle, with the string as the hypotenuse. Here's the question: How high up is the balloon?

A. Identify the parts of the right triangle in this situation and substitute the known values into the Pythagorean Theorem. Refer to the following picture.

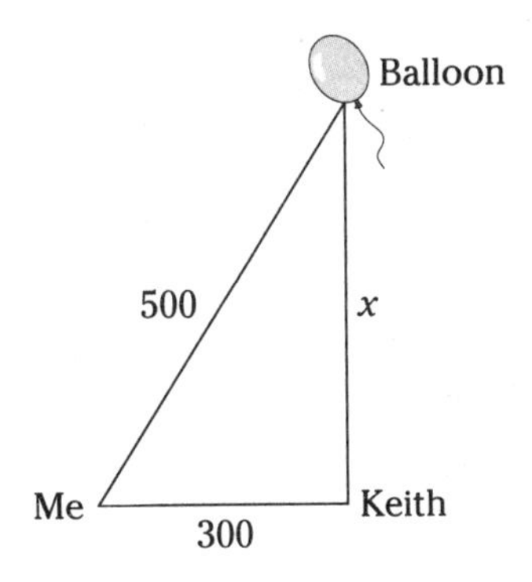

The 500-foot string forms the hypotenuse, so the equation reads $300^2 + x^2 = 500^2$.

$$90{,}000 + x^2 = 250{,}000$$

Solving for x,

$$x^2 = 250{,}000 - 90{,}000 = 160{,}000$$

$$x = \pm\sqrt{160{,}000} = 400$$

You only use the positive solution to this equation because a negative answer doesn't make any sense.

YOUR TURN

1. Jack's recliner is in the corner of a rectangular room that measures 10 feet by 24 feet. His television is in the opposite corner. How far is Jack (his head, when the recliner is wide open) from the television? (***Hint:*** Assume that the distance across the room is the hypotenuse of a right triangle.)

2. Sammy's house is 1,300 meters from Tammy's house — straight across the lake. The paths they need to take to get to one another's houses (and not get wet) form a right angle. If the path from Sammy's house to the corner is 1,200 meters, then how long is the path from the corner to Tammy's house?

3. A ladder from the ground to a window that's 24 feet above the ground is placed 7 feet from the base of the building, forming a right triangle. How long is the ladder, if it just reaches the window?

4. Calista flew 400 miles due north and then turned due east and flew another 90 miles. How far is she from where she started?

Using Geometry to Solve Story Problems

Geometry is a subject that has something for everyone. It has pictures, formulas, proofs, and practical applications for the homeowner. The perimeter, area, and volume formulas are considered to be a part of geometry. But geometry also deals in angle measures, parallel lines, congruent triangles, polygons and similar figures, and so forth. The different properties that you find in the study of geometry are helpful in solving many story problems.

Story problems that use geometry are some of the more popular (if you can call any story problem popular) because they come with ready-made equations from the formulas. Also, you can draw a picture to illustrate the problem. I'm a very visual person, and I find pictures and labels on the pictures to be very helpful.

EXAMPLE

Q. The opposite angles in a parallelogram are equal, and the adjacent angles in the parallelogram are supplementary (add up to 180 degrees). If one of the angles of a parallelogram measures 20 degrees more than three times another angle, then how big is the larger of the two angles?

A. The two angles have different measures, so they must be adjacent to one another. The sum of adjacent angles is 180. Let x represent the measure of the smaller angle. Then, because the larger angle is 20 degrees more than three times the smaller angle, you can write its measure as $20+3x$. Represent this in an equation in which the two angle measures are added together and the sum is 180: $x+20+3x=180$. Simplifying and subtracting 20 from each side, you get $4x=160$. Divide by 4 to get $x=40$. Putting 40 in for x in the measure $20+3x$, you get $20+120=140$.

So the two angles measure 40 and 140. Their sum is 180 degrees, which is your "check," because this makes them supplementary angles.

Q. An isosceles triangle has two sides that have the same measure. In a particular isosceles triangle that has a perimeter of 27 inches, the base (the side with a different measure) is 1 foot less than twice the measure of either of the other two sides. How long is the base?

A. The perimeter of the isosceles triangle is $x+x+2x-1=27$. The x represents the lengths of the two congruent sides, and $2x-1$ is the measure of the base. Simplifying and adding 1 to each side of the equation, you get $4x=28$.

Dividing by 4, you get $x=7$ inches.

The base is equal to $2(7)-1=14-1=13$ inches. To check this out, see that the perimeter is 27 inches; add the three sides together: $7+7+13=27$. Checks!

YOUR TURN

5. The sum of the measures of the angles of a triangle is 180 degrees. In a certain triangle, one angle is 10 degrees greater than the smallest angle, and the biggest angle is 15 times as large as the smallest. What is the measure of that biggest angle?

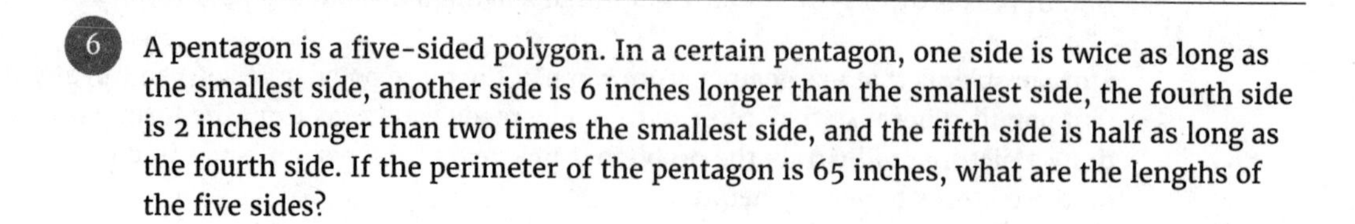

6. A pentagon is a five-sided polygon. In a certain pentagon, one side is twice as long as the smallest side, another side is 6 inches longer than the smallest side, the fourth side is 2 inches longer than two times the smallest side, and the fifth side is half as long as the fourth side. If the perimeter of the pentagon is 65 inches, what are the lengths of the five sides?

7. The sum of the measures of all the angles in any polygon can be found with the formula $A = 180(n - 2)$, where n is the number of sides that the polygon has. How many sides are there on a polygon where the sum of the measures is 1,080 degrees?

8. Two figures are similar when they're exactly the same shape — their corresponding angles are exactly the same measure, but the corresponding sides don't have to be the same length. When two figures are similar, their corresponding sides are proportional (all have the same ratio to one another). In the figure, triangle *ABC* is similar to triangle *DEF*, with *AB* corresponding to *DE* and *BC* corresponding to *EF*.

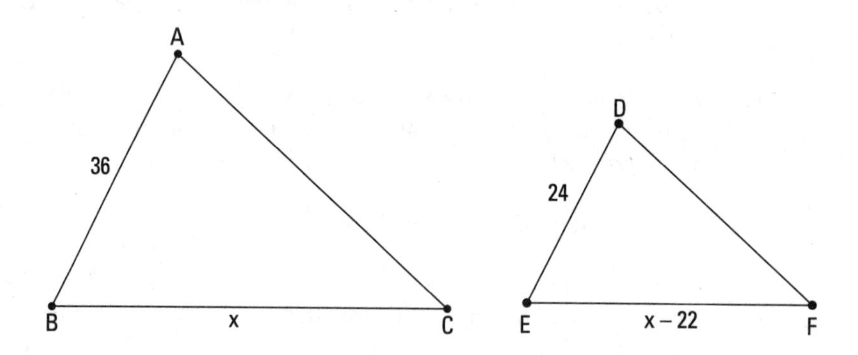

If side *EF* is 22 units smaller than side *BC*, then what is the measure of side *BC*?

Working around Perimeter, Area, and Volume

Perimeter, area, and volume problems are some of the most practical of all story problems. It's hard to avoid situations in life where you have to deal with one or more of these measures. For example, someday you may want to put up a fence and need to find the perimeter of your yard to help determine how much material you need to buy. Maybe you're expecting a baby and you want to add a room to your home; you can use an area formula to figure how much space your new room will take up. Finding a box to contain your present for your Grandma Erma's 100th birthday may require calculating the volume of standard box sizes and then constructing your own box for that special gift. Lucky for you, standard formulas to deal with all these situations are available, and many of them are in this section.

Parading out perimeter and arranging area

Perimeter is the measure around the outside of a region or area. Perimeter is used when you want to put a fence around a yard or some baseboards around a room. The police put yellow crime-scene tape around the perimeter of an accident or crime scene.

To find the perimeter of a rectangle, add twice the length, l, and twice the width, w. The formula for the perimeter of a rectangle is $P = 2(l + w) = 2l + 2w$. To find the area, A, of a rectangle, multiply the length times the width: $A = lw$. (You can find these formulas and many more in Chapter 20.)

EXAMPLE

Q. Juan wants to fence in a rectangular field along the river for his flock of sheep. He won't need any fence along the side of the field next to the river, just the other three sides. Juan wants his field to be twice as long as it is wide, and he'd like it to have a total area of 80,000 square feet. What should the dimensions of his field be, and how much fencing will he need?

A. This problem is a classic example of needing a picture. Figure 21-1 shows a possible sketch of the situation. Juan is assuming that sheep don't swim, so he thinks he can save money by not fencing along the river. (Have you ever smelled wet wool?)

The first issue has to do with the area. The formula for the area of a rectangle is $A = lw$. The area is to be 80,000 square feet, so $80{,}000 = lw$.

There are two variables. To change the equation so that it has one variable, go back to the problem where it says Juan wants the length to be twice the width. That means $l = 2w$. Replacing the l with $2w$ in the area formula, you get

$$80{,}000 = 2w \cdot w$$
$$80{,}000 = 2w^2$$

You need to solve the equation for w. First, divide by 2:

$$40{,}000 = w^2$$

Then take the square root of each side:

$$w = 200$$

The width is 200. The length is twice that, or 400. If the three sides that need fencing are $200 + 400 + 200$, then the amount of fencing needed is 800 feet.

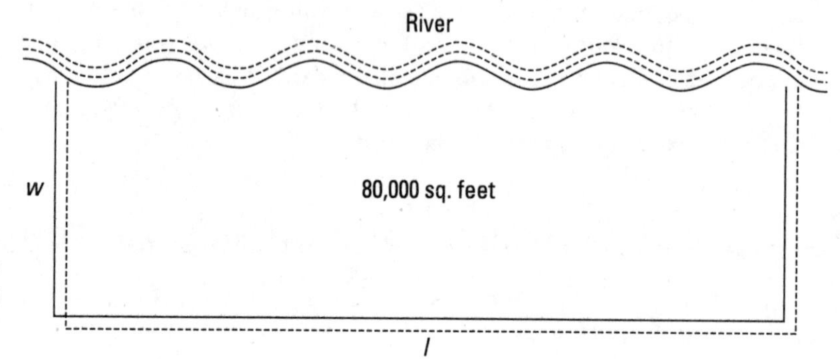

FIGURE 21-1: Fencing three sides of the field.

YOUR TURN

9. You've dug up a triangular garden in your backyard. You hadn't counted on there being so many rabbits and groundhogs, and now have to put a fence around the garden. The triangle is a right triangle with an hypotenuse of 205 feet and one leg measuring 45 feet. You can't measure the other leg right now, because it's really muddy from last night's rainfall. How many feet of fencing will you need?

10. A friend is building a rectangular corral along the side of their barn, so they'll only need to fence in three sides. They want the length of the corral to be three times the width, and the side of the barn is 60 feet — all of which can be used for a side of the corral. How much fencing is needed? (***Hint:*** To minimize the amount of fencing, let the side along the barn be the length of the corral.)

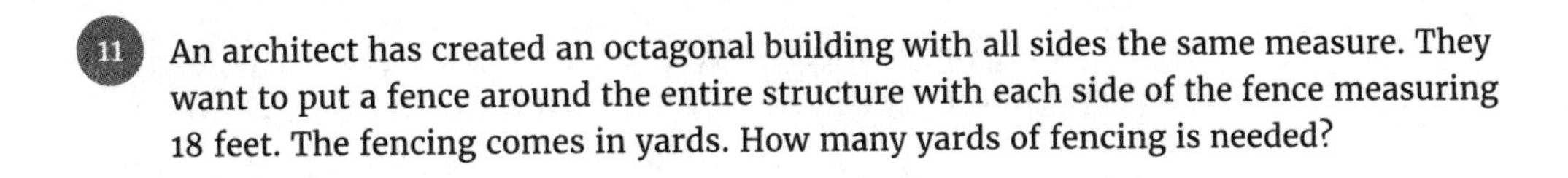

11 An architect has created an octagonal building with all sides the same measure. They want to put a fence around the entire structure with each side of the fence measuring 18 feet. The fencing comes in yards. How many yards of fencing is needed?

---

12 Your patio is trapezoidal in shape — formed from a rectangle with a right triangle added to the end. The two parallel bases of the trapezoid measure 60 feet and 100 feet, and the height of the trapezoid is 30 feet. You need to create a border around the outside of the patio to keep the concrete blocks in place. How long does the border need to be?

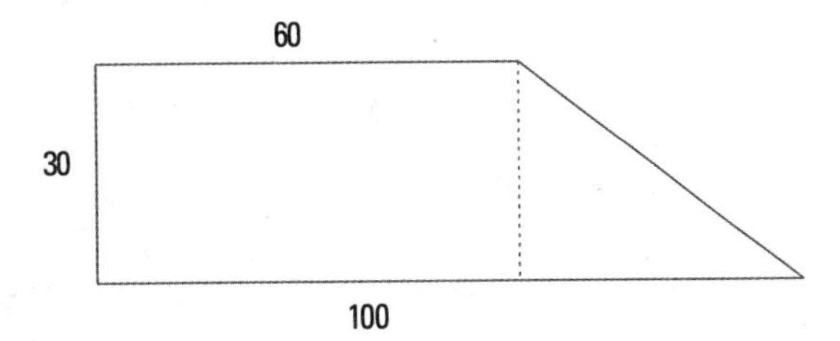

Adjusting the area

You may want to buy an area rug. You may meet someone who lives in your area. In both cases, area can be interpreted as some measured-off region or surface that has a shape or size. When doing area problems, you can find the area if you know what the shape is because there are so many nice formulas to use. You just have to match the shape with the formula.

EXAMPLE

Q. Eli and Esther are thinking of enlarging their family room. Right now, it's a rectangle with an area of 120 square feet. If they increase the length by 4 feet and the width by 5 feet, the new family room will have an area of 240 square feet. What are the dimensions of the family room now, and what will the new dimensions be?

A. Draw a rectangle, labeling the shorter sides as w and the longer sides as l. The area of a rectangle is $A = lw$, so, in this case, because you know the area, you write the equation $120 = lw$.

The length is going to increase by 4 feet, so you write $l+4$ to represent the new length; and the width is increasing by 5 feet, so write $w+5$ for the new width.

The new area is 240 square feet, so $240=(l+4)(w+5)$.

In the original room, $120=lw$, so you can solve for l and substitute that into the new equation. Then you'll have just one variable in the equation.

$$l=\frac{120}{w}$$

$$\left(\frac{120}{w}+4\right)(w+5)=240$$

Using FOIL (refer to Chapter 9) to simplify the left side,

$$120+\frac{600}{w}+4w+20=240$$
$$\frac{600}{w}+4w=100$$

To solve this, get rid of the fraction by multiplying both sides by w:

$$600+4w^2=100w$$

Now you have a quadratic equation that can be solved:

$$4w^2-100w+600=0$$

Divide through by 4 to make the coefficients and constant smaller:

$$w^2-25w+150=0$$

The quadratic factors using unFOIL:

$$(w-15)(w-10)=0$$

Now use the multiplication property of zero (MPZ), where $w-15=0$ or $w-10=0$, to get the solutions $w=15$ and $w=10$:

If $w=15$, then $l=\frac{120}{15}=8$; the width is increased by 5 and the length by 4, giving you the new dimensions of 20 by 12.

If $w=10$, then $l=\frac{120}{10}=12$; the width is increased by 5 and the length by 4, giving you new dimensions of 15 by 16.

Technically, these both work. Both are acceptable answers if you can accept a width that is greater than the length. In the case of $w=15$, the width is 15 and the length 8. If you're going to hold fast to width being less than length, then only the second solution works: original dimensions of 10 by 12 and new dimensions of 15 by 16.

13 You're putting 81 feet of molding along the walls around a rectangular room that's 12 feet wide; what is the area of that room (if you subtracted 3 feet for the width of the door)?

14 You're painting a large right triangle on the side of a building. The triangle has an hypotenuse measuring 61 feet and a base measuring 60 feet. If a gallon of paint covers 300 square feet, then how many gallons of paint will you need?

15 Your square patio currently measures 8 feet on each side. You want to increase each side by 4 feet. How much does this increase the area of the patio?

16 What is the area of an isosceles trapezoid if the 10-foot-shorter base is half as long as the longer base, and the height is $\frac{3}{5}$ the longer base?

Pumping up the volume

An area is a flat measurement. It can be shown on a floor or sports field, in two dimensions. Volume adds a third dimension, as Figure 21-2 shows. Take a room that is 10 by 12 feet and make the ceiling 8 feet high. You're talking about an area of 10×12, or 120 square feet, and with the height, a volume of 120×8, or 960 cubic feet. Volume is measured in cubic units. The amount of gas in a balloon is a cubic measure. The amount of cement in a sidewalk is a cubic measure.

A cube is a box that has equal length, width, and height. Picture a sugar cube, or a pair of dice. The volume of a cube is the cube (third power) of the length of a side: $V = s^3$.

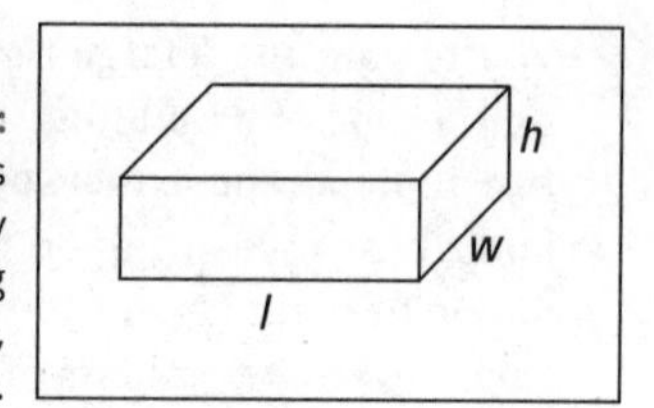

FIGURE 21-2: Volume is determined by multiplying length, width, and height.

EXAMPLE

Q. Aunt Sadie got a wonderful deal on some chocolate candies. You're the favorite of all her nieces and nephews, so Aunt Sadie wants to send all the candies to you. The candies came in a huge plastic bag, but she wants to ship them in a box. The candies take up 900 cubic inches of space. If the box she's going to use to ship them must have a 9-x-9-inch bottom, then how high does the box have to be to fit the candy?

A. The volume of a prism (in this case, the box) is found by multiplying the length of the box times its width times its height: $V = lwh$.

In this case, the bottom is square, and each side of the bottom is 9 inches, so, substituting into the formula,

$$900 = 9(9)(h)$$

Simplifying, you get:

$$900 = 81h$$

Solving for h, divide each side by 81 to get

$$\frac{900}{81} = h$$

$$h = 11\frac{1}{9} \text{ inches}$$

Building a pyramid

Pyramids are among the more recognizable geometric figures. Children are introduced to the pyramids of Egypt early in their schooling. You see the pyramid shape in everything from tents to meditation sites to Figure 21-3. If your tent has a pyramid shape, you can find its volume to see if you and your three friends can all fit. You'll want breathing room.

FIGURE 21-3: Some people believe pyramids have preservation powers.

The formula for the volume of a pyramid with a square base is $V = \frac{1}{3}x^2h$. The x^2 represents the area of the base. In general, the volume of a pyramid is one-third of the product of the area of the base and the height.

EXAMPLE

Q. The Great Pyramid of Cheops is a solid mass of limestone blocks. It's estimated to contain 2.3 million blocks of stone. Originally, the pyramid had a square base of 756 feet by 756 feet and was 480 feet high, but wind and sand have eroded it over time. Pretend it still has its original dimensions. If each of the blocks is a cube, what are the dimensions of the cubes?

A. You have only one measure to name — the measure of each edge — so call it *x*. First, find the volume of the Great Pyramid in cubic feet:

$$V = \frac{1}{3}\left(756^2\right)\left(480\right) = 91{,}445{,}760 \text{ cubic feet}$$

If each block were a cube 1 foot by 1 foot by 1 foot, there would be over 91 million of them. But, according to the estimate, there are 2.3 million blocks of stone, not 91 million, so

$$91{,}445{,}760 \div 2{,}300{,}000 \approx 39.759$$

That means that each of the 2.3 million blocks of stone measures more than 39 cubic feet. To find the measure of an edge, which gives you the dimensions, look at the formula for the volume of a cube, $V = s^3$. In this case, assign *s* to be the length of a side of any of the cubes. So, if $V = 39.759 = s^3$, then

$$s = \sqrt[3]{39.759} \approx 3.413$$

So each cube would be about $3\frac{1}{2}$ feet on each edge.

Picture a huge block of stone longer than a yardstick on each side (some of the stones are reportedly larger than this). Now picture lifting that stone up to the top of the Great Pyramid.

Circling Jupiter

Figuring out how much air you have to expel to blow up a 9-inch balloon involves cubic inches of air, force, propulsion, and all sorts of complicated physics, and in the end, do you really care? You just blow until the balloon is full. But that's not to say that you may not want to figure out how many balloons you need to fill up the big balloon net you rented for your 5-year-old's party.

The example in this section involves much larger spheres — a couple of planets, in fact — but I do my best to keep your feet on the ground. Figure 21-4 shows you a sphere.

The volume of a sphere is found with the formula $V = \frac{4}{3}\pi r^3$. The only dimension you need is the radius, *r*.

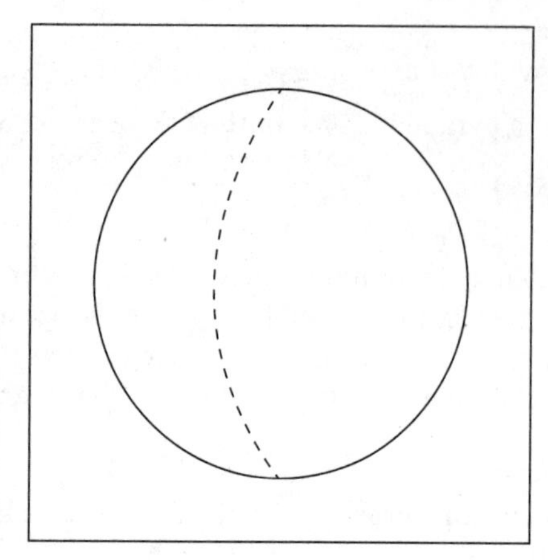

FIGURE 21-4: Basketballs, globes, planets, and sometimes oranges are spheres.

EXAMPLE

Q. Dan's fraternity is planning an elaborate prank to impress the ladies of the neighboring sorority. They plan on filling a spherical balloon with water and then bursting the balloon at an appropriate moment. The balloon they bought expands to a diameter of 20 feet. How many gallons of water will it take to fill the balloon? (Disregard the possible warping of the balloon's shape and the weight of the water — Dan is just dreaming that this will work, anyway.)

A. 31,336.33799 gallons. Using the formula for the volume of a sphere, you replace the *r* with 10; if the diameter is 20 feet, then the radius is 10 feet.

$$V = \frac{4}{3}\pi\left(10^3\right) \approx 4{,}188.790 \text{ cubic feet of water}$$

You now need the conversion equation from cubic feet to gallons. One cubic foot is equal to approximately 7.481 gallons. To find the total number of gallons necessary, multiply the number of cubic feet by the number of gallons and you get $(4{,}188.790)(7.481) = 31{,}336.33799$ gallons of water. I think they'd best scrap this bright idea.

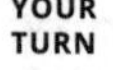

YOUR TURN

17 What is the increase in volume when a cube that's 4 inches on an edge has its edges each grow by 1 inch?

18 How many 2-by-3-by-4-foot cartons can you fit in a truck measuring 20 by 30 by 40 feet?

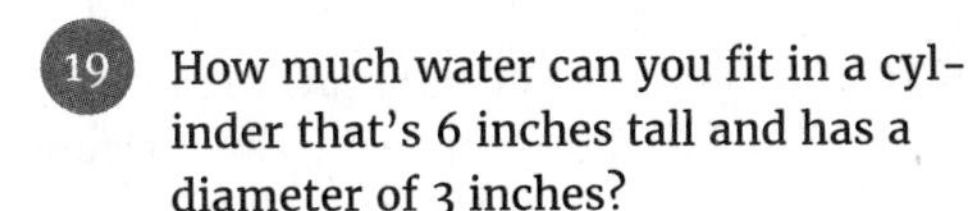

19 How much water can you fit in a cylinder that's 6 inches tall and has a diameter of 3 inches?

20 Wolford is constructing a greenhouse that's a hemisphere with a radius of 40 feet. How many cubic feet of volume will the greenhouse contain?

Going 'Round in Circles

The circle, as Figure 21-5 shows, is a very nice, efficient shape, although using a circular shape isn't always practical in buildings. Circles don't fit together well. There are always gaps between them, so they don't make good shapes for fields, yards, or areas shared with other circles. But even though circles don't fit in, circles are useful, letting you consider situations involving their area: circular rugs and race tracks, fields and swimming pools.

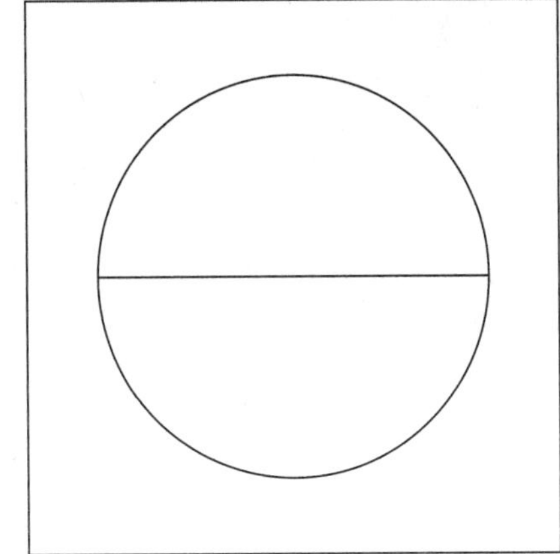

FIGURE 21-5: The diameter is the longest distance across a circle.

The area of a circle can be determined if you know the radius or the diameter.

EXAMPLE

Q. Grace decided to get an 18-foot-diameter, above-ground pool instead of a 12-foot-diameter pool. How much more area (of her yard) will this bigger pool cover?

A. Approximately 141.4 square feet. The area of a circle is found using $A = \pi r^2$.

The diameter of a circle is twice the radius, so an 18-foot-diameter pool has a radius of 9 feet, and the 12-foot-diameter pool has a 6-foot radius.

difference in area = area of bigger pool − area of smaller pool

$$\text{difference} = \pi(9)^2 - \pi(6)^2 = 81\pi - 36\pi = 45\pi \approx 141.4 \text{ square feet}$$

Q. If you have a certain amount of fencing, you can enclose more area with a circular shape than you can with any other shape. To prove this point, let me show you how much bigger a circular yard enclosed by 314 feet of fencing is than a square yard enclosed by the same amount of fencing.

A. The area of a circle is found with $A = \pi r^2$, and the area of a square is found with $A = s^2$. It looks like this will be fairly simple; you just have to find the difference between the two values.

$$\text{difference} = \text{area of circle} - \text{area of square}$$

The challenge comes in when you need the value of *r*, the radius of the circle, and the value of *s*, the length of a side of the square. You don't have the value of *r* or the value of *s*. You just have the distance around the outside called the *perimeter* or, in the case of a circle, the *circumference*, and there's a formula for each figure.

The circumference of a circle is found with $C = 2\pi r$, and the perimeter of a square is found with $P = 4s$.

If 314 is the circumference of the circle, then $314 = 2\pi r$, or $r = \frac{314}{2\pi} \approx 50$.

So the area of the circle is $A = \pi \cdot 50^2 = 2{,}500\pi \approx 7{,}854$ square feet.

The perimeter of a square is just four times the measure of the side. Because 314 is the perimeter of the square, $314 = 4s$, or $s = 78.5$. That means that the area of the square is $78.52 = 6{,}162.25$.

$$\text{difference} = 7{,}854 - 6{,}162.25 = 1{,}691.75 \text{ square feet}$$

That's quite a bit more area in the circle than in the square.

YOUR TURN

21 Your circular pool has a sprinkler mounted in the center that sprays water all the way to the edges. The radius of the pool is 18 feet. To keep more water in the pool, you're going to install some new edging to catch what the sprinkler puts out. How much edging do you need?

You are renting a circular tent for an outdoor wedding. You're going to put a short fence along the edge of the tent to keep out the chipmunks. (They are so annoying with their chatter.) If the tent is 40 feet across the center, then how much fencing will you need?

23 You see some irrigation circles from the air and wonder what area they cover. You estimate that the diameter is about 3,000 feet. How much area is watered in one of the circles?

24 After calculating the amount of area watered in an irrigation circle with a diameter of 3,000 feet, you wonder how much land is left unwatered in a square containing four adjacent circles. (See the following figure.)

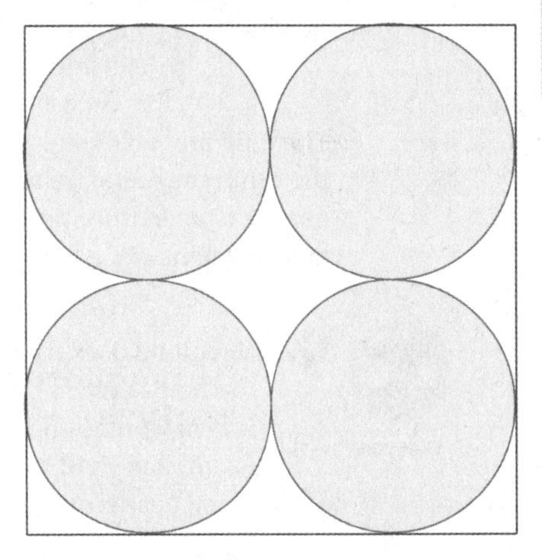

Putting Distance, Rate, and Time in a Formula

There are two basic *distance* formulas used in algebra. One of them has to do with the distance between two points on the coordinate plane. You find information on that formula in Chapter 24. The other distance relationship has to do with how far something moves from a starting point given a particular rate of speed and a certain amount of time.

Going the distance with the distance-rate-time formula

You travel, I travel, everybody travels, and at some point everybody asks, "Are we there yet?" Algebra can't answer that question for you, but it can help you estimate how long it takes to get there — wherever "there" is.

The distance formula, $d = rt$, says that distance is equal to the rate of speed multiplied by the time it takes to get from the starting point to the destination. You can apply this formula and its variations to determine how long, how far, and how fast you travel.

One type of distance problem involves setting two distances *equal to* one another. The usual situation is that one person is traveling at one speed and another person is traveling at another speed, and they end up at the same place at the same time. The other traditional distance problem involves *adding* two distances together, giving you a total distance apart. That's it. All you have to do is determine whether you're equating or adding!

The only real challenge in using the distance-rate-time formula is to be sure that the units in the different parts are the same. (For example, if the rate is in miles per hour, then you can't use the time in minutes or seconds.) If the units are different, you first have to convert them to an equivalent value before you can use the formula.

EXAMPLE

Q. How long does it take for light from the Sun to reach the Earth? (***Hint:*** The Sun is 93 million miles from the Earth, and light travels at 186,000 miles per second.)

A. $8\frac{1}{3}$ minutes. Using $d = rt$ and substituting the distance (93 million miles) for d and 186,000 for r, you get $93{,}000{,}000 = 186{,}000t$. Dividing each side by 186,000, t comes out to be 500 seconds. Divide 500 seconds by 60 seconds per minute, and you get $8\frac{1}{3}$ minutes.

Q. How fast do you go to travel 200 miles in 300 minutes?

A. 40 mph. Assume you're driving over the river and through the woods, and you need to get to grandmother's house by the time the turkey is done, which is in 300 minutes. It's 200 miles to grandmother's house. Because your speedometer is in miles per hour, change the 300 minutes to hours by dividing by 60, which gives you 5 hours. Fill the values into the distance formula: $200 = 5r$. Dividing by 5, it looks like you have to average 40 miles per hour.

YOUR TURN

25 How long will it take you to travel 600 miles if you're averaging 50 mph?

What was your average speed (just using the actual driving time) if you left home at noon, drove 200 miles, stopped for an hour to eat, drove another 130 miles, and arrived at your destination at 7 p.m.?

Figuring distance plus distance

One of the two basic distance problems involves one object traveling a certain distance, a second object traveling another distance, and the two distances getting added together. There could be two kids on walkie-talkies, going in opposite directions to see how far apart they'd have to be before they couldn't communicate anymore. Another instance would be when two cars leave different cities heading toward each other on the same road and you figure out where they meet.

EXAMPLE

Q. Deirdre and Donovan are in love and will be meeting in Kansas City to get married. Deirdre boarded a train at noon traveling due east toward Kansas City. Two hours later, Donovan boarded a train traveling due west, also heading for Kansas City, and going at a rate of speed 20 mph faster than Deirdre. At noon, they were 1,100 miles apart. At 9 p.m., they both arrived in Kansas City. How fast were they traveling?

A. Deirdre was travelling at 60 mph; Donovan, 80 mph. You can write the following formulas:

$$\text{distance of Deirdre from Kansas City} + \text{distance of Donovan from Kansas City} = 1{,}100$$
$$(\text{rate} \times \text{time}) + (\text{rate} \times \text{time}) = 1{,}100$$

Let the speed (rate) of Deirdre's train be represented by r. Donovan's train was traveling 20 mph faster than Deirdre's, so the speed of Donovan's train is $r + 20$.

Let the time traveled by Deirdre's train be represented by t. Donovan's train left two hours after Deirdre's, so the time traveled by Donovan's train is $t - 2$. Substituting the expressions into the first equation,

$$rt + (r + 20)(t - 2) = 1{,}100$$

Deirdre left at noon and arrived at 9, so $t = 9$ hours for Deirdre's travels and $t - 2 = 7$ hours for Donovan's. Replacing these values in the equation,

$$r(9) + (r + 20)(7) = 1{,}100$$

Now distribute the 7:

$$9r + 7r + 140 = 1{,}100$$

Combine the two terms with r:

$$16r = 960$$

Divide each side by 16:

$$r = 60$$

Deirdre's train is going 60 mph; Donovan's is going $r + 20 = 80$ mph.

The distance-rate-time formula is probably the formula you're most familiar with from daily life — even though you may not think of it as using a formula all the time. The distance formula is $d = rt$. The d is the distance traveled, the r is the speed at which you're traveling, and the t is the amount of time spent traveling.

EXAMPLE

Q. One train leaves Kansas City traveling due east at 45 mph. A second train leaves Kansas City three hours later, traveling due west at 60 mph. When are they 870 miles apart?

A. 10 hours. This type of problem is where you add two distances together. Let t represent the time that the first train traveled and $45t$ represent the distance that the first train traveled. This is rate times time, which equals the distance. The second train didn't travel as long; its time will be $t-3$, so represent its distance as $60(t-3)$. Set the equation up so that the two distances are added together and the sum is equal to 870: $45t+60(t-3)=870$. Distribute the 60 on the left and simplify the terms to get $105t-180=870$. Add 180 to each side to get $105t=1{,}050$. Divide each side by 105, and you have $t=10$ hours.

Equating distances

The other standard distance formula involves two people or objects traveling the same route but at different speeds or starting at different times.

EXAMPLE

Q. Angelina left home traveling at an average of 40 mph. Brad left the same place an hour later using the same route, traveling at an average of 60 mph. How long did it take for Brad to catch up to Angelina?

A. This type of problem is where you set the distances equal to one another. You don't know what the distance is, but you know that the rate times the time of each must equal the same thing. So let t represent the amount of time that Angelina traveled and set Angelina's distance, $40t$, equal to Brad's distance, $60(t-1)$. (***Remember:*** He traveled one less hour than Angelina did.) Then solve the equation $40t=60(t-1)$. Distribute the 60 on the right, which gives you $40t=60t-60$. Subtract $60t$ from each side, which results in $-20t=-60$. Divide by -20, and you get that $t=3$. Angelina traveled for 3 hours at 40 mph, which is 120 miles. Brad traveled for 2 hours at 60 mph, which is also 120 miles.

YOUR TURN

27 Kelly left school at 4 p.m., traveling at 25 mph. Ken left at 4:30 p.m., traveling at 30 mph, following the same route as Kelly. At what time did Ken catch up with Kelly?

A Peoria Charter Coach bus left the bus terminal at 6 a.m. heading due north and traveling at an average of 45 mph. A second bus left the terminal at 7 a.m., heading due south and traveling at an average of 55 mph. When were the buses 645 miles apart?

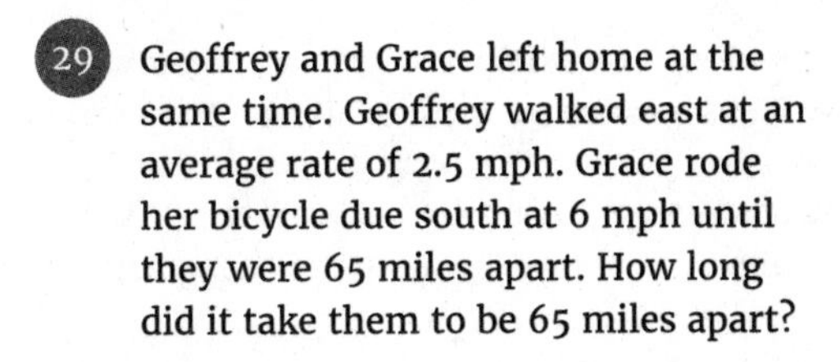

29 Geoffrey and Grace left home at the same time. Geoffrey walked east at an average rate of 2.5 mph. Grace rode her bicycle due south at 6 mph until they were 65 miles apart. How long did it take them to be 65 miles apart?

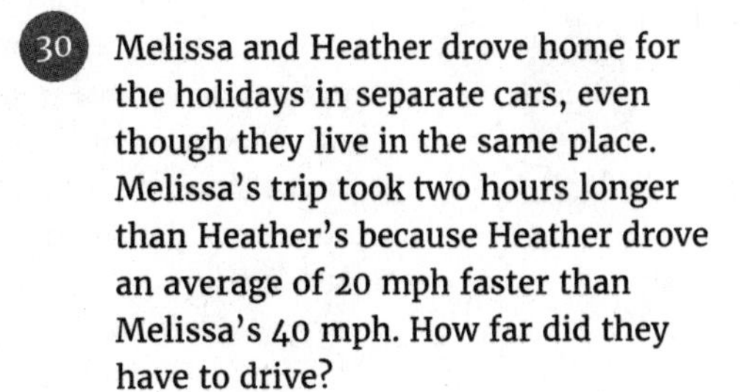

30 Melissa and Heather drove home for the holidays in separate cars, even though they live in the same place. Melissa's trip took two hours longer than Heather's because Heather drove an average of 20 mph faster than Melissa's 40 mph. How far did they have to drive?

Figuring distance and fuel

My son, Jim, sent me this problem when he was stationed in Afghanistan with the Marines. He was always a whiz at story problems — doing them in his head and not wanting to show any work. He must have been listening to me, because, at the end of this contribution, he added, "Don't forget to show your work!"

EXAMPLE

Q. A CH-47 troop-carrying helicopter can travel 300 miles if there aren't any passengers. With a full load of passengers, it can travel 200 miles before running out of fuel. If Camp Tango is 120 miles away from Camp Sierra, can the CH-47 carry a full load of Special Forces members from Tango to Sierra, drop off the troops, and return safely to Tango before running out of fuel? If so, what percentage of fuel will it have left?

A. I felt a little nervous, working on this problem, with so much at stake. So I took my own advice and drew a picture, tried some scenarios with numbers, and assigned a variable to an amount.

Let x represent the number of gallons of fuel available in the helicopter, and write expressions for the amount used during each part of the operation.

When the helicopter is loaded, it can travel 200 miles on a full tank of fuel. The camps are 120 miles apart, so the helicopter uses $\frac{120}{200}x$ gallons for that part of the trip.

When there are no passengers, the helicopter can travel 300 miles on a full tank. So it uses $\frac{120}{300}x$ gallons for the return flight.

Adding the two amounts together,

$$\frac{120}{200}x+\frac{120}{300}x=\frac{3}{5}x+\frac{2}{5}x=\frac{5}{5}x=x$$

It looks like there's no room for a scenic side trip. And I haven't figured in the fuel needed for landing and taking off. Hopefully, there's a reserve tank.

YOUR TURN

31 Your new car gets an average of 28 mpg, and the old one got only 22 mpg. If both cars have tanks that hold 18 gallons of fuel, then how much farther can you go in your new car on one tank of gas?

32 You're planning a trip of 1,088 miles, and your car gets an average of 28 mpg. If fuel costs $3.80 per gallon, then how much will it cost in fuel for you to make the trip?

Counting on Interest and Percent

It's all about the money. Do you buy the item at the reduced price? Should you invest in the program? Is this a fair deal? Understanding how the different interest and percentage formulas work will help you answer questions such as these — and protect your savings.

EXAMPLE

Q. You can invest your $10,000 at 1.5% interest compounded quarterly or at 2% simple interest. You're willing to invest the money for 5 years. Which is the better investment plan?

A. Using the compound interest formula,

$$A = 10{,}000\left(1 + \frac{0.015}{4}\right)^{4\cdot 5} = 10{,}000(1.00375)^{20} = 10{,}777.33$$

Using simple interest, you add the interest onto the initial investment:

$$A = 10{,}000 + 10{,}000(0.015)(5) = 10{,}000 + 750 = 10{,}750$$

You get a walloping $27 more by investing at the compounded interest rate.

Q. Chris has been exercising regularly and has increased the amount of weight she can lift by 25%. If she can now lift 210 pounds, what was her starting amount?

A. Letting her starting amount be m, use the formula $210 = m + 0.20m = m(1 + 0.20)$.

Solving for m in $210 = m(1.20)$, divide each side of the equation by 1.20 to get that $m = 175$ pounds.

YOUR TURN

33 If 31% of Tracy's gross monthly salary results in deductions totaling $1,302, then what is her monthly salary?

34 And if Tracy gets a 15% raise in her monthly salary, then what will her new deductions be?

35 You see that $12 has been deposited in your savings account after 3 months. This account earns 2% interest, compounded monthly. How much do you now have in your account?

36 If the balance on Chuck's credit card is $500 and the account has a 12.9% interest rate, compounded daily, what will the new balance be at the end of the month, if he doesn't charge any more to the account?

Practice Questions Answers and Explanations

1. **26 feet.** Using the Pythagorean Theorem where the 10 and 24 are legs of the triangle:

$$\begin{aligned} 10^2 + 24^2 &= c^2 \\ 100 + 576 &= c^2 \\ c^2 &= 676 = 26^2 \\ c &= 26 \end{aligned}$$

2. **500 meters.** The distance across the lake is the hypotenuse of a right triangle.

$$\begin{aligned} 1{,}200^2 + b^2 &= 1{,}300^2 \\ 1{,}440{,}000 + b^2 &= 1{,}690{,}000 \\ b^2 &= 1{,}690{,}000 - 1{,}440{,}000 = 250{,}000 = 500^2 \\ b &= 500 \end{aligned}$$

3. **25 feet.** Check out the following figure.

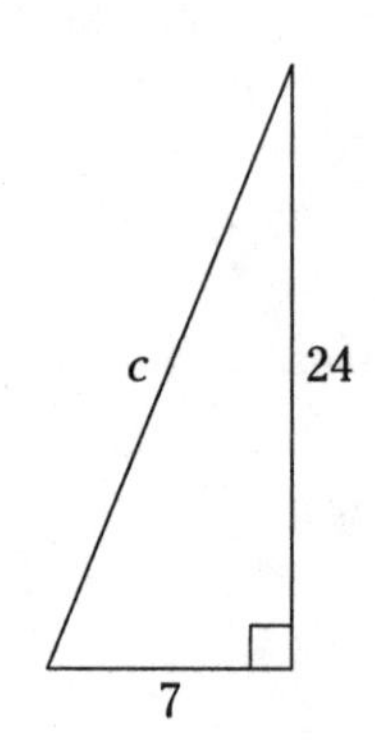

$$\begin{aligned} 7^2 + 24^2 &= c^2 \\ 49 + 576 &= c^2 \\ c^2 &= 625 = 25^2 \\ c &= 25 \end{aligned}$$

4. **410 miles.** See the following figure.

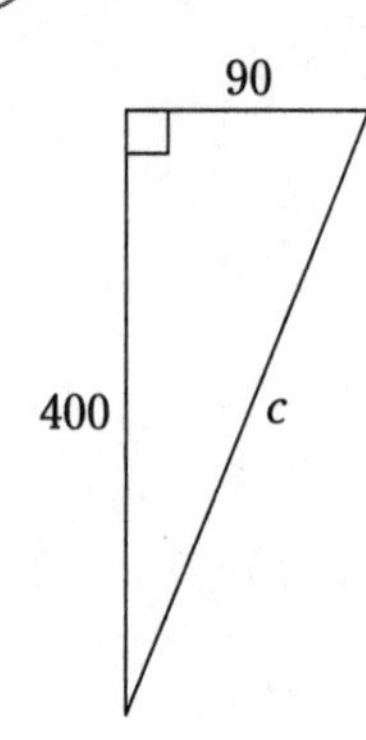

$$400^2 + 90^2 = c^2$$
$$160{,}000 + 8{,}100 = c^2$$
$$c^2 = 168{,}100 = 410^2 \text{ because } \sqrt{168{,}100} = 410$$
$$c = 410$$

5. **150 degrees.** Look at the following figure for help.

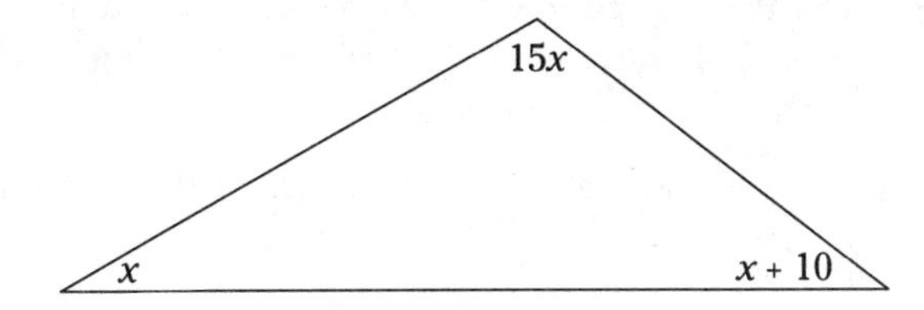

Let x = the measure of the smallest angle in degrees. Then the other two angles measure $x+10$ and 15x. Add them to get $x + x + 10 + 15x = 180$. Simplifying on the left, you get $17x + 10 = 180$. Subtract 10 from each side: $17x = 170$. Then, dividing by 17, $x = 10$ degrees, and the largest angle, which is 15 times as great, is 150 degrees.

To check, find the measure of the other angle, $x + 10 = 10 + 10 = 20$. Adding up the three angles, you get $10 + 20 + 150 = 180$.

6. **8 inches, 16 inches, 14 inches, 18 inches, 9 inches.** Let x represent the length of the smallest side. Then the second side is 2x long, the third side is $x+6$, the fourth is $2+2x$, and the fifth is $\frac{1}{2}(2+2x) = 1+x$. Add up all the sides and set the sum equal to 65: $x + 2x + (x+6) + (2+2x) + (1+x) = 65$. Simplifying, you get $7x + 9 = 65$. Subtract 9 from each side and divide by 7 to get $x = 8$.

7. **8 sides.** Use the formula $A = 180(n-2)$, where n is that number you're trying to find. Replace the A with 1,080 to get $1{,}080 = 180(n-2)$. Divide each side by 180 to get $6 = n - 2$. Add 2 to each side, and $n = 8$. So it's an eight-sided polygon that has that sum for the angles.

8. **66 units.**

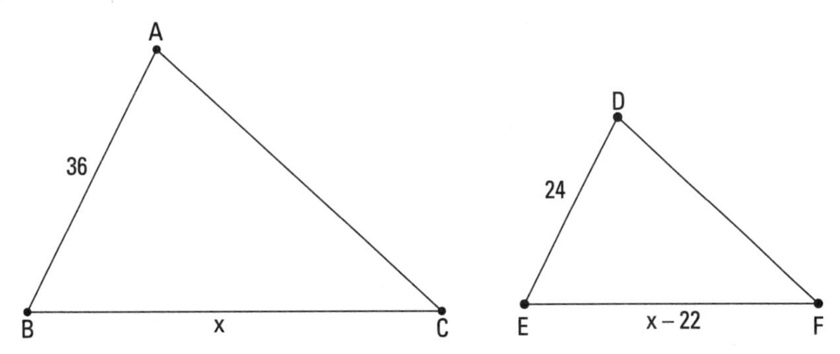

A proportion that represents the relationship between sides of the triangle is $\frac{AB}{DE} = \frac{BC}{EF}$.

Replacing the names of the segments with the respective labels, you get $\frac{36}{24} = \frac{x}{x-22}$.

Cross-multiply to get $36(x-22) = 24x$. Distributing on the left, you get $36x - 792 = 24x$. Subtract 24x from each side and add 782 to each side for $12x = 792$. Dividing by 12, you find that $x = 66$. (***Hint:*** The numbers would have been smaller if I had reduced the fraction on the left to $\frac{3}{2}$ before cross-multiplying. The answer is the same, of course.)

9. **450 feet.** You need to find the length of the second leg of the right triangle. Using the Pythagorean Theorem, $a^2+b^2=c^2$, let $a=45$ and $c=205$, giving you $45^2+b^2=205^2$. Solving for b^2, you have $b^2=205^2-45^2=42{,}025-2{,}025=40{,}000$. The square root of 40,000 is 200, so $b=200$. Now add up the measures of the three sides: $45+205+200=450$ feet of fencing needed.

10. **100 feet.** If the length of the corral is to be three times the width, then you write $l=3w$. Because the length can be the full 60 feet, replace the l in the equation and write $60=3w$. Dividing by 3, you have $w=20$. Now add up the length and two times the width, because you don't need fencing along the barn. The total fencing is $60+20+20=100$ feet.

11. **48 yards.** An octagon has eight sides. The perimeter is 8 times 18 feet: $P=8(18)=144$ feet. Because the fencing comes in yards, and there are three feet in a yard, divide the 144 feet by 3 to get 48 yards.

12. **240 feet.** You're missing the length of the slant-side of the trapezoid. This slant-side is also the hypotenuse of a right triangle with one leg measuring 30 feet (the height of the trapezoid) and the other leg measuring 40 feet (subtracting 60 from 100 feet). Using the Pythagorean Theorem, $a^2+b^2=c^2$, let $a=30$ and $b=40$, giving you $30^2+40^2=c^2$. Solving for c^2, $900+1{,}600=2{,}500=c^2$. The square root of 2,500 is 50, so the missing length is 50 feet. Add up the measures of the four sides: $100+30+60+50=240$ feet of border is needed.

13. **360 square feet.** Adding the 3 feet back into the 81 feet, you have a perimeter of 84 feet. The perimeter is twice the length plus twice the width. So, using $P=2l+2w$, you have $84=2l+2(12)$, which simplifies to $84=2l+24$. Solving for l, $2l=84-24=60$. Dividing by 2, $l=30$ feet. Now find the area of the rectangle using $A=lw=30(12)=360$ square feet.

14. **2 gallons.** To find the area of a triangle, you need to use $A=\frac{1}{2}bh$. The base of the triangle measures 60 feet. You find the height using the Pythagorean Theorem, $a^2+b^2=c^2$. In this case, $a^2+60^2=61^2$. Solving for a^2, you have $a^2=61^2-60^2=3{,}721-3{,}600=121$. The square root of 121 is 11, so the height of the triangle is 11 feet. The area of the triangle is then $A=\frac{1}{2}(60)(61)=330$ square feet. It looks like you'll need to buy two gallons of paint, because one gallon will only cover the first 300 square feet. You'll have lots left over for touch-ups.

15. **80 square feet.** The area of a square is found with $A=s^2$. The current patio has an area of $A=8^2=64$ square feet. Increasing each side by 4 feet gives you sides measuring 12 feet. The area of the new patio will be $A=12^2=144$ square feet. The difference between the two areas is $144-64=80$ square feet. Quite a change!

16. **180 square feet.** You find the area of a trapezoid with $A=\frac{1}{2}h(b_1+b_2)$. If the 10-foot-shorter base is half as long as the longer base, then the longer base measures 20 feet. If the height is $\frac{3}{5}$ the measure of the longer base, then the height is $\frac{3}{5}(20)=12$ feet. Using the formula for area, $A=\frac{1}{2}(12)(10+20)=180$ square feet.

17. **61 cubic inches.** The volume of a cube is determined by using the formula $V=e^3$. So a cube measuring 4 inches on an edge has a volume of $V=4^3=64$ cubic inches. Increasing each edge by one inch creates a cube with a volume of $V=5^3=125$ cubic inches. The difference in the volume is $125-64=61$ cubic inches; it almost doubles in volume!

18. **1,000.** Placing the cartons in the truck (and assuming no space is needed along the sides of the truck or between the boxes), you can place 10 boxes with the 2-foot edge along the 20-foot side and 10 boxes with the 3-foot edge along the 30-foot side, giving you a layer of 10 by 10, or 100 boxes. These can be stacked 10 high, fitting the 4-foot edge along the 40-foot side. So, 10 layers of 100 is 1,000 boxes.

(19) **42.39 cubic inches.** The volume of a cylinder is found with $V = \pi r^2 h$. The diameter is 3 inches, so the radius is 1.5 inches. The volume is then $V = \pi(1.5)^2(6)$, which is about 42.39 cubic inches, using 3.14 for the value of π.

(20) **133,973.33 cubic feet.** The volume of a sphere is found with $V = \frac{4}{3}\pi r^3$. A hemisphere is half a sphere, so the volume of a hemisphere is $V = \frac{1}{\cancel{2}} \cdot \frac{\cancel{4}^2}{3}\pi r^3 = \frac{2}{3}\pi r^3$. The volume of Wolford's hemisphere is $V = \frac{2}{3}\pi(40)^3$, which is about 133,973.33 cubic feet, using 3.14 for the value of π.

(21) **113.04 feet.** The edge or perimeter of a circular pool is called the *circumference*. You find the circumference with $C = \pi d$ or $C = 2\pi r$. If the radius is 18 feet, then the circumference is $C = 2\pi(18) = 113.04$ feet, using 3.14 for the value of π.

(22) **125.6 feet.** The edge or perimeter of a circular tent is called the *circumference*. You find the circumference with $C = \pi d$ or $C = 2\pi r$. If the diameter is 40 feet, then the circumference is $C = \pi(40) = 125.6$ feet, using 3.14 for the value of π.

(23) **7,065,000 square feet.** The area of a circle is found with $A = \pi r^2$. If the diameter of the area is 3,000 feet, then the radius is 1,500 feet, and the area is $A = \pi(1{,}500)^2 = 7{,}065{,}000$ square feet, using 3.14 for the value of π.

(24) **7,740,000 square feet.** A square containing four adjacent irrigation circles, lined up two-by-two, measures 6,000 feet on a side, because the diameters of the circles are each 3,000 feet. The area of the square is found with $A = e^2$ and is $A = 6{,}000^2 = 36{,}000{,}000$ square feet. If one irrigation circle has a diameter of 3,000 feet, then its area is $A = \pi(1{,}500)^2 = 7{,}065{,}000$ square feet, using 3.14 for the value of π. Four circles are four times that area, or $4(7{,}065{,}000) = 28{,}260{,}000$. Subtract that area from the area of the entire square, and you have $36{,}000{,}000 - 28{,}260{,}000 = 7{,}740{,}000$ square feet unwatered.

(25) **12 hours.** Use the distance-rate-time formula ($d = rt$), and replace the d with 600 and the r with 50 to get $600 = 50t$. Divide each side by 50 to get $12 = t$.

(26) **55 mph.** Use the formula $d = rt$, where $d = 200 + 130$ total miles, and the time is $t = 7 - 1 = 6$ hours. Substituting into the formula, you get $330 = r(6)$. Divide each side by 6 to get the average rate of speed: $55 = r$.

(27) **7 p.m.** Let t = time in hours after 4 p.m. Kelly's distance is $25t$, and Ken's distance is $30\left(t - \frac{1}{2}\right)$, using $t - \frac{1}{2}$ because he left a half hour after 4 p.m.

Ken will overtake Kelly when their distances are equal. So the equation to use is $25t = 30\left(t - \frac{1}{2}\right)$.

Distributing on the right gives you $25t = 30t - 15$. Subtract $30t$ from each side to get $-5t = -15$. Divide each side by -5, and $t = 3$. If t is the time in hours after 4 p.m., then $4 + 3 = 7$; so Ken caught up with Kelly at 7 p.m.

(28) **1 p.m.** Let t = time in hours after 6 a.m. The distance the first bus traveled is $45t$, and the distance the second bus traveled is $55(t - 1)$, which represents 1 hour less of travel time. The sum of their distances traveled gives you their distance apart. So $45t + 55(t - 1) = 645$. Distribute the 55 to get $45t + 55t - 55 = 645$. Combine the like terms on the left and add 55 to each side: $100t = 700$. Dividing by 100, $t = 7$. The buses were 645 miles apart 7 hours after the first bus left. Add 7 hours to 6 a.m., and you get 1 p.m.

(29) **10 hours.** Look at the following figure, showing the distances and directions that Geoffrey and Grace traveled.

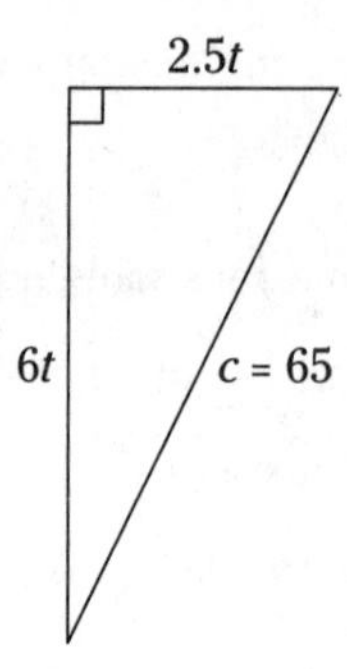

Geoffrey walks $d = rt = 2.5t$ miles, and Grace rides $d = rt = 6t$ miles. Use the Pythagorean Theorem:

$$(6t)^2 + (2.5t)^2 = 65^2 \rightarrow 36t^2 + 6.25t^2 = 65^2$$

Multiply each side by 4 to get rid of the decimal:

$$144t^2 + 25t^2 = 4 \times 65^2 \rightarrow 169t^2 = 4 \times 65^2 \rightarrow t^2 = \frac{4 \times 65^2}{169}$$

$$t = \sqrt{\frac{4 \times 65^2}{169}} = \frac{2 \times 65}{13} = \frac{130}{13} = 10 \text{ hours}$$

In 10 hours, Geoffrey walked $2.5(10) = 25$ miles, and Grace rode $6(10) = 60$ miles. Plug these values into the Pythagorean Theorem: $25^2 + 60^2 = 625 + 3,600 = 4,225 = 65^2$.

30 **240 miles.** Because Heather's time is shorter, let t = Heather's time in hours. Heather's distance is $60t$, using $40 + 20$ for her speed. Melissa's distance is $40(t+2)$, because she took 2 hours longer. Their distances are equal, so $60t = 40(t+2)$. Distribute the 40 to get $60t = 40t + 80$. Subtract $40t$ from each side: $20t = 80$. Divide by 20 for $t = 4$. Heather's distance is $60(4) = 240$ miles. Melissa's distance is $40(4+2) = 40(6) = 240$ miles. It's the same, of course.

31 **108 miles.** A car averaging 28 miles per gallon can travel $28(18) = 504$ miles on 18 gallons of fuel. A car averaging 22 miles per gallon can travel $22(18) = 396$ miles on 18 gallons of fuel. The difference in miles is $504 - 396 = 108$ miles.

32 **$147.66.** Traveling 1,088 miles in a car averaging 28 mpg, you divide 1,088 by 28 to get the number of gallons needed: $\frac{1,088}{28} \times 38.857$ gallons. At $3.80 per gallon, that's $38.857(\$3.80) = \$147.66$.

33 **$4,200.** Let Tracy's gross salary be t and solve: $0.31t = 1,302$. Dividing by 0.31, you have $t = 4,200$.

34 **$1,497.30.** You can either find her new salary and compute 31%, or just increase the deductions she now has by 15%. I'll take the easier route. New deductions $= 1,302(1.15) = 1,497.3$.

35 **$2,412.** A full-blown equation to solve this problem would look like this:
$A = P\left(1 + \frac{0.02}{4}\right)^{4\left(\frac{1}{4}\right)} = P + 12$, where A is the new amount in your account, P is what the $12 was earned on, and the exponent, $4\left(\frac{1}{4}\right)$, reflects the fact that this was over a three-month period, or $\frac{1}{4}$ of a year. Instead of solving for P in this equation, I will take the less-complicated route and determine what the initial principal was. Using $I = Prt$, $12 = P(0.02)\left(\frac{1}{4}\right)$, and solving for P, you have $P = 2,400$. The new balance is $P + 12$, or $2,412.

36 **\$505.33.** Using $A = 500\left(1+\frac{0.129}{365}\right)^{365\left(30/365\right)}$, I used 30 days for the days in a month (which is pretty standard). So $A = 500\left(1+\frac{0.129}{365}\right)^{365\left(30/365\right)} \approx 505.3286274$. That rate sounded pretty scary, but it adds up slowly.

If you're ready to test your skills a bit more, take the following chapter quiz that incorporates all the chapter topics.

Whaddya Know? Chapter 21 Quiz

Quiz time! Complete each problem to test your knowledge on the various topics covered in this chapter. You can then find the solutions and explanations in the next section.

1. You paid \$51.75 for new jeans that were on sale at a 25% discount. What was the original price of the jeans? (No tax is included here.)

2. A cook wants to cut the largest possible circular pie crust from a square piece of dough that's 10 inches on a side. If the largest possible crust is cut out, then how much dough is left over?

3. It cost you \$75 for gas when you made your 500-mile trip. If your car gets 25 miles per gallon, then how much per gallon did the gas cost?

4. On a recent trip, you drove an average of 65 mph for the first 2 hours and then 35 mph for the next 3 hours. What was your average rate of speed for the entire trip?

5. A yard has a right triangular shape and the perpendicular sides are 11 feet and 60 feet. What is the area of the triangle, and what is its perimeter?

6. You want to build a rectangular corral, and you won't need fencing on the side of the corral that runs along a river. You have 600 feet of fencing available. Which length of the corral opposite the river, x, will create the corral of the greatest area if your choices are $x = 100, 200, 300, 400, 500$? (See Figure 21-6.)

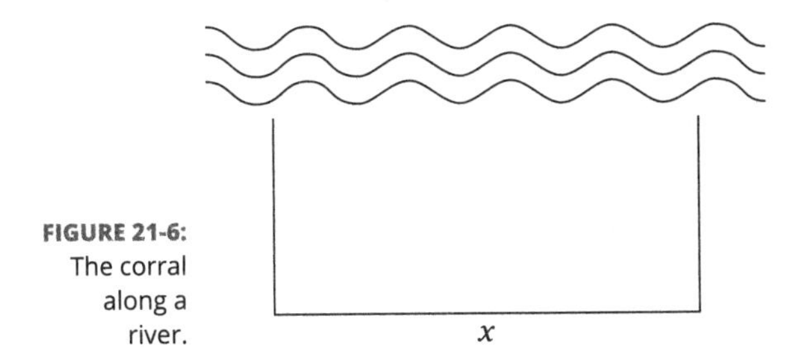

FIGURE 21-6: The corral along a river.

7. You are building a tree house in the shape of a pyramid with a square base. If the volume of the tree house is 400 cu. ft. and the sides of the base are 10 feet, then how tall is the tree house?

8. A circular race track has an outside fence that is 1,570 feet long. How far across is the race track (diameter)?

9. Jim left Tampa headed for Naples at 2:00, traveling an average of 70 mph. Janet left Naples headed for Tampa at 3:00, using the same route, and averaging 45 mph. If the distance between the two starting points is 185 miles, then at what time did they meet?

10. Triangles *ABC* and *DEF* are similar triangles, as shown in Figure 21-7. What is the measure of side *BC*?

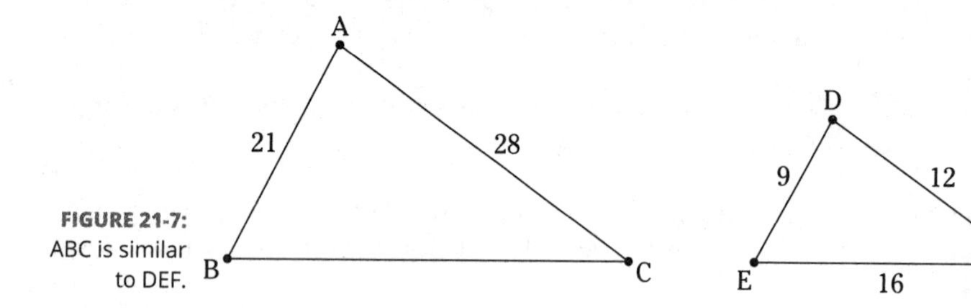

FIGURE 21-7: ABC is similar to DEF.

11. How much will there be in an account that starts with $12,000 if it earns 7% interest compounded monthly for five years and there are no additional deposits or withdrawals?

12. The largest angle in a triangle measures 3 more degrees than twice the smallest angle, and the third angle is 21 degrees smaller than the largest. What are the measures of the angles?

13. It was noted that 50% of the girls and 40% of the boys in a class brought their own lunch from home rather than purchasing it at the cafeteria. If there are 10 more boys than girls in this class, and if a total of 30 students brought their own lunch, then how many students bought theirs at the cafeteria?

14. What is the area of the trapezoidal room shown in Figure 21-8? The wall on the left side is perpendicular to the two bases, and the measures shown are in feet.

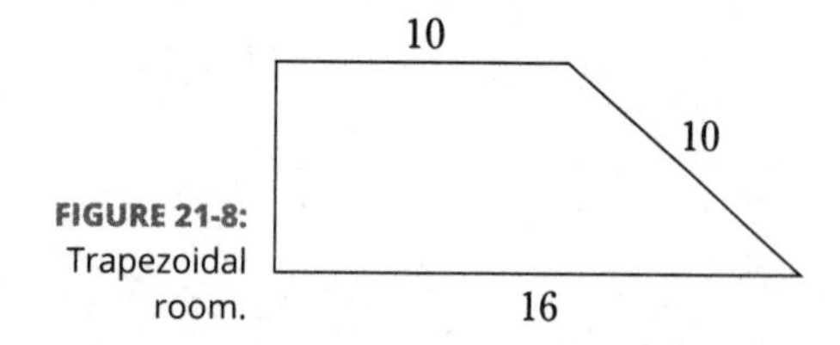

FIGURE 21-8: Trapezoidal room.

Answers to Chapter 21 Quiz

1. **\$69.** Let the sales price $51.75 = p - 0.25p = 0.75p$, where p is the original price. Dividing each side of the equation by 0.75, you have $p = 69$.

2. **22.5 sq. in.** If the circular crust touches all four sides, then the diameter of the circle is 10 inches. The radius is then 5 inches. Using $A = \pi r^2$, $A = \pi 5^2 = 25\pi = 25(3.14) = 78.5$. The area of the 10-inch-square dough is 100 square inches. So $100 - 78.5 = 22.5$. That's almost a fourth of the dough! I'm sure the cook will save it for the next "rollout."

3. **\$3.75.** If you get 25 miles per gallon, then 500 miles takes 20 gallons: $\frac{500 \text{ miles}}{25 \text{ gallons}} =$ 20 miles per gallon. If 20 gallons cost \$75, then $\frac{75 \text{ dollars}}{20 \text{ gallons}} = 3.75$ dollars per gallon.

4. **47 mph.** Using $d = rt$, the distance traveled for the first part of the trip was $d = 65 \cdot 2 = 130$ miles and the second part was $d = 35 \cdot 3 = 105$ miles. That's a total of 235 miles in 5 hours. So $135 = r \cdot 5$, giving you $r = 47$.

5. **$A = 330$ sq. ft., $P = 132$ ft.** The area of a right triangle is found with $A = \frac{1}{2}bh$, where the *base* and *height* are the lengths of the two legs. So $A = \frac{1}{2}(11)(60) = 330$. To find the perimeter, you need the length of the hypotenuse. Using the Pythagorean Theorem, $11^2 + 60^2 = c^2$, $3{,}721 = c^2$, so $c = 61$. Add up the three sides for the perimeter: $11 + 60 + 61 = 132$.

6. **300 feet.** Let the measures of the two sides each be w. The total of the three sides needs to be 600, so $x + 2w = 600$. Solving for w, you have $w = 300 - \frac{1}{2}x$. Make a chart with the given values of x, w, and the area, xw.

x	$w = 300 - \frac{1}{2}x$	$A = xw$
100	250	25,000
200	200	40,000
300	150	45,000
400	100	40,000
500	50	25,000

You see that the largest area occurs when x is 300 feet. Some good news: this problem is much easier when you use calculus! Now I bet you can't wait!

7. **12 ft.** Using $V = \frac{1}{3}Bh$, where B is the area of the base, you have $400 = \frac{1}{3}(100)h$, because the base is a square that's 10 feet on each side. Solving for h, you have $h = 12$.

8. **500 ft.** The fence is the circumference. Using $C = \pi d$, and 3.14 for π, you have $1{,}570 = 3.14d$. Dividing by 3.14, $d = 500$.

9. **4:00.** Let the time that Jim traveled be t and the time that Janet traveled be $t - 1$. Then, using their respective rates, $70t + 45(t - 1) = 185$. This simplifies to $115t = 230$, giving you $t = 2$. Because t is the time Jim traveled, and he left at 2:00, they met at 4:00.

10. **$37\frac{1}{3}$.** Set up the proportion for $\frac{AB}{BC} = \frac{DE}{EF}$ and you have $\frac{21}{x} = \frac{9}{16}$. (Note that there are several other proportions you can use.) Reducing, first, you have $\frac{7}{x} = \frac{3}{16}$, and cross-multiplying, $3x = 112$. Divide each side by 3 to get $x = \frac{112}{3} = 37\frac{1}{3}$.

11. **$17,011.50.** Using $A = P\left(1+\frac{r}{n}\right)^{nt}$, you find that $A = 12{,}000\left(1+\frac{0.07}{12}\right)^{12\cdot 5} = 12{,}000(1.0058333)^{60} =$ 17,011.50.

12. **39, 81, and 60 degrees.** Let the smallest angle measure x, so the largest is $2x+3$, and the third angle is $2x+3-21=2x-18$. The sum of the measures of a triangle is 180 degrees, so $x+2x+3+2x-18=180$, which simplifies to $5x=195$ and $x=39$. Using this value for x, $2x+3=81$ and $2x-18=60$.

13. **17.** If there are 10 more boys than girls, then let g represent the number of girls and $g+10$ be the number of boys. Then $g+g+10=30$, giving you $2g=20$ and $g=10$. The number of boys is $g+10$, which is 20. If 50% of the girls brought their lunch, that's 5 girls. If 40% of the boys brought their lunch, that's 40% of 20, or 8 boys. With 5 girls and 8 boys bringing their own lunch, $30-13=17$ bought their lunch at the cafeteria.

14. **104 sq. ft.** To find the area of a trapezoid, you need the height between the two parallel sides. You can find this by isolating the right triangle to the right of the room. Drop a perpendicular from the upper-right corner, and you have a right triangle whose hypotenuse is 10 and one side that is 6. Using the Pythagorean Theorem, $10^2=6^2+h^2$, giving you $h^2=64$ and $h=8$. Now, using the formula for area, $A=\frac{1}{2}(8)(10+16)=104$.

IN THIS CHAPTER

» Dealing with age problems

» Counting up consecutive integers

» Working on sharing the workload

Chapter 22

Relating Values in Story Problems

Yes, this is another chapter on story problems. Just in case you're one of those people who are less than excited about the prospect of problems made up of words, I've been breaking these problems down into specific types to help deliver the material logically. Chapter 20 deals with formulas, Chapter 21 gets into geometry and distances, and now this chapter focuses on age problems, consecutive integers, and work problems.

Problems dealing with age and consecutive integers have something in common: You use one or more base values or ages, assigning variables, and you keep adding the same number to each base value. As you work through these problems, you find some recurring patterns, and the theme is to use the same format — pick a variable for a number and add something on to it.

As for work problems, they're a completely different bird. In work problems, you need to divvy up the work and add it all together to get the whole job done. The portions usually aren't equal, and the number of participants can vary.

This chapter offers plenty of opportunities for you to tackle these story problems and overcome any trepidation.

Tackling Age Problems

Age problems in algebra don't have anything to do with wrinkles or thinning hair. Algebra deals with age problems very systematically and with an eye to the future. A story problem involving ages usually includes something like "in four years" or "ten years ago." The trick is to have everyone in the problem age the same amount (for example, add four years to each person's age, if needed).

When establishing your equation, make sure you keep track of how you name your variables. The letter x can't stand for Joe. The letter x can stand for Joe's *age.* If you keep in mind that the variables stand for numbers, the problem — and the process — will all make more sense.

EXAMPLE

Q. Joe's father is twice as old as Joe is. Twelve years ago, Joe's father was three years older than three times Joe's age. How old is Joe's father now?

A. Joe's father is 42 years old. Now, I'll show you how to set this up and solve it.

Joe's father is twice as old as Joe is. (He hasn't always been and won't be again — think about it: When Joe was born, was his father twice as old as he was?)

1. **Assign a variable to Joe's age.**

   Let Joe's age be x. Joe's father is twice as old, so Joe's father's age is $2x$.

2. **Continue to read through the problem.**

   To reflect the text, "Twelve years ago," both Joe's age and his father's age have to be backed up by 12 years. Their respective ages, 12 years ago, are $x-12$ and $2x-12$.

3. **Take the rest of the sentence where "Joe's father was three years older than" and change it into an equation, putting an equal sign where the verb is.**

   "Twelve years ago, Joe's father" becomes $2x-12$.

   "was" becomes =.

   "three years older than" becomes 3 +.

   "three times Joe's age (12 years ago)" becomes $3(x-12)$.

4. **Put this information all together in an equation.**

   $2x-12=3+3(x-12)$

5. **Solve for x.**

   $2x-12=3+3x-36=3x-33, \text{ so } x=21$

   That's Joe's age, so his father is twice that, or 42.

YOUR TURN

1. Jack is three times as old as Chloe. Ten years ago, Jack was five times as old as Chloe. How old are Jack and Chloe now?

2. Linda is ten years older than Luke. In ten years, Linda's age will be 30 years less than twice Luke's age. How old is Linda now?

3. Avery is six years older than Patrick. In four years, the sum of their ages will be 26. How old is Patrick now?

4. Jon is three years older than Jim, and Jim is two years older than Jane. Ten years ago, the sum of their ages was 40. How old is Jim now?

Tackling Consecutive Integer Problems

When items are *consecutive,* they follow along one after another. An *integer* is a positive or negative whole number, or 0. So you put these two things together to get consecutive integers. *Consecutive integers* have patterns — they're evenly spaced. The following three lists are examples of consecutive integers.

Consecutive integers: 5, 6, 7, 8, 9,

Consecutive odd integers: 11, 13, 15, 17, 19,

Consecutive multiples of 8: 48, 56, 64, 72, 80,

After you get one of the integers in a list and are given the rule, you can pretty much get all the rest of the integers. When doing consecutive integer problems, let x represent one of the integers (usually the first in your list) and then add on 1, 2, or whatever the spacing is to the next number; then add that amount on again to the new number, and so on until you have as many integers as you need.

EXAMPLE

Q. The sum of six consecutive integers is 255. What are they?

A. The first integer in my list is x. The next is $x+1$, the one after that is $x+2$, and so on. The equation for this situation reads: $x+(x+1)+(x+2)+(x+3)+(x+4)+(x+5)=255$. (*Note:* The parentheses aren't necessary. I just include them so you can see the separate terms.) Adding up all the x's and numbers, the equation becomes $6x+15=255$. Subtracting 15 from each side and dividing each side by 6, you get $x=40$. Fill the 40 into your original equations to get the six consecutive integers. (If $x=40$, then $x+1=41$, $x+2=42$, and so on.) The integers are 40, 41, 42, 43, 44, and 45.

Q. The sum of four consecutive odd integers is 8. What are they?

A. The equation for this problem is $x+(x+2)+(x+4)+(x+6)=8$. It becomes $4x+12=8$. Subtracting 12 and then dividing by 4, you get $x=-1$. You may have questioned using the +2, +4, +6 when dealing with odd integers. The first number is −1. Replace the x with −1 in each case to get the rest of the answers. The integers are −1, 1, 3, and 5. The problem designates x as an odd integer, and the other integers are all two steps away from one another. It works!

YOUR TURN

5 The sum of three consecutive integers is 57. What are they?

6 The sum of four consecutive even integers is 52. What is the largest of the four?

7 The sum of three consecutive odd integers is 75. What is the middle number?

8 The sum of five consecutive multiples of 4 is 20. What are they?

9. The sum of the smallest and largest of three consecutive integers is 126. What is the middle number of those consecutive integers?

10. The product of two consecutive integers is 89 more than their sum. What are they?

Working Together on Work Problems

Work problems in algebra involve doing jobs alone and together. Together is usually better, unless the person you're working with distracts you. I take the positive route and assume that two heads are better than one.

The general format for these problems is to let x represent how long it takes to do the job working together. Follow these steps and you won't even break a sweat when solving work problems:

1. **Write the amount that a person can do in one time period as a fraction.**

   If they can do the job in 6 hours, then they can do $\frac{1}{6}$ in one hour.

2. **Multiply that amount by the x, the length of time it takes to do the job working together.**

   You've multiplied each fraction that each person can do by the time it takes to do the whole job. So the person doing $\frac{1}{6}$ does $\frac{x}{6}$.

3. **Add the portions of the job that are completed in one time period together and set the sum equal to 1.**

   Setting the amount to 1 is 100 percent of the job.

EXAMPLE

Q. Meg can clean out the garage in five hours. Mike can clean out the same garage in three hours. How long will the job take if they work together?

A. Let x represent the amount of time it takes to do the cleaning when Meg and Mike work together. Meg can do $\frac{1}{5}$ of the job in one hour, and Mike can do $\frac{1}{3}$ of the job in one hour. The equation to use is $\frac{x}{5}+\frac{x}{3}=1$. Multiply both sides of the equation by the common denominator and add the two fractions together: $15\left(\frac{x}{5}+\frac{x}{3}\right)=15(1)$, or $3x+5x=15$, giving you $8x=15$. Divide by 8, and you have $x=\frac{15}{8}$. With the two of them working together, it'll take just under two hours.

Q. Carlos can wash the bus in seven hours, and when Carlos and Carol work together, they can wash the bus in three hours. How long would it take Carol to wash the bus by herself?

A. Instead of having x in the numerators, you already know that it'll take 3 hours working together, so put 3's in the numerators. Let y represent the amount of time it takes Carol to wash the bus by herself. This time, you have the time that it takes working together, so your equation is $\frac{3}{7}+\frac{3}{y}=1$. The common denominator of the two fractions on the left is $7y$. Multiply both sides of the equation by $7y$, simplify, and solve for y.

$$7y\left(\frac{3}{7}+\frac{3}{y}\right)=7y(1)$$

$$\cancel{7}y\left(\frac{3}{\cancel{7}}\right)+7\cancel{y}\left(\frac{3}{\cancel{y}}\right)=7y$$

$$3y+21=7y$$

$$21=4y$$

$$5\frac{1}{4}=y$$

It would take Carol $5\frac{1}{4}$ hours working by herself.

YOUR TURN

11 Alissa can do the job in three days, and Alex can do the same job in four days. How long will it take if they work together?

12 George can paint the garage in five days, Geanie can paint it in eight days, and Greg can do the job in ten days. How long will painting the garage take if they all work together?

13 Working together, Jon and Helen wrote a company organizational plan in $1\frac{1}{3}$ days. Working alone, it would have taken Jon four days to write that plan. How long would it have taken Helen if she had written it alone?

14 Rancher Biff needs his new fence put up in four days — before the herd arrives. Working alone, it'll take him six days to put up all the fencing. He can hire someone to help. How fast does the hired hand have to work in order for them to complete the job before the herd arrives?

15 When hose A is running full-strength to fill the swimming pool, it takes 8 hours; when hose B is running full-strength, filling the pool takes 12 hours. How long would it take to fill the pool if both hoses were running at the same time?

16 Elliott set up hose A to fill the swimming pool, and planned on it taking eight hours. But he didn't notice that water was leaking out of the pool through a big crack in the bottom. With just the water leaking, the pool would be empty in 12 hours. How long would filling the pool take with Elliott adding water with hose A and the leak emptying the pool at the same time?

Throwing an Object into the Air

A well-known formula used to determine the height of an object that has been dropped, thrown up in the air, or thrown downward from an elevated position is $h = -16t^2 + v_0 t + h_0$. The letter h (without the subscript) stands for the height of the object, t stands for how much time has elapsed in seconds, v_0 stands for the initial velocity of the object in feet per second, and h_0 stands for the initial or beginning height. The multiplier -16 on the squared term is based on the pull of gravity. You find this formula in physics and calculus problems, where one of the goals is to find out how high the object goes or how fast it's traveling at a particular point in time. In this section, I concentrate on the algebra aspect that determines when the object hits the ground.

EXAMPLE

Q. A rocket is shot upward from ground level at an initial speed of 128 feet per second. How long does it take for the rocket to come back and hit the ground?

A. Using $h=-16t^2+v_0t+h_0$, let $h=0$, because the height of the object will be 0 feet when it hits the ground. The initial velocity, $v_0=128$, and the initial height, $h_0=0$, because it starts at ground level. The equation to solve is $0=-16t^2+128t$, which factors into $0=-16t(t-8)$. Using the multiplication property of zero, you have $t=0$ or $t=8$. The solution $t=0$ tells you that at time 0 (the instant of launch), the rocket was at ground level. The solution $t=8$ tells you that 8 seconds later, the rocket has hit the ground. The answer is 8 seconds.

Q. A man is standing on a ladder that's 48 feet above the ground and throws a water balloon up into the air at a speed of 32 feet per second. How long will it take for the balloon to hit the ground?

A. Using $h=-16t^2+v_0t+h_0$, let $h=0$, because the height of the object will be 0 feet when it hits the ground. The initial velocity, $v_0=32$, and the initial height, $h_0=48$. The equation to solve is $0=-16t^2+32t+48$, which factors into $0=-16(t^2-2t-3)=-16(t-3)(t+1)$. Using the multiplication property of zero, you have $t=3$ or $t=-1$. The solution $t=3$ tells you that 3 seconds after being thrown, the balloon hits the ground. But what about $t=-1$? You can't go back in time. What this represents is when the balloon would have started its ascent if it had started from the ground, not up on the ladder. The answer is 3 seconds.

YOUR TURN

17 A projectile is launched upward at an initial speed of 160 feet per second. How long does it take for the projectile to hit the ground?

18 A baseball is thrown upward from a bridge that's 240 feet above ground at a speed of 32 feet per second. When does the baseball hit the ground?

19 A target is dropped from a balloon 576 feet above the ground. How long does it take for it to reach the ground (and for those shooting at the target to hit it)?

20 An egg is thrown by a man into the air at a speed of 24 feet per second. When the egg leaves the man's hand, it is 7 feet above ground. How long will it take the egg to hit the ground?

Practice Questions Answers and Explanations

1. **Chloe is 20, and Jack is 60.** Let x = Chloe's present age and 3x = Jack's present age. Ten years ago, Chloe's age was $x-10$, and Jack's was $3x-10$. Also ten years ago, Jack's age was five times Chloe's age. You can write this equation as $3x-10=5(x-10)$. Distribute the 5 to get $3x-10=5x-50$. Subtract 3x from each side and add 50 to each side, and the equation becomes $40=2x$. Divide by 2 to get $x=20$.

2. **Linda is 40.** Let x = Luke's age now. That makes Linda's present age = $x+10$. In ten years, Luke will be $x+10$, and Linda will be $(x+10)+10=x+20$. But at these new ages, Linda's age will be 30 years less than twice Luke's age. This is written as $x+20=2(x+10)-30$. Distribute the 2 to get $x+20=2x+20-30$. Simplifying, $x+20=2x-10$. Subtract x from each side and add 10 to each side to get $x=30$. Luke is 30, and Linda is 40. In ten years, Luke will be 40, and Linda will be 50. Twice Luke's age then, minus 30, is $80-30=50$. It checks.

3. **Patrick is 6.** Let x = Patrick's age now. Then Avery's age = $x+6$. In four years, Patrick will be $x+4$ years old, and Avery will be $x+6+4=x+10$ years old. Write that the sum of their ages in four years: $(x+4)+(x+10)=26$. Simplify on the left to get $2x+14=26$. Subtract 14 from each side: $2x=12$. Divide by 2, and you get $x=6$. Patrick is 6, and Avery is 12. In four years, Patrick will be 10, and Avery will be 16. The sum of 10 and 16 is 26.

4. **Jim is 23.** Let x represent Jane's age now. Jim is two years older, so Jim's age is $x+2$. Jon is three years older than Jim, so Jon's age is $(x+2)+3=x+5$. Ten years ago, Jane's age was $x-10$, Jim's age was $x+2-10=x-8$, and Jon's age was $x+5-10=x-5$. Add the ages ten years ago together to get 40: $x-10+(x-8)+(x-5)=40$. Simplifying on the left, $3x-23=40$. Add 23 to each side to get $3x=63$. Dividing by 3, $x=21$. Jane's age is 21, so Jim's age is $21+2=23$.

5. **18, 19, and 20.** Let x = the smallest of the three consecutive integers. Then the other two are $x+1$ and $x+2$. Adding the integers together to get 57, $x+(x+1)+(x+2)=57$, which simplifies to $3x+3=57$. Subtract 3 from each side to get $3x=54$. Dividing by 3, $x=18$. Checking, you get $18+19+20=57$.

6. **16.** The four integers are 10, 12, 14, and 16. Let the smallest integer = x. The other integers will be 2, 4, and 6 larger, so they can be written as $x+2$, $x+4$, and $x+6$. Adding them, you get $x+(x+2)+(x+4)+(x+6)=52$. Simplifying gives you $4x+12=52$. Subtract 12 from each side to get $4x=40$. Divide each side by 4, and $x=10$. Checking, you get $10+12+14+16=52$. The largest number is 16.

7. **25.** Let the smallest odd integer = x. Then the other two are $x+2$ and $x+4$. Add them together: $x+(x+2)+(x+4)=75$. Simplifying, $3x+6=75$. Subtract 6 and divide by 3 to get $x=23$. That's the smallest. The other two are 25 and 27: $23+25+27=75$. The middle number is 25.

8. **−4, 0, 4, 8, and 12.** Let x = the first of the consecutive multiples of 4. Then the other four are $x+4$, $x+8$, $x+12$, and $x+16$. Add them together: $x+(x+4)+(x+8)+(x+12)+(x+16)=20$. Simplifying on the left, $5x+40=20$. Subtract 40 from each side to get $5x=-20$. Dividing by 5, $x=-4$. Then add 4's to find the other integers.

9 **63.** Let x = the smallest of the consecutive integers. Then the other two are $x+1$ and $x+2$. Because the sum of the smallest and largest integers is 126, you can write it as $x+(x+2)=126$. Simplifying the equation, $2x+2=126$. Subtract 2 from each side to get $2x=124$. Dividing by 2, $x=62$, which is the smallest integer. The middle one is one bigger, so it's 63, and the largest is 64.

10 **10 and 11 or −9 and −8.** Let the smaller of the integers = x. The other one is then $x+1$. Their product is written $x(x+1)$ and their sum is $x+(x+1)$. Now, to write that their product is 89 more than their sum, the equation is $x(x+1)=89+x+(x+1)$. Distributing the x on the left and simplifying on the right, $x^2+x=2x+90$. Subtract 2x and 90 from each side to set the quadratic equation equal to 0: $x^2-x-90=0$. The trinomial on the left side of the equation factors to give you $(x-10)(x+9)=0$. So, $x=10$ or $x=-9$. If $x=10$, then $x+1=11$. The product of 10 and 11 is 110. That's 89 bigger than their sum, 21. What if $x=-9$? The next bigger number is then −8. Their product is 72. The difference between their product of 72 and sum of −17 is $72-(-17)=89$. So this problem has two possible solutions.

11 $\mathbf{\frac{12}{7}}$ **days.** Let x = the number of days to do the job together. Alissa can do $\frac{1}{3}$ of the job in one day and $\frac{1}{3}(x)$ of the job in x days. In x days, as they work together, they're to do 100 percent of the job. $\frac{1}{3}(x)+\frac{1}{4}(x)=1$, which is 100 percent. Multiply by 12: $4x+3x=12$, or $7x=12$. Divide by 7: $x=\frac{12}{7}=1\frac{5}{7}$ days to do the job.

Alissa's share is $\frac{1}{3}\left(\frac{12}{7}\right)=\frac{4}{7}$, and Alex's share is $\frac{1}{4}\left(\frac{12}{7}\right)=\frac{3}{7}$. Together, $\frac{4}{7}+\frac{3}{7}=1$.

12 $\mathbf{2\frac{6}{17}}$ **days.** Let x = the number of days to complete the job together. In x days, George will paint $\frac{1}{5}(x)$ of the garage, Geanie $\frac{1}{8}(x)$ of the garage, and Greg $\frac{1}{10}(x)$ of the garage. The equation for completing the job is $\frac{1}{5}(x)+\frac{1}{10}(x)+\frac{1}{8}(x)=1$. Multiplying through by 40, which is the least common denominator of the fractions, you get $8x+4x+5x=40$. Simplifying, you get $17x=40$. Dividing by 17 gives you $x=\frac{40}{17}=2\frac{6}{17}$ days to paint the garage. Checking the answer, George's share is $\frac{1}{5}\left(\frac{40}{17}\right)=\frac{8}{17}$, Geanie's share is $\frac{1}{8}\left(\frac{40}{17}\right)=\frac{5}{17}$, and Greg's share is $\frac{1}{10}\left(\frac{40}{17}\right)=\frac{4}{17}$. Together, $\frac{8}{17}+\frac{5}{17}+\frac{4}{17}=\frac{17}{17}$, or 100 percent.

13 **2 days.** Let y = the number of days for Helen to write the plan alone. So Helen writes $\frac{1}{y}$ of the plan each day. Jon writes $\frac{1}{4}$ of the plan per day. In $1\frac{1}{3}=\frac{4}{3}$ days, they complete the job together. The equation is $\left(\frac{1}{y}\right)\left(\frac{4}{3}\right)+\left(\frac{1}{4}\right)\left(\frac{4}{3}\right)=1$. Multiplying each side by 12y, the least common denominator, gives you $\left(\frac{4}{3y}\right)(12y)+\left(\frac{4}{12}\right)(12y)=1(12y), 16+4y=12y$. Subtract 4y from each side to get $16=8y$. Dividing by 8, $y=2$ days. Helen will complete the job in 2 days. To check this, in $\frac{4}{3}$ days, Jon will do $\frac{1}{4}\left(\frac{4}{3}\right)=\frac{1}{3}$ of the work, and Helen will do $\frac{1}{2}\left(\frac{4}{3}\right)=\frac{2}{3}$. Together, $\frac{1}{3}+\frac{2}{3}=1$ or 100 percent.

14. **12 days.** Let y = the number of days the hired hand needs to complete the job alone. The hired hand does $\frac{1}{y}$ of the fencing each day, and Biff puts up $\frac{1}{6}$ of the fence each day. In four days, they can complete the project together, so $\left(\frac{1}{y}\right)(4)+\left(\frac{1}{6}\right)(4)=1$. Multiply by the common denominator $6y$: $\left(\frac{4}{y}\right)(6y)+\left(\frac{4}{6}\right)(6y)=(1)(6y)$. Simplifying, $24+4y=6y$. Subtracting $4y$ from each side, $24=2y$. Dividing by 2, $y=12$. The hired hand must be able to do the job alone in 12 days. To check this, in 4 days, Biff does $\left(\frac{1}{6}\right)(4)=\frac{2}{3}$ of the fencing, and the hired hand $\left(\frac{1}{12}\right)(4)=\frac{1}{3}$ of the job.

15. **4.8 hours.** Let x represent the amount of time to complete the job. Your equation is $\frac{x}{8}+\frac{x}{12}=1$. Multiplying both sides of the equation by 24, you get $3x+2x=24$. Simplifying and then dividing by 5, you get $5x=24$, $x=4.8$.

16. **24 hours.** This time, one of the terms is negative. The leaking water takes away from the completion of the job, making the time longer. Using the equation $\frac{x}{8}-\frac{x}{12}=1$, multiply both sides of the equation by 24, giving you $3x-2x=24$ or $x=24$.

17. **10 seconds.** Using $0=-16t^2+160t$, the equation factors into $0=-16t(t-10)$. The two solutions are $t=0$ and $t=10$. The solution $t=0$ is the height at time 0, the beginning. The solution $t=10$ tells you that the projectile hits the ground after 10 seconds.

18. **5 seconds.** Using $0=-16t^2+32t+240$, the equation factors into $0=-16(t^2-2t-15)=-16(t-5)(t+3)$. The two solutions are $t=5$ and $t=-3$. The solution $t=5$ tells you that the baseball hits the ground in 5 seconds. The solution $t=-3$ tells you that it would have started its ascent 3 seconds earlier, if it hadn't started from the bridge.

19. **6 seconds.** Using $0=-16t^2+576$, the equation factors into $0=-16(t^2-36)=-16(t-6)(t+6)$. The two solutions are $t=6$ and $t=-6$. The solution $t=6$ tells you that the target hits the ground in 6 seconds. The solution $t=-6$ tells you that it would have started its descent 6 seconds earlier, if it hadn't started from the balloon.

20. $1\frac{3}{4}$ **seconds.** Using $0=-16t^2+24t+7$, the equation factors into $0=-(4t+1)(4t-7)$, but sometimes it's easier to use the quadratic formula (see Chapter 16) and get

$$t=\frac{-24\pm\sqrt{24^2-4(-16)(7)}}{2(-16)}=\frac{-24\pm\sqrt{576+448}}{-32}=\frac{-24\pm\sqrt{1024}}{-32}=\frac{-24\pm 32}{-32}=\frac{-3\pm 4}{-4}$$

The first solution, using +4 in the numerator, is $t=\frac{-3+4}{-4}=-\frac{1}{4}$, and the second solution, using −4 in the numerator, is $t=\frac{-3-4}{-4}=\frac{7}{4}$. The first solution, $t=-\frac{1}{4}$, tells you when the egg would have started its ascent, if it hadn't been at the end of the man's outstretched arm. The solution $t=\frac{7}{4}$ tells you when the egg hits the ground, after $1\frac{3}{4}$ seconds.

If you're ready to test your skills a bit more, take the following chapter quiz that incorporates all the chapter topics.

Whaddya Know? Chapter 22 Quiz

Quiz time! Complete each problem to test your knowledge on the various topics covered in this chapter. You can then find the solutions and explanations in the next section.

1. Wolf is 4 years older than Blake. Twice the sum of their ages is three times Wolf's age. How old is Wolf?
2. The sum of four consecutive integers is 86. What are they?
3. If Donna can weed the garden in 8 hours, and Dave can weed it in 5 hours, then how long will it take to weed the garden if they work together?
4. The sum of the smallest and largest of three consecutive odd integers is 62. What is the largest odd number?
5. A toy rocket was launched in the air at 80 feet per second. How long did it take for the rocket to hit the ground?
6. The product of two consecutive numbers is 4 more than four times their sum. What are the numbers?
7. Twins Ryan and Ron are three years older than triplets Tom, Tim, and Terry. If the sum of their ages is 56, then how old is Ryan?
8. If Hank can clean the pool in 3 hours, Eddie can clean the pool in 4 hours, and Freddie can clean the pool in 6 hours, then how long would it take for all three of them to clean the pool working together?
9. Fiona has a wonderful surprise for Oakley's birthday. She's dropping his gift from a helicopter that's 400 feet above the ground. How long will it take for the gift to reach the ground? (Hope it's not breakable.)
10. In 1776, George was 3 years older than John. Thomas was 8 years younger than John. And James was 3 years younger than George. If the sum of their ages was 143, then how old was George?
11. Stella can paint the garage in 18 hours. But if Stella and Eva can paint it in $7\frac{1}{5}$ hours when working together, then how long would it take Eva to do the job by herself?
12. Standing in a building's window 80 feet off the ground, when you throw a softball in the air at 64 feet per second, how long does it take for the softball to reach the ground?

Answers to Chapter 22 Quiz

1. **8 years old.** Let Blake's age = b and Wolf's age = $b+4$. Twice the sum of their ages is $2(b+b+4)=2(2b+4)$. Set this equal to three times Wolf's age and solve for b. $2(2b+4)=3(b+4)$ becomes $4b+8=3b+12$ or $b=4$. If Blake is 4, then Wolf is 8.

2. **20, 21, 22, 23.** Using $n+(n+1)+(n+2)+(n+3)=86$, you get $4n=80$ or $n=20$.

3. **$3\frac{1}{13}$ hours.** Using $\frac{x}{8}+\frac{x}{5}=1$, multiply through by 40 to get $5x+8x=40$ or $x=\frac{40}{13}$.

4. **33.** Three consecutive odd integers is written: n, $n+2$, $n+4$. Add the smallest and largest to get 62, and you have $n+(n+4)=62$ or $2n=58$. If n is 29, then the three consecutive odd numbers are 29, 31, 33.

5. **5 seconds.** Using $-16t^2+80t+0=0$, you have $-16t(t-5)=0$, giving you the solutions 0 and 5.

6. **8 and 9 or −1 and 0.** Solve $n(n+1)=4+4(n+(n+1))$, which simplifies to $n^2+n=4+8n+4$ and is written as the quadratic equation $n^2-7n-8=0$. This factors into $(n-8)(n+1)=0$, giving you the solutions $n=8$ and $n=-1$. When $n=8$, the next number is 9, and their product is 72, which is 4 more than 4(17). When $n=-1$, the next number is 0. The product of −1 and 0 is 0, which is 4 more than $4(-1)$.

7. **13 years old.** Let the triplets be t years old, making the twins $t+3$. The sum of all five ages is $2(t+3)+3t=56$, giving you $5t+6=56$ or $t=10$. Add 3 to the age, and Ryan is 13.

8. **$1\frac{1}{3}$ hours.** Using $\frac{x}{3}+\frac{x}{4}+\frac{x}{6}=1$, multiply through by 12 to get $4x+3x+2x=12$ or $9x=12$, which is $x=\frac{4}{3}$ hours.

9. **5 seconds.** Using $-16t^2+400=0$, you have $-16(t^2-25)=0$, which factors into $-16(t-5)(t+5)=0$. The $t=-5$ refers back to if the gift had been tossed up from the ground. Your answer is $t=5$.

10. **40 years old.** Let n represent John's age; thus, George is $n+3$, Thomas is $n-8$, and James is n years old. Find the sum and solve for n: $n+n+3+n-8+n=143$ gives you $4n=148$ or $n=37$. George's age is $n+3$, so he was 40.

11. **12 hours.** Using $\frac{7\frac{1}{5}}{18}+\frac{7\frac{1}{5}}{y}=1$, multiply through by $18y$ to get $7\frac{1}{5}y+18\left(7\frac{1}{5}\right)=18y$ or $\frac{36}{5}y+\frac{18\cdot 36}{5}=18y$. You don't need to multiply that second numerator yet. You're hoping it'll reduce nicely later. Now multiply through by 5 to get rid of the denominators. You have $36y+18\cdot 36=90y$ or $18\cdot 36=54y$. Divide each side by 8 to get $36=3y$ or $12=y$.

12. **5 seconds.** Using $-16t^2+64t+80=0$, you have $-16(t^2-4t-5)=0$, which factors into $-16(t-5)(t+1)=0$. The $t=-1$ refers back to if the softball had been thrown from the ground. Your answer is $t=5$.

IN THIS CHAPTER

- » Making the most of mixtures
- » Understanding the strength of a solution
- » Keeping track of money in piggy banks and interest problems

Chapter 23

Measuring Up with Quality and Quantity Story Problems

The story problems in this chapter have a common theme to them: they deal with *quality* (the strength or worth of an item) and *quantity* (the measure or count), and adding up to a total amount. (Chapters 21, 22, and 23 have other types of story problems.) You encounter quality and quantity problems almost on a daily basis. For instance, if you have four dimes, you know that you have 40 cents. How do you know? You multiply the quantity, four dimes, times the quality, ten cents each, to get the total amount of money. And if you have a fruit drink that's 50 percent real juice, then a gallon contains one-half gallon of real juice (and the rest is who-knows-what) — again, multiplying quality times quantity.

In this chapter, take time to practice with these story problems. Just multiply the amount of something, the *quantity*, times the strength or worth of it, *quality*, in order to solve for the total value.

Achieving the Right Blend with Mixture Problems

Mixture problems can take on many different forms. There are the traditional types, in which you can actually mix one solution with another, such as water and antifreeze. There are the types in which different solid ingredients are mixed, such as in a salad bowl or candy dish. Another type is where different investments at different interest rates are mixed together. I lump all these types of problems together in this chapter because you use basically the same process to solve them.

Drawing a picture helps with all mixture problems. The same picture can work for all: liquid, solid, and investments. Figure 23-1 shows three sample containers — two added together to get a third (the mixture). In each case, the containers are labeled with the quality and quantity of the contents. These two values get multiplied together before adding. The quality is the strength of the antifreeze or the percentage of the interest or the price of the ingredient. The quantity is the amount in quarts or dollars or pounds. You can use the same picture for the containers in every mixture problem, or you can change to bowls or boxes. It doesn't matter — you just want to visualize the way the mixture is going together.

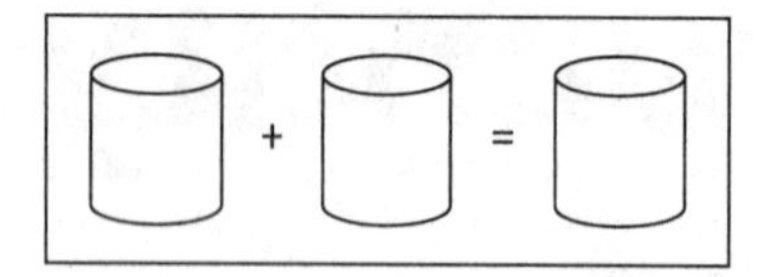

FIGURE 23-1: Visualizing containers can help with mixture problems.

Do you buy cans of mixed nuts? When I do, I always pick out and eat the cashews first. Do you wonder why there seem to be so few of your favorite type and so many peanuts? Well, some types of nuts are more expensive than others, and some are more popular than others. The nut folks take these factors into account when they devise the proportions for a mixture that is both desirable and affordable.

How many pounds of cashews that cost $5.50 per pound should be mixed with 3 pounds of peanuts that cost $2 per pound to create a mixture that costs $3 per pound? (You can use this formula to save your budget for your next big party.)

Using containers makes sense here. Let x represent the number of pounds of cashews. The quality is the cost of the nuts and the quantity is the number of pounds. Put the cost (quality) on the top of each container and the number of pounds (quantity) on the bottom.

Refer to Figure 23-2. The first container has $5.50 on the top and x pounds on the bottom. The second container has $2 on the top and 3 pounds on the bottom. The third container, with the mixture, has $3 on the top and $x+3$ on the bottom. Now write the equation reflecting this relationship.

$$5.5x+2(3)=3(x+3)$$
$$5.5x+6=3x+9$$

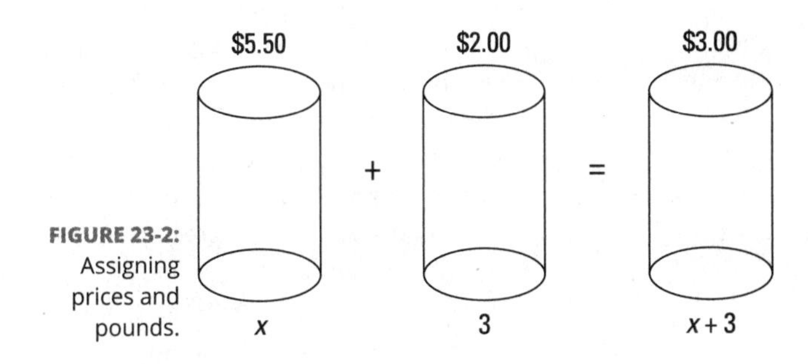

FIGURE 23-2: Assigning prices and pounds.

Subtracting $3x$ from each side and 6 from each side, I get $2.5x = 3$ or $x = 1.2$ pounds of the expensive nuts. I mix that with 3 pounds of peanuts to create a mixture of 4.2 pounds of nuts that costs \$3 per pound.

EXAMPLE

Q. A health store is mixing up some granola that has many ingredients, but three of the basics are oatmeal, wheat germ, and raisins. Oatmeal costs \$1 per pound, wheat germ costs \$3 per pound, and raisins cost \$2 per pound. The store wants to create a base granola mixture of those three ingredients that will cost \$1.50 per pound. (These items serve as the base of the granola; the rest of the ingredients and additional cost will be added later.) The granola is to have nine times as much oatmeal as wheat germ. How much of each ingredient is needed?

A. To start this problem, let *x* represent the amount of wheat germ in pounds. Because you need nine times as much oatmeal as wheat germ, you have 9*x* pounds of oatmeal. How much in raisins? The raisins can have whatever's left of the pound after the wheat germ and oatmeal are taken out: $1-(x+9x)$ or $1-10x$ pounds. Now multiply each of these amounts by their respective price: $\$3(x)+\$1(9x)+\$2(1-10x)$. Set this equal to the \$1.50 price multiplied by its amount, as follows: $\$3(x)+\$1(9x)+\$2(1-10x)=\$1.50(1)$. Simplify and solve for *x*: $3x+9x+2-20x=-8x+2=\$1.50$. Subtracting 2 from each side and then dividing each side by 8, gives you $-8x=0.50$, $x=\frac{-0.50}{-8}=\frac{1}{16}$.

Every pound of mixed granola will need $\frac{1}{16}$ pound of wheat germ, $\frac{9}{16}$ pound of oatmeal, and $\frac{6}{16}$ or $\frac{3}{8}$ pound of raisins.

YOUR TURN

1. Kathy's Kandies features a mixture of chocolate creams and chocolate-covered caramels that sells for \$9 per pound. If creams sell for \$6.75 per pound and caramels sell for \$10.50 per pound, how much of each type of candy should be in a 1-pound mix?

2. Solardollars Coffee is trying new blends to attract more customers. The premium Colombian costs \$10 per pound, and the regular blend costs \$4 per pound. How much of each should the company use to make 100 pounds of a coffee blend that costs \$5.50 per pound?

3. Peanuts cost $2 per pound, almonds cost $3.50 per pound, and cashews cost $6 per pound. How much of each should you use to create a mixture that costs $3.40 per pound, if you have to use twice as many peanuts as cashews?

4. A mixture of jellybeans is to contain twice as many red as yellow, three times as many green as yellow, and twice as many pink as red. Red jelly beans cost $1.50 per pound, yellow cost $3.00 per pound, green cost $4.00 per pound, and pink only cost $1.00 per pound. How many pounds of each color jellybean should be in a 10-pound canister that costs $2.20 per pound?

5. A Very Berry Smoothie calls for raspberries, strawberries, and yogurt. Raspberries cost $3 per cup, strawberries cost $1 per cup, and yogurt costs $0.50 per cup. The recipe calls for twice as many strawberries as raspberries. How many cups of strawberries are needed to make a gallon of this smoothie that costs $10.10? (*Hint:* 1 gallon = 16 cups).

6. A supreme pizza contains five times as many ounces of cheese as mushrooms, twice as many ounces of peppers as mushrooms, twice as many ounces of onions as peppers, and four more ounces of sausage than mushrooms. Mushrooms and onions cost 10 cents per ounce, cheese costs 20 cents per ounce, peppers are 25 cents per ounce, and sausage is 30 cents per ounce. If the toppings are to cost no more than a total of $5.80, then how many ounces of each ingredient can be used?

Concocting the Correct Solution 100% of the Time

A traditional solutions-type problem is where you mix water and antifreeze. When the liquids are mixed, the strengths of the two liquids average out.

How many quarts of 80% antifreeze have to be added to 8 quarts of 20% antifreeze to get a mixture of 60% antifreeze?

First, label your containers. The first would be labeled 80% on the top and x on the bottom. (I don't know yet how many quarts have to be added.) The second container would be labeled with 20% on the top and 8 quarts on the bottom. The third container, which represents the final mixture, would be labeled 60% on the top and $x+8$ quarts on the bottom. To solve this, multiply each "quality" or percentage strength of antifreeze times its "quantity" and put these in the equation:

$$80\%(x \text{ quarts}) + 20\%(8 \text{ quarts}) = 60\%(x+8)\text{quarts}$$
$$(0.8)x + 0.2(8) = 0.6(x+8)$$
$$0.8x + 1.6 = 0.6x + 4.8$$

Subtracting 0.6x from each side and subtracting 1.6 from each side, you get $0.2x = 3.2$. Dividing each side of the equation by 0.2, you get $x = 16$. So, 16 quarts of 80% antifreeze have to be added.

You can use the liquid mixture rules with salad dressings, mixed drinks, and all sorts of sloshy concoctions.

Solutions problems are sort of like mixtures problems (explained in the preceding section). The main difference is that solutions usually deal in percents — 30%, $27\frac{1}{2}\%$, 0%, or even 100%. These last two numbers indicate, respectively, that none of that ingredient is in the solution (0%) or that it's *pure* for that ingredient (100%). You've dealt with these solutions if you've had to add antifreeze or water to your radiator. Or how about adding that frothing milk to your latte mixture?

The general format for these solutions problems is

$$(\%\ \text{A} \times \text{amount A}) + (\%\ \text{B} \times \text{amount B}) = (\%\ \text{C} \times \text{amount C})$$

TIP

If you're adding pure alcohol or pure antifreeze or something like this, use 1 (which is 100%) in the equation. If there's no alcohol, chocolate syrup, salt, or whatever in the solution, use 0 (which is 0%) in the equation.

EXAMPLE

Q. How many gallons of 60% apple juice mix need to be added to 20 gallons of mix that's currently 25% apple juice to bring the new mix up to 32% apple juice?

A. To solve this problem, let x represent the unknown amount of 60% apple juice. Using the format of all the percents times the respective amounts, you get $(60\% \times x) + (25\% \times 20 \text{ gallons}) = [32\% \times (x + 20 \text{ gallons})]$. Change the percents to decimals, and the equation becomes $0.60x + 0.25(20) = 0.32(x + 20)$. (If you don't care for decimals, you could multiply each side by 100 to change everything to whole numbers.) Now distribute the 0.32 on the right and simplify so you can solve for x:

$$\begin{aligned} 0.60x + 5 &= 0.32x + 6.4 \\ 0.28x &= 1.4 \\ x &= 5 \end{aligned}$$

You need to add 5 gallons of 60% apple juice.

Q. How many quarts of water do you need to add to 4 quarts of lemonade concentrate in order to make the drink 25% lemonade?

A. To solve this problem, let x represent the unknown amount of water, which is 0% lemonade. Using the format of all the percents times the respective amounts, you get $(0\% \cdot x) + (100\% \cdot 4 \text{ quarts}) = [25\% \cdot (x + 4) \text{ quarts}]$. Change the percents to decimals, and the equation becomes $0x + 1.00(4) = 0.25(x + 4)$. When the equation gets simplified, the first term disappears.

$$\begin{aligned} 4 &= 0.25x + 1 \\ 3 &= 0.25x \\ \frac{3}{0.25} &= \frac{0.25x}{0.25} \\ 12 &= x \end{aligned}$$

You need to add 12 quarts of water.

YOUR TURN

7 How many quarts of 25% solution do you need to add to 4 quarts of 40% solution to create a 31% solution?

8 How many gallons of a 5% fertilizer solution have to be added to 2 gallons of a 90% solution to create a fertilizer solution that has 15% strength?

9. How many quarts of pure antifreeze need to be added to 8 quarts of 30% antifreeze to bring it up to 50%?

10. How many cups of chocolate syrup need to be added to 1 quart of milk to get a mixture that's 25% syrup?

11. What concentration should 4 quarts of salt water have so that, when it's added to 5 quarts of 40% solution salt water, the concentration goes down to $33\frac{1}{3}\%$?

12. What concentration and amount of solution have to be added to 7 gallons of 60% alcohol to produce 16 gallons of $37\frac{1}{2}\%$ alcohol solution?

Dealing with Money Problems

Story problems involving coins, money, or interest earned are all solved with a process like that used in solutions problems: You multiply a quantity times a quality. In these cases, the qualities are the values of the coins or bills, or they're the interest rate at which money is growing.

Investigating investments and interest

You can invest your money in a safe CD or savings account and get one interest rate. You can also invest in riskier ventures and get a higher interest rate, but you risk losing money. Most financial advisors suggest that you diversify — put some money in each type of investment — to take advantage of each investment's good points.

Use the simple interest formula in the following sample problem to simplify the process. With simple interest, the interest is figured on the beginning amount only. In practice, financial institutions are more likely to use the compound interest formula. Compound interest is figured on the changing amounts as the interest is periodically added into the original investment.

Khalil had $20,000 to invest last year. He invested some of this money at $3\frac{1}{2}\%$ interest and the rest at 8% interest. His total earnings in interest, for both of the investments, were $970. How much did he have invested at each rate?

Use containers again. Let x represent the amount of money invested at $3\frac{1}{2}\%$. The first container has $3\frac{1}{2}\%$ on top and x on the bottom. The second container has 8% on top and $20{,}000 - x$ on the bottom. The third container, the mixture, has $970 right in the middle. That's the result of multiplying the mixture percentage times the total investment of $20,000. You don't need to know the mixture percentage — just the result.

$$\begin{aligned} 3\tfrac{1}{2}\text{ percent}(x) + 8\text{ percent}(20{,}000 - x) &= 970 \\ 0.035(x) + 0.08(20{,}000 - x) &= 970 \\ 0.035x + 1{,}600 - 0.08x &= 970 \end{aligned}$$

Subtract 1,600 from each side and simplify on the left side:

$$-0.045x = -630$$

Dividing each side by −0.045, you get

$$x = 14{,}000$$

That means that $14,000 was invested at $3\frac{1}{2}\%$ and the other $6,000 was invested at 8%.

EXAMPLE

Q. Kathy wants to withdraw only the interest on her investment each year. She's going to put money into the account and leave it there, just taking the interest earnings. She wants to take out and spend $10,000 each year. If she puts two-thirds of her money where it can earn 5% interest and the rest at 7% interest, how much should she put at each rate to have the $10,000 spending money?

A. Let x represent the total amount of money Kathy needs to invest. The first container has 5% on top and $\frac{2}{3}x$ on the bottom. The second container has 7% on top and $\frac{1}{3}x$ on the bottom. The third container, or mixture, has $10,000 in the middle; this is the result of the "mixed" percentage and the total amount invested.

$$5\%\left(\frac{2}{3}x\right) + 7\%\left(\frac{1}{3}x\right) = 10{,}000$$

Change the decimals to fractions and multiply:

$$\begin{aligned} 0.05\left(\frac{2}{3}x\right) + 0.07\left(\frac{1}{3}x\right) &= 10{,}000 \\ \frac{1}{30}x + \frac{7}{300}x &= 10{,}000 \end{aligned}$$

Find a common denominator and add the coefficients of x:

$$\frac{17}{300}x = 10{,}000$$

Divide each side by $\frac{17}{300}$:

$$x \approx 176{,}470.59$$

Kathy needs over $176,000 in her investment account. Two-thirds of it, about $117,647, has to be invested at 5% and the rest, about $58,824, at 7%.

Q. Jon won the state lottery and has $1 million to invest. He invests some of it in a highly speculative venture that earns 18% interest. The rest is invested more wisely, at 5% interest. If he earns $63,000 in simple interest in one year, how much did he invest at 18%?

A. Let x represent the amount of money invested at 18%. Then the remainder, $1{,}000{,}000 - x$, is invested at 5%. The equation to use is $0.18x + .05x(1{,}000{,}000 - x) = 63{,}000$. Distributing the 0.05 on the left, and combining terms, you get $0.13x + 50{,}000 = 63{,}000$. Subtract 50,000 from each side, and you get $0.13x = 13{,}000$. Dividing each side by 0.13 gives you x = 100,000, the amount invested at 18%. It's kind of mind-boggling.

YOUR TURN

13 Blake invested $10,000 in two different funds for one year. She invested part at 2% and the rest at 3%. She earned $240 in simple interest. How much did she invest at each rate? (**Hint:** Use the simple interest formula: $I = Prt$.)

14 Elliott got a bonus check for $4,000. He invested it in two different funds for one year. Some was invested in a rather risky fund promising to earn 8% and the rest in a safer fund at 2%. If all goes well, he will earn $230 in simple interest. How much did he invest at each rate? (**Hint:** Use the simple interest formula: $I = Prt$.)

15 Fiona made some investments putting part of her money at 4% and twice as much as that amount at 5%. If she earned $1,400 in simple interest at the end of the year, how much did she invest in each fund? (***Hint:*** Use the simple interest formula: $I = Prt$.)

16 Wolf had $24,000 to invest. If he invested twice as much at 2% as he did at 1% and three times as much at 3% as at 1%, then how much interest can he expect to earn with simple interest by the end of the year? (***Hint:*** Use the simple interest formula: $I = Prt$.)

Going for the green: Money

Money is everyone's favorite topic; it's something everyone can relate to. It's a blessing and a curse. When you're combining money and algebra, you have to consider the number of coins or bills and their worth or denomination. Other situations involving money can include admission prices, prices of different pizzas in an order, or any commodity with varying prices. For the purposes of this book, U.S. coins and bills are used in the examples and practice problems in this section. I don't want to get fancy by including other countries' currencies.

EXAMPLE

Q. Chelsea has five times as many quarters as dimes, three more nickels than dimes, and two fewer than nine times as many pennies as dimes. If she has $15.03 in coins, how many of them are quarters?

A. The containers work here, too. There will be four of them added together: dimes, quarters, nickels, and pennies. The quality is the value of each coin. Every coin count refers to dimes in this problem, so let the number of dimes be represented by *x* and compare everything else to it.

The first container contains dimes, so put 0.10 on top and *x* on the bottom. The second container contains quarters, so put 0.25 on top and 5*x* on the bottom. The third container contains nickels, so put 0.05 on top and $x+3$ on the bottom. The fourth container contains pennies, so put 0.01 on top and $9x-2$ on the bottom. The mixture container, on the right, has $15.03 right in the middle.

$$0.10(x)+0.25(5x)+0.05(x+3)+0.01(9x-2)=15.03$$
$$0.10x+1.25x+0.05x+0.15+0.09x-0.02=15.03$$

Simplifying on the left, you get

$$1.49x+0.13=15.03$$

Subtracting 0.13,

$$1.49x = 14.90$$

And, after dividing by 1.49,

$$x = 10$$

Because x is the number of dimes, there are 10 dimes, 5 times as many (or 50) quarters, 3 more (or 13) nickels, and 2 less than 9 times (or 88) pennies. The question was, "How many quarters?" There were 50 quarters; use the other answers to check to see if this comes out correctly.

Q. Gabriella is counting the bills in her cash drawer before the store opens for the day. She has the same number of \$10 bills as \$20 bills. She has two more \$5 bills than \$10 bills, and ten times as many \$1 bills as \$5 bills. She has a total of \$300 in bills. How many of each does she have?

A. You can compare everything, directly or indirectly, to the \$10 bills. Let x represent the number of \$10 bills. The number of \$20 bills is the same, so it's also x. The number of \$5 bills is two more than the number of \$10 bills, so let the number of fives be represented by $x+2$. Multiply $x+2$ by 10 for the number of \$1 bills, $10(x+2)$. Now take each *number* of bills and multiply by the quality or value of that bill. Add them to get \$300:

$$\begin{aligned}10x + 20x + 5(x+2) + 1(10(x+2)) &= 300\\ 10x + 20x + 5x + 10 + 10x + 20 &= 300\\ 45x + 30 &= 300\\ 45x &= 270\\ x &= 6\end{aligned}$$

Using $x = 6$, the number of \$10 bills and \$20 bills, you get $x + 2 = 8$ for the number of \$5 bills, and $10(x+2) = 80$ for the number of \$1 bills.

YOUR TURN

17 Carlos has twice as many quarters as nickels and a total of \$8.25. How many quarters does he have?

18 Gregor has twice as many \$10 bills as \$20 bills, five times as many \$1 bills as \$10 bills, and half as many \$5 bills as \$1 bills. He has a total of \$750. How many of each bill does he have?

19 Stella has 100 coins in nickels, dimes, and quarters. She has 18 more nickels than dimes and a total of \$7.40. How many of each coin does she have?

20 Hawkeye has \$3.50 in coins in his pocket. He has twice as many nickels as fifty-cent pieces, two more quarters than nickels, and twice as many dimes as nickels. How many of each coin does he have?

Practice Questions Answers and Explanations

1. **0.4 pound of creams and 0.6 pound of caramels.** Let x represent the amount of chocolate creams in pounds. Then $1-x$ is the pounds of chocolate caramels. In a pound of the mixture, creams cost $6.75x and caramels, $\$10.50(1-x)$. Together, the mixture costs $9. So,

$$6.75x+10.5(1-x)=9 \to 6.75x+10.5-10.5x=9 \to -3.75x+10.5=9$$

Subtract 10.5 from each side and then divide by −3.75:

$$-3.75x=-1.5 \to x=\frac{-1.5}{-3.75}=0.4 \text{ pound of creams}$$

To get the amount of caramels, $1-x=1-0.4=0.6$ pound of caramels.

Checking this, the cost of creams is $\$6.75(0.4)=\$2.70$. The cost of caramels is $\$10.50(0.6)=\$6.30$. Adding these together, you get $2.70+6.30=9$ dollars.

2. **25 pounds of Colombian and 75 pounds of regular blend.** Let x = the pounds of Colombian coffee at $10 per pound. Then $100-x$ = pounds of regular blend at $4 per pound. The cost of 100 pounds of the mixture blend is $\$5.50(100)=\$550$. Use

$$10x+4(100-x)=550 \to 10x+400-4x=550 \to 6x+400=550$$

Subtract 400 from each side and then divide each side by 6:

$$6x=150 \to x=\frac{150}{6}=25$$

To check, multiply $10(25) and $4(75) to get $\$250+\$300=\$550$, as needed.

3. **$\frac{2}{5}$ pound almonds, $\frac{2}{5}$ pound peanuts, $\frac{1}{5}$ pound cashews.** Let x = the pounds of cashews at $6 per pound. Then $2x$ = pounds of peanuts at $2 per pound. The almonds are then $1-(x+2x)=1-3x$ pounds at $3.50 per pound. Combine all this and solve:

$$\begin{aligned}6x+2(2x)+3.5(1-3x)&=3.4\\ 6x+4x+3.5-10.5x&=3.4\\ -0.5x+3.5&=3.4\\ -0.5x&=-0.1\\ x&=\frac{-0.1}{-0.5}=\frac{1}{5}\text{ pound of cashews}\end{aligned}$$

Use this amount for x and substitute in to get the other weights. You get $2\left(\frac{1}{5}\right)=\frac{2}{5}$ pound of peanuts and $1-3\left(\frac{1}{5}\right)=1-\frac{3}{5}=\frac{2}{5}$ pound of almonds.

4. **1 pound yellow, 2 pounds red, 3 pounds green, and 4 pounds pink jellybeans.** Let x represent the number of pounds of yellow jellybeans. Then $2x$ is the pounds of red jellybeans, $3x$ is the pounds of green jellybeans, and $2(2x)=4x$ is the pounds of pink jellybeans. Multiply each quantity times its price and solve:

$$\begin{aligned}x(3.00)+2x(1.50)+3x(4.00)+4x(1.00)&=10(2.20)\\ 3x+3x+12x+4x&=22\\ 22x&=22,\ x=1\end{aligned}$$

With 1 pound of yellow jellybeans, you then know that the mixture has $2(1)=2$ pounds of red jellybeans, 3(1) pounds of green jellybeans, and 4(1) pounds of pink jellybeans. The $1+2+3+4$ pounds add up to 10 pounds of jellybeans.

5 **1.2 cups.** Let x = the cups of raspberries at $3 per cup. Then $2x$ = the cups of strawberries at $1 per cup. The remainder of the 16 cups of mixture are for yogurt, which comes out to be $16-(x+2x)=16-3x$ cups of yogurt at $0.50 per cup. The equation you need is

$$\begin{aligned}3(x)+1(2x)+0.5(16-3x)&=10.10\\5x+8-1.5x&=10.1\\3.5x+8&=10.1\\3.5x&=2.1\\x&=\frac{2.1}{3.5}=0.6\end{aligned}$$

The mixture has 0.6 cup of raspberries and $2(0.6)=1.2$ cups of strawberries.

6 **2 ounces of mushrooms, 10 ounces of cheese, 4 ounces of peppers, 8 ounces of onions, and 6 ounces of sausage.** Let x represent the number of ounces of mushrooms. Then $5x$ is the ounces of cheese, $2x$ is the ounces of peppers, $2(2x)=4x$ is the ounces of onions, and $x+4$ is the ounces of sausage. Multiplying each quantity times its cost (quality) and setting that equal to 580 cents, you get $x(10)+5x(20)+2x(25)+4x(10)+(x+4)(30)=580$. Simplifying, you get $10x+100x+50x+40x+30x+120=580$. Solving for x gives you $230x+120=580$, or $230x=460$. Divide each side by 230 to get $x=2$. So you need 2 ounces of mushrooms, $5(2)=10$ ounces of cheese, $2(2)=4$ ounces of peppers, $4(2)=8$ ounces of onions, and $2+4=6$ ounces of sausage.

7 **6 quarts.** Let x = the quarts of 25% solution needed. Add x quarts of 25% solution to 4 quarts of 40% solution to get $(x+4)$ quarts of 31% solution. Write this as $x(0.25)+4(0.40)=(x+4)(0.31)$. Multiply through by 100 to get rid of the decimals: $25x+4(40)=(x+4)(31)$. Distribute the 31 and simplify on the left: $25x+160=31x+124$. Subtract $25x$ and 124 from each side to get $36=6x$. Divide by 6 to get $x=6$. The answer is 6 quarts of 25% solution.

8 **15 gallons.** Let x = the gallons of 5% solution. Then $(0.05)x+(0.90)(2)=(0.15)(x+2)$ $(0.05)x+(0.90)(2)=(0.15)(x+2)$ for the $x+2$ gallons. Multiply through by 100: $5x+(90)(2)=(15)(x+2)$. Simplify each side: $5x+180=15x+30$. Subtract $5x$ and 30 from each side: $150=10x$ or $x=15$ gallons of 5% solution.

9 **3.2 quarts.** Let x = the quarts of pure antifreeze to be added. Pure antifreeze is 100% antifreeze $(100\%=1)$. So, $1(x)+(0.30)(8)=(0.50)(x+8)$. Simplify on the left and distribute on the right: $x+2.4=0.5x+4$. Subtract $0.5x$ and 2.4 from each side: $0.5x=1.6$. Divide by 0.5 to get $x=3.2$ quarts of pure antifreeze.

10 $\mathbf{1\frac{1}{3}}$ **cups.** Let x = the cups of chocolate syrup needed. (I only use the best-quality chocolate syrup, of course, so you know that the syrup is pure chocolate.) 1 quart $=4$ cups, and the milk has no chocolate syrup in it. So

$$\begin{aligned}1(x)+(0)(4)&=(0.25)(x+4)\\x&=0.25x+1\\0.75x&=1\\x&=\frac{1}{0.75}=\frac{100}{75}=\frac{4}{3}\text{ cups of chocolate syrup}\end{aligned}$$

11 **25%.** Let x% = the percent of the salt solution in the 4 quarts:

$$(x\%)(4)+(40\%)(5)=\left(33\tfrac{1}{3}\%\right)(9)$$
$$4x+200=\left(33\tfrac{1}{3}\right)(9)$$
$$4x+200=300$$
$$4x=100$$
$$x=25$$

The 4 quarts must have a 25% salt solution.

12 **9 gallons of 20% solution.** To get 16 gallons, 9 gallons must be added to the 7 gallons. Let x% = the percent of alcohol in the 9 gallons:

$$(x\%)(9)+(60\%)(7)=\left(37\tfrac{1}{2}\%\right)(16)$$
$$9x+420=600$$
$$9x=180$$
$$x=\frac{180}{9}=20$$

13 **Blake has $6,000 invested at 2% and the other $4,000 invested at 3%.** Let x = the amount invested at 2%. Then the amount invested at 3% is 10,000 – x. Blake earns interest of 2% on x dollars and 3% on 10,000 – x dollars. The total interest is $240, so

$$(0.02)(x)+(0.03)(10{,}000-x)=240$$
$$0.02x+300-0.03x=240$$
$$-0.01x+300=240$$
$$-0.01x=-60$$
$$x=\frac{-60}{-0.01}=6{,}000$$

14 **Elliott invested $2,500 at 8% and $1,500 at 2%.** Let x = the amount invested at 8%. Then the amount invested at 2% is $\$4{,}000-x$. He earns 8% interest on x dollars and 2% interest on $4,000 dollars. The total interest is $230, so $0.08(x)+0.02(4{,}000-x)=230$. Multiplying, you have $0.08x+80-0.02x=230$. Simplifying on the left, the equation becomes $0.06x+80=230$. Subtract 80 from each side to get $0.06x=150$. Divide each side of the equation by 0.06, and $\frac{\cancel{0.06}x}{\cancel{0.06}}=\frac{\cancel{150}^{2{,}500}}{\cancel{0.06}}$ or $x=2{,}500$. So $2,500 was invested at 8% and $\$4{,}000-\$2{,}500=\$1{,}500$ was invested at 2%.

15 **Fiona invested $10,000 at 4% and $20,000 at 5%.**

Let x = the amount of money invested at 4%. Then twice that is 2x, the amount invested at 5%. The total interest at the end of the year was $1,400, so $0.04x+0.05(2x)=1{,}400$. Simplifying on the left, first, $0.04x+0.10x=1{,}400$ giving you $0.14x=1{,}400$. Dividing each side of the equation by 0.14, $\frac{\cancel{0.14}x}{\cancel{0.14}}=\frac{\cancel{1{,}400}^{10{,}000}}{\cancel{0.14}}$ or $x=10{,}000$. So $10,000 was invested at 4%, and twice that, $20,000, was invested at 5%.

16. **Wolf invested $4,000 at 1%, $8,000 at 2%, and $12,000 at 3% and earned $560 interest.**

    Let x = the amount invested at 1%. Then 2x = the amount invested at 2%, and 3x = the amount invested at 3%. The total amount invested was $24,000, so you can determine how much was invested at each interest level using $x+2x+3x=24{,}000$. The equation simplifies to $6x=24{,}000$. Dividing each side by 6, you have $x=4{,}000$. Then $2x=8{,}000$ and $3x=12{,}000$. To compute the amount of interest,
    $0.01x+0.02(2x)+0.03(3x)=0.01(4{,}000)+0.02(8{,}000)+0.03(12{,}000)=40+160+360=560$.

17. **30 quarters.** Let x = the number of nickels. Then 2x = the number of quarters. These coins total $8.25 or 825 cents. So, in cents,

$$\begin{aligned}5(x)+25(2x)&=825\\5x+50x&=825\\55x&=825\\x&=\frac{825}{55}=15\end{aligned}$$

    Because Carlos has 15 nickels, he must then have 30 quarters.

18. **10 $20 bills, 20 $10 bills, 100 $1 bills, and 50 $5 bills.** Let x = the number of $20 bills. Then 2x = the number of $10 bills, $5(2x)=10x$ = the number of $1 bills, and $\frac{1}{2}(10x)=5x$ = the number of $5 bills. The total in dollars is 750. So, the equation should be $20(x)+10(2x)+1(10x)+5(5x)=750$. Simplify on the left to get $75x=750$ and $x=10$. Gregor has 10 $20 bills, 20 $10 bills, 100 $1 bills, and 50 $5 bills.

19. **40 dimes, 58 nickels, and 2 quarters.** Let x = the number of dimes. Then $x+18$ = the number of nickels, and the number of quarters is $100-\left[x+(x+18)\right]=82-2x$. These coins total $7.40 or 740 cents:

$$\begin{aligned}10(x)+5(x+18)+25\left(100-\left[x+(x+18)\right]\right)&=740\\10x+5(x+18)+25(82-2x)&=740\\10x+5x+90+2{,}050-50x&=740\\2{,}140-35x&=740\\-35x&=-1{,}400\\x&=\frac{-1{,}400}{-35}=40\end{aligned}$$

    So Stella has 40 dimes, 58 nickels, and 2 quarters.

20. **6 quarters, 8 dimes, 2 fifty-cent pieces, and 4 nickels.** Everything seems to compare to nickels, but, to avoid a fractional coefficient, start with x = the number of fifty-cent pieces. Then twice the number of fifty-cent pieces = 2x = the number of nickels. Two more than the number of nickels is $2+2x$ = the number of quarters, and twice the number of nickels is 4x = the number of dimes. Multiplying each number of coins by what they're worth, $0.50(x)+0.05(2x)+0.25(2+2x)+0.10(4x)=3.50$. Multiplying on the left, you have $0.50x+0.10x+0.50+0.50x+0.40x=3.50$. Adding like terms, the equation becomes $1.50x+0.50=3.50$. Now subtract 0.50 from each side, and $1.50x=3.00$. Divide each side by 1.50, and $x=2$. So there are $x=2$ fifty-cent pieces, $2x=4$ nickels, $2+2x=6$ quarters, and $4x=8$ dimes.

If you're ready to test your skills a bit more, take the following chapter quiz that incorporates all the chapter topics.

Whaddya Know? Chapter 23 Quiz

Quiz time! Complete each problem to test your knowledge on the various topics covered in this chapter. You can then find the solutions and explanations in the next section.

1. Dried cranberries cost $4.00 per pound, and raisins cost $2.50 per pound. How much of each is needed to make a $3.00-per-pound mixture?

2. How many quarts of 70% solution do you need to add to 40 quarts of 40% solution to create a 50% solution?

3. You have half as many dimes as nickels, for a total of $2.00. How many dimes do you have?

4. How many ounces of ginger ale do you add to 6 ounces of orange juice to create a mixture that's 30% orange juice?

5. You have 10 pounds of dark chocolate that costs $6.00 per pound. You want to add chocolate-covered nuts that cost $9.00 per pound. How many pounds of nuts do you need to add to create a mixture that costs $8.00 per pound?

6. A collection box had two more $10 bills than $5 bills and twice as many $1 bills as $5 bills. How many of each bill were there, if the amount totaled $122.

7. Greg spent $14 on his lunch at the fair. Hot dogs cost $2 more than cheese fries, and a soft drink costs $1 less than a hot dog. If he had 2 hot dogs and 3 servings of cheese fries with his soft drink, how much did the soft drink cost?

8. What percent solution do you need to add to 20 pints of 35% solution to create 50 pints of a 26% solution?

9. Henry has four times as many $1 bills as quarters and ten more dimes than quarters. If he has 610 bills and coins in all, then how much money does he have in total?

10. If walnuts cost $3.00 per pound and pecans cost $4.50 per pound, what is the cost of a 6-pound mixture if you have twice as many pounds of walnuts as pecans?

11. You have a total of 90 coins, all in quarters and dimes. They total $18. How many of each coin do you have?

12. One cup of breakfast cereal is 30% sugar. How much sugar should be added to make it 40% sugar?

Answers to Chapter 23 Quiz

1. **$\frac{1}{3}$ pound cranberries, $\frac{2}{3}$ pound raisins.** Let c represent the amount of cranberries and $1-c$ the amount of raisins. Then, using $4.00(c)+2.50(1-c)=3(1)$, you get $1.50c=0.50$ or $c=\frac{1}{3}$.

2. **20 quarts.** Use $0.70(x)+0.40(40)=0.50(x+40)$, which becomes $0.7x+16=0.5x+20$ or $0.2x=4$. This gives you $x=20$.

3. **10 dimes.** Let d represent the number of dimes and $2d$ the number of nickels. Using $0.10(d)+0.05(2d)=2.00$, you get $0.20d=2.00$ or $d=10$.

4. **14 ounces.** Because there's no orange juice in ginger ale, the percentage for the ginger ale is 0%, and the percentage for the orange juice is 100%. Using $0(x)+1.00(6)=0.30(x+6)$, you have $6=0.3x+1.8$ or $4.2=0.3x$. Because $x=14$, you need 14 ounces of ginger ale.

5. **20 pounds of nuts.** Let x represent the number of pounds of nuts and use $6.00(10)+9.00(x)=8.00(10+x)$, which simplifies to $60+9x=80+8x$ or $x=20$. You need 20 pounds of nuts, creating a 30-pound mixture.

6. **8 \$10's, 6 \$5's, 12 \$1's.** Let x represent the number of \$5's. Then $x+2$ is the number of \$10's, and $2x$ is the number of \$1's. Using $10(x+2)+5(x)+1(2x)=122$, this simplifies to $17x+20=122$ and then $x=6$. That's 6 \$5's, 8 \$10's, and 12 \$1's.

7. **\$2.50.** Let x represent the cost of a serving of cheese fries, $x+2$ the cost of a hot dog, and $x+1$ the cost of a soft drink. Using $x(3)+(x+2)(2)+(x+1)(1)=14$, you then have $6x+5=14$, which is $6x=9$ or $x=1.5$. The cost of a soft drink is $x+1=2.5$.

8. **20% solution.** Because you need a total of 50 pints and already have 20 pints, you want to add 30 pints of the new solution. Use $x(30)+0.35(20)=0.26(50)$, which becomes $30x+7=13$ or $x=\frac{6}{30}=\frac{1}{5}=0.20$.

9. **\$436.** Let x represent the number of quarters. That means he has $4x$ \$1's and $10+x$ dimes. The number of coins and bills is 610, so $x+4x+10+x=610$. That means $6x=600$, giving you $x=100$. That's 100 quarters, 400 in \$1's, and 110 dimes: $0.25(100)+1(400)+0.10(110)=25+400+11=436$.

10. **\$3.50.** If the total amount is to be 6 pounds, and you need twice as many walnuts as pecans, then, solving $2x+x=6$, where x is the amount of pecans, you have $x=2$. That's 2 pounds of pecans and 4 pounds of walnuts. Now, using $3.00(4)+4.50(2)=x(6)$, where x is the cost of the mixture, $21=6x$, meaning that the cost will be \$3.50 per pound.

11. **60 quarters, 30 dimes.** Let x represent the number of quarters. Then $90-x$ is the number of dimes. Using $0.25(x)+0.10(90-x)=18$, this becomes $0.25x+9-0.10x=18$ or $0.15x=9$. Dividing by 0.15, $x=60$. So there are 60 quarters and 30 dimes.

12. **$\frac{1}{6}$ cup.** If you're adding pure sugar, then the percentage is 100%. Use $0.30(1)+1.00(x)=0.40(1+x)$, which becomes $0.3+x=0.4+0.4x$ or $0.6x=0.1$. Because $x=\frac{1}{6}$, you need to use $\frac{1}{6}$ cup of sugar.

8 Getting a Grip on Graphing

Contents at a Glance

IN THIS CHAPTER

- » Plotting points and lines
- » Computing distances, midpoints, and slopes
- » Intercepting and intersecting
- » Writing equations of lines
- » Finding parallels and perpendiculars

Chapter 24

Getting a Handle on Graphing

Graphs are as important to algebra as pictures are to books and magazines. A graph can represent data that you've collected, or it can represent a pattern or model of an occurrence. A graph illustrates what you're trying to demonstrate or understand.

Algebraic equations match up with their graphs. With algebraic operations and techniques applied to equations to make them more usable, the equations can be used to predict, project, and figure out various problems.

The standard system for graphing in algebra is to use the *Cartesian coordinate system*, where points are represented by ordered pairs of numbers; connected points can be lines, curves, or disjointed pieces of graphs.

This chapter can help you sort out much of the graphing mystery and even perfect your graphing skills. Just watch out! The slope may be slippery.

Thickening the Plot with Points

You do the type of point-finding needed to do graphing when you find the whereabouts of Peoria, Illinois, at G7 on a road atlas. You move your finger so it's down from the G and across from the 7. Graphing in algebra is just a bit different because numbers replace the letters, and you start in the middle at a point called the *origin*.

Graphing on the Cartesian coordinate system begins by constructing two perpendicular axes, the x-axis (horizontal) and the y-axis (vertical). The Cartesian coordinate system identifies a point by an ordered pair, (x,y). The order in which the coordinates are written matters. The first coordinate, the x, represents how far to the left or right the point is from the *origin*, or where the axes intersect. A positive x is to the right; a negative x is to the left. The second coordinate, the y, represents how far up or down from the origin the point is.

Points are dots on a piece of paper or blackboard that represent positions or places with respect to the axes of a graph. The coordinates of a point tell you its exact position on the graph (unlike maps, where G7 is a big area and you have to look around for the city).

The axes of an algebraic graph are usually labeled with integers, but they can be labeled with any rational numbers, as long as the numbers are the same distance apart from each other, such as the one-quarter distance between $\frac{1}{4}$, $\frac{1}{2}$, and $\frac{3}{4}$.

Interpreting ordered pairs

An *ordered pair* is a set of two numbers called *coordinates* that are written inside parentheses with a comma separating them. Some examples are (2,3), (−1,4), and (5,0). The point for the ordered pair (3,2) is 3 units to the right of the origin, and 2 units up from the x-axis. Everything starts at the origin — the intersection of the two axes. The ordered pair for the origin is (0,0). The numbers in this ordered pair tell you that the point didn't go left, right, up, or down. Its position is at the starting place.

The point (2,0) lies to the right on the x-axis. Whenever 0 is a coordinate within the ordered pair, the point must be located on an axis.

Table 24-1 gives you the names of the quadrants, their positions in the coordinate plane, and the characteristics of coordinate points in the various quadrants. Table 24-2 describes what's happening on the axes as they radiate out from the origin.

TABLE 24-1 Quadrants

Quadrant	Position	Coordinate Signs	How to Plot
Quadrant I	Upper-right side	(positive, positive)	Move right and up.
Quadrant II	Upper-left side	(negative, positive)	Move left and up.
Quadrant III	Lower-left side	(negative, negative)	Move left and down.
Quadrant IV	Lower-right side	(positive, negative)	Move right and down.

TABLE 24-2 Axes

Position	Coordinate Signs	How to Plot
Right axis	(positive, 0)	Move right and sit on the x-axis.
Left axis	(negative, 0)	Move left and sit on the x-axis.
Upper axis	(0, positive)	Move up and sit on the y-axis.
Lower axis	(0, negative)	Move down and sit on the y-axis.

Actually Graphing Points

To plot a point, look at the coordinates — the numbers in the parentheses. The first number tells you which way to move, horizontally, from the origin. Place your pencil on the origin and move right if the first number is positive; move left if the first number is negative. Next, from that position, move your pencil up or down — up if the second number is positive and down if it's negative.

Cartesian coordinates designate where a point is in reference to the two perpendicular axes. To the right and up is positive, to the left and down is negative. Any point that lies on one of the axes has a 0 for one of the coordinates, such as (0,2) or (−3,0). The coordinates for the *origin*, the intersection of the axes, are (0,0).

EXAMPLE

Q. Use the following figure to graph the points (2,6), (8,0), (5,−3), (0,−7), (−4,−1), and (−3,4).

A. Notice that the points that lie on an axis have a 0 in their coordinates.

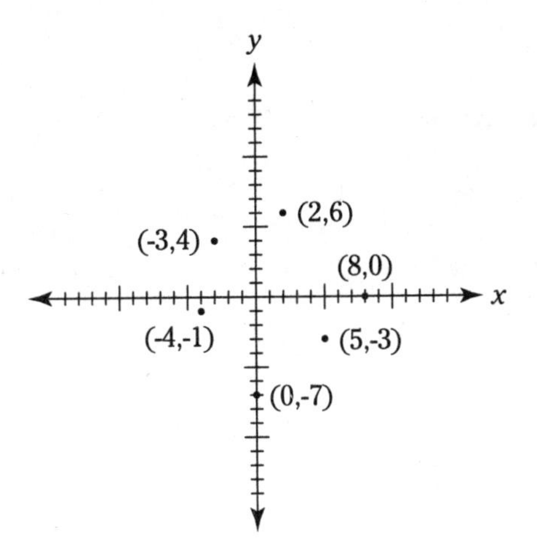

YOUR TURN

Graph the points.

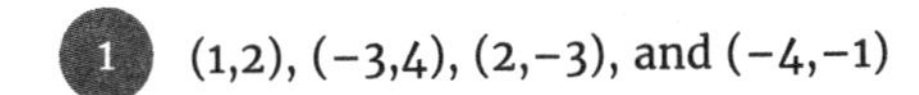

1. (1,2), (−3,4), (2,−3), and (−4,−1)

2. (0,3), (−2,0), (5,0), and (0,−4)

Sectioning Off by Quadrants

Another description for a point is the *quadrant* that the point lies in. The quadrants are referred to in many applications because of the common characteristics of points that lie in the same quadrant. The quadrants are numbered one through four, usually with Roman numerals. Check out Figure 24-1 or Table 24-1, in the previous section, to see how the quadrants are identified.

EXAMPLE

Q. Referring to Figure 24-1, describe which coordinates are positive or negative in the different quadrants.

A. In Quadrant I, both the x and y coordinates are positive numbers. In Quadrant II, the x coordinate is negative, and the y coordinate is positive. In Quadrant III, both the x and y coordinates are negative. In Quadrant IV, the x coordinate is positive, and the y coordinate is negative.

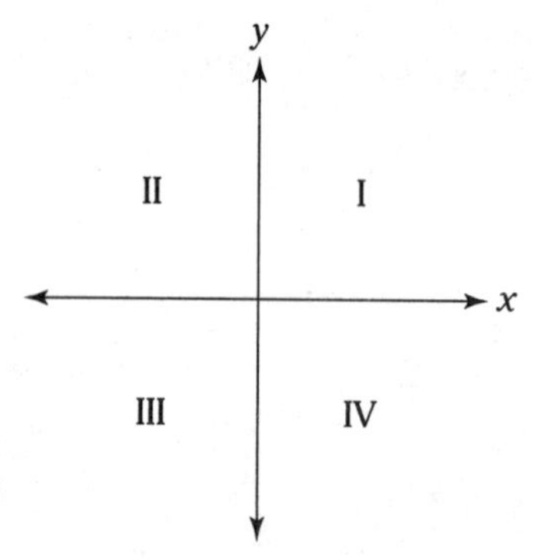

FIGURE 24-1: Identifying quadrants.

YOUR TURN

Referring to Figure 24-1, in which quadrant do the points lie?

 3 (−3,2) and (−4,11)

 4 (−4,−1) and (−2,−2)

Graphing Lines

One of the most basic graphs you can construct by using the coordinate system is the graph of a straight line. You may remember from geometry that only two points are required to determine a particular line.

When graphing lines using points, though, it's a good idea to plot three points to be sure that you've graphed the points correctly and put them in the correct positions. You can think of the third point as a sort of check (like the *check digit* in a UPC barcode). The third point can be anywhere, but try to spread out the three points and not have them clumped together. If the three points aren't in a straight line, you know that at least one of them is wrong.

Using points to lay out lines

An equation whose graph is a straight line is said to be *linear*. A linear equation has a standard form of $ax + by = c$, where x and y are variables and a, b, and c are real numbers. The equation of a line usually has an x or a y (often both), which refers to all the points (x,y) that make the equation true. The x and y both have a power of 1. (If the powers were higher or lower than 1, the graph would curve.)

When graphing a line, you can find some pairs of numbers that make the equation true and then connect them. Connect the dots!

What does the equation of a line look like? It looks like any of the following examples. Notice that the first three equations are written in the standard form, and the fourth has you solve for y. The last two have only one variable; this situation happens with horizontal and vertical lines.

$x + y = 10 \qquad 2x + 3y = 4 \qquad -5x + y = 7$

$y = \frac{1}{2}x + 3 \qquad x = 3 \qquad y = -2$

Graphing lines from their equations just takes finding enough points on the line to convince you that you've done the graph correctly.

Find a point on the line $y - x = 3$.

1. **Choose a random value for one of the variables, either x or y.**

   To make the arithmetic easy for yourself, pick a large-enough number so that, when you subtract x from that number, you get a positive 3. In $y - x = 3$, you can let $y = 8$, so $8 - x = 3$.

2. **Solve for the value of the other variable.**

   You probably can tell the answer just by looking: $x = 5$. But you can solve this by subtracting 8 from both sides to get $-x = -5$. You then multiply each side by the number -1 to get $x = 5$. (For a review of solving linear equations, turn to Chapter 15.)

3. **Write an ordered pair for the coordinates of the point.**

   You chose 8 for y and solved to get $x = 5$, so your first ordered pair is $(5,8)$.

The graph of a linear equation in two variables is a line. For example, the graph of the linear equation $y - x = 3$ is a line that appears to move upward as the x-coordinates increase. Here's how to do a basic graph.

The graph of $y - x = 3$ goes through all the points in the coordinate plane that make the equation a true statement. You already found the point $(5,8)$. For another point, if $x = 2$, then $y - 2 = 3$, which gives you $x = 5$, and you have the point $(2,5)$. Here are some of the points that make the equation true:

$$\begin{array}{llll} (-4,-1) & (-3,0) & (-2,1) & (-1,2) \\ (0,3) & (1,4) & (2,5) & (3,6) \end{array}$$

The number of points that satisfy the equation is infinitely large. You just need a few to draw a decent graph. (Actually, you only need two points to draw a particular line, but I like to graph more than that for accuracy's sake.) Figure 24-2 shows you the points graphed and then connected to form the line.

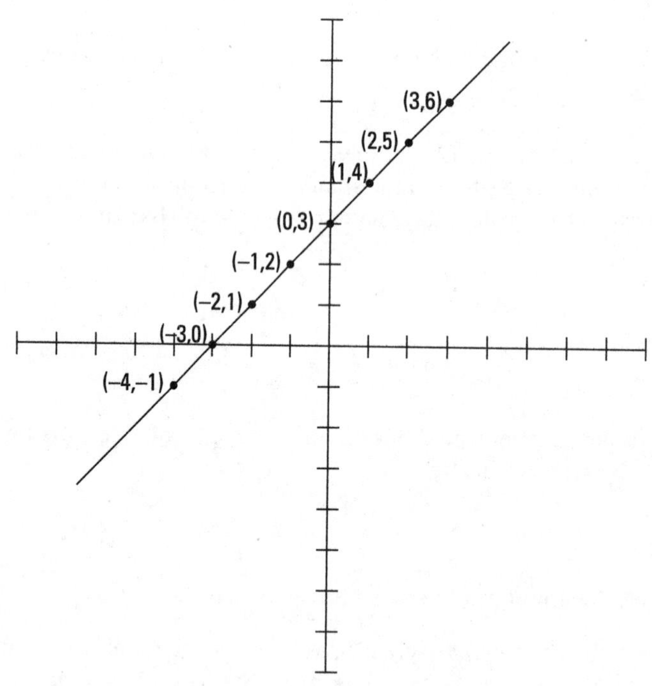

FIGURE 24-2: Graphing $y - x = 3$.

You can find more ordered pairs by choosing another number to substitute for either x or y.

For more of a challenge, find points that lie on a line with coefficients on x and y other than 1. The multipliers (2 and 3 in the next example) make this just a little trickier. You may find one or two points fairly easily, but others could be more difficult because of fractions. A good plan in a case like this is to solve for x or y and then plug in numbers.

Find points that lie on the line $2x+3y=12$.

Use these steps.

1. **Solve the equation for one of the variables.**

   Solving for y in the sample problem $2x+3y=12$, you get $3y=12-2x$, or

   $$y=\frac{12-2x}{3}$$

   With multipliers involved, you often get a fraction.

2. **Choose a value for the other variable and solve the equation.**

   Try to pick values so that the result in the numerator is divisible by the 3 in the denominator, giving you an integer.

   For example, let $x=3$. Solving the equation,

   $$y=\frac{12-2\cdot 3}{3}=\frac{6}{3}=2$$

   So, the point (3,2) lies on the line.

Going with the horizontal and vertical

Whenever you have an equation where y equals a constant number, you have a *horizontal* line going through all those y values. Conversely, if you have an equation where x equals a constant number, then all the x values are the same, and you have a *vertical* line. Horizontal lines are all parallel to the x-axis; their equations look like $y=3$ or $y=-2$. Vertical lines are all parallel to the y-axis; their equations all look like $x=5$ or $x=-11$. Figure 24-3 shows a graph of $y=4$, using four points: (−4,4), (0,4), (1,4), and (3,4).

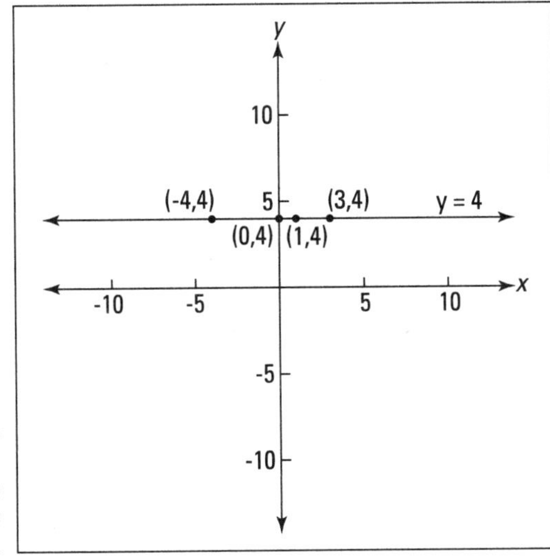

FIGURE 24-3: Horizontal lines are parallel to the x-axis.

Finding the points that lie on the line $x = 4$ may look like a really tough assignment, with only an x showing in the equation. But this actually makes the whole thing much easier. You can write down anything for the y value, as long as x is equal to 4. Some points are (4,9), (4,−2), (4,0), (4,3.16), (4,−11), and (4,4).

Notice that the 4 is always the first number. The point (4,9) is not the same as the point (9,4). The order counts in ordered pairs.

Graphing these points gives you a nice, vertical line, as Figure 24-4 shows. On the other hand, if all the y-coordinates are the same point, the line is — you guessed it — horizontal.

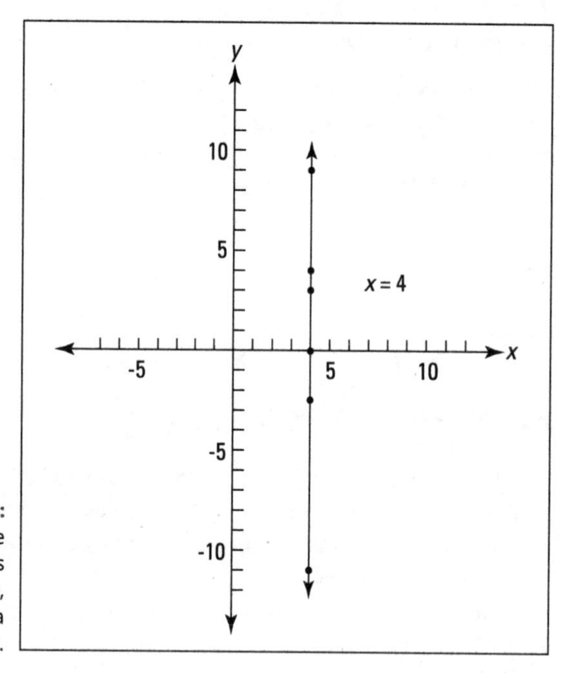

FIGURE 24-4: When all the x-coordinates are the same, you get a vertical line.

EXAMPLE

Q. Graph the line represented by the equation $2x + 3y = 10$.

A. To graph the line, first find three sets of coordinates that satisfy the equation. Three points that work for this line are (5,0), (2,2), and (−1,4). Plot the three points and then draw a line through them.

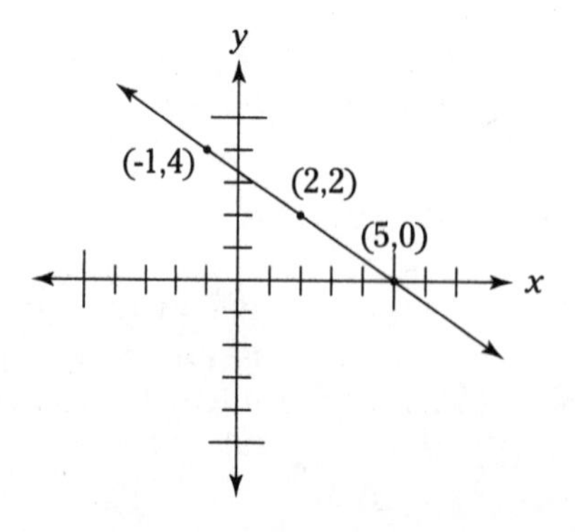

These aren't the only three points you could choose. I'm just demonstrating how to spread the points out so that you can draw a better line.

Find three points that lie on the line, plot them, and draw the line through them.

5 $2x - y = 3$	6 $x + 3y + 6 = 0$
7 $x = 4$	8 $y = -2$

Getting a Handle on Graphing

Graphing Lines Using Intercepts

An *intercept* of a line is a point where the line crosses an axis. Unless a line is vertical or horizontal, it crosses both the x and y axes, so it has two intercepts: an x-intercept and a y-intercept. Horizontal lines have just a y-intercept, and vertical lines have just an x-intercept. The exceptions are when the horizontal line is actually the x-axis or the vertical line is the y-axis.

Intercepts are quick and easy to find and can be a big help when graphing. The reason they're so useful is that one of the coordinates of every intercept is a 0. Zeros in equations cut down on the numbers and the work, and it's nice to take advantage of zeros when you can.

The x-intercept of a line is where the line crosses the x-axis. To find the x-intercept, let the y in the equation equal 0 and solve for x.

For example, find the x-intercept and y-intercept of the line $4x - 7y = 8$.

First, let $y = 0$ in the equation. Then

$$\begin{aligned} 4x - 0 &= 8 \\ 4x &= 8 \\ x &= 2 \end{aligned}$$

The x-intercept of the line is (2,0); the line goes through the x-axis at that point.

The y-intercept of a line is where the line crosses the y-axis. To find the y-intercept, let the x in the equation equal 0 and solve for y. You start with $0 - 7y = 8$, which gives you $y = -\frac{8}{7}$. The y-intercept is $\left(0, -\frac{8}{7}\right)$.

EXAMPLE

Q. Find the intercepts of the line $9x - 4y = 18$.

A. First, to find the x intercept, let $y = 0$ in the equation of the line to get $9x = 18$. Solving that, $x = 2$, and so the intercept is (2,0). Next, for the y intercept, let $x = 0$ to get $-4y = 18$.

Solving for y, $y = \frac{18}{-4} = -\frac{9}{2}$. So, the y-intercept is $\left(0, -\frac{9}{2}\right)$.

Intercepts are also especially helpful when graphing the line. Use the two intercepts you found to graph the line, and then you can check with one more point. For instance, in the following figure, you can check to see whether the point $\left(\frac{2}{9}, -4\right)$ is on the line.

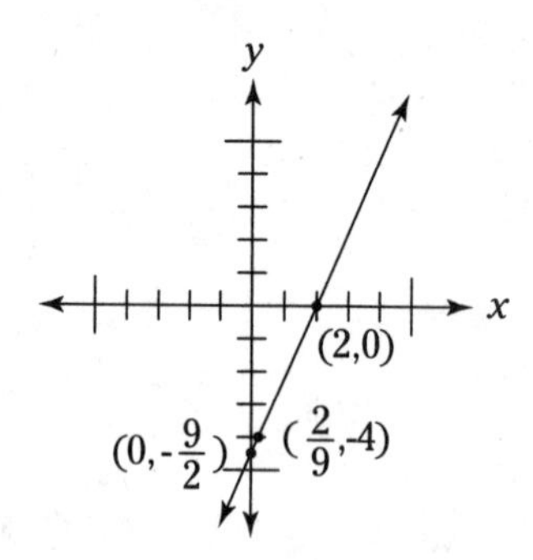

Use the intercepts to graph the line.

9 $3x + 4y = 12$	**10** $x - 2y = 4$

Computing Slopes of Lines

The *slope* of a line is simply a number that describes the steepness of the line and whether the line is rising or falling as it moves from left to right in a graph. When referring to how steep a line is, when you're given its slope, the general rule is that the farther the number is from 0, the steeper the line. A line with a slope of 7 is much steeper than a line with a slope of 2. And a line whose slope is −6 is steeper than a line whose slope is −3.

Sighting the slope

Knowing the slope of a line beforehand helps you graph the line. You can find a point on the line and then use the slope and that point to graph it. A line with a slope of 6 goes up steeply. If you know what the line should look like (that is, whether it should go up or down) — information you get from the slope — then you'll have an easier time graphing it correctly.

The value of the slope is important when the equation of the line is used in modeling situations. For example, in equations representing the cost of so many items, the value of the slope is called the *marginal cost.* In equations representing depreciation, the slope is the *annual depreciation.*

Figure 24-5 shows some lines with their slopes. The lines are all going through the origin just for convenience.

What about a horizontal line — one that doesn't go upward or downward? A horizontal line has a 0 slope. What about a vertical line? It has no slope; the slope of a vertical line (it's so steep) is undefined. Figure 24-6 shows graphs of lines that have a 0 slope or an undefined slope.

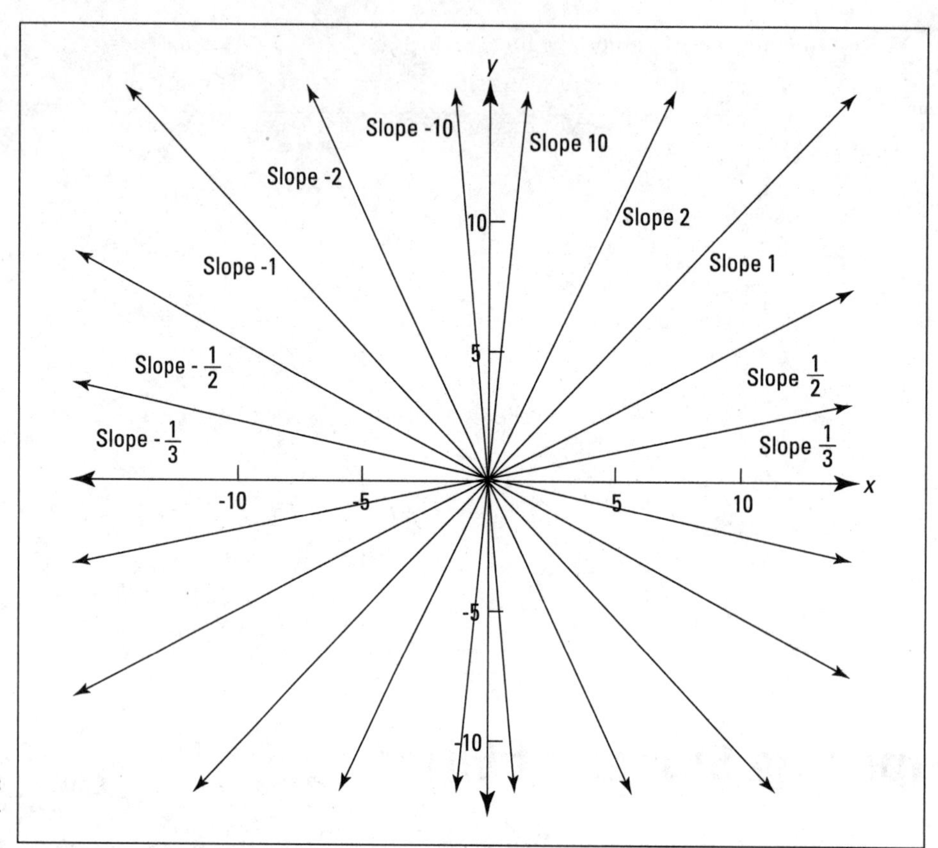

FIGURE 24-5: Pick a line — see its slope.

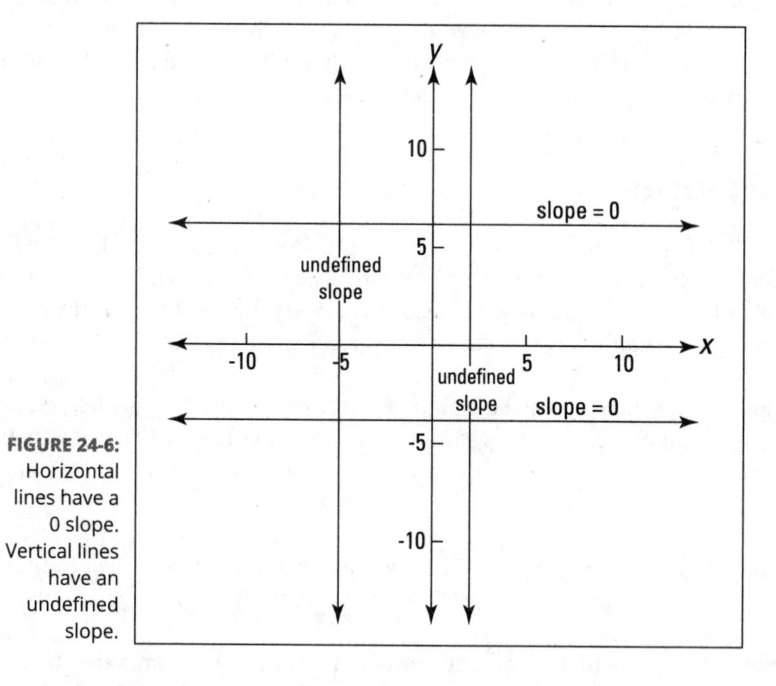

FIGURE 24-6: Horizontal lines have a 0 slope. Vertical lines have an undefined slope.

One way of referring to the slope, when it's written as a fraction, is *rise over run.* If the slope is $\frac{3}{2}$, it means that for every 2 units the line runs along the x-axis, it rises 3 units along the y-axis. A slope of $-\frac{1}{8}$ indicates that as the line runs 8 units horizontally, parallel to the x-axis, it drops (negative rise) 1 unit vertically.

Formulating slope

If you know two points on a line, you can compute the number representing the slope of the line.

The slope of a line, denoted by the small letter m, is found when you know the coordinates of two points on the line, (x_1,y_1) and (x_2,y_2):

$$m = \frac{y_2 - y_1}{x_2 - x_1}$$

Subscripts are used here to identify which is the first point and which is the second point. There's no rule as to which is which; you can name the points any way you want. It's just a good idea to identify them to keep things in order. Reversing the points in the formula gives you the same slope (when you subtract in the opposite order):

$$m = \frac{y_1 - y_2}{x_1 - x_2}$$

Note that you can't mix them and do $x_1 - y_2$ over $x_2 - y_1$.

Now, I'll show you how to compute slope using the following examples.

Find the slope of the line going through (3,4) and (2,10).

Let (3,4) be (x_1,y_1) and (2,10) be (x_2,y_2). Substitute into the formula:

$$m = \frac{y_2 - y_1}{x_2 - x_2} = \frac{10 - 4}{2 - 3}$$

Simplify:

$$m = \frac{6}{-1} = -6$$

This line is pretty steep as it falls from left to right.

Here's another example. Find the slope of the line going through (4,2) and (−6,2).

Let (4,2) be (x_1,y_1) and (−6,2) be (x_2,y_2). Substitute into the formula:

$$m = \frac{y_2 - y_1}{x_2 - x_2} = \frac{2 - 2}{-6 - 4}$$

Simplify:

$$m = \frac{0}{-10} = 0$$

These points are both 2 units above the x-axis and form a horizontal line. That's why the slope is 0. And the equation of the line is $y = 2$.

Here's one more example: Find the slope of the line going through (2,4) and (2,−6).

Let (2,4) be (x_1,y_1) and (2,−6) be (x_2,y_2). Substitute into the formula:

$$m = \frac{y_2 - y_1}{x_2 - x_2} = \frac{-6-4}{2-2}$$

Simplify:

$$m = \frac{-10}{0}$$

Oops! You can't divide by 0, as there is no such number. The slope doesn't exist or is undefined. These two points are on a vertical line, and the equation of the line is $x = 2$.

WARNING

Watch out for these common errors when working with the slope formula:

- **Be sure that you subtract the *y* values on the top of the division formula.** A common error is to subtract the *x* values on the top.
- **Be sure to keep the numbers in the same order when you subtract.** Decide which point is first and which point is second. Then take the second *y* minus the first *y* and the second *x* minus the first *x*. Don't do the top subtraction in a different order from the bottom.

To find the slope of a line, you can use two points on the graph of the line and apply the formula $m = \frac{y_2 - y_1}{x_2 - x_1}$. Here, the letter *m* is the traditional symbol for slope, and the (x_1,y_1) and (x_2,y_2) are the coordinates of any two points on the line. The point you choose to go first in the formula doesn't really matter. Just be sure to keep the order the same — from the same point — because you can't mix and match.

A *horizontal* line has a slope of 0, and a vertical line has no slope. To help you remember, picture the sun coming up on the *horizon* — that 0 is just peeking out at you.

EXAMPLE

Q. Find the slope of the line that goes through the two points (−3,4) and (1,−8) and graph it.

A. To find the slope, use $m = \frac{4-(-8)}{-3-1} = \frac{12}{-4} = -3$. The following figure shows a graph of that line. It's fairly steep — any slope greater than 1 or less than −1 is steep. The negative part indicates that the line's falling as you go from left to right. Another description of slope is that the bottom number is the *change in x*, and the top is the *change in y*. Here's how you read a slope of −3: *For every 1 unit you move to the right parallel to the x-axis, you drop down 3 units parallel to the y-axis.*

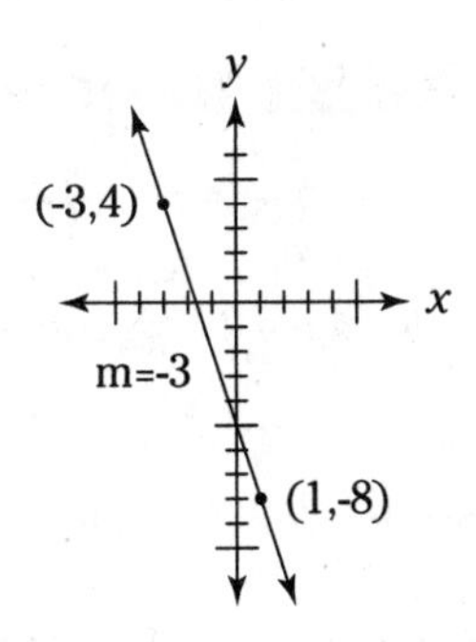

YOUR TURN

Find the slope of the line through the points and graph the line.

11 $(3,2)$ and $(-4,-5)$	**12** $(-1,7)$ and $(1,3)$
13 $(3,-4)$ and $(5,-4)$	**14** $(2,3)$ and $(2,-8)$

Getting a Handle on Graphing

Graphing with the Slope-Intercept Form

Equations of lines can take many forms, but one of the most useful is called the *slope-intercept form.* The numbers for the slope and y-intercept are part of the equation. When you use this form to graph a line, you just plot the y-intercept and use the slope to find another point from there.

REMEMBER

The slope intercept form is $y = mx + b$, where the m represents the slope of the line, and the b is the y-coordinate of the intercept where the line crosses the y-axis. For example, a line with the equation $y = -3x + 2$ has a slope of −3 and a y-intercept of (0,2).

Having the equation of a line in the *slope-intercept* form makes graphing the line an easy chore. Follow these steps:

1. **Plot the y-intercept on the y-axis.**
2. **Write the slope as a fraction.**

   Using the equation $y = -3x + 2$, the fraction would be $\frac{-3}{1}$. (If the slope is negative, you put the negative part in the numerator.) The slope has the change in y in the numerator and the change in x in the denominator.
3. **Starting with the y-intercept, count the amount of the change in x (the number in the denominator) to the right of the intercept, and then count up or down from that point (depending on whether the slope is positive or negative), using the number in the numerator.**

   Wherever you end up is another point on the line.
4. **Mark that point and draw a line through the new point and the y-intercept.**

EXAMPLE

Q. Graph $y = -3x + 2$, using the method in the previous steps.

A. Graph the intercept (0,2). Then count 1 unit to the right and 3 units down, and graph the second point.

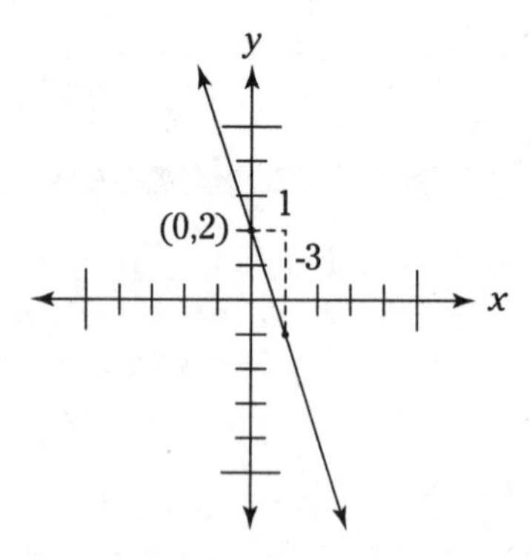

Q. Graph $y = \frac{2}{5}x - 1$.

A. Graph the intercept (0,−1). Then count 5 units to the right and 2 units up, and graph the second point.

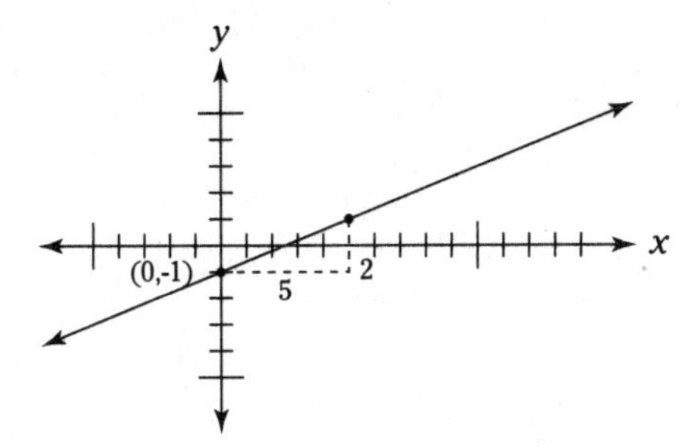

YOUR TURN

Graph the line using the y-intercept and slope.

 **15** $y = -\frac{2}{3}x + 4$

 **16** $y = 5x - 2$

Changing to the Slope-Intercept Form

Graphing lines by using the slope-intercept form is a piece of cake. But what if the equation you're given isn't in that form? Are you stuck with substituting in values and finding coordinates of points that work? Not necessarily. Changing the form of the equation using algebraic manipulations — and then graphing using the new form — is often easier.

Remember: To change the equation of a line to the slope-intercept form, $y = mx + b$, first isolate the term with y in it on one side of the equation and then divide each side by any coefficient of y. You can rearrange the terms so the x term, with the slope multiplier, comes first.

EXAMPLE

Q. Change the equation $3x - 4y = 8$ to the slope-intercept form.

A. First, subtract $3x$ from each side: $-4y = -3x + 8$. Then divide each term by -4:

$$\frac{-4y}{-4} = \frac{-3x}{-4} + \frac{8}{-4}$$

$$y = \frac{3}{4}x - 2$$

This line has a slope of $\frac{3}{4}$ and a y-intercept at $(0,-2)$.

Q. Change the equation $y - 3 = 0$ to the slope-intercept form.

A. There's no x term, so the slope must be 0. If you want a complete slope-intercept form, you can write the equation as $y = 0x + 3$ to show a slope of 0 and an intercept of $(0,3)$.

YOUR TURN

Change the equation to the slope-intercept form.

17 $8x + 2y = 3$

 $4x - y - 3 = 0$

Writing Equations of Lines

Up until now, you've been given the equation of a line and have been told to graph that line using two points, the intercepts, or the slope and y-intercept. But how do you re-create the line's equation if you're given either two points (which could be the two intercepts) or the slope and some other point?

Given a point and a slope

When given the slope and any point on the line, you use the *point-slope* form, $y - y_1 = m(x - x_1)$, to write the equation of the line. The letter m represents the slope of the line, and (x_1, y_1) is any point on the line. After filling in the information, simplify the form.

EXAMPLE

Q. Find the equation of the line that has a slope of 3 and goes through the point (−4,2).

A. Using the point-slope form, you replace the m with 3, the y_1 with the 2, and the x_1 with the −4. Your equation becomes $y - 2 = 3(x - (-4))$. Simplifying, you get $y - 2 = 3(x + 4)$. Distributing, you have $y - 2 = 3x + 12$. And, adding 2 to each side, $y = 3x + 14$.

Given two points

When given two points, you have just one more step than the procedure when you're given a point and the slope: you have to find the slope!

EXAMPLE

Q. Find the equation of the line that goes through the points (5,−2) and (−4,7).

A. First, find the slope of the line: $m = \frac{7-(-2)}{-4-5} = \frac{9}{-9} = -1$. Now use the point-slope form with the slope of −1 and with the coordinates of one of the points. It doesn't matter which point, so I choose (5,−2). Filling in the values, $y - (-2) = -1(x - 5)$. Simplifying, you get $y + 2 = -x + 5$ or $y = -x + 3$.

YOUR TURN

19 Find an equation of the line with a slope of $\frac{1}{3}$ that goes through the point (0,7).

20 Find an equation of the line that goes through the points (−3,−1) and (−2,5).

Picking on Parallel and Perpendicular Lines

The slope of a line gives you information about a particular characteristic of the line. It tells you if it's steep or flat and if it's rising or falling as you read from left to right. The slope of a line can also tell you if one line is parallel or perpendicular to another line. Figure 24-7 shows parallel and perpendicular lines.

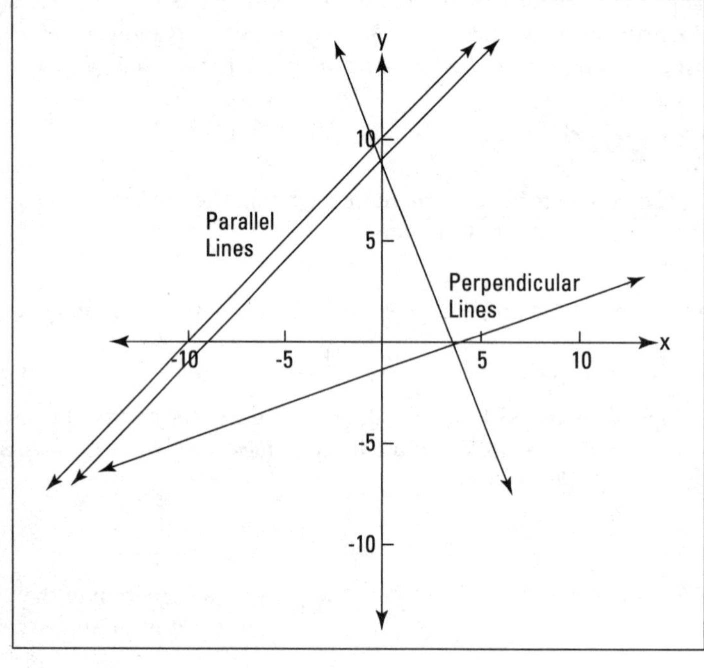

FIGURE 24-7: Parallel lines are like railroad tracks; perpendicular lines meet at a right angle.

Parallel lines never touch. They're always the same distance apart and never share a common point. They have the same slope.

Perpendicular lines form a 90-degree angle (a *right angle*) where they cross. They have slopes that are negative reciprocals of one another. The *x* and *y* axes are perpendicular lines.

REMEMBER

Two numbers are reciprocals if their product is the number 1. The numbers $\frac{3}{4}$ and $\frac{4}{3}$ are reciprocals. Two numbers are negative reciprocals if their product is the number −1. The numbers $\frac{3}{4}$ and $-\frac{4}{3}$ are negative reciprocals.

If line y_1 has a slope of m_1, and if line y_2 has a slope of m_2, then the lines are parallel if $m_1 = m_2$. If line y_1 has a slope of m_1, and if line y_2 has a slope of m_2, then the lines are perpendicular if $m_1 = -\frac{1}{m_2}$.

The following examples show you how to determine whether lines are parallel or perpendicular by just looking at their slopes:

Q. Are the lines $3x+2y=8$ and $6x+4y=7$ parallel or perpendicular?

EXAMPLE **A.** The line $3x+2y=8$ is parallel to the line $6x+4y=7$ because their slopes are both $-\frac{3}{2}$. Write each line in the slope-intercept form to see this: $3x+2y=8$ can be written $y=-\frac{3}{2}x+4$ and $6x+4y=7$ can be written $y=-\frac{3}{2}x+\frac{7}{4}$.

Q. Are the lines $y=\frac{3}{4}x+5$ and $y=-\frac{4}{3}x+6$ parallel or perpendicular?

A. The line $y=\frac{3}{4}x+5$ is perpendicular to the line $y=-\frac{4}{3}x+6$ because their slopes are negative reciprocals of one another.

YOUR TURN

21 What is the slope of a line that's parallel to the line $2x-3y=4$?

22 What is the slope of a line that's perpendicular to the line $4x+2y+7=0$?

Finding Distances between Points

A segment can be drawn between two points that are plotted on the coordinate axes. You can determine the distance between those two points by using a formula that actually incorporates the Pythagorean Theorem — it's like finding the length of a hypotenuse of a right triangle. (Check Chapter 21 for more practice with the Pythagorean Theorem.) If you want to find the distance between the two points (x_1,y_1) and (x_2,y_2), use the formula $d=\sqrt{(x_2-x_1)^2+(y_2-y_1)^2}$.

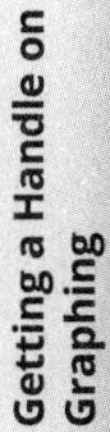

EXAMPLE

Q. Find the distance between the points (−8,2) and (4,7).

A. Use the distance formula and plug in the coordinates of the points:

$$\begin{aligned} d &= \sqrt{\left(4-(-8)\right)^2+(7-2)^2} \\ &= \sqrt{(12)^2+(5)^2} \\ &= \sqrt{144+25} \\ &= \sqrt{169} \\ &= 13 \end{aligned}$$

Of course, not all the distances come out nicely with a perfect square under the radical. When it isn't a perfect square, either simplify the expression or give a decimal approximation (refer to Chapter 6).

Q. Find the distance between the points (4,−3) and (2,11).

A. Using the distance formula, you get

$$\begin{aligned} d &= \sqrt{(2-4)^2+\left(11-(-3)\right)^2} \\ &= \sqrt{(-2)^2+(14)^2} \\ &= \sqrt{4+196} \\ &= \sqrt{200} = \sqrt{100}\sqrt{2} \\ &= 10\sqrt{2} \end{aligned}$$

If you want to estimate the distance, just replace the $\sqrt{2}$ with 1.4 and multiply by 10. The distance is about 14 units.

YOUR TURN

23 Find the distance between (3,−9) and (−9,7).

Find the distance between (4,1) and (−2,2). Round the decimal equivalent of the answer to two decimal places.

Finding Midpoints of Segments

Another useful function that's available on a graph is finding the midpoint of a segment when you're given two points. This is helpful when working with polygons and circles. To find the midpoint, M, of the segment with endpoints (x_1, y_1) and (x_2, y_2), use the following formula: $M = \left(\frac{x_1 + x_2}{2}, \frac{y_1 + y_2}{2}\right)$. You see that you're just averaging the values of the x-coordinates and y-coordinates.

EXAMPLE

Q. What is the midpoint of the segment whose endpoints are $(4,-3)$ and $(8,2)$?

A. Applying the formula, $M = \left(\frac{4+8}{2}, \frac{-3+2}{2}\right) = \left(6, -\frac{1}{2}\right)$.

YOUR TURN

25 Find the midpoint of the segment with endpoints $(-6,11)$ and $(10,5)$.

Find the coordinates of the center of a circle if the coordinates of its diameter are $(5,1)$ and $(-6,-13)$.

Practice Questions Answers and Explanations

1 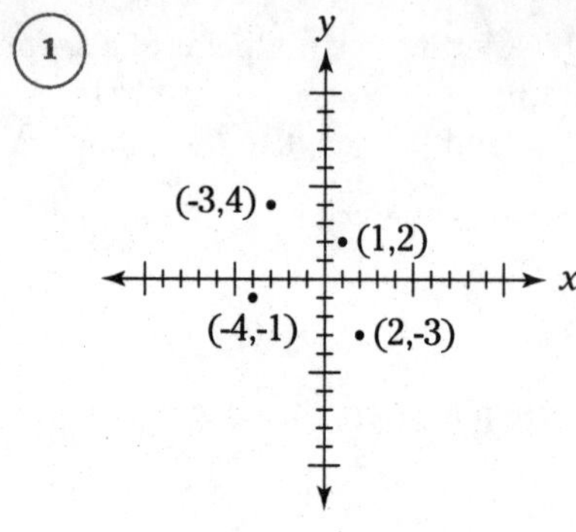

2 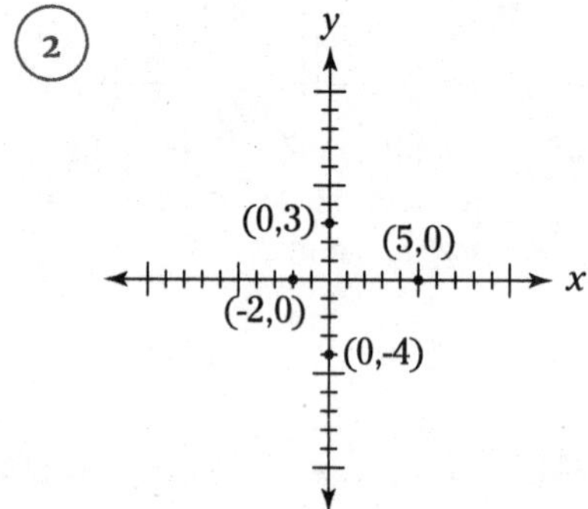

3 **Quadrant II,** the upper-left quadrant.

4 **Quadrant III,** the lower-left quadrant.

5 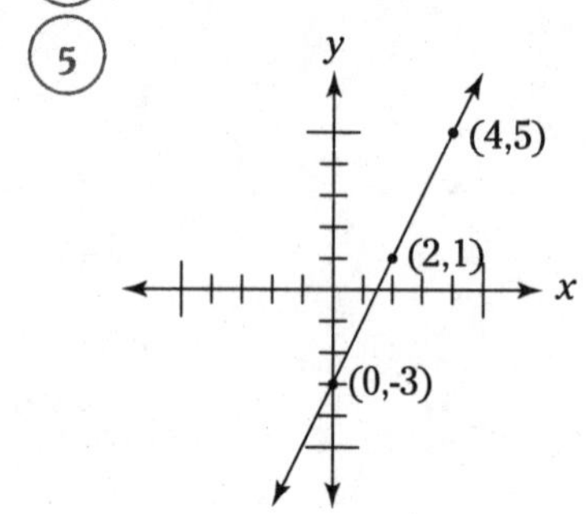

6 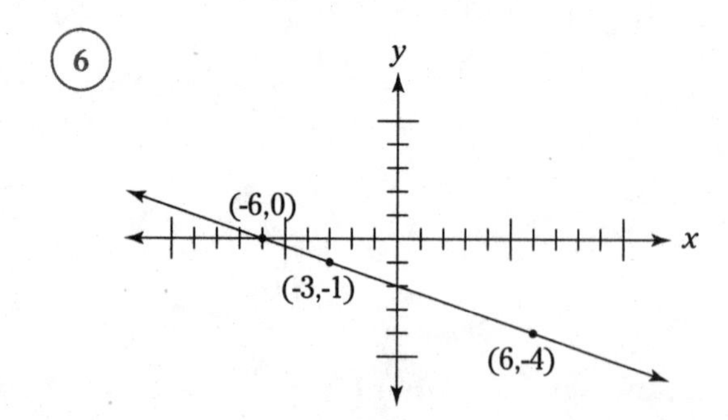

7

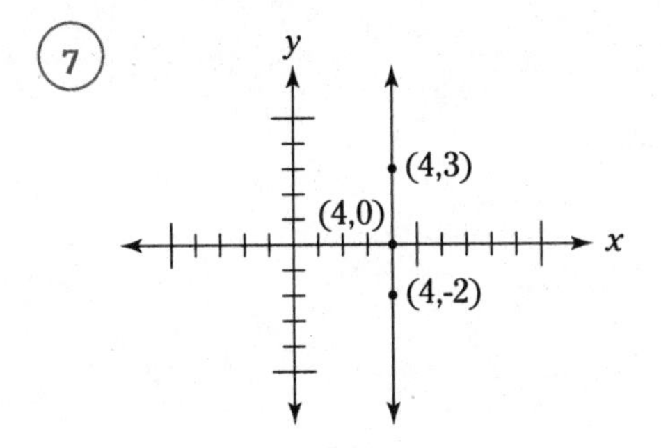

8

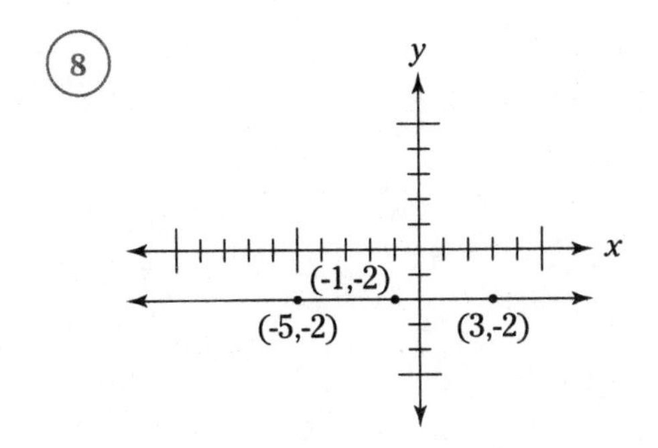

9 **(0,3), (4,0).** Let $y = 0$ to get $3x + 0 = 12$, $x = 4$; the intercept is (4,0). Let $x = 0$ to get $0 + 4y = 12$, $y = 3$; the intercept is (0,3).

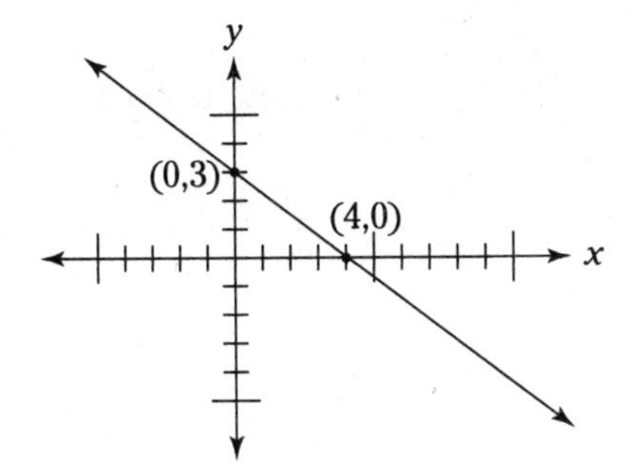

10 **(0,−2), (4,0).** Let $y = 0$ to get $x - 0 = 4$, $x = 4$; the intercept is (4,0). Let $x = 0$ to get $0 - 2y = 4$, $y = -2$; the intercept is (0,−2).

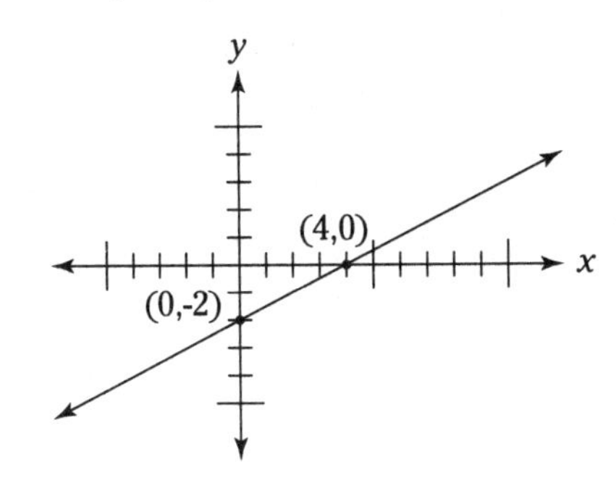

(11) **1.** $m = \frac{2-(-5)}{3-(-4)} = \frac{7}{7} = 1$

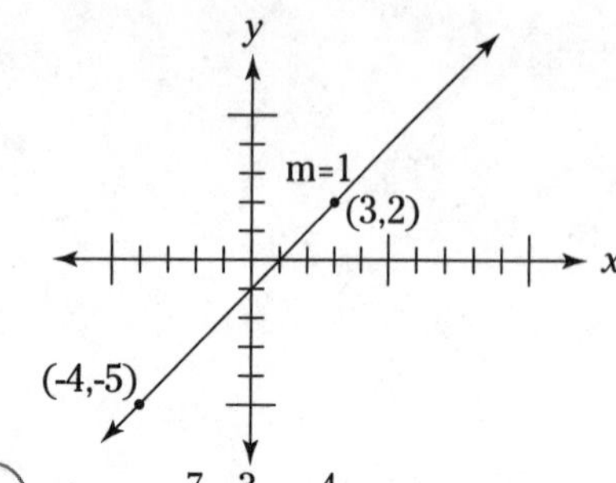

(12) **−2.** $m = \frac{7-3}{-1-1} = \frac{4}{-2} = -2$

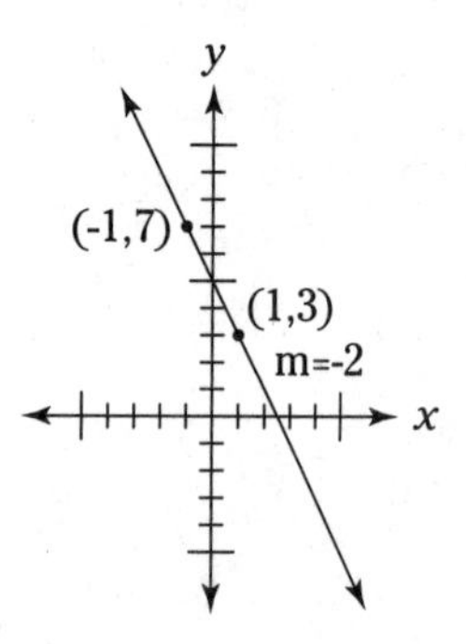

(13) **0.** $m = \frac{-4-(-4)}{3-5} = \frac{0}{-2} = 0$. This fraction is equal to 0, meaning that the line through the points is a horizontal line with the equation $y = -4$.

(14) **No slope.** $m = \frac{3-(-8)}{2-2} = \frac{11}{0}$. This fraction has no value; it's undefined. The line through these two points has no slope. It's a vertical line with the equation $x = 2$.

(15) **The slope is $-\frac{2}{3}$ and the y-intercept is 4.** Place a point at (0,4). Then count three units to the right of that intercept and two units down. Place a point there and draw a line through the intercept and the new point.

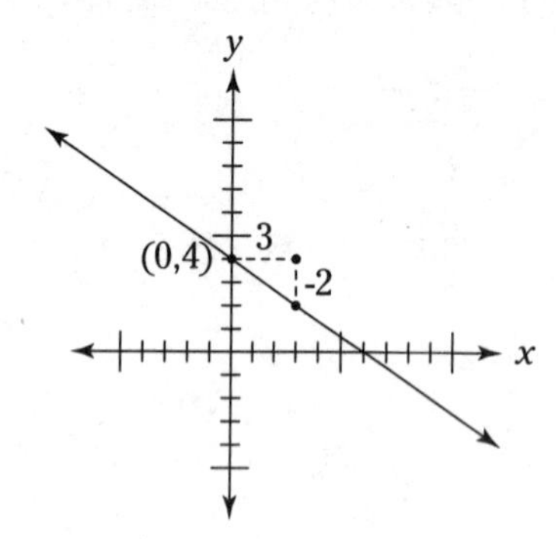

16 **The y-intercept is −2, and the slope is 5.** Place a point at (0,−2). Then count one unit to the right and five units up from that intercept. Place a point there and draw a line.

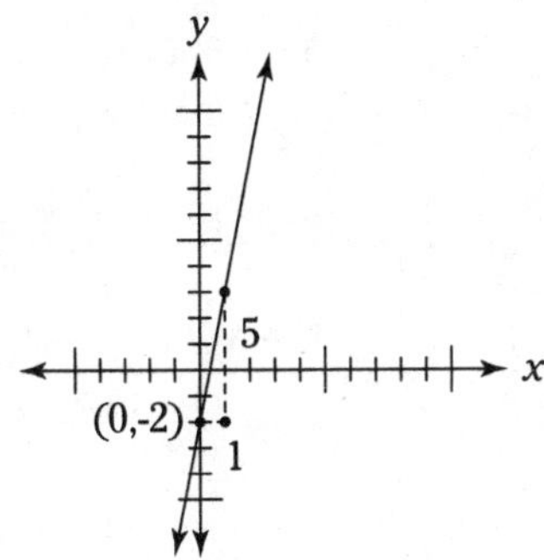

17 $\mathbf{y = -4x + \frac{3}{2}}$. Subtract 8x from each side: $2y = -8x + 3$. Then divide every term by 2: $y = -4x + \frac{3}{2}$. The slope is −4, and the y-intercept is $\frac{3}{2}$.

18 $\mathbf{y = 4x - 3}$. Add y to each side to get $4x - 3 = y$. Then use the symmetric property to turn the equation around: $y = 4x - 3$. (This is easier than subtracting $4x$ from each side, adding 3 to each side, and then dividing by −1.)

19 $\mathbf{y = \frac{1}{3}x + 7}$. You're given the slope and the y-intercept, so you can just put in the 7 for the value of b. If you want this in a form without fractions, just multiply through by 3 to get $3y = x + 21$.

20 $\mathbf{y = 6x + 17}$. First, find the slope by using the two points: $m = \frac{5-(-1)}{-2-(-3)} = \frac{6}{1} = 6$. Using the point-slope form and the coordinates of the point (−3,−1), you get $y-(-1) = 6(x-(-3))$ or $y + 1 = 6x + 18$. Subtracting 1 from each side gives you $y = 6x + 17$.

21 $\mathbf{\frac{2}{3}}$. You can find the slope of a line that's parallel to the line $2x - 3y = 4$ by changing the equation to the slope-intercept form and then subtracting 2x from each side and dividing each term by −3. The equation is then $-3y = -2x + 4$, $y = \frac{-2}{-3}x + \frac{4}{-3}$, $y = \frac{2}{3}x - \frac{4}{3}$. The slope is $\frac{2}{3}$, and the y-intercept is $-\frac{4}{3}$.

22 $\mathbf{\frac{1}{2}}$. Change this equation to the slope-intercept form to find the slope of the line: $4x + 2y + 7 = 0$, $2y = -4x - 7$, $y = -2x - \frac{7}{2}$. The negative reciprocal of −2 is $\frac{1}{2}$, so that's the slope of a line perpendicular to this one.

23 **20 units.** Using the distance formula, $d = \sqrt{[3-(-9)]^2 + (-9-7)^2} = \sqrt{12^2 + (-16)^2} = \sqrt{144 + 256} = \sqrt{400} = 20$.

24 **6.08 units.** $d = \sqrt{[4-(-2)]^2 + (1-2)^2} = \sqrt{6^2 + (-1)^2} = \sqrt{36 + 1} = \sqrt{37}$. Use a calculator and round to two decimal places.

25 **(2,8).** Using the formula, $M = \left(\frac{-6+10}{2}, \frac{11+5}{2}\right) = \left(\frac{4}{2}, \frac{16}{2}\right) = (2,8)$.

26 $\mathbf{\left(-\frac{1}{2}, -6\right)}$. Using the formula, $M = \left(\frac{5+(-6)}{2}, \frac{1+(-13)}{2}\right) = \left(\frac{-1}{2}, \frac{-12}{2}\right) = \left(-\frac{1}{2}, -6\right)$.

Whaddya Know? Chapter 24 Quiz

Quiz time! Complete each problem to test your knowledge on the various topics covered in this chapter. You can then find the solutions and explanations in the next section.

1. Identify the coordinates of the points shown on the graph.

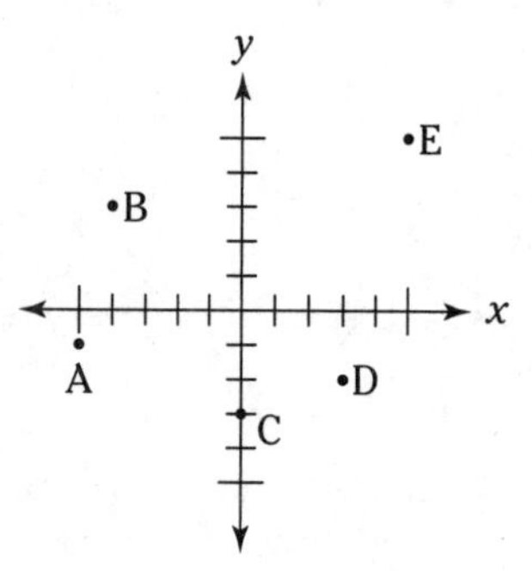

2. Refer to the graph in Problem 1 and determine in which quadrant each point lies.

3. Find the midpoint of the points $(6,-2)$ and $(4,1)$.

4. Determine the intercepts of the line $x-4y=8$.

5. Find the slope of the line $10x-5y=13$.

6. What are the slope and y-intercept of the line $4x-9y=36$?

7. Write the equation of the line whose slope is -3 and y-intercept is 4.

8. Find the distance between the points $(6,-2)$ and $(4,1)$.

9. Given the equation of the line $3x-4y=11$, fill in the missing values a, b, and c in the coordinates of points that lie on that line: $(a,-2)$, $(5,b)$, $(c,0)$.

10. Are the lines $3x+2y=4$ and $2x-3y=6$ parallel or perpendicular?

11. Write the equation of the line whose slope is $\frac{1}{2}$ and which goes through the point $(3,-5)$.

12 Match the equations $x - 3y = 6$ and $y = 4x + 4$ with the correct graphed line.

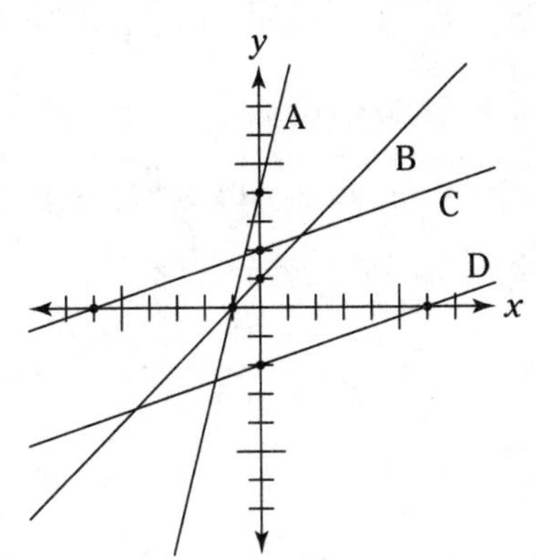

13 Determine the intercepts of the line $x = 7$.

14 What are the slope and y-intercept of the line $y = -3x + 7$?

15 Write the equation of the line that goes through the points $(-2, 6)$ and $(8, -4)$.

16 Find the slope of the line $y = -4$.

17 Write the equation of the line $x - 3y = 8$ in slope-intercept form.

Getting a Handle on Graphing

Answers to Chapter 24 Quiz

1. **A: (–5,–1), B: (–4,3), C: (0,–3), D: (3,–2), E: (5,5)**

2. **A: Quadrant III, B: Quadrant II, C: no quadrant — on axis, D: Quadrant IV, E: Quadrant I**

3. $\left(5,-\frac{1}{2}\right)$. Using the formula, $M=\left(\frac{6+4}{2},\frac{-2+1}{2}\right)=\left(\frac{10}{2},\frac{-1}{2}\right)=\left(5,-\frac{1}{2}\right)$.

4. **(8,0) and (0,–2).** When $y=0$, $x-4(0)=8$ and $x=0$. When $x=0$, $0-4y=8$ and $y=-2$.

5. $\boldsymbol{m=2}$. Subtract 10*x* from each side and then divide by –5. $-5y=-10x+13$ becomes $y=2x-\frac{13}{5}$. The slope is 2.

6. $\boldsymbol{m=\frac{4}{9}, b=-4}$. Writing the equation in slope-intercept form, you have $4x-9y=36$, which becomes $-9y=-4x+36$ and then $y=\frac{4}{9}x-4$. The slope is $\frac{4}{9}$, and the *y*-intercept is –4.

7. $\boldsymbol{y=-3x+4}$

8. **3.61.** $d=\sqrt{(6-4)^2+(-2-1)^2}=\sqrt{2^2+(-3)^2}=\sqrt{4+9}=\sqrt{13}\approx 3.61$

9. $\boldsymbol{a=1,\ b=1,\ c=\frac{11}{3}}$. The points on $3x-4y=11$ are $(1,-2)$, $(5,1)$, $\left(\frac{11}{3},0\right)$.

10. **perpendicular.** The slope of $3x+2y=4$ is $-\frac{3}{2}$, and the slope of $2x-3y=6$ is $\frac{2}{3}$. These are negative reciprocals, so the lines are perpendicular.

11. $\boldsymbol{y=\frac{1}{2}x-\frac{13}{2}}$. Use the slope-intercept form to solve for the *y*-intercept. $-5=\frac{1}{2}(3)+b$ gives you $b=-\frac{13}{2}$. Now you have $y=\frac{1}{2}x-\frac{13}{2}$, which can be written as $x-2y=13$.

12. $\boldsymbol{x-3y=6}$**: line D and** $\boldsymbol{y=4x+4}$**: line A.** The line $x-3y=6$ has intercepts (0,–2) and (6,0). The line $y=4x+4$ has intercepts (0,4) and (–4,0).

13. **(7,0).** This is a vertical line, and the only intercept is (7,0).

14. $\boldsymbol{m=-3, b=7}$. Using the slope-intercept form, $y=-3x+7$, the slope is –3, and the *y*-intercept is (0,7).

15. $\boldsymbol{y=-x+4}$. First, find the slope of the line: $m=\frac{-4-6}{8-(-2)}=\frac{-10}{10}=-1$. Then, using the point (–2,6), solve for *b* with the slope-intercept form: $6=-1(-2)+b$, which gives you $b=4$. So the equation is $y=-1x+4$.

16. **0.** The line $y=-4$ is horizontal, so its slope is 0.

17. $\boldsymbol{y=\frac{1}{3}x-\frac{8}{3}}$. First, subtract *x* from each side to get $-3y=-x+8$. Then divide each term by –3, which gives you $\frac{-3y}{-3}=\frac{-x}{-3}+\frac{8}{-3}$ or $y=\frac{1}{3}x-\frac{8}{3}$.

IN THIS CHAPTER

» Meeting up with intersections of lines

» Circling around with circles and other conics

» Plotting points on polynomials

» Taking on inequalities and absolute-value graphs

» Transforming basic equations

Chapter 25
Extending the Graphing Horizon

The graphs in algebra are unique because they reveal relationships that you can use to model a situation: A line can model the depreciation of the value of your boat; parabolas can model daily temperatures; and a flat, S-shaped curve can model the number of people infected with the flu. All these and other models are useful for illustrating what's happening and predicting what can happen in the future.

In Chapter 24, you find the basic techniques necessary to either read or create the graph of a function. In this chapter, I introduce you to some of the curves that belong to *conic sections* (figures formed from slicing a cone in various directions). You also find specialty graphs that require shading for completion and some graphs involving inequalities instead of equal signs. And, finally, I fill you in on the method of graphing using transformations.

Finding the Intersections of Lines

If two lines *intersect*, or cross one another, then they intersect exactly once and only once. The place they cross is the point of intersection, and that common point is the only one both lines share. Careful graphing can sometimes help you to find the point of intersection.

The point (5,1) is the point of intersection of the two lines $x+y=6$ and $2x-y=9$ because the coordinates make each equation true:

- If $x+y=6$, then substituting the values $x=5$ and $y=1$ gives you $5+1=6$, which is true.
- If $2x-y=9$, then substituting the values $x=5$ and $y=1$ gives you $2\times 5-1=10-1=9$, which is also true.

This is the only point that works for both of these lines.

Graphing for intersections

Two lines will intersect at exactly one point — unless they're parallel to one another. You can find the intersection of lines by careful graphing of the lines or by using simple algebra. You'll see how to do it algebraically later in this section and in Chapter 26. Graphing is quick and easy, but it's hard to tell the exact answer if there's a fraction in one or both of the coordinates of the point of intersection.

Careful graphing can give you the intersection of two lines. The only problem is that if your graph is even a little off, you can get the wrong answer. Also, if the answer has a fraction in it, it's difficult to figure out what that fraction is. That's where the algebraic techniques are a big help.

Suppose you need to find the intersection of the lines $3x-y=5$ and $x+y=-1$.

Look at the graphs in Figure 25-1. The lines appear to cross at the point (1,−2). Replace the coordinates in the equations to check this out:

- If $3x-y=5$, then substituting the values gives you $3\times 1-(-2)=3+2=5$, which is true.
- If $x+y=-1$, then substituting the values gives you $1+(-2)=-1$, which is also true.

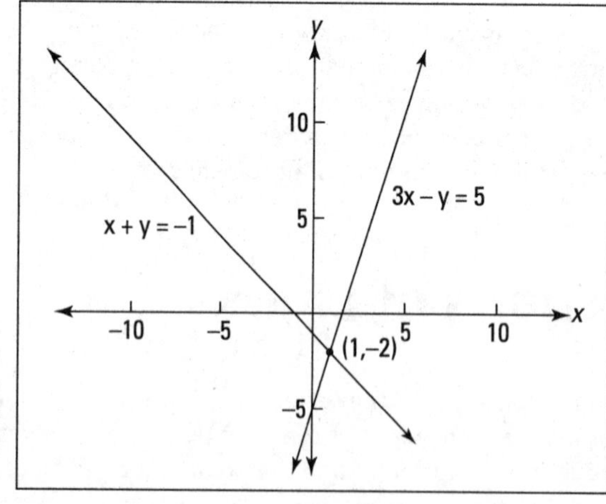

FIGURE 25-1: The intersection of two lines at a point (1,−2).

REMEMBER

Graphing is an inexact way to find the intersection of lines. You have to be super-careful when plotting the points and lines, and you should always check your answer in the equations of the lines.

Substituting to find intersections

Another way to find the point where two lines intersect is to use a technique called *substitution* — you substitute the *y* value from one equation for the *y* value in the other equation and then solve for *x*. Because you're looking for the place where *x* and *y* of each line are the same — that's where they intersect — you can write the equation $y = y$, meaning that the *y* from the first line is equal to the *y* from the second line. Replace the *y*'s with what they're equal to in each equation, and solve for the value of *x* that works. Here's an example:

Find the intersection of the lines $3x - y = 5$ and $x + y = -1$. (This is the same problem that is graphed in the preceding section.)

Follow these steps:

1. **Put each equation in the slope-intercept form, which is a way of solving each equation for y.**

   $3x - y = 5$ is written as $y = 3x - 5$, and $x + y = -1$ is written as $y = -x - 1$. The lines are not parallel, and their slopes are different, so there will be a point of intersection.

2. **Set the y points equal and solve.**

   From $y = 3x - 5$ and $y = -x - 1$, you substitute what *y* is equal to in the first equation with the *y* in the second equation: $3x - 5 = -x - 1$.

3. **Solve for the value of x.**

   Add *x* to each side and add 5 to each side:

   $$\begin{aligned} 3x + x - 5 + 5 &= -x + x - 1 + 5 \\ 4x &= 4 \\ x &= 1 \end{aligned}$$

   Substitute that 1 for *x* into either equation to find that $y = -2$. The lines intersect at the point $(1, -2)$.

This technique shows you how the solution can be found without even graphing. If the lines are parallel, it's apparent immediately because their slopes are the same. If that's the case, stop — there's no solution. Also, if the two equations are just two different ways of representing the same line, then this will be apparent: The equations will be exactly the same in the slope-intercept form.

EXAMPLE

Q. Find the intersection of the lines $y = 3x - 2$ and $y = -2x - 7$.

A. Set $3x - 2 = -2x - 7$ and solve for x. Adding $2x$ to each side and adding 2 to each side, you get $5x = -5$. Dividing by 5 gives you $x = -1$. Now substitute -1 for x in either of the original equations. You get $y = -5$. It's really a good idea to do that substitution back into *both* of the equations as a check. The answer is $(-1, -5)$.

Q. Find the intersection of the lines $x + y = 6$ and $2x - y = 6$.

A. First, write each equation in the slope-intercept form. The line $x + y = 6$ becomes $y = -x + 6$, and the line $2x - y = 6$ becomes $y = 2x - 6$. Now, setting $-x + 6 = 2x - 6$, you get $-3x = -12$ or $x = 4$. Substituting 4 for x in either equation, you get $y = 2$. The answer is $(4, 2)$.

YOUR TURN

Find the intersection of the lines.

1 $y = -4x + 7$ and $y = 5x - 2$

2 $3x - y = 1$ and $x + 2y + 9 = 0$

3 $y = 4x - 3$ and $8x - 2y = 7$

4 $3x + y = 4$ and $3y = 12 - 9x$

Graphing Parabolas and Circles

A *parabola* is a sort of U-shaped curve. It's one of the conic sections. A *circle* is the most easily recognized conic (the other conics are hyperbolas and ellipses). The equations and graphs of parabolas are used to describe all sorts of natural phenomena. For instance, headlight reflectors are formed from parabolic shapes. Circles — well, circles are just circles: handy, easy to deal with, and so symmetric.

An equation for parabolas that open upward or downward is $y = ax^2 + bx + c$, where a isn't 0. If a is a positive number, then the parabola opens upward; a negative a gives you a downward parabola.

Here are the standard forms for parabolas and circles.

- **Parabola:** $y = a(x-h)^2 + k$ or $x = a(y-k)^2 + h$
- **Circle:** $(x-h)^2 + (y-k)^2 = r^2$

Notice the h and k in both forms. For the parabola, the coordinates (h,k) tell you where the *vertex* (bottom or top or left or right of the U-shape) is. For the circle, (h,k) gives you the coordinates of the center of the circle. The r part of the circle's form gives you the radius of the circle.

Curling Up with Parabolas

Parabolic curves are the graphs of quadratic equations where either an x term is squared or a y term is squared, but both are not squared at the same time. The reflectors in headlights have parabola-like curves running through them. The McDonalds golden arches are parabolas. The abundance of manufactured parabolas points to the fact that the properties responsible for creating a parabola often occur naturally. Mathematicians are able to put an equation to this natural phenomenon.

Parabolas have a highest point or a lowest point (or the farthest left or the farthest right), called the *vertex.* The curve is lower on the left and right of a vertex that is the highest point, and it's higher to the left and right of a vertex that is the lowest point.

Trying out the basic parabola

My favorite equation for the parabola is $y = x^2$, the basic parabola. Figure 25-2 shows a graph of this formula. This equation says that the y-coordinate of every point on the parabola is the square of the x-coordinate. Notice that whether x is positive or negative, the y is a square of it and is positive.

The vertex of the parabola in Figure 25-2 is at the origin, (0,0), and it curves upward.

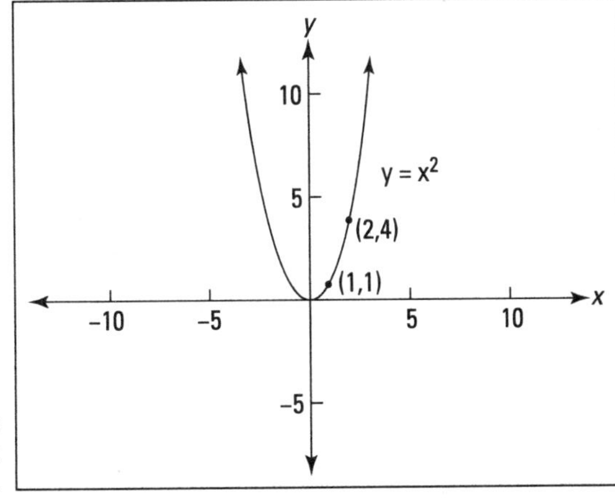

FIGURE 25-2: The simplest parabola.

You can make this parabola steeper or flatter by multiplying the x^2 by certain numbers. If you multiply the squared term by numbers bigger than 1, it makes the parabola steeper. If you multiply by numbers between 0 and 1 (proper fractions), it makes the parabola flatter (examples for both types are shown in Figure 25-3). Making it steeper or flatter than the basic parabola helps the parabola fit different applications. The flatter ones are more like the curve of a headlight reflector. The steeper ones could be models for the time it takes to swim a certain distance, depending on your age.

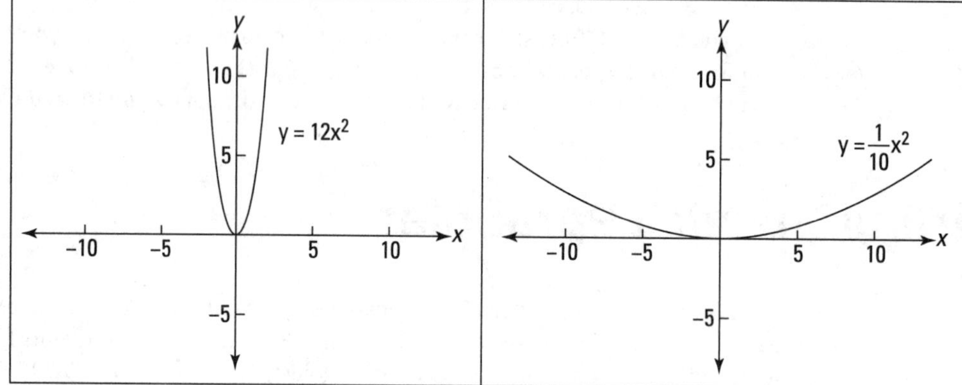

FIGURE 25-3: A steeper parabola and a flatter parabola.

You can make the parabola open downward by multiplying the x^2 by a negative number, and make it steeper or flatter than the basic parabola — in a downward direction.

Putting the vertex on an axis

The basic parabola, $y = x^2$, can be slid around — left, right, up, down — placing the vertex somewhere else on an axis and not changing the general shape.

If you change the basic equation by adding a constant number to the x^2 — such as $y = x^2 + 3$, $y = x^2 + 8$, $y = x^2 - 5$, or $y = x^2 - 1$ — then the parabola moves up and down the y-axis. Note that adding a negative number is also part of this rule. These manipulations help make a parabola fit the model of a certain situation.

Note: Not everything starts at 0. Figure 25-4 shows several parabolas, only one of which starts at 0.

If you change the basic parabolic equation by adding a number to the x first and then squaring the expression — as in $x = (x+3)^2$, $y = (x+8)^2$, $y = (x-5)^2$, or $y = (x-1)^2$ — then you move the graph to the left or right of where the basic parabola lies. Using +3 as in the equation $y = (x+3)^2$ moves the graph to the left, and using −3 as in the equation $y = (x-3)^2$ moves the graph to the right. It's the opposite of what you may expect, but it works this way consistently (see Figure 25-5).

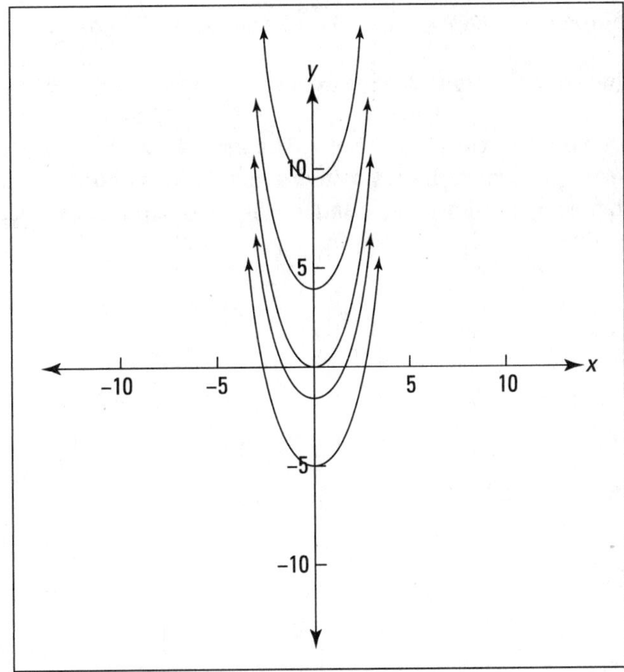

FIGURE 25-4: Parabolas spooning.

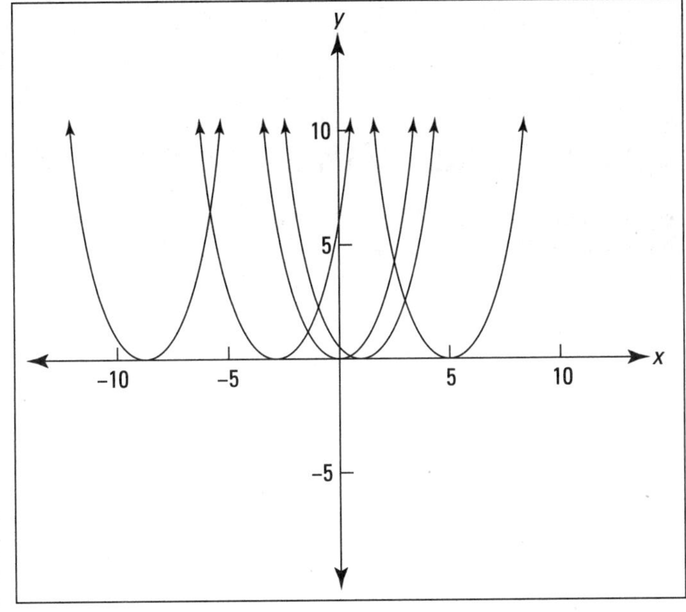

FIGURE 25-5: Pretty parabolas all in a row.

The standard equation of a parabola is $y = a(x-h)^2 + k$. You find the vertex of the parabola with (h,k), and the a tells you whether the parabola opens upward or downward and how steep the parabola is. If a is positive, the parabola opens upward; negative means downward.

EXAMPLE

Q. Find the vertex of the parabola $y = -3(x+3)^2 + 7$ and sketch its graph.

A. Vertex: (−3,7); the parabola opens downward.

To find the vertex, you need the opposite of 3. The general form has $(x - h)$, which is the same as $(x - (-h))$ to deal with the positive sums in the parentheses. Because a is negative, the parabola opens downward; and a being −3 introduces a steepness, much like the slope of a line. You can plot a few extra points to help with the shape.

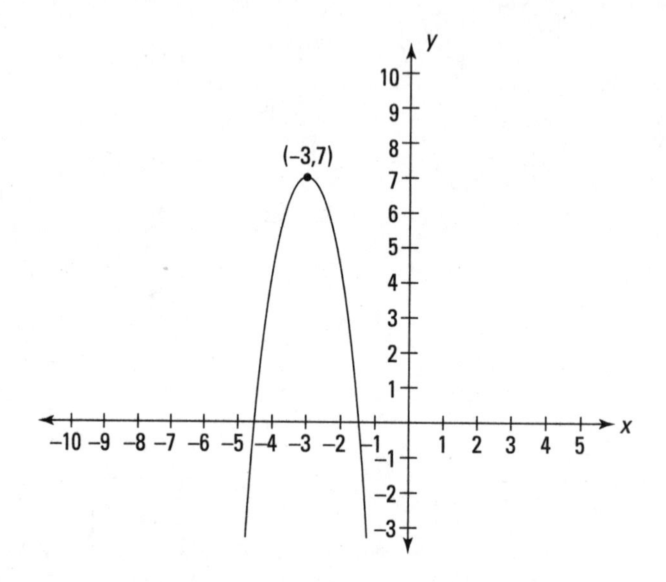

Going around in circles with a circular graph

An example of an equation of a circle is $x^2 + y^2 = 25$. The circle representing this equation goes through an infinite number of points. Here are just some of those points: (0,5), (0,−5), (5,0), (−5,0), (3,4), (−3,4), (4,3), (4,−3), (−3,4), (−3,−4), (−4,−3), and many more.

I haven't finished all the possible points with integer coordinates, let alone points with fractional coordinates, such as $\left(\frac{25}{13}, \frac{60}{13}\right)$.

When graphing an equation, you don't expect to find all the points. You just want to find enough points to help you sketch in all the others without naming them.

The general equation for a circle is $(x-h)^2 + (y-k)^2 = r^2$, where (h,k) are the coordinates of the center of the circle, and r is the length of the radius.

In Figure 25-6, I show you the graph of the circle and some of the named points that make up the graph of the circle.

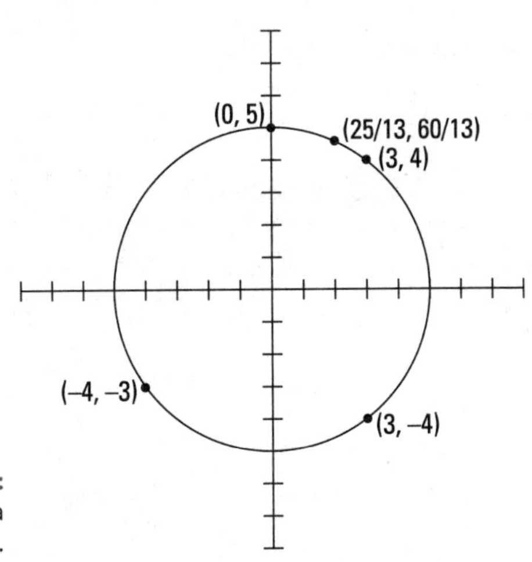

FIGURE 25-6: The circle has a radius of 5.

EXAMPLE

Q. Find the center and radius of the circle $(x-4)^2+(y+2)^2=36$ and sketch its graph.

A. Center: (4,−2); radius = 6. First, plot the center. Then find four or more points that are 6 units from the center; usually counting up, down, left, and right are enough.

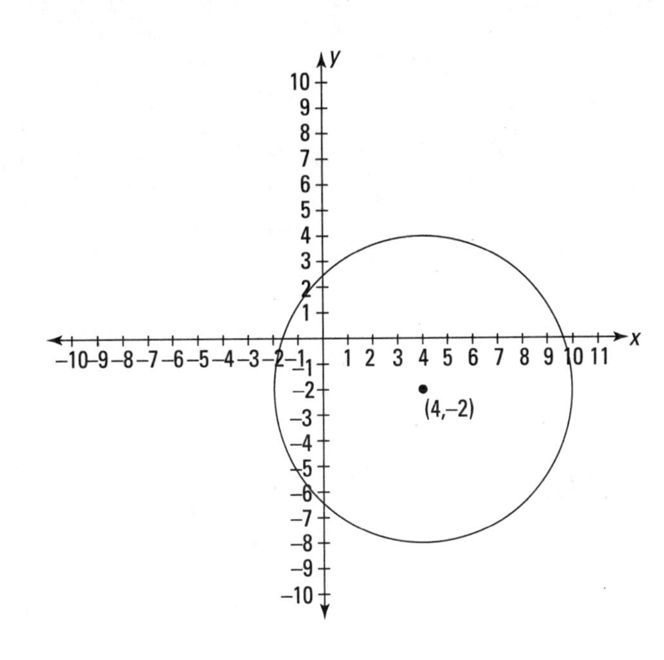

YOUR TURN

Find the vertex and sketch the graph of the equation.

5 $y = \frac{1}{2}(x-3)^2 - 2$

6 $y = 2(x+2)^2$

Find the center and radius, and sketch the graph of the equation.

7 $(x+3)^2 + (y-4)^2 = 25$

 $(x-5)^2 + (y-2)^2 = 9$

Plotting and Plugging in Polynomial Graphs

The graph of a polynomial function is a smooth curve. You find methods for factoring polynomial expressions in Chapter 13. This is where all that factoring pays off, because you want to find the x-intercepts or zeros of a polynomial function when sketching its graph.

The standard equation of a polynomial function is: $P(x) = a_n x^n + a_{n-1}x^{n-1} + a_{n-2}x^{n-2} + \cdots + a_1 x^1 + a_0$, where n is a positive integer, and the a_n coefficients and constant are real numbers.

The following are steps you can use to sketch a polynomial curve.

1. **Let $x = 0$, then $P(0) = a_0$, which gives you the y-intercept of the curve, (0,ao).**
2. **Factor P(x) and set the factored form equal to 0.**

If $P(x) = 0$ when $x = b$, then $(b,0)$ is an x-intercept of the curve.

3. **Place the intercepts on the coordinate axes.**

   The curve will either touch or cross at each x-intercept and will cross the y-axis at the y-intercept. If the exponent on the factor of an x-intercept is an even number, then the curve will just touch at that point and not cross the axis. If the exponent is odd, then the curve will cross the x-axis at that point.

4. **Determine if the function P(x) is positive or negative between the x-intercepts and to the right of the right-most intercept and the left of the left-most intercept. Use the results to help you sketch the graph.**

EXAMPLE

Q. Sketch the graph of $y = x(x-3)(x+4)^2$.

A. When $x = 0$, $y = 0$, so the y-intercept is (0,0), the origin.

Setting $y = 0$, you have $0 = x(x-3)(x+4)^2$, which has solutions $x = 0$, $x = 3$, and $x = -4$. (Refer to Chapters 16 and 17 for more on the multiplication property of zero and solving equations.) The x-intercepts are (0,0), (3,0), and (−4,0). The intercept (0,0) has appeared twice — it's both the y-intercept and an x-intercept.

Place the intercepts on coordinate axes. The exponent 2 on the factor $(x+4)$ means that the curve just touches at that intercept. The left side of Figure 25-7 shows you the intercepts in their places.

Now test the sign of the function for values smaller than −4, between −4 and 0, between 0 and 3, and greater than 3. Just find the signs of the test values — positive means above the x-axis, and negative means below the x-axis.

Smaller than −4: Try $x = -5$, which gives you $y = -5(-5-3)(-5+4)^2 = -5(-8)(1)$, which is positive.

Between −4 and 0: Try $x = -1$, which gives you $y = -1(-1-3)(-1+4)^2 = -1(-4)(9)$, which is positive.

Between 0 and 3: Try 1, which gives you $y = 1(1-3)(1+4)^2 = 1(-2)(25)$, which is negative.

Greater than 3: Try 4, which gives you $y = 4(4-3)(4+4)^2 = 4(1)(64)$, which is positive.

The test values aren't unique or special. I just try to pick numbers that are easiest to compute with in the equation.

Now sketch in a curve that is above the x-axis when $x < -4$, touches at (−4,0), is above the x-axis when $-4 < x < 0$, crosses at (0,0), is below the x-axis when $0 < x < 3$, crosses at (3,0), and is above the x-axis when $x > 3$. The right side of Figure 25-7 shows you just such a graph.

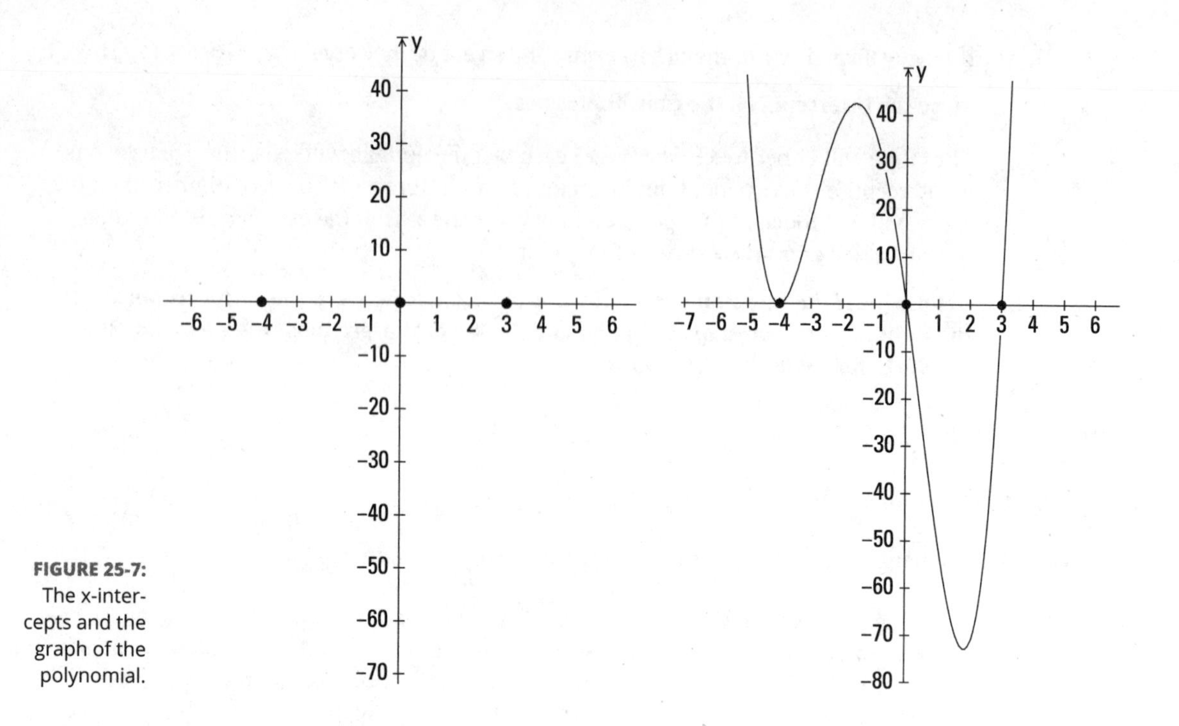

FIGURE 25-7: The x-intercepts and the graph of the polynomial.

Q. Sketch the graph of $y = x^4 - 17x^2 + 16$.

A. When $x = 0$, $y = 16$, so the y-intercept is (0,16). Now factor the polynomial using the pattern in quadratic-like expressions. (See Chapter 16 for more on this.) You get $y = (x^2 - 16)(x^2 - 1) = (x+4)(x-4)(x+1)(x-1)$. Setting $y = 0$, you have $(x+4)(x-4)(x+1)(x-1) = 0$, which has solutions $x = -4$, $x = 4$, $x = -1$, and $x = 1$. (Refer to Chapters 16 and 17 for more on the multiplication property of zero and solving equations.) The x-intercepts are (−4,0), (4,0), (−1,0), and (1,0). Place these intercepts on the x-axis, and the y-intercept (0,16) on the y-axis. All the factors have exponents of 1, so the curve crosses at each intercept. The left side of Figure 25-8 shows you the placement of the intercepts.

Now test the sign of the function for values smaller than −4, between −4 and −1, between −1 and 1, between 1 and 4, and greater than 4. Just find the signs of the test values — positive means above the x-axis, and negative means below the x-axis.

Smaller than −4: Try $x = -5$, which gives you $y = (-5+4)(-5-4)(-5+1)(-5-1) = (-1)(-9)(-4)(-6)$, which is positive.

Between −4 and 1: Try $x = -2$, which gives you $y = (-2+4)(-2-4)(-2+1)(-2-1) = (2)(-6)(-1)(-3)$, which is negative.

Between −1 and 1: Try 0, which gives you $y = (0+4)(0-4)(0+1)(0-1) = (4)(-4)(1)(-1)$, which is positive.

Between 1 and 4: Try 2, which gives you $(2+4)(2-4)(2+1)(2-1) = (6)(-2)(3)(1)$, which is negative.

Greater than 4: Try 5, which gives you $(5+4)(5-4)(5+1)(5-1)=(9)(1)(6)(4)$, which is positive.

The test values aren't unique or special. I just try to pick numbers that are easiest to compute with in the equation.

Now sketch in a curve that is above the x-axis when $x<-4$, crosses at (−4,0), is below the x-axis when $-4<x<-1$, crosses at (−1,0), is above the x-axis when $-1<x<1$, crosses at (1,0), is below the x-axis when $1<x<4$, crosses at (4,0) and is above the x-axis when $x>4$. The right side of Figure 25-8 shows you just such a graph.

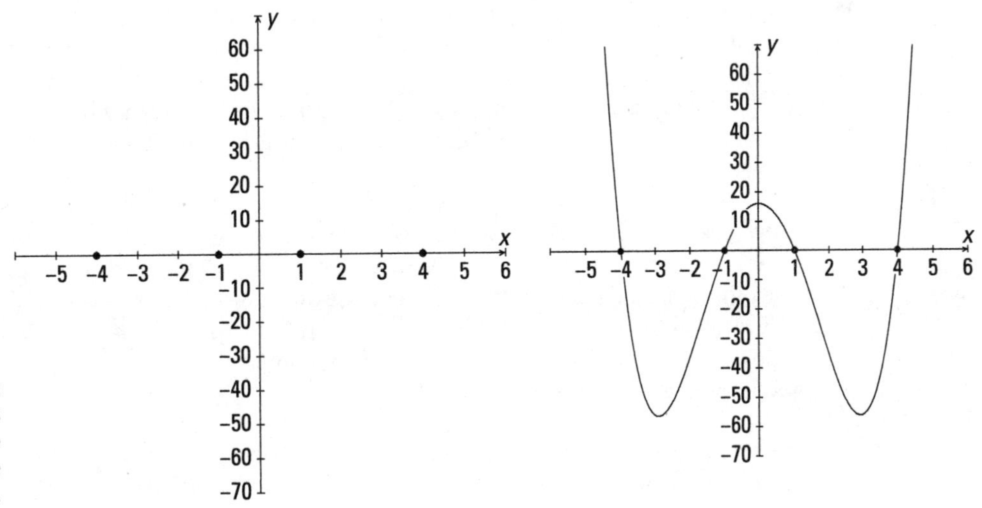

FIGURE 25-8: The intercepts help with the graph of the polynomial.

YOUR TURN

Sketch the graph of the equation.

9 $y=x^2(x+1)(x-5)$

10 $y=x^3+7x^2-9x-63$

Investigating Graphs of Inequality Functions

An *inequality function* has the general format $y > f(x)$, $y \geq f(x)$, $y < f(x)$, or $y \leq f(x)$. The graph of an inequality function consists of an infinite number of points, all displayed by shading in one side of a line or curve. To graph an inequality function, you graph the corresponding function $y = f(x)$ and then determine which side of the curve is shaded by trying a test point to see if it belongs in the solution.

EXAMPLE

Q. Sketch the graph of $x + y > 3$.

A. First, rewrite the inequality by subtracting x from each side to get $y > -x + 3$. The corresponding function is $y = -x + 3$, which is a line with a slope of −1 and a y-intercept of 3. (Refer to Chapter 24 for more on graphing lines.) Next, sketch the graph of the line $y = -x + 3$, but use a dotted or dashed line, because the inequality reads "$>$"; you don't include any of the points on that line in the solution. The left side of Figure 25-9 illustrates how you graph that line.

Now pick a *test point.* This is a point you use to determine which side of the line to shade. Whenever possible, I use the origin, (0,0) — as long as it's clearly on one side or the other. Placing (0,0) in the original inequality, $x + y > 3$, you have $0 + 0 > 3$. That's not true! So (0,0) is not in the solution; you can't shade that side of the line. The right side of Figure 25-9 shows you the correct shading.

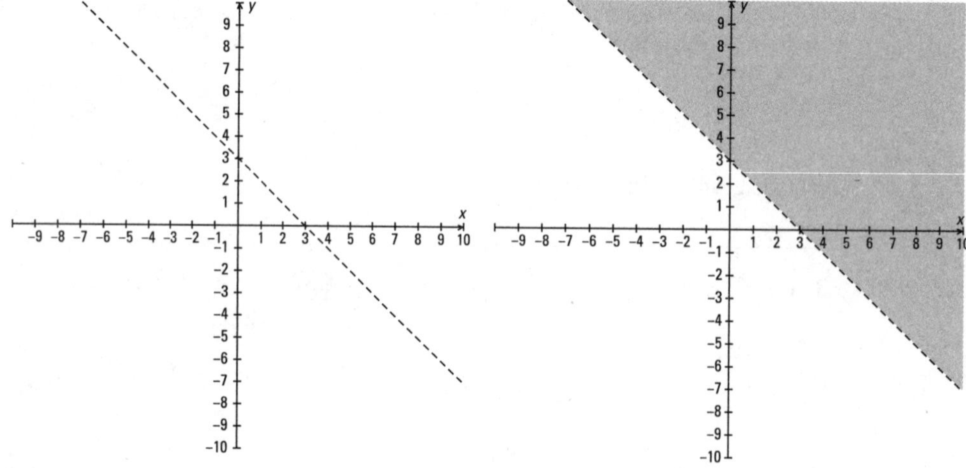

FIGURE 25-9: Use a dashed line when the line is not included.

Q. Sketch the graph of $4x - y \leq 8$.

A. First, rewrite the inequality by adding y to each side and subtracting 8 from each side to get $4x - 8 \leq y$. The corresponding function is $y = 4x - 8$, which is a line with a slope of 4 and a y-intercept of −8. (Refer to Chapter 24 for more on graphing lines.) Next, sketch the graph of the line $y = 4x - 8$, using a solid line, because the inequality reads "$\leq$", meaning you include all the points on that line in the solution. The left side of Figure 25-10 illustrates how you graph that line.

Now pick a *test point*. This is a point you use to determine which side of the line to shade. Whenever possible, I use the origin, (0,0) — as long as it's clearly on one side or the other. Placing (0,0) in the original inequality, $4x - y \leq 8$, you have $0 - 0 \leq 8$, which is true. So (0,0) is in the solution; you shade everything on that side of the line. The right side of Figure 25-10 shows you the correct shading.

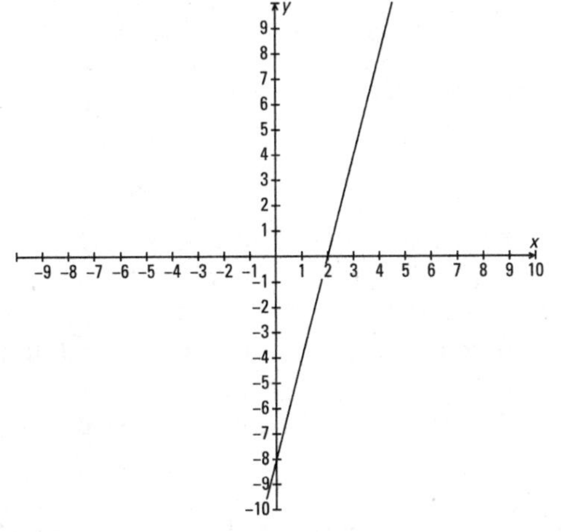

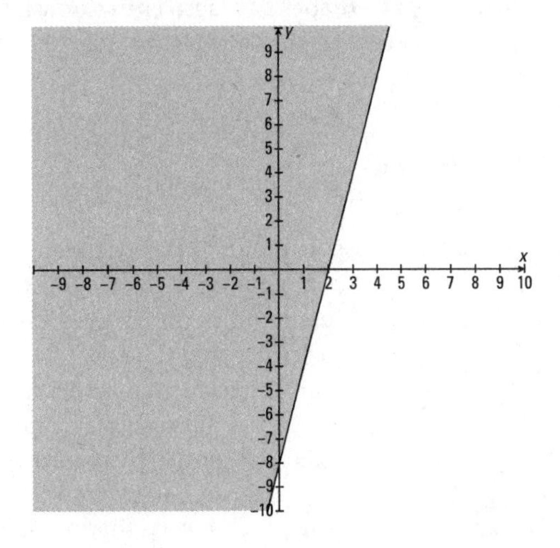

FIGURE 25-10: The origin is included in the solution.

YOUR TURN

Sketch the graph of the inequality.

 $x + y \leq 2$

 $2x - y > -3$

Taking on Absolute-Value Function Graphs

An *absolute-value function* has the general format $y = a|f(x)| + b$. The $|f(x)|$ values are never negative, but the multiplier a can flip the positive values to below the x-axis, and a negative b can drag the y-values downward. The graph of an absolute-value function when $f(x)$ is linear is a V-shape that faces either upward or downward. When the a is positive, it's upward — when negative, it's downward.

EXAMPLE

Q. Sketch the graph of $y = |x-2|$.

A. The graph of this function is never below the x-axis, because the absolute value of any real number is always positive or 0. (See Chapter 2 for more on absolute value.) The lowest point on the graph is (2,0), because letting $x = 2$ gives you $y = 0$. One way to describe what the absolute-value function does to the graph of a function is to say it flips the negative portion up above the x-axis. In Figure 25-11, I show you the graph of $y = x - 2$ and then what happens when the negative portion is flipped up above the x-axis to form the graph of $y = |x-2|$.

The graph has its distinguishing V-shaped feature with the lowest point where you have the absolute value of 0.

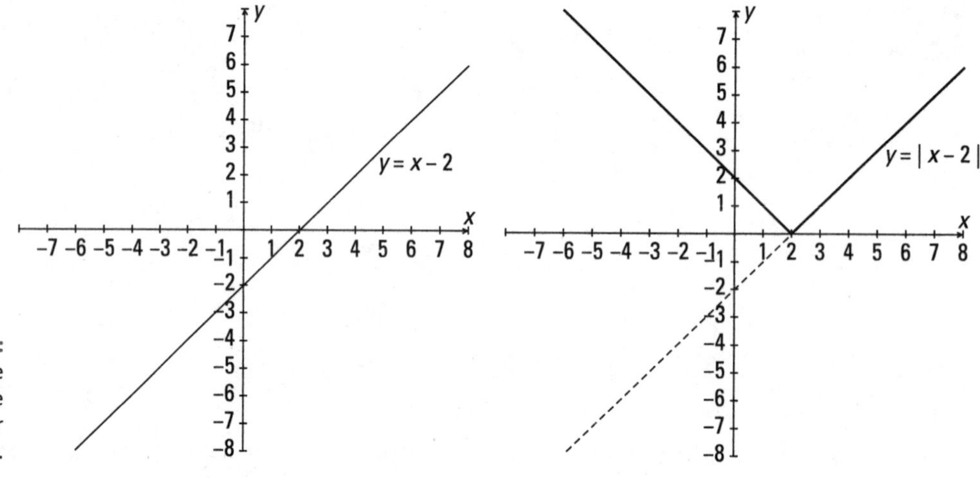

FIGURE 25-11: The negative values are flipped over the x-axis.

Q. Sketch the graph of $y = 2|x+3| - 4$.

A. The graph of $y = |x+3|$ is never below the x-axis, because the absolute value of any real number is always positive or 0. (See Chapter 2 for more on absolute value.) The lowest point on the graph is (−3,0), because letting $x = -3$ gives you $y = 0$. But subtracting 4 from the absolute value drops the basic graph below the x-axis to the point (−3,−4), and the multiplier of 2 makes the V-shape steeper than usual. In Figure 25-12, I first show you the graph of $y = |x+3|$, and then I show you the results of performing the two transformations. (For more on transformations, refer to the next section, "Graphing with Transformations.")

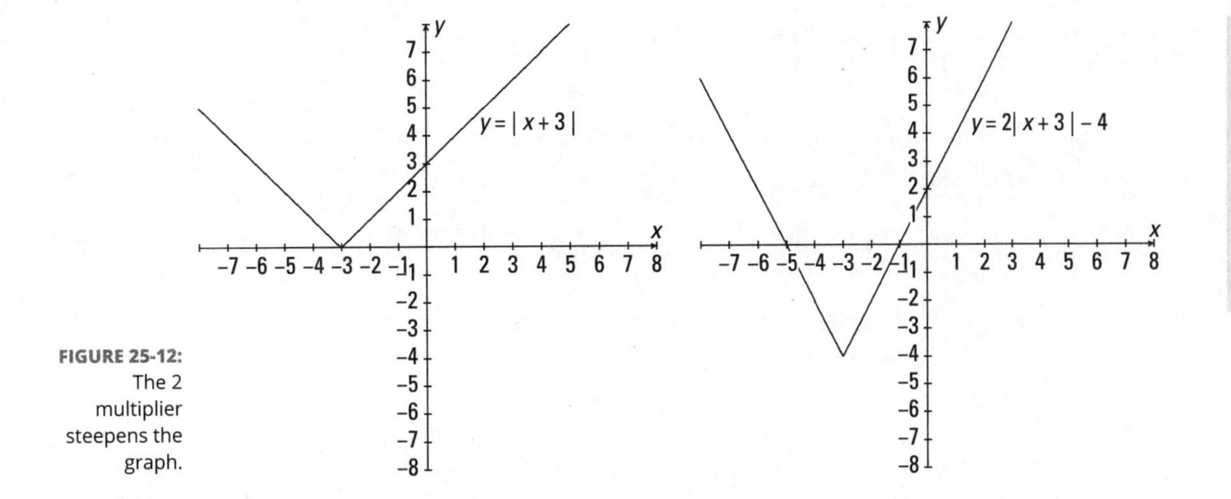

FIGURE 25-12: The 2 multiplier steepens the graph.

YOUR TURN

Sketch the graph of the equation.

13 $y = |x + 9|$

14 $y = -3|x - 4| + 2$

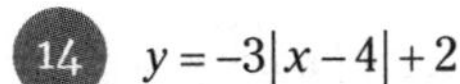

Graphing with Transformations

You can graph the curves and lines associated with different equations in many different ways. Intercepts are helpful when graphing lines or parabolas; the vertex of a parabola and center of a circle are critical to the graphing. But many curves can be quickly sketched when they're just a slight variation on the basic form of the particular graph. Two *transformations* used in graphing are *translations* (slides) and *reflections* (flips). Neither of these transformations changes the basic shape of the graph; they just change its position or orientation. Using these transformations can save you a lot of time when graphing figures.

When a curve is changed by a translation or slide, it slides to the left, to the right, or up or down. For instance, you can take the basic parabola with the equation $y = x^2$ and slide it around, using the following rules. The C represents some positive number.

» $y = x^2 + C$: Raises the parabola by C units.

» $y = x^2 - C$: Lowers the parabola by C units.

» $y = (x + C)^2$: Slides the parabola left by C units.

» $y = (x - C)^2$: Slides the parabola right by C units.

When a curve is changed by a reflection or flip, you have a symmetry that is vertical or horizontal.

» $y = -x^2$: Flips the parabola over a horizontal line.

» $y = (-x)^2$: Flips the parabola over a vertical line.

You can change the steepness of a graph by multiplying it by a number. This is closely related to the slopes of lines — the larger the number, the steeper the slope. If you multiply the basic function or operation by a positive number greater than 1, then the graph becomes steeper. If you multiply it by a positive number smaller than 1, then the graph becomes flatter. Multiplying a basic function by a negative number results in a flip or reflection over a horizontal line. Use these rules when using the parabola.

» $y = kx^2$: The parabola becomes steeper when k is positive and greater than 1.

» $y = kx^2$: The parabola becomes flatter when k is positive and smaller than 1.

Sliding and multiplying

You can combine the two operations of changing the steepness of a parabola and moving the vertex. These operations change the basic parabola to suit your purposes.

The equation used to model a situation may require a steep parabola because the changes happen rapidly, and it may require that the starting point be at 8 feet, not 0 feet. By moving the parabola around and changing the shape, you can get a better fit for the information you want to demonstrate.

The following equations and their graphs are shown in Figure 25-13.

» $y = 3x^2 - 2$: The 3 multiplying the x^2 makes the parabola steeper, and the –2 moves the vertex down to (0,–2).

» $y = \frac{1}{4}x^2 + 1$: The $\frac{1}{4}$ multiplying the x^2 makes the parabola flatter, and the +1 moves the vertex up to (0,1).

» $y = -5x^2 + 3$: The –5 multiplying the x^2 makes the parabola steeper and causes it to go downward, and the +3 moves the vertex to (0,3).

» $y = 2(x - 1)^2$: The 2 multiplier makes the parabola steeper, and subtracting 1 moves the vertex right to (1,0).

» $y = -\frac{1}{3}(x + 4)^2$: The $-\frac{1}{3}$ multiplier makes the parabola flatter and causes it to go downward, and adding 4 moves the vertex left to (–4,0).

» $y = -\frac{1}{20}x^2 + 5$: The $-\frac{1}{20}$ multiplying the x^2 makes the parabola flatter and causes it to go downward, and the +5 moves the vertex to (0,5).

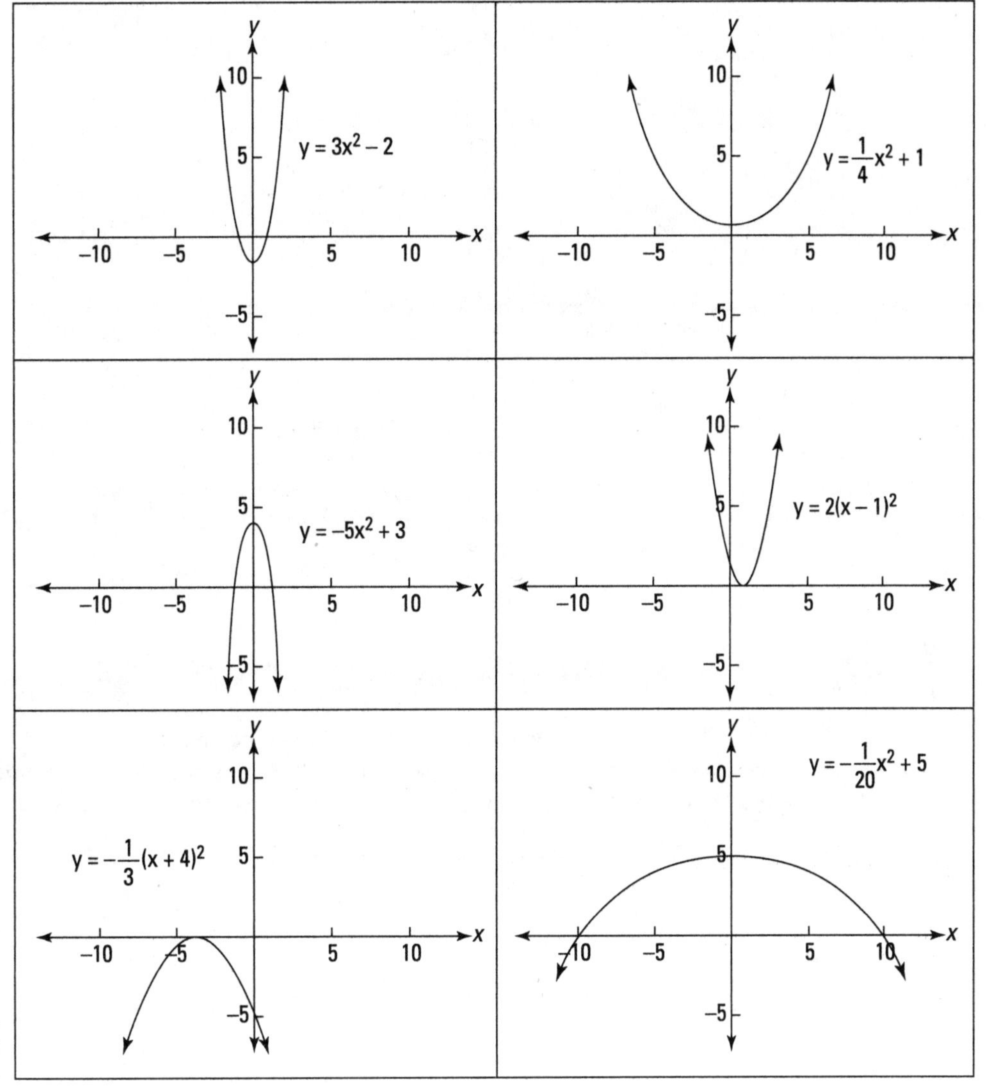

FIGURE 25-13: Parabolas galore.

EXAMPLE

Q. Use the basic graph of $y = x^2$ to graph $y = -3x^2 + 1$.

A. The multiplier -3 flips the parabola over the x-axis and makes it steeper. The $+1$ raises the vertex up one unit.

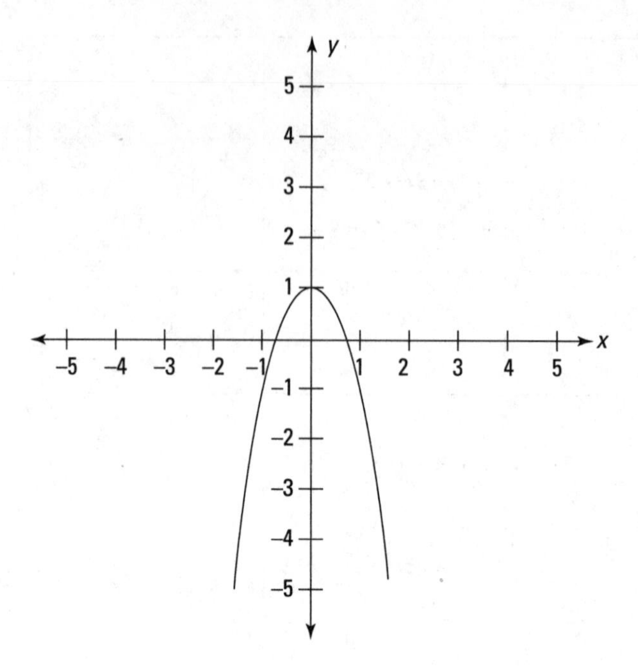

Q. Use the basic graph of $y = x^2$ to graph $y = (x+1)^2 - 3$.

A. The −3 lowers the parabola by three units. The +1 inside the parentheses slides the parabola one unit to the left. Compare this graphing technique with the method of graphing using the vertex, explained in the earlier section, "Graphing Parabolas and Circles."

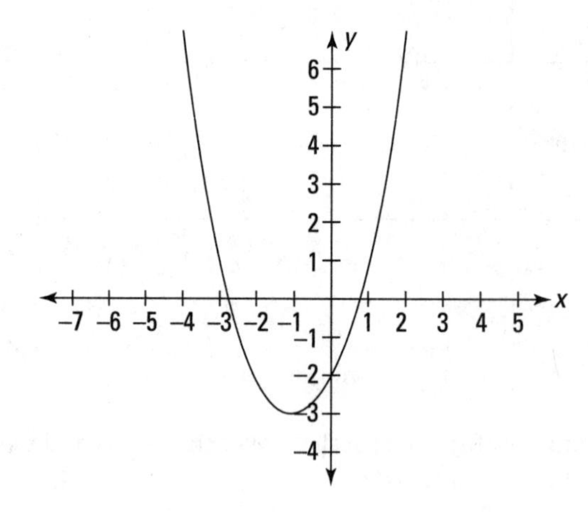

YOUR TURN

Sketch the graph of the equation.

15 $y = (x-4)^2$

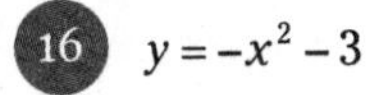

16 $y = -x^2 - 3$

Practice Questions Answers and Explanations

1. **(1,3).** Set $-4x + 7 = 5x - 2$. Solving for x, you get $-9x = -9$ or $x = 1$. Substituting back into either equation, $y = 3$.

2. **(−1,−4).** First, write the equations in slope-intercept form: $3x - y = 1$ becomes $y = 3x - 1$ and $x + 2y + 9 = 0$ becomes $y = -\frac{1}{2}x - \frac{9}{2}$. Setting $3x - 1 = -\frac{1}{2}x - \frac{9}{2}$, you solve for x and get $\frac{7}{2}x = -\frac{7}{2}$ or $x = -1$. Substitute back into either equation to get $y = -4$.

3. **No common solution.** First, rewrite the second equation in the slope-intercept form by moving the y term to the right and the 7 to the left: $8x - 7 = 2y$. Now divide each term by 2, and you have $y = 4x - \frac{7}{2}$. You should stop right there. The slope of the first line, $y = 4x - 3$, is the same as the second line. The two lines are parallel, so there's no intersection. But I'll continue to show you how their being parallel is confirmed. Setting $4x - 3 = 4x - \frac{7}{2}$, when you subtract 4x from each side, you get $-3 = -\frac{7}{2}$. That's just not true! It's a false statement, so there is no point of intersection — no common solution.

4. **$(k, \ -3k + 4)$.** Rewrite each equation in the slope-intercept form: $y = -3x + 4$ and $y = -3x + 4$. The two equations are just two different ways of writing the equation of the same line. You can stop there, but, instead, I'll show you two more facets of the problem. First, if you were to continue solving and set $-3x + 4 = -3x + 4$ and solve for x, you would end up with $0 = 0$. When is that true? It's always true for any ordered pairs that satisfy the equations. And what are those ordered pairs? You can list them, or you can give a general form for the ordered pairs. Let x be represented by some parameter (representation of a number), such as k. Then $y = -3k + 4$, and the ordered pairs are all of the form $(k, -3k + 4)$.

5. **Vertex: (3,−2).** The multiplier of $\frac{1}{2}$ in front of the parentheses makes the graph open wider (more flat).

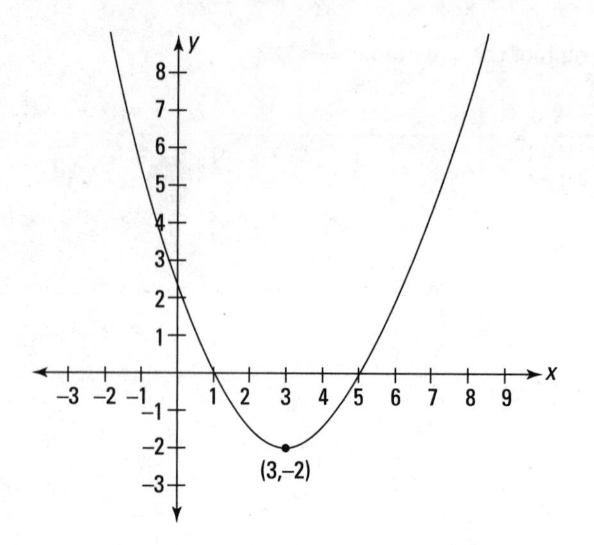

6 **Vertex: (−2,0).** The vertex is on the x-axis. The multiplier of 2 makes the parabola steeper.

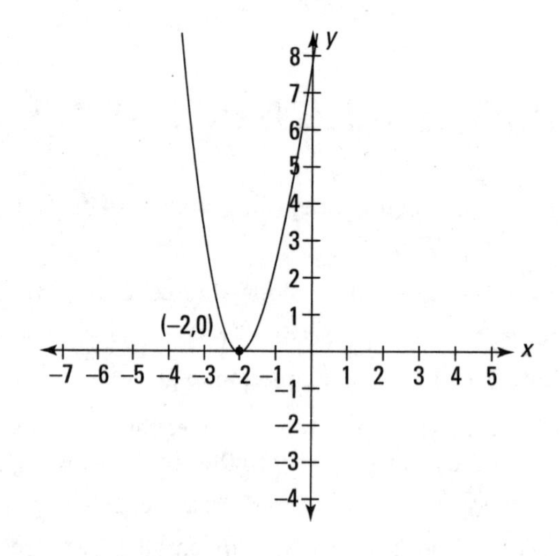

7. **Center: (−3,4); radius: 5.**

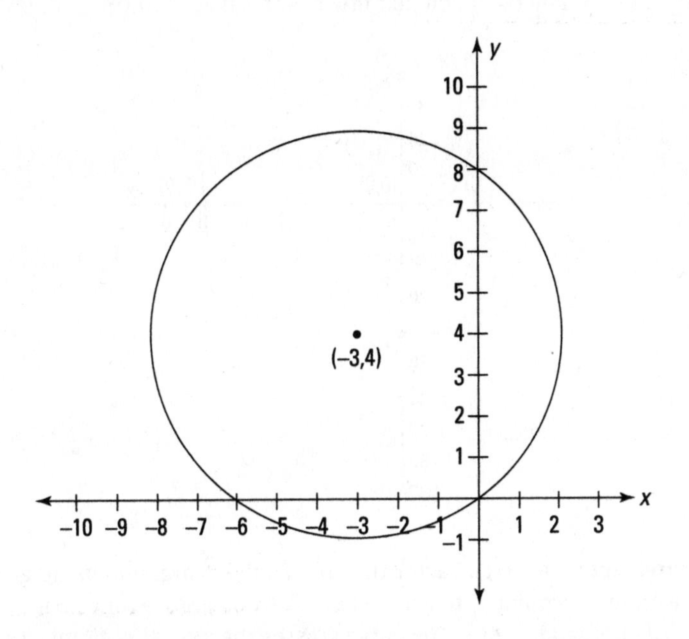

8. **Center: (5,2); radius: 3.**

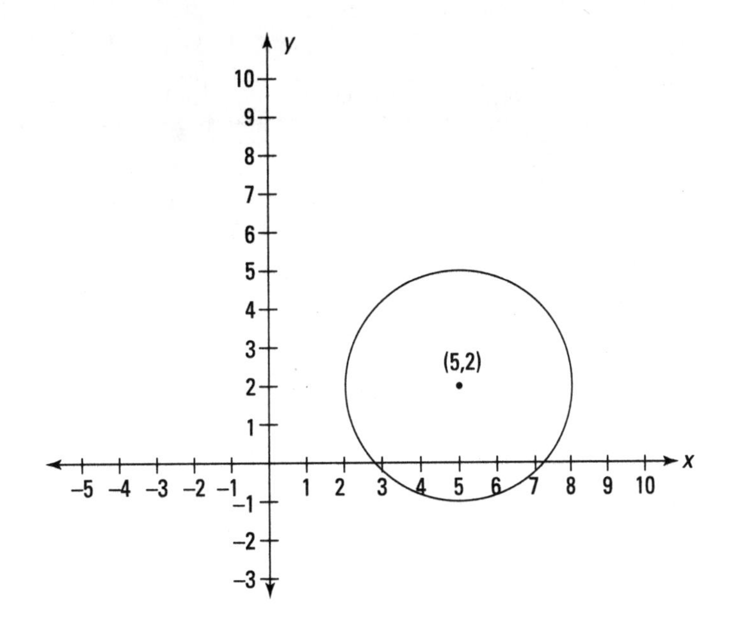

Extending the Graphing Horizon

(9) **The y-intercept is (0,0), and the x-intercepts are (0,0), (−1,0), and (5,0).** The curve crosses the axes at (−1,0) and (5,0) and just touches the axes at (0,0).

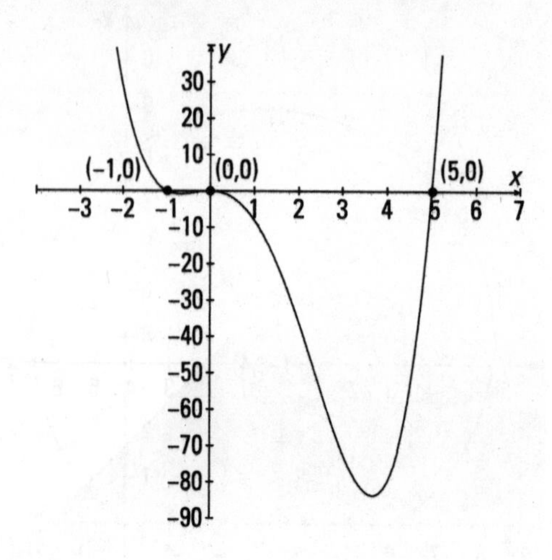

(10) **The y-intercept is (0,−63).** Factor the polynomial expression using grouping (see Chapter 13 for more on this technique) to get $y = (x^2 - 9)(x + 7) = (x + 3)(x - 3)(x + 7)$. The x-intercepts are (−3,0), (3,0), and (−7,0). The curve crosses the axes at each intercept.

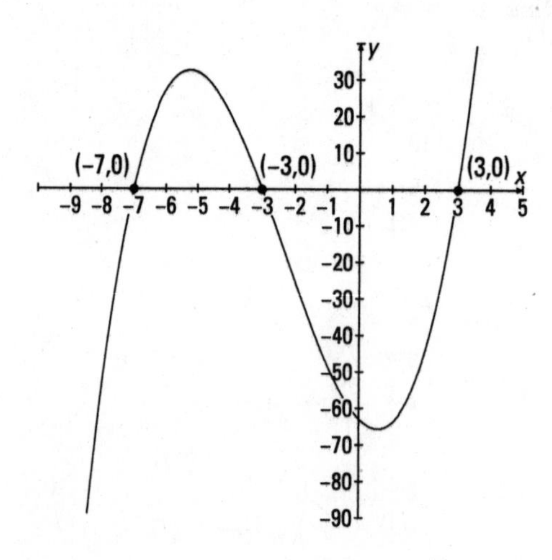

11 **Use a solid line.** The shading goes below the line to the left.

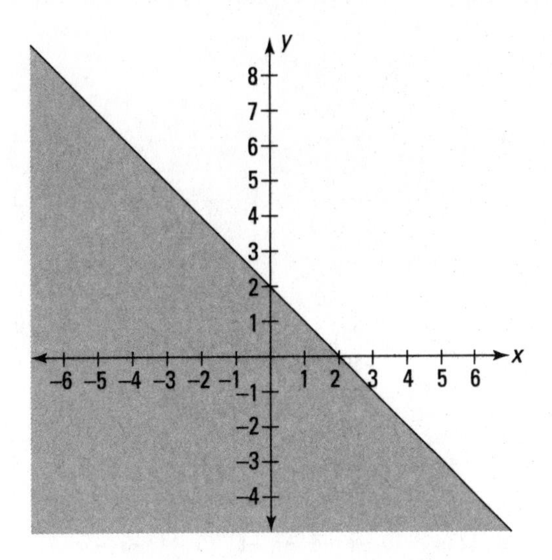

12 **Use a dashed line.** The shading goes below the line to the right.

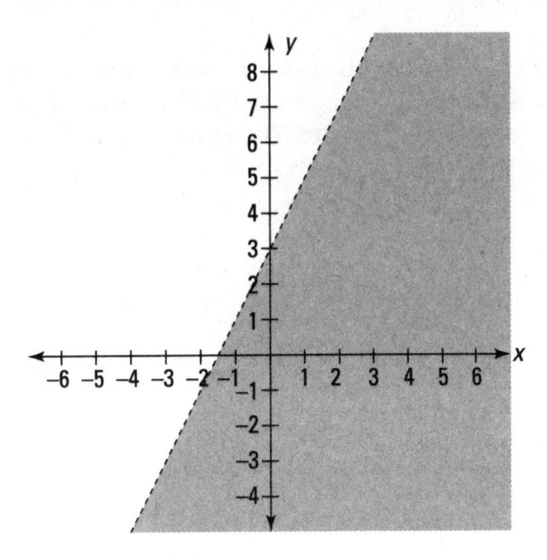

(13) $y=|x+9|$. The lowest point is at $(-9,0)$. The graph has the classic V-shape with the lowest point at that intercept.

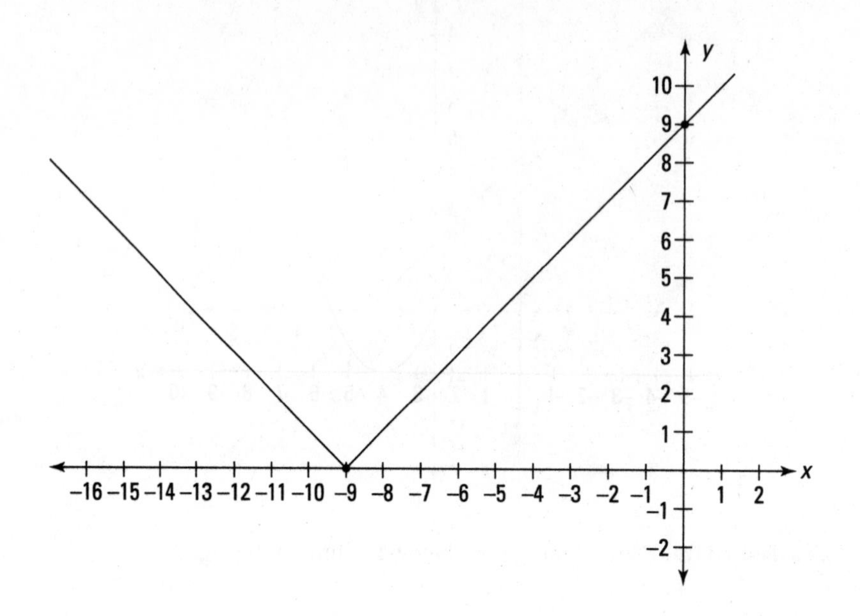

(14) $y=-3|x-4|+2$. The basis of this absolute-value graph is $y=|x-4|$, which has its lowest point at (4,0). The multiplier −3 makes the V-shape steeper, and the negative sign flips it over so the V is now pointing upward. The +2 moves the tip of the V up two units to the point (4,2).

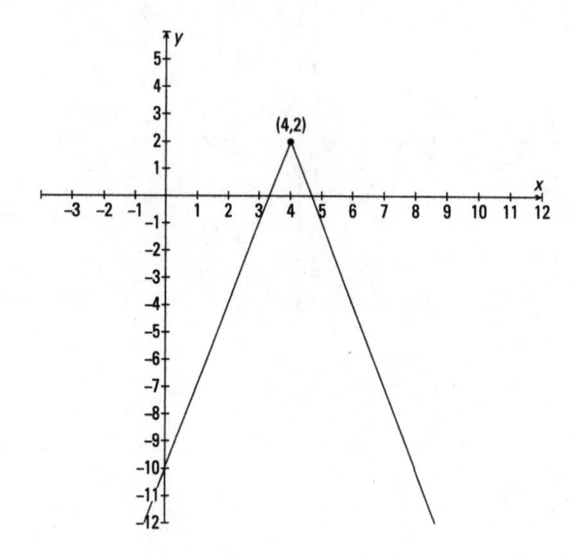

15 **The parabola has been translated 4 units to the right.**

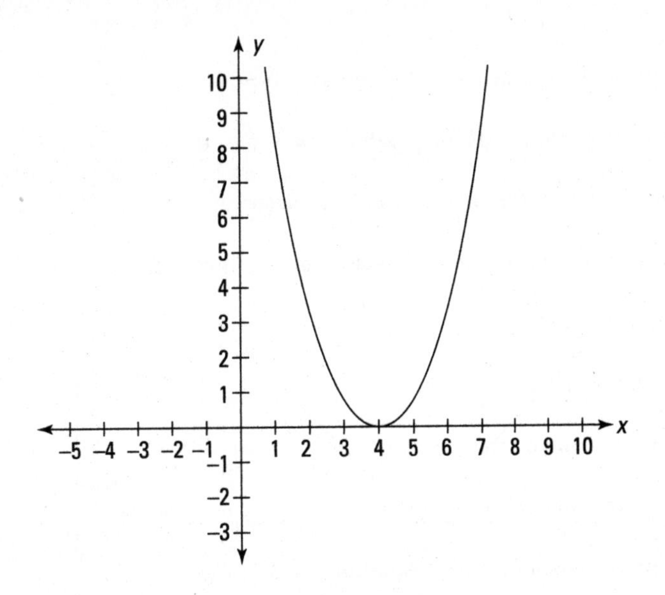

16 **The parabola has been lowered by 3 units and flipped over to face downward.**

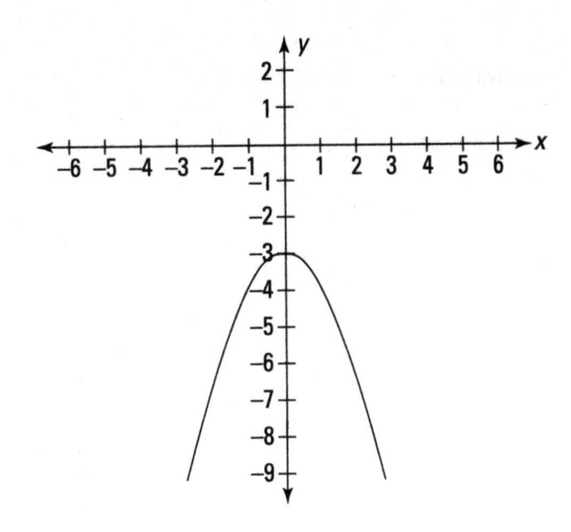

Extending the Graphing Horizon

Whaddya Know? Chapter 25 Quiz

1. Identify the intercepts of the line $4x-5y=20$.
2. Identify the vertex of the parabola $y=3(x+2)^2+7$.
3. Identify the center and radius of the circle $(x-3)^2+(y+4)^2=16$.
4. Use a graph to find the intersection of the lines $y=2x+1$ and $y=-x+4$.
5. Graph the parabola $y=x^2-9$.
6. Graph the circle $x^2+y^2=1$.
7. Graph the polynomial $y=x(x-2)(x+3)$.
8. Graph the inequality $2x+y>6$.
9. Graph the absolute-value equation $y=3|x-2|$.
10. Use a graph to find the intersection of the lines $3x+y=5$ and $x=4y+6$.
11. Graph the parabola $y=-2(x+3)^2+1$.
12. Graph the polynomial $y=x^4-9x^2$.
13. Graph the inequality $x-y\le -1$.
14. Graph the absolute-value equation $y=-|x+1|-3$.
15. Graph the circle $(x+3)^2+(y-1)^2=36$.

Answers to Chapter 25 Quiz

1. **(5,0), (0,−4).** Set y equal to 0 to solve for the x-intercept and x equal to 0 to solve for the y-intercept.

2. **(−2,7).**

3. **(3,−4), $r = 4$.**

4. **(1,3).**

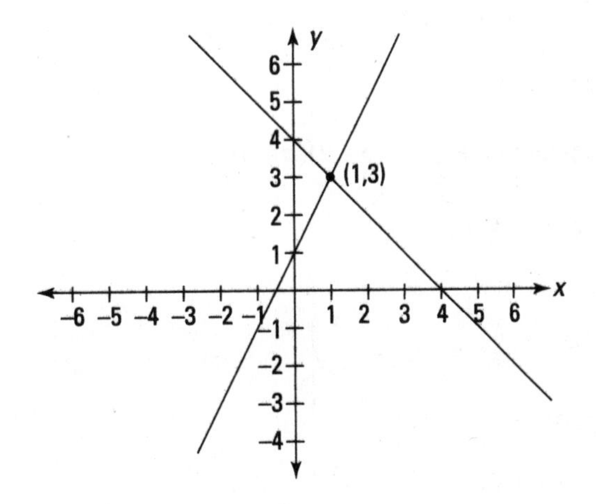

5. **The vertex is at (0,−9).**

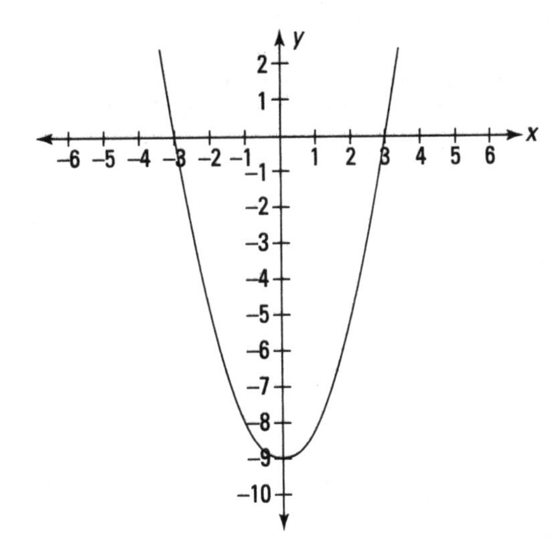

6. **The center is at (0,0) and the radius is 1.**

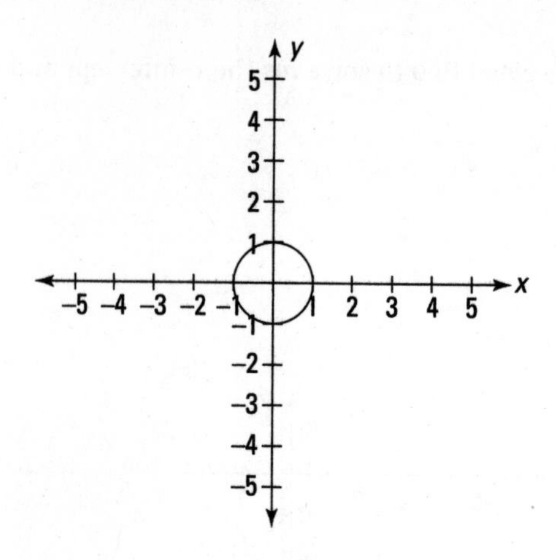

7. **The x-intercepts are at $x = -3, 0, 2$.**

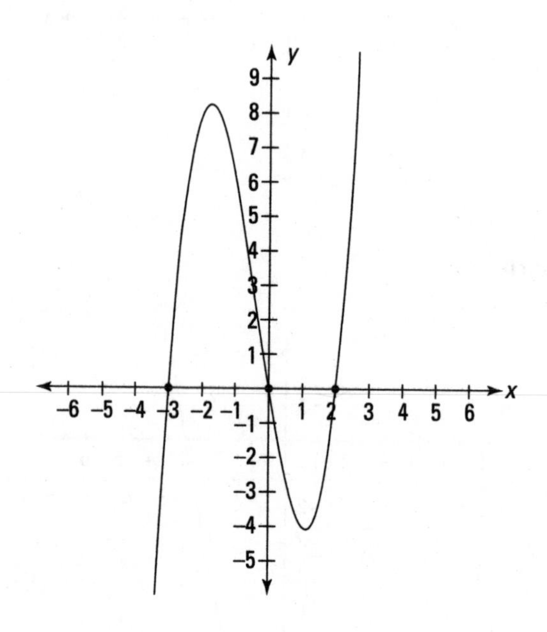

Extending the Graphing Horizon

8 **The two intercepts of the dashed line are at (3,0) and (0,6).** The shading is to the upper right of the dashed line.

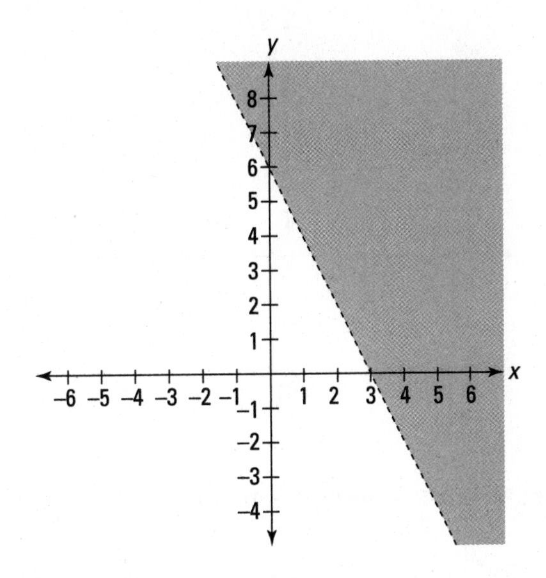

9 **The lowest point is at (2,0).**

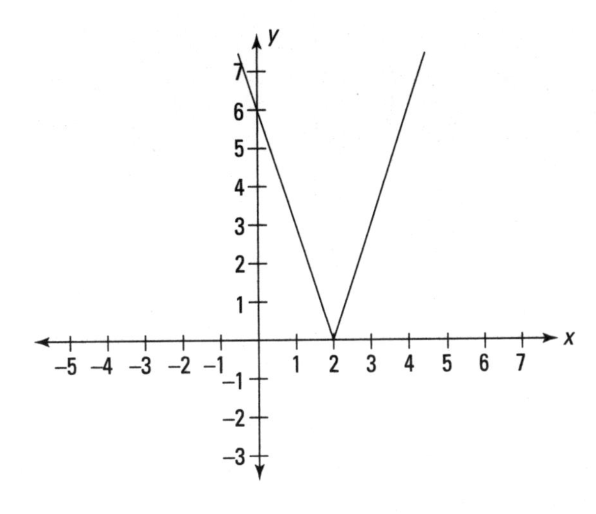

(10) (2,−1).

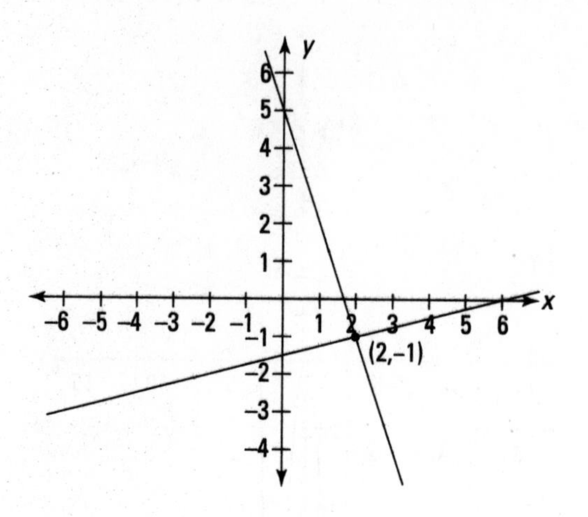

(11) **The vertex is at (−3,1), and it opens downward.**

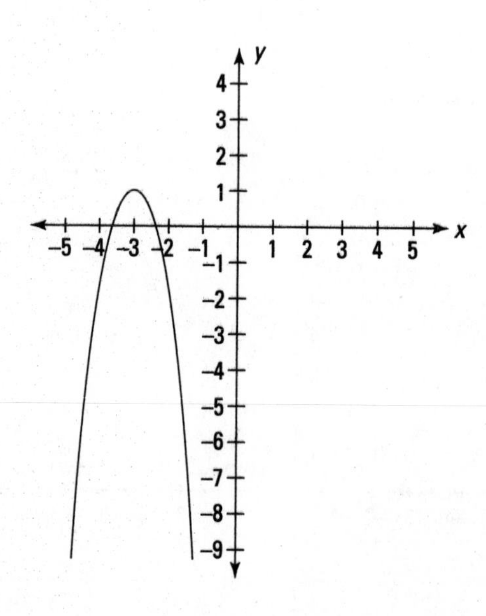

12 First, factor the binomial: $y = x^4 - 9x^2 = x^2(x^2 - 9) = x^2(x+3)(x-3)$. The double root at $x = 0$ means that the graph will just touch the intercept at that point.

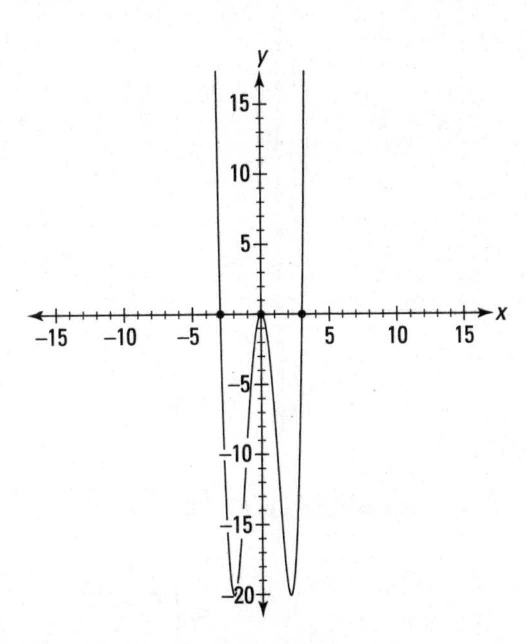

13 **The intercepts of the solid line are at (−1, 0) and (0,1).** The shading is above and to the left.

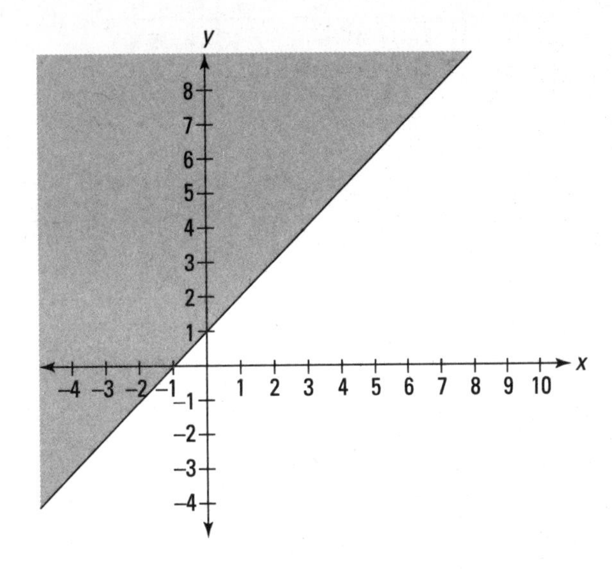

(14) **The highest point is at $(-1,-3)$, and the only intercept is the y-intercept at $(0,-4)$.**

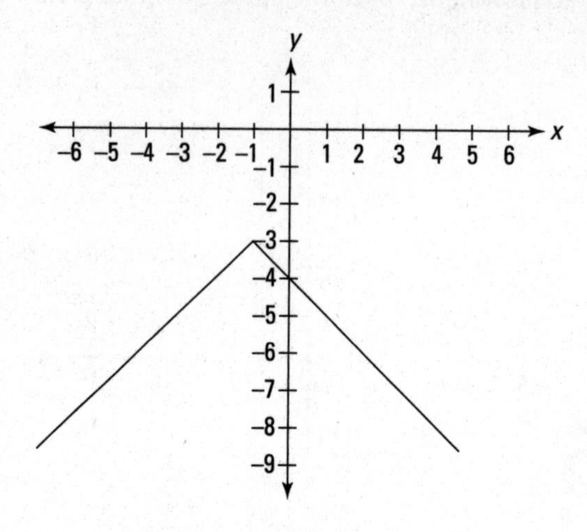

(15) **The center is at $(-3,1)$, and the radius is 6.**

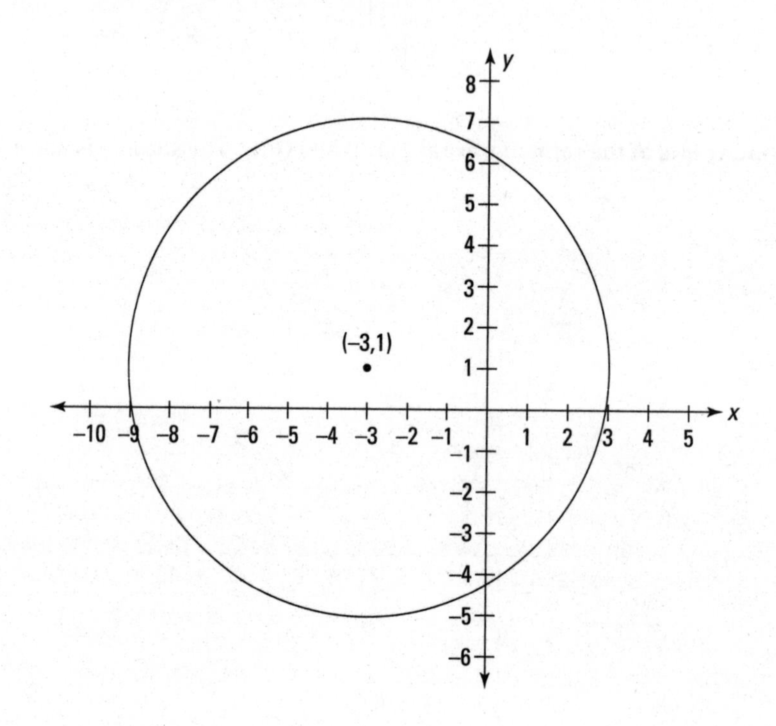

IN THIS CHAPTER

» Creating and analyzing solutions of systems of equations

» Choosing a method to solve systems

» Taking on non-linear systems

» Tackling systems of three equations

Chapter 26

Coordinating Systems of Equations and Graphing

A system of equations consists of two or more equations representing functions or other curves. A system of equations has a solution or solutions if they share variable values that make their statements true. Solutions such as these are important in business, when you want to know the break-even point and have income and cost functions. If you're in a control tower, you want to know where the plane and the radar should intersect. There are all sorts of wonderful applications.

In this chapter, you'll see several options for recognizing these common solutions and for finding them yourself — either with a graph or some algebra, or just some really good guessing.

Defining Solutions of Systems of Equations

The solution or solutions of a system of equations can be represented by the points they share on a graph or the variables that make their statements true. For example, you may have the equations of a cubic polynomial and a parabola. You want to determine if the points you have lie on both curves. In other words, you want to be sure that the points $(-2,-28)$, $(1,8)$, $(3,22)$ satisfy the system of equations.

$$\begin{cases} y = x^3 - 3x^2 + 6x + 4 \\ y = -x^2 + 11x - 2 \end{cases}$$

To do this, start by replacing each x with –2 and each y with –28, you have:

$$\begin{cases} -28 = (-2)^3 - 3(-2)^2 + 6(-2) + 4 \\ -28 = -(-2)^2 + 11(-2) - 2 \end{cases} \text{ or } \begin{cases} -28 = -8 - 12 - 12 + 4 \\ -28 = -4 - 22 - 2 \end{cases}$$, which are true.

Replacing each x with 1 and each y with 8, you have:

$$\begin{cases} 8 = (1)^3 - 3(1)^2 + 6(1) + 4 \\ 8 = -(1)^2 + 11(1) - 2 \end{cases} \text{ or } \begin{cases} 8 = 1 - 3 + 6 + 4 \\ 8 = -1 + 11 - 2 \end{cases}$$, which are true.

Replacing each x with 3 and each y with 22, you have:

$$\begin{cases} 22 = (3)^3 - 3(3)^2 + 6(3) + 4 \\ 22 = -(3)^2 + 11(3) - 2 \end{cases} \text{ or } \begin{cases} 22 = 27 - 27 + 18 + 4 \\ 22 = -9 + 33 - 2 \end{cases}$$, which are true.

These are the points are where they intersect — where they have common solutions.

EXAMPLE

Q. Determine if the pair of variables $x = 4$ and $y = -1$ satisfy the system of equations,

$$\begin{cases} x + y = 3 \\ 2x - y = 9 \end{cases}$$

A. Replacing x with 4 and y with –1 in each of the equations, you see that they create true statements in both cases.

$$\begin{cases} 4 + (-1) = 3 \\ 2(4) - (-1) = 9 \end{cases} \text{ or } \begin{cases} 4 - 1 = 3 \\ 8 + 1 = 9 \end{cases}$$

This pair of variables can also be written as the point $(4, -1)$.

Q. Determine if the variables in the point $(-2, 5)$ satisfy the system of equations,

$$\begin{cases} y = 2x^2 - 3 \\ y - 1 = x^2 \end{cases}$$

A. Replacing each x with –2 and each y with 5, you have:

$$\begin{cases} 5 = 2(-2)^2 - 3 \\ 5 - 1 = (-2)^2 \end{cases} \text{ or } \begin{cases} 5 = 8 - 3 \\ 5 - 1 = 4 \end{cases}$$

Both equations produce true statements using those values. And, as an added bonus, the point (2,5) also satisfies both equations.

YOUR TURN

1 Determine which of the following equations have a solution when $x = -3$ and $y = 5$:

A. $x + 2y = 7$

B. $2x + 3y = 1$

C. $3y = 27 + 4x$

D. $y - x = 8$

2 Find the value of y in the solution $(2, y)$ for the system of equations,

$$\begin{cases} x + y = 3 \\ 2x - y = 3 \end{cases}$$

Solving Systems of Linear Equations

A system of linear equations consists of equations whose variables all have exponents of 1. You can have systems of two, three, or more linear equations. There are several ways of determining the solution of a system of equations: by-guess-or-by-golly, algebraically using elimination, algebraically using substitution, and graphing. You'll find descriptions of all but the guessing part, although making a guess or an estimate is always a good plan to help with the checking.

Using elimination

The *elimination* system involves just what the name suggests: You eliminate one of the variables and solve for the other. This allows you to solve the system of equations for the values of all the variables. The elimination method essentially involves adding the equations together and having one of the variables disappear because you happen to have an exact opposite of the other variable. If the original set of equations doesn't have opposites, then you create them!

EXAMPLE

Q. Use elimination to solve the system of equations.

$$\begin{cases} 2x + y = 5 \\ x - y = 4 \end{cases}$$

A. You note that the two y-variables are opposites of one another. Add the two equations together.

$$\begin{array}{rcrcr} 2x & + & y & = & 5 \\ x & - & y & = & 4 \\ \hline 3x & & & = & 9 \end{array}$$

Now divide each side of the resulting equation by 3 and you get that $x = 3$. Substitute that into either equation to solve for y.

Using $x - y = 4$, $3 - y = 4$, or $y = -1$.

And, of course, you check by putting $x = 3$ and $y = -1$ in the other equation.

$2x + y = 5$ becomes $2(3) + (-1) = 5$, and, yes, $6 - 1 = 5$.

Q. Use elimination to solve the system of equations.

$$\begin{cases} x + 5y = 13 \\ 2x + 3y = 5 \end{cases}$$

A. This time, there isn't a pair of opposite terms. You could multiply the top equation through by 3 and the bottom equation through by −5 to get 15*y* and $-15y$, but there's an easier choice. Multiply the top equation through by −2, and the two *x* terms will be opposites.

$$\begin{array}{rcrcr} -2x & - & 10y & = & -26 \\ 2x & + & 3y & = & 5 \\ \hline & - & 7y & = & -21 \end{array}$$

Dividing both sides by −7 gives you that $y = 3$. Substitute that into the original first equation, and you have that $x + 5(3) = 13$ or $x = -2$. Does the solution $(-2, 3)$ work in the second equation? Substituting, you have $2(-2) + 3(3) = 5$, and, yes, $-4 + 9 = 5$. It checks.

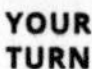

YOUR TURN

3 Solve the system using elimination.

$$\begin{cases} x+y=3 \\ x-y=5 \end{cases}$$

4 Solve the system using elimination.

$$\begin{cases} 3x-2y=1 \\ x-y=1 \end{cases}$$

Using substitution

The *substitution* system is also discussed in Chapter 25. And, again, the name describes exactly what the method involves. You substitute a portion of one equation into the other. You use this method when it's handy to solve for one variable in an equation and substitute what you've found into the other. This method finishes off pretty much like the elimination method in that you find the value of the one variable and then go back to the original equations to find the value of the other.

EXAMPLE

Q. Use substitution to solve the system of equations.

$$\begin{cases} y=4x-1 \\ y=2x+3 \end{cases}$$

A. The two equations are both solved for y. You can substitute either equation into the other. If you substitute the second equation into the first, you replace the y in the first equation with what y is equal to in the second equation. Put the $2x+3$ in for the y in the first equation:

$2x+3=4x-1$

Now, solving for x, you have $4=2x$ or $x=2$.

Solve for y by putting the 2 in for x in the first equation. You have $y=4(2)-1=7$. So your solution is $(2,7)$.

Check this solution in the second equation: $7=2(2)+3$. Yes, $7=4+3$.

Q. Use substitution to solve the system of equations.

$$\begin{cases} 4x-3y=14 \\ 3x-y=8 \end{cases}$$

A. This time, neither equation is solved for y, but this can be easily done in the second equation. If you solve for y, you have $y=3x-8$.

Substitute this into the first equation, replacing the y with $3x-8$.

$4x-3(3x-8)=14$

Multiplying and simplifying, you have $4x-9x+24=14$, and then $-5x=-10$. This gives you $x=2$.

Substitute that into $y=3x-8$, and you have that $y=3(2)-8=-2$. So your solution is $(2,-2)$. Check this answer by substituting into the first equation.

$4x-3y=14$ becomes $4(2)-3(-2)=14$, and, yes, $8+6=14$.

YOUR TURN

5 Solve the system of equations using substitution.

$$\begin{cases} y = -2x + 9 \\ y = 3x + 4 \end{cases}$$

Solve the system of equations using substitution.

$$\begin{cases} 3x - 4y = 3 \\ x - 3y = 6 \end{cases}$$

Introducing intersections of lines

The final way I'll be showing you how to solve a system of equations is to use graphing. This is also illustrated in Chapter 25. Sometimes graphing is the handiest way to solve a system of linear equations, but there's a big drawback: the solutions have to consist of integers. It's really difficult to estimate fractional solutions on a graph. Is that $\frac{1}{2}$ or $\frac{1}{3}$? You don't want to be guessing.

When you graph two lines and find their point of intersection, you've found the solution to the system of equations made up of those lines.

EXAMPLE

Q. Find the solution to the system of equations by graphing the lines.

$$\begin{cases} y = -2x + 3 \\ 3x + y = 6 \end{cases}$$

A. The first line has a y-intercept of 3 and a slope of -2. Plot the point $(0,3)$ and move 1 unit to the right and 2 down to find a second point and graph the line. The second line has intercepts of $(2,0)$ and $(0,6)$. Plot the two points and draw the line through them. Refer back to Chapter 24 for more on using intercepts and slope-intercept forms. You see the point of intersection at $(3,-3)$. This is the solution of the system of equation.

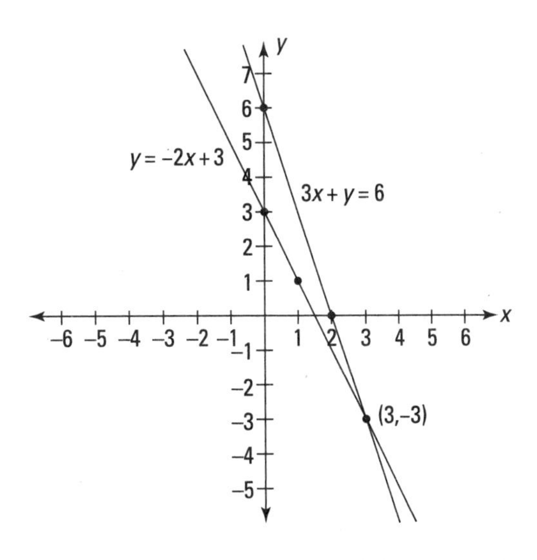

Coordinating Systems of Equations and Graphing

YOUR TURN

7. Solve the system of equations by graphing.

$$\begin{cases} x+2y=2 \\ y=3x+8 \end{cases}$$

8. Solve the system of equations by graphing.

$$\begin{cases} 3x+4y=7 \\ x=1 \end{cases}$$

Solving Systems Involving Non-Linear Equations

Non-linear equations have graphs that are not lines. The exponents on the variables in non-linear equations are not all 1's. When you have an equation with exponents of 0, 1, and 2, you may have the graph of a parabola or a circle. An exponent of 3 indicates a cubic. And it gets more and more interesting with higher powers, fractional powers, and negative powers! You won't see all the possibilities here, but I'll show you some techniques that will help you with most of what there is to offer.

EXAMPLE

Q. Solve the system of equations by graphing and then checking the answer with substitution.

$$\begin{cases} y=x^2+3x+2 \\ y=x+5 \end{cases}$$

A. The equations represent a parabola and a line. The equation of the parabola factors into $y=(x+1)(x+2)$, so the x-intercepts are $(-1,0)$ and $(-2,0)$; it opens upward. The line has a y-intercept at $(0,5)$, and the slope is 1. Graph the two functions.

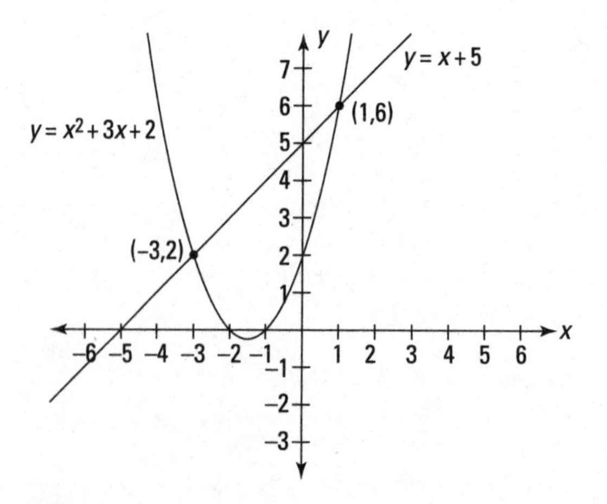

There are two points of intersection: (–3,2) and (1,6). Instead of checking by inserting the coordinates of the points into the equations, I will use the substitution method for solving systems, substituting the second equation's y equivalence into the first equation. When $x+5=x^2+3x+2$, you subtract the x and the 5 from both sides and get $0=x^2+2x-3$. You factor this and apply the MPZ (see Chapter 16 for more on quadratic equations). When $0=(x+3)(x-1)$, $x=-3$ or $x=1$. Substitute these values into the line of the equation to solve for y, and you get $y=-3+5=2$ and $y=1+5=6$. Those are the points of intersection on the graph.

Q. Solve the system of equations by graphing and then checking the answer with substitution.

$$\begin{cases} x^2+y^2=25 \\ y=x+1 \end{cases}$$

A. These are the equations representing a circle and a line. If a circle and line intersect, then they intersect in one point, a tangent, or in two points. The circle has its center at (0,0), and its radius is 5. (You find more on graphing circles in Chapter 25.) The line has a y-intercept of 1 and a slope of 1.

The line and circle intersect in two points: (3,4) and (–4,–3).

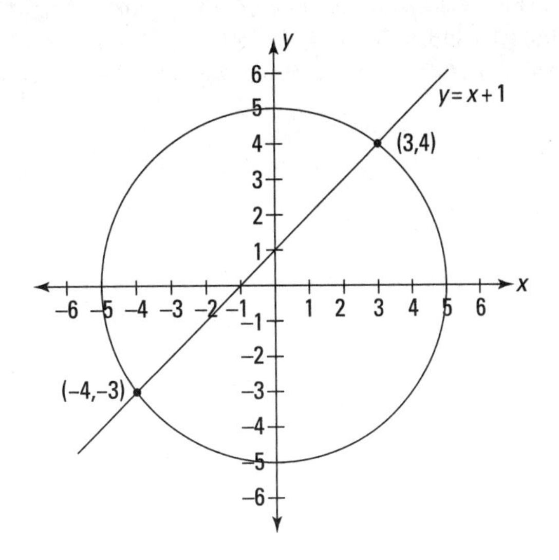

These points are the solutions of the system of equations. To check this, substitute the y value of the second equation into the first equation and solve for x. Here, $x^2+(x+1)^2=25$ becomes $x^2+x^2+2x+1=25$, which simplifies to $x^2+x-12=0$. Factoring, you get $(x+4)(x-3)=0$, and the MPZ provides you with $x=-4$ or $x=3$. Substituting into the equation for the line, you get those same two points that were obtained with the graphing.

Q. Solve the system of equations using elimination.

$$\begin{cases} (x-2)^2+y^2=25 \\ \quad x^2+y^2=16 \end{cases}$$

A. Sometimes, graphing a system of equations for the solution just doesn't work. This is especially true when the solutions involve fractions. These two circles intersect in two places, but the coordinates aren't integers.

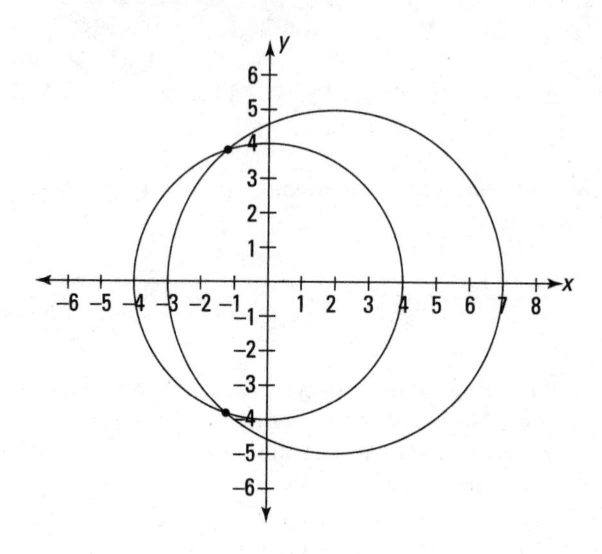

To solve the system using elimination, first simplify the top equation by squaring the binomial and combining like terms: $x^2 - 4x + 4 + y^2 = 25$ further simplifies to $x^2 - 4x + y^2 = 21$. Now subtract the second equation from this new equation.

$$\begin{array}{rrrrr} & x^2 & -\ 4x & +\ y^2 & = & 21 \\ - & x^2 & & +\ y^2 & = & 16 \\ \hline & & -\ 4x & & = & 5 \end{array}$$

Solving for x, you have $x = -\frac{5}{4}$. Put that value in the second equation and solve for y.

$\left(-\frac{5}{4}\right)^2 + y^2 = 16$ becomes $y^2 = 16 - \frac{25}{16} = \frac{256 - 25}{16} = \frac{231}{16}$ and $y = \pm\frac{\sqrt{231}}{4} \approx \pm 3.80$

The two points of intersection are $\left(-\frac{5}{4}, 3.8\right)$ and $\left(-\frac{5}{4}, -3.8\right)$.

YOUR TURN

9 Solve the system of equations.

$$\begin{cases} y = x^2 - 4 \\ y = 2x - 4 \end{cases}$$

10 Solve the system of equations.

$$\begin{cases} y = 2x^2 - 3x - 1 \\ y = x + 5 \end{cases}$$

11 Solve the system of equations.

$$\begin{cases} x^2 + y^2 = 1 \\ y = x + 1 \end{cases}$$

Taking on Systems of Three Linear Equations

Coordinating Systems of Equations and Graphing

When you have a system of three linear equations that involve three different variables, you have graphs of planes. No, these aren't the flying type. These planes are the flat surfaces like floors and streets. Your piece of paper is a plane on which you can draw things. When two planes cross one another, their intersection is a line. When three planes cross, you can get several lines or, when they all agree, you get a single point.

One of the most efficient ways of solving systems of three linear equations is to use the elimination method — three times. You pick a variable to eliminate in two pairs of equations, and then you eliminate a second variable and substitute in backwards.

EXAMPLE

Q. Solve the system of equations for the values of x, y, and z.

$$\begin{cases} 2x + y + z = 5 \\ x + 2y + z = 1 \\ 3x - y + z = 11 \end{cases}$$

A. First, choose a variable to eliminate in the first round. Because all the z variables have a coefficient of 1, this is a good candidate.

Pairing up the first and second equations, subtract the second equation from the first equation.

$$\begin{array}{rrrrrrrr} & 2x & + & y & + & z & = & 5 \\ - & x & + & 2y & + & z & = & 1 \\ \hline & x & - & y & & & = & 4 \end{array}$$

Now, pairing up the second and third equations, subtract the third from the second.

$$\begin{array}{rrrrrrrr} & x & + & 2y & + & z & = & 1 \\ - & 3x & - & y & + & z & = & 11 \\ \hline & -2x & + & 3y & & & = & -10 \end{array}$$

Now you have two equations in just x and y. To use elimination, you need two of the coefficients to be the same or opposites. Multiplying the equation $x - y = 4$ through by 2 will accomplish this. Now add the two equations.

$$\begin{array}{rcrcr} 2x & - & 2y & = & 8 \\ -2x & + & 3y & = & -10 \\ \hline & & y & = & -2 \end{array}$$

Substitute this value of y into the equation $x - y = 4$, and you have $x - (-2) = 4$ or $x = 2$.

Take the values $y = -2$ and $x = 2$ and substitute them into any of the original equations to solve for z. Using the first equation, you have $2(2) + (-2) + z = 5$, which becomes $2 + z = 5$, giving you $z = 3$. The solution can be written as an ordered triple: $(2, -2, 3)$. And, yes, you need to check this by substituting these values into the other two equations.

YOUR TURN

12 Solve the system of equations for the values of x, y, and z.

$$\begin{cases} 4x + 2y + 3z = 1 \\ \quad -2y + 5z = -1 \\ \qquad\quad 6z = -6 \end{cases}$$

13 Solve the system of equations for the values of x, y, and z.

$$\begin{cases} 2x + y - z = 1 \\ x - y + 2z = -1 \\ 3x + y - 2z = 5 \end{cases}$$

Practice Problems Answers and Explanations

1. A. **Yes.** $-3 + 2(5) = 7$

   B. **No.** $2(-3) + 3(5) \neq 1$

   C. **Yes.** $3(5) = 27 + 4(-3)$

   D. **Yes.** $5 - (-3) = 8$

2. $\mathbf{y = 1}$. Replacing the x with 2 in the first equation gives you $2 + y = 3$, so $y = 1$. Replacing the x with 2 in the second equation results in $2(2) - y = 3$ or $4 - y = 3$. Again, $y = 1$.

3. $\mathbf{(4, -1)}$. Adding the two equations together to eliminate the y-term you have $2x = 8$. Dividing by 2, you have that $x = 4$. Substitute 4 for x in the first equation to get $4 + y = 3$, giving you $y = -1$.

4. $\mathbf{(-1, -2)}$. Multiply the second equation through by -2, and you then have the equation $-2x + 2y = -2$. Add that to the first equation to eliminate the y-term and you get $x = -1$. Substitute that value into the first equation to get $3(-1) - 2y = 1$ or $-3 - 2y = 1$. Add 3 to each side to get $-2y = 4$. Dividing each side by -2 gives you $y = -2$.

5. **(1,7).** Substitute $-2x+9$ for y in the second equation and you have $-2x+9=3x+4$. Adding $2x$ to each side and subtracting 4 from each side, $5=5x$ or $x=1$. Substitute 1 for x in the second equation and you have $y=3(1)+4=7$.

6. **(−3,−3).** Solve for x in the second equation to get $x=6+3y$. Substitute this equivalence for x in the first equation and you have $3(6+3y)-4y=3$, which simplifies to $18+5y=3$. Subtract 18 from each side and divide by 5 to get $y=-3$. Replace the y with −3 in the second equation to get $x-3(-3)=6$ or $x+9=6$. Solving for x you get $x=-3$.

7. **(−2, 2)** Graphing the lines, you find their intersection at the point $(-2,2)$, as shown in the graph.

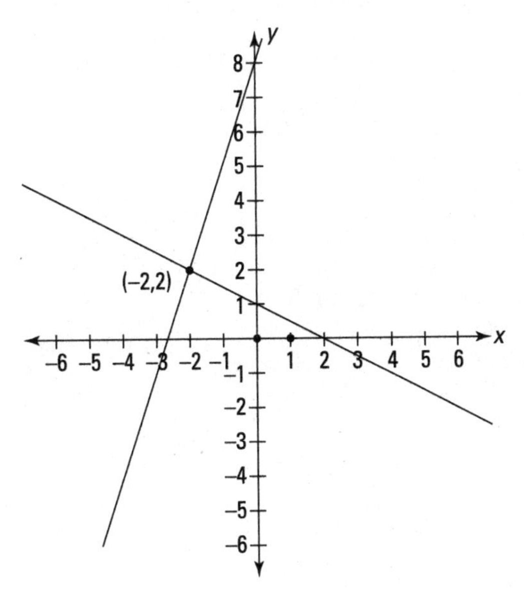

8. **(1, 1)** Graphing the lines, you find their intersection at the point $(1,1)$, as shown in the graph. The graph of $x=1$ is a vertical line.

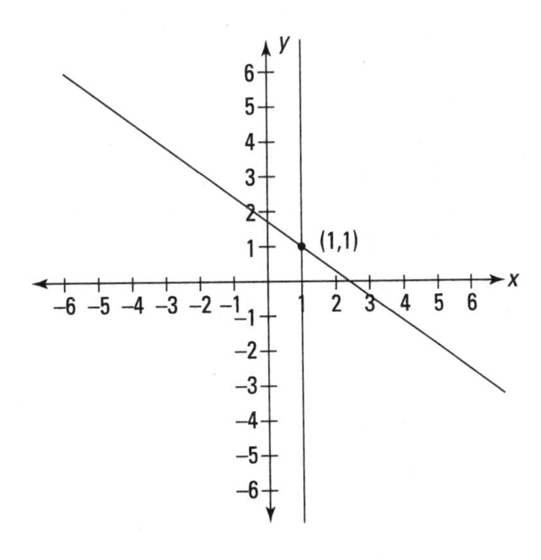

Coordinating Systems of Equations and Graphing

9. **(2,0) and (0,–4).** Graphing the parabola and line, you find the intersections at (2,0) and (0,–4). Using substitution, you first have $2x-4=x^2-4$, which simplifies to $0=x^2-2x=x(x-2)$. The two x-values are 0 and 2. Substitute them into one of the equations to find the y-values.

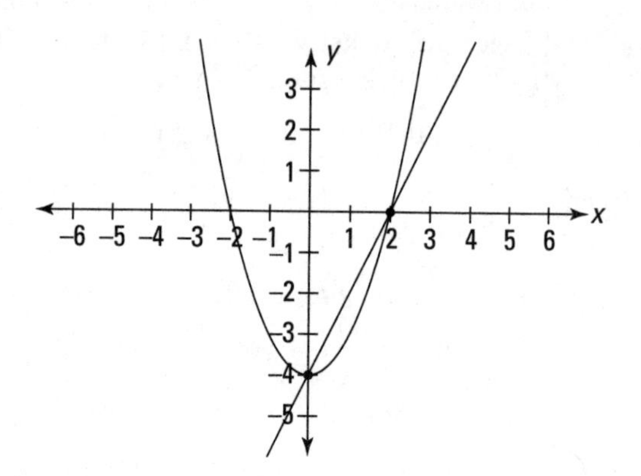

10. **(3,8) and (–1,4).** Graphing the parabola and line, you find the intersections at (3,8) and (–1,4). Using substitution, you first have $x+5=2x^2-3x-1$, which simplifies to $0=2x^2-4x-6=2(x-3)(x+1)$. The two x-values are 3 and –1. Substitute them into the equation of the line to find the y-values.

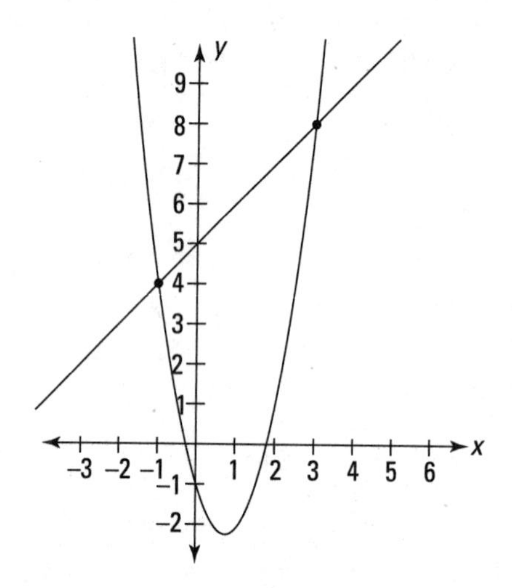

11. **(0,1) and (–1,0).** Graphing the circle and line, you find the intersections at (0,1) and (–1,0). Using substitution, you first have $x^2+(x+1)^2=1$, which simplifies to $2x^2+2x=2x(x+1)=0$. The two x-values are 0 and –1. Substitute them into the equation of the line to find the y-values.

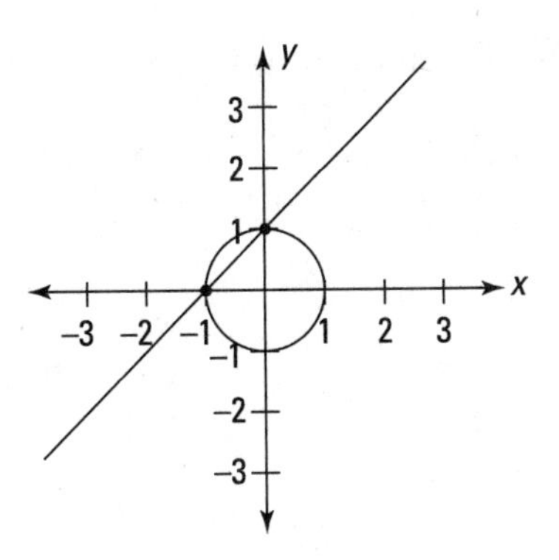

12. $\boldsymbol{x = 2,\ y = -2,\ z = -1}$. Work backwards starting with the third equation. You find that $z = -1$. Substituting into the second equation, $-2y - 5 = -1$, giving you $-2y = 4$ or $y = -2$. Substituting both values into the first equation, $4x - 4 - 3 = 1$, so $4x = 8$ and $x = 2$.

13. $\boldsymbol{x = 1,\ y = -4,\ z = -3}$. Eliminate the y term in each equation. First, add the top two equations together to get $3x + z = 0$, and then add the second and third equations together to get $4x + 0 = 4$. This gives you $x = 1$. Substitute that into $3x + z = 0$, and $z = -3$. Finally, put $x = 1$ and $z = -3$ into the first equation and you have $2 + y - (-3) = 1$ or $y = -4$. Be sure to check these in the second and third equations.

If you're ready to test your skills a bit more, take the following chapter quiz that incorporates all the chapter topics.

Whaddya Know? Chapter 26 Quiz

Quiz time! Complete each problem to test your knowledge on the various topics covered in this chapter. You can then find the solutions and explanations in the next section.

1. Use substitution to solve the system of equations.

$$\begin{cases} 5x - y = 2 \\ y = 2x + 1 \end{cases}$$

2. Use elimination to solve the system of equations.

$$\begin{cases} 4x - 5y = 5 \\ 4x + 3y = -3 \end{cases}$$

3. Use a graph to solve the system of equations.

$$\begin{cases} 3x - 2y = 12 \\ -2x + y = -7 \end{cases}$$

4. Use substitution to solve the system of equations.

$$\begin{cases} y = 9 - x^2 \\ y = x + 7 \end{cases}$$

5. Solve for x, y, and z.

$$\begin{cases} x + y + z = 3 \\ x \qquad + z = 1 \\ \qquad y - z = 5 \end{cases}$$

6. Use substitution to solve the system of equations.

$$\begin{cases} 2x + y = 5 \\ x - 3y = 6 \end{cases}$$

7. Use a graph to solve the system of equations.

$$\begin{cases} y = \frac{1}{4}x - 3 \\ y = x - 6 \end{cases}$$

8. Use elimination to solve the system of equations.

$$\begin{cases} 3x - 5y = 7 \\ x + 7y = -15 \end{cases}$$

9. Solve the system of equations.

$$\begin{cases} y = x^2 - 4x + 3 \\ x + y = 1 \end{cases}$$

10. Solve the system of equations.

$$\begin{cases} 2x + y - z = 9 \\ x - 2y + z = 4 \\ x + 3y - z = 3 \end{cases}$$

Answers to Chapter 26 Quiz

1. **(1,3).** Replace the y term in the first equation with the y equivalence in the second. You get the equation $5x - (2x + 1) = 2$, which simplifies to $3x - 1 = 2$ and then $3x = 3$. You now have $x = 1$. Substitute into the second equation to get $y = 2(1) + 1 = 3$.

2. **(0,–1).** To make the first terms opposites, multiply the first equation through by –1 and then add the two equations together.

Index

Symbol

A

B

C

D

E

F

H

I

L

M

P

Q

R

S

T

About the Author

Mary Jane Sterling is the author of many *For Dummies* products, including *Algebra I, Algebra II, Math Word Problems, Business Math, Linear Algebra, Finite Math,* and *Pre-Calculus.* She is currently in a not-really-retired state, after forty-plus years of teaching mathematics. She loves sharing math-related topics through workshops and Zoom sessions to school-aged through senior audiences. She and her husband, Ted, enjoy spending their leisure time with their children and grandchildren, traveling, and seeking out new adventures.

Dedication

I want to dedicate *Algebra I All-In-One* to all the doctors, nurses, hospital aides, and other healthcare workers who are getting us through this very challenging time. Bless you for all your hard work and dedication.

Author's Acknowledgments

A big thank you to Chrissy Guthrie, my project editor, for pulling together all the elements of this project. This has been a big challenge, but she is, as always, up to the challenge! Thank you, also, to Michelle Hacker, my managing editor, for all her support and assistance. Another big thank you to Marylouise Wiack, my copyeditor, for her great catches and her way with words. And Amy Nicklin gets a big thanks as technical editor, being sure I have my numbers right! And, of course, I can't say thank you enough to Lindsay Lefevere, who made this project possible and included me, yet again!

Publisher's Acknowledgments

Executive Editor: Lindsay Sandman Lefevere

Project Manager and Development Editor: Christina N. Guthrie

Managing Editor: Michelle Hacker

Copy Editor: Marylouise Wiack

Technical Editor: Amy Nicklin

Production Editor: Mohammed Zafar Ali

Cover Image: © Zebra Finch / Shutterstock